Applied Mathematical Sciences

Volume 49

Applied Mathematical Sciences

A Selection

21. Courant/Friedrichs: **Supersonic Flow and Shock Waves.**
22. Rouche/Habets/Laloy: **Stability Theory by Liapunov's Direct Method.**
23. Lamperti: **Stochastic Processes: A Survey of the Mathematical Theory.**
24. Grenander: **Pattern Analysis: Lectures in Pattern Theory, Vol. II.**
25. Davies: **Integral Transforms and Their Applications,** 2nd ed.
26. Kushner/Clark: **Stochastic Approximation Methods for Constrained and Unconstrained Systems.**
27. de Boor: **A Practical Guide to Splines.**
28. Keilson: **Markov Chain Models—Rarity and Exponentiality.**
29. de Veubeke: **A Course in Elasticity.**
30. Sniatycki: **Geometric Quantization and Quantum Mechanics.**
31. Reid: **Sturmian Theory for Ordinary Differential Equations.**
32. Meis/Markowitz: **Numerical Solution of Partial Differential Equations.**
33. Grenander: **Regular Structures: Lectures in Pattern Theory, Vol. III.**
34. Kevorkian/Cole: **Perturbation Methods in Applied Mathematics.**
35. Carr: **Applications of Centre Manifold Theory.**
36. Bengtsson/Ghil/Källén: **Dynamic Meterology: Data Assimilation Methods.**
37. Saperstone: **Semidynamical Systems in Infinite Dimensional Spaces.**
38. Lichtenberg/Lieberman: **Regular and Stochastic Motion.**
39. Piccinini/Stampacchia/Vidossich: **Ordinary Differential Equations in R^n.**
40. Naylor/Sell: **Linear Operator Theory in Engineering and Science.**
41. Sparrow: **The Lorenz Equations: Bifurcations, Chaos, and Strange Attractors.**
42. Guckenheimer/Holmes: **Nonlinear Oscillations, Dynamical Systems and Bifurcations of Vector Fields.**
43. Ockendon/Tayler: **Inviscid Fluid Flows.**
44. Pazy: **Semigroups of Linear Operators and Applications to Partial Differential Equations.**
45. Glashoff/Gustafson: **Linear Optimization and Approximation: An Introduction to the Theoretical Analysis and Numerical Treatment of Semi-Infinite Programs.**
46. Wilcox: **Scattering Theory for Diffraction Gratings.**
47. Hale et al.: **An Introduction to Infinite Dimensional Dynamical Systems — Geometric Theory.**
48. Murray: **Asymptotic Analysis.**
49. Ladyzhenskaya: **The Boundary-Value Problems of Mathematical Physics.**
50. Wilcox: **Sound Propagation in Stratified Fluids.**
51. Golubitsky/Schaeffer: **Bifurcation and Groups in Bifurcation Theory, Vol. I.**
52. Chipot: **Variational Inequalities and Flow in Porous Media.**
53. Majda: **Compressible Fluid Flow and Systems of Conservation Laws in Several Space Variables.**
54. Wasow: **Linear Turning Point Theory.**
55. Yosida: **Operational Calculus: A Theory of Hyperfunctions.**
56. Chang/Howes: **Nonlinear Singular Perturbation Phenomena: Theory and Applications.**

O. A. Ladyzhenskaya

The Boundary Value Problems of Mathematical Physics

Translated by Jack Lohwater†

Springer-Verlag
New York Berlin Heidelberg Tokyo

O. A. Ladyzhenskaya
Mathematical Institute
Fontanka 27
191011 Leningrad D-11
U.S.S.R.

AMS Subject Classifications: 35-01, 35F15, 35G15, 35J65, 35K60, 35L35, 35R10, 35R35

Library of Congress Cataloging in Publication Data
Ladyzhenskaiâ, O. A. (Ol'ga Aleksandrovna)
The Boundary-value problems of mathematical physics.
(Applied mathematical sciences; v. 49)
Translation of: Kraevye zadachi matematicheskoĭ fiziki.
Bibliography: p.
1. Boundary value problems. 2. Mathematical physics. I. Title. II. Series: Applied mathematical sciences (Springer-Verlag New York Inc.); v. 49.
QA1.A647 vol. 49 510 s [530.1'5535] 84-1293
[QC20.7.B6]

Original Russian edition: *Craevie Zadachi Matematicheskoi Phiziki.* Moscow: Nauka, 1973.

ISBN 978-1-4419-2824-5

Printed in the United States of America.

9 8 7 6 5 4 3 2 1

In fond respectful memory of
Professor K. O. Friedrichs, Professor A. J. Lohwater,
and my mother, A. M. Ladyzhenskaya

Preface to the English Edition

In the present edition I have included "Supplements and Problems" located at the end of each chapter. This was done with the aim of illustrating the possibilities of the methods contained in the book, as well as with the desire to make good on what I have attempted to do over the course of many years for my students—to awaken their creativity, providing topics for independent work.

The source of my own initial research was the famous two-volume book *Methods of Mathematical Physics* by D. Hilbert and R. Courant, and a series of original articles and surveys on partial differential equations and their applications to problems in theoretical mechanics and physics. The works of K. O. Friedrichs, which were in keeping with my own perception of the subject, had an especially strong influence on me.

I was guided by the desire to prove, as simply as possible, that, like systems of n linear algebraic equations in n unknowns, the solvability of basic boundary value (and initial-boundary value) problems for partial differential equations is a consequence of the uniqueness theorems in a "sufficiently large" function space. This desire was successfully realized thanks to the introduction of various classes of general solutions and to an elaboration of the methods of proof for the corresponding uniqueness theorems. This was accomplished on the basis of comparatively simple integral inequalities for arbitrary functions and of *a priori* estimates of the solutions of the problems without enlisting any special representations of those solutions.

In this present edition I included some explanations of the basic text, and corrected misprints and inaccuracies that I noticed.

In conclusion, I want to express my deep gratitude to Professor A. J. Lohwater, who, regardless of the demands of his own scientific and pedagogical work, expressed the desire to acquaint himself with my book in detail and translate it into English. He translated all six chapters, but was not able to edit the book. The tragic, untimely death of Professor Lohwater cut short work on the book. The translation of "Supplements and Problems" I completed myself, and the translation of the "Introduction" and of this preface was done under the supervision of Springer-Verlag.

I thank all who have worked on this edition, especially the editorial and production staff of Springer-Verlag.

July 1984 O. A. LADYZHENSKAYA

Contents

Introduction

This book originated as a course of lectures that I have delivered since 1949 at the Departments of Mathematics and Physics of Leningrad University in areas of mathematical physics and differential equations with partial derivatives.

The content of these lectures varied as my own understanding of the subject developed. However, the pivotal idea, which determined the style of the lectures, was clearly formulated from the very beginning. This idea consists of replacing classical formulations of boundary value problems by generalized formulations. The generalized formulation, not being unique, is determined by the indication of the functional space where the solution is being sought.

The idea of introducing generalized solutions first started to penetrate mathematical physics in the 1920's. It came from two sources. The first source was that of two-dimensional variational problems. Investigation of these necessitated the extension of classes of functions among which a minimum is sought, and admission for consideration not only of continuously differentiable functions but also of continuous functions possessing so-called generalized derivatives (Tonelli classes). This approach led to a revision of one of the most cardinal concepts of calculus—the concept of partial derivatives. A number of different generalizations was proposed (see G. C. Evans [1], Tonelli's papers [1], C. Morrey's papers [1], etc.), including a definition of a generalized derivative which subsequently received general recognition, and which we will use in this book (see §4, Chapter 1).

The second source of generalized solutions was that of non-stationary problems, first, the wave equation $u_{tt} = c^2 \Delta u$ and then the equations of hydrodynamics. Discontinuous solutions were introduced for both a long time ago: for the former, flat and spherical waves with a strong discontinuity on the front moving at velocity c; and for the latter, solutions describing shock waves. In the 1920's, researchers were trying to understand which discontinuous solutions should be considered as "admissible." In the U.S.S.R. these questions were raised by A. A. Friedman and were investigated by him and his co-workers (I. A. Kibbel, N. E. Kochin *et al.*) as problems of kinematic and dynamic compatibility conditions. It was realized that "admissible" solutions of the linear equation $\mathscr{L}u = f$ (or a system) are those which satisfy the integral identity

$$\int u\mathscr{L}^*\eta \, dx = \int f\eta \, dx \tag{1}$$

for any sufficiently smooth function $\eta(x)$ with a compact support (in (1) $\mathscr{L}^*$ is a differential operator conjugate with $\mathscr{L}$ in the sense of Lagrange). Identity (1) also appears in Wiener's paper [1], which concerns the one-dimensional hyperbolic equation.

The 1930's led to a further development of the above-mentioned trends. In K. O. Friedrich's papers [2], [3], devoted to finding the minimum of the quadratic functional

$$J(u) = \int_\Omega (a_{ij}u_{x_i}u_{x_j} + au^2 + 2uf) \, dx \tag{2}$$

under the boundary conditions $u|_{\partial\Omega} = 0$ and the condition of ellipticity

$$a_{ij}\xi_i\xi_j \geq \nu \sum_i \xi_i^2, \qquad \nu > 0,$$

the considerations were carried out in the class of functions which subsequently was designated by $\overset{\circ}{W}{}_2^1(\Omega)$. The elements of $\overset{\circ}{W}{}_2^1(\Omega)$ are not continuous functions but have so-called generalized derivatives of first order which, along with the functions themselves, are square-summable over Ω. In this class of functions, which is a complete Hilbert space, it is relatively easy to find the function $u(x)$ which realizes $\inf J(u)$ if $a(x) \geq 0$, $f \in L_2(\Omega)$, and the coefficients a_{ij} and a are bounded functions on Ω. On this function, the first variation J vanishes, that is

$$\delta J(u) = 2\int_\Omega (a_{ij}u_{x_j}\eta_{x_i} + au\eta + f\eta) \, dx = 0 \tag{3}$$

for all $\eta \in \overset{\circ}{W}{}_2^1(\Omega)$, where such a function $u(x)$ is unique.

This solution of the variational problem is found directly with the help of so-called direct methods of variational calculus originating in Hilbert's work. This method does not use any information about the solvability of the Dirichlet problem

$$\mathscr{L}u = \frac{\partial}{\partial x_i}(a_{ij}u_{x_j}) - au = f, \qquad u|_S = 0, \tag{4}$$

in which $\mathscr{L}u = f$ is the Euler equation for the functional $J(u)$. On the contrary, the variational problem is used to investigate problem (4) (in this part of the study a_{ij} are considered to be smooth functions). Friedrichs demonstrates that if the differential operator $\mathscr{L}$ is considered as an unbounded operator in the Hilbert space $L_2(\Omega)$—with an original domain $\mathscr{D}(\mathscr{L})$ consisting of twice continuously differentiable functions in $\overline{\Omega}$ equal to zero on $\partial\Omega$ ($\mathscr{L}$, as is easy to verify, is symmetric on $\mathscr{D}(\mathscr{L})$)—and if we add to $\mathscr{D}(\mathscr{L})$ the solutions $u(x)$ of the variational problem described earlier for all f from $L_2(\Omega)$, and if we define $\tilde{\mathscr{L}}u = f$, then a self-conjugate extension $\tilde{\mathscr{L}}$ of the operator $\mathscr{L}$ is obtained. This extension subsequently came to be called the rigid extension or the Friedrichs extension, and functions $u(x)$ realizing $\inf J(u)$ on $\overset{\circ}{W}{}_2^1(\Omega)$ were called the generalized solutions of problem (4) from class $W_2^1(\Omega)$. The question of whether such generalized solutions have second-order derivatives and whether thereby they satisfy the equation $\mathscr{L}u = f$ remains open at this time.

Another group of ideas and methods were employed to investigate discontinuous solutions of non-stationary equations. S. L. Sobolev bases the definition of generalized solutions $u(x)$ on identity (1), supposing that u is a summable function. He shows, for differential operators $\mathscr{L}$ with constant coefficients, that such solutions are limits in L_1 of classical solutions of equation $\mathscr{L}u = f$ (more precisely, he investigated equations $u_{tt} - c^2\Delta u = 0$, $\Delta u = 0$ and $u_t - \Delta u = 0$). Accordingly, he arrived at another definition of a generalized solution of the equation $\mathscr{L}u = f$ as a limit (in one sense or another) of the classical solutions of such equations and he works with this definition. Following this approach, S. L. Sobolev obtains generalized solutions of the Cauchy problem for second-order hyperbolic equations with sufficiently smooth coefficients and homogeneous initial conditions but with rather "poor" free terms; he also goes beyond the classes of the function of a point, introducing generalized solutions as linear functionals over $\overset{\circ}{C}{}^l(\overline{\Omega})$ ([SO 1]). Investigation of the above-mentioned questions made it necessary to research, in more detail, various classes of functions possessing generalized derivatives of one kind or another. In the 1930's, the spaces $W_p^l(\Omega)$ were subjected to the most complete analysis; they consisted of functions $u(x)$ possessing generalized derivatives up to the order l inclusively, summable together with $u(x)$ over Ω with degree p. S. L. Sobolev and his associate, V. I. Kondrashov, obtained the most complete results, for that time, for the spaces $W_p^l(\Omega)$. These were preceded by results obtained by F. Rellich on the compactness of imbedding $W_2^1(\Omega)$ in $L_2(\Omega)$ and a number of inequalities (Poincaré inequalities, etc.), which, in modern terminology, are various imbedding theorems. This whole complex of analytic facts became the basis for the subsequent development of new approaches to investigating boundary value problems.

Also, in the 1930's, J. Leray went beyond the scope of classical solutions for boundary value problems in [1]; this paper concerns non-stationary problems of hydrodynamics for viscous, incompressible fluids. The same applies to Günter's work [1]–[4] on the Dirichlet problem for the equation $\Delta u = f$, and on the initial-boundary value problem for the equation of the oscillation of a non-homogeneous string. N. M. Günter was actively campaigning to abandon the classical formulations and to introduce generalized solutions which would reflect more closely the physical phenomena described by differential equations. He introduced and investigated generalized solutions of the Dirichlet problem for the equation $\Delta u = f$ which are functionals over the space $C(\overline{\Omega})$. Another class of generalized solutions was introduced by him when he was investigating initial-boundary value problems for the string equation.

(Note that in this book we shall deal only with generalized solutions that are functions of a point. Papers discussing generalized solutions in the form of linear functionals over classes of smooth functions will not be considered in either the body of this book or in this Introduction. Most of these papers are devoted to the Cauchy problem or to the general theory of equations in the total space. These papers mainly investigate equations with constant coefficients or coefficients from some special classes. The theory of boundary value problems has received little attention in this group of papers. Broad interest in generalized solutions—functionals—arose after the publication of L. Schwartz' book [1]. This book served as a foundation for numerous studies, among which we mention monographs by I. M. Gelfand, G. E. Šilov [1], and L. Hörmander [1].)

In the late 1940's I proposed that generalized solutions of boundary value and initial-boundary value problems for various types of equations (elliptic, parabolic, and hyperbolic) should be determined with the help of integral identities replacing the equation and, sometimes, a part of the initial and boundary value conditions. Furthermore, the importance of the fact that for each problem one can introduce various classes of generalized solutions, defined by that functional space W, to which the generalized solution which is sought should belong, was also duly noted. All the requirements of the problem should be restated in accordance with the indication of the space W. The choice of the space W is up to the researcher and could be limited only by smoothness of data. The only requirement which must be satisfied is that of the admissibility of the introduced extension, i.e., the requirement that the uniqueness theorem be preserved in the class W provided that this theorem is in agreement with the "spirit" of the problem and occurs in the class of classical solutions. In addition, there should be a natural subordination of classes of generalized solutions: solutions from a narrow class W are generalized solutions from a wider functional class W.

Thus, the definition of a generalized solution of a problem was separated from any method of obtaining it (as it was earlier in the papers by K. Friedrichs and S. L. Sobolev) and, even more, from any analytic representation of it (as in the works of N. M. Günter and J. Leray). The freedom that arose in choosing a class W of generalized solutions made it possible to transfer some of the difficulties which occurred when a theorem of existence is proved over to the second part of the investigation—the establishment of a corresponding uniqueness theorem. It is obvious that upon the extension of W these difficulties show a tendency to move towards proving theorems of uniqueness: the wider the W the easier the proof of existence of solutions from W, and the harder the confirmation of their uniqueness. In the class of classical solutions the principal weight falls on the proof of existence theorems, whereas the proof of uniqueness theorems is comparatively simple. For elliptic equations in the 1940's a "golden mean" was found—the space $W_2^1(\Omega)$ for second-order equations—where the difficulties of the whole investigation are distributed evenly. For non-stationary problems I shifted them towards uniqueness theorems by introducing classes for their generalized solutions, the existence of which is proved simply enough: for hyperbolic second-order equations it is the class $W_2^1(Q_T)$, $Q_T = \Omega \times (0, T)$, and for parabolic second-order equations it is the class $W_2^{1,0}(Q_T)$. Proofs of uniqueness theorems in these classes required some ingenuity. It was done by using only some of the general properties of considered problems (see [LA 4]).

Somewhat later (in 1953, see [LA 6]) theorems of uniqueness for non-stationary problems were successfully proved in the class of generalized solutions from $L_2(Q_T)$ (i.e., solutions not having even generalized derivatives) from which theorems of existence follow almost "automatically," roughly similar to what takes place in the theory of linear algebraic systems.

At first I established the existence of admissible generalized solutions using the method of finite differences. This method subsequently made it possible to investigate the increase in the smoothness of these generalized solutions depending on the increase in the smoothness of the data of the problem. The entire program under discussion was realized in the book [LA 4], which is mainly devoted to initial-boundary value problems for hyperbolic equations that were the most difficult and the least investigated at that time. During the years when that book was being written (1949–1951) the method of finite differences held the most prominent place in my lectures; the solvability of all the principal boundary value and initial-boundary value problems for equations (and certain classes of systems) of the elliptic, parabolic, and hyperbolic types was proved by means of this method.

For elliptic equations of second order, as was mentioned above, the more convenient space W proved to be the space $W_2^1(\Omega)$. In this space Fredholm solvability of the principal boundary value problems is established without complicated and long analytic considerations. Initially it was demonstrated by K. O. Friedrichs and after that in other forms by S. G. Mihlin, M. I. Vishik, and the author for different boundary conditions.

It appears that the method proposed by me (described in §§2–5, Chapter II) has a number of advantages due, to a great extent, to the form of treatment of generalized solutions with the help of integral identities. It is this form that insured its applicability to more difficult problems such as stationary linear and non-linear problems of hydrodynamics for viscous incompressible fluids [LA 13, 15] and many others.

Further, for elliptic operators of second order, I established a so-called "second principal inequality" (see §§6 and 7, Chapter II) that permits us to answer one of the most important questions that arose on the boundary between functional analysis and the theory of differential equations; namely, what does the domain of definition consist of for the closure of the elliptic operator $\mathscr{L}$ as an unbounded operator in Hilbert space $L_2(\Omega)$ originally defined on smooth functions satisfying some homogeneous boundary condition? This question was raised by I. M. Gelfand in 1944 at his well-known seminar where a program for studying boundary problems for elliptic equations from the point of view of functional analysis was first outlined. From the second principal inequality it follows, in particular, that for the extension $\tilde{\mathscr{L}}$, constructed by K. O. Friedrichs, the domain $\mathscr{D}(\tilde{\mathscr{L}})$ is the entire space $W_2^2(\Omega) \cap \overset{\circ}{W}{}_2^1(\Omega)$, provided that the boundary of Ω and the coefficients $\mathscr{L}$ possess some smoothness. For this reason, solutions of variational problems for $J(u)$ have generalized derivatives of second order and satisfy equation (4) for almost all x in Ω. The second principal inequality is a particular case of more general inequalities established by me in the process of proving the Fourier method for hyperbolic equations. This material was published in full in the second chapter of [LA 4]. In the present book I confined myself to presenting only the most simple of them (§6, Chapter II). Later, in 1955, there occurred to me a fairly simple way of investigating the solvability of Dirichlet problems for elliptic equations based on this inequality [LA 7]. This is the method of continuation in a parameter connecting the given elliptic operator with some simple reversible elliptic operator; for example, the Laplace operator (it is presented in §7, Chapter II). However, later on I became convinced that I had just "discovered America" all over again. It turned out that exactly the same method was proposed by J. Schauder in 1934 ([1]) to prove the solvability of the Dirichlet problem in a Hölder space $H^{2+\alpha}(\overline{\Omega})$.

This remarkable work was unknown in Russian and it did not begin to influence Soviet mathematicians until the late 1950's. Also, from this paper came Korn's idea (which was later applied widely, both in Russia and abroad) of reducing a complex problem to elementary problems for equations with constant ("frozen") coefficients in the entire space or half-space. The problem under investigation is then "glued together" from such elementary problems. In view of its great technical complexity J. Schauder's work is not included in the present book (for details, see §§2, 3 [LU 1]).

In addition to the above-mentioned methods, both in the lectures and here, Galerkin's method is presented. For elliptic equations, numerous papers were devoted to this method. For general equations of this type its convergence in $W_2^1(\Omega)$ was proved by S. G. Mihlin (see [1]–[3]). For non-stationary problems, Galerkin's method was first used by S. Faedo [1] to construct classical solutions (which complicated the matter considerably) of hyperbolic equations with a single space variable and then by J. W. Green [1] for parabolic equations with a single space variable. In E. Hopf's paper

[3], with the use of this method, a weak solution of the initial-boundary value problem for general non-stationary Navier–Stokes equations was found.

For linear equations of parabolic and hyperbolic type, Galerkin's method in its classical form permits us to prove the existence of generalized solutions in the class of functions that is somewhat broader than "solutions with a finite energy norm." Such solutions were introduced by me and the uniqueness theorem was proved for them. I presented this method both in my lectures and here, with respect to equations of all three classical types. In the process, I cite some of its variants that make it possible to prove the existence of smoother solutions and also give better convergence than the standard Galerkin method. For equations of parabolic type, I combine Galerkin's method with another, "functional," method, which permits us to prove the unique solvability in the class of "generalized solutions with a finite energy norm" under minimal assumptions on the smoothness of all data of the problem. Let us discuss in some more detail the above-mentioned "functional" method of proving existence theorems. I give this (not very "felicitous") name to methods which do not use any analytic constructions for building up solutions (i.e., neither exact formulas, nor integral equations, nor any approximations), but, are only based on some theorems or other arguments of functional analysis.

For equations of elliptic type, such a method is the method mentioned earlier in connection with the Fredholm solvability of the Dirichlet problem; it is presented in §§2–5, Chapter II. It is essentially based on the positiveness of the principal square form, i.e., on the property of ellipticity of the equation and it was believed that such reasoning was not applicable to equations of other types, especially hyperbolic equations, to which there corresponds an indefinite metric. Nevertheless, it was successfully shown that the unique solvability of initial-boundary value problems may be obtained using the following naive idea. Let us state the problem in the form of the operator equation $Au = f$, where the operator A acts in the Hilbert space H (in our case, it is the space $L_2(Q_T)$). We shall assume initial and boundary conditions to be reduced to homogeneous. The operator A is an unbounded linear operator. Let us take as its domain $\mathscr{D}(A)$ the "not too poor" functions satisfying homogeneous boundary and initial conditions and let us attempt to prove by "a frontal attack" that the set $\mathscr{R}(A)$ of its values is dense in H. This means that the identity

$$(Av, \mathscr{F}) = 0, \qquad \forall v \in \mathscr{D}(A), \tag{5}$$

guarantees the disappearance of the element $\mathscr{F} \in H$. From the point of view of functional analysis, identity (5) signifies that $\mathscr{F} \in \mathscr{D}(A^*)$ and $A^*\mathscr{F} = 0$. In my terminology (5) means that $\mathscr{F}$ is a generalized solution from H of a problem conjugate with the problem being investigated. But such a problem is a problem of the same type as the original one and, consequently, the proof of the disappearance of the element $\mathscr{F} \in H$ is also a proof of the uniqueness theorem for the problem of the type under consideration in the class of generalized solutions from H.

If this theorem can successfully be proved in a direct manner, then the theorem of existence follows from it fairly simply, since in all the problems under investigation the operator A admits the closure $\bar{A}$ and there exists on $\mathscr{R}(\bar{A})$ a bounded reverse operator $\bar{A}^{-1}$, hence the equation $\bar{A}u = f$ will have a unique solution from $\mathscr{D}(\bar{A})$ for all f from H. This entire scheme has one non-trivial part, namely, the proof of the uniqueness theorem of generalized solutions from H (i.e., from $L_2(Q_T)$).

Such theorems of uniqueness and, with them, theorems of existence, were proved by me in 1953 and were published in summary form in the note [6]. Essentially, they did not make use of the fact that A is a differential operator and were found to be applicable to broad classes of operator equations

$$S_i(t)\frac{d^2u}{dt^2} + S_2(t)\frac{du}{dt} + S_3(t)u = f(t), \tag{6}$$

where $S_i(t)$ are unbounded linear operators in the Hilbert space $\mathscr{H}$ depending on parameter t and possessing some common properties. It was found possible to regard many initial-boundary value problems and the Cauchy problem for non-stationary differential equations, and systems of different types, as particular cases of the Cauchy problem for equations of form (6) admitting of the method of investigation just described [(LA 8]). Other methods for the solution of the Cauchy problem for equation (6) were proposed by M. I. Vishik [5], [6]. He applied Galerkin's method to them; in addition, he gave, for parabolic equations, the functional method conceptually close to (but more complicated than) the functional method by which equations of elliptic type have been investigated and which were mentioned above.

The results just described on non-stationary problems, including results on the Cauchy problem for equations of the form (6) are presented in fairly great detail in the review article [LV 1]. The work in this direction was continued by many mathematicians including J. L. Lions [1], [2], S. G. Krein, M. A. Krasnoselskii, and P. E. Sobolevskii [KKS 1]. Another method having its origin in the theory of semigroups was developed by T. Kato for investigating the Cauchy problem for equation (6) of first order ([KA 1]). This work was also continued by a number of authors and its development was applied to problems of mathematical physics (see T. Kato [2], [3], M. Z. Solomjak [1], [2], S. G. Krein [1], P. E. Sobolevskii [2], etc.).

As I was writing my papers [8], I was telling my students about them during my lectures. However, this required a presentation of the principal facts from the theory of differentiation and integration of functions $u(\cdot)$ with values in the Hilbert space and a proof of a number of assertions concerning unbounded bilinear forms. In the present book I confine myself to the presentation of the functional method described above on examples of the heat equation and the wave equation; the results obtained for the heat equation were subsequently used to obtain more subtle results on general parabolic equations.

It can be seen from the above that during different periods preference was given first to some methods of investigation, and then to others.

The only thing that did not change was the chief object of investigation, that is, boundary value problems (the first, second, and third) for linear equations of elliptic, parabolic, and hyperbolic type and of the Schrödinger type with variable coefficients in arbitrary areas of variation of space variables (lying in the Euclidean space R^n) and the chief approach to their investigation, namely, studying generalized rather than classical solutions. The question of when generalized solutions possess some degree of smoothness was touched upon only partially, namely, we investigate only when generalized solutions have generalized derivatives entering into equations and thereby satisfy equations almost everywhere. A more detailed analysis of generalized solutions, in particular, determining when they become classical, was not discussed at the lectures in view of the technical complexity of the question. We shall now proceed to describe the version of the lectures that is presented in this book.

The book contains six chapters. The first chapter is of an auxiliary character. In it we remind the reader of some facts from the theory of Hilbert spaces and the specific function spaces that will be used in the sequel. Special attention is given to one of the central concepts of non-classical calculus, namely, the concept of a generalized derivative and the description of properties of such derivatives, as well as to the function spaces $W_2^1(\Omega)$ and $\mathring{W}_2^1(\Omega)$ to which we have allotted a central place in the book. For the latter, we give proofs of criteria of strong compactness of families of functions, belonging to these spaces, in the spaces $L_2(\Omega)$ and $L_2(\partial\Omega)$. The chapter also contains a preliminary discussion of the behavior of the elements $W_2^1(\Omega)$ on a surface of dimension $n - 1$ (n is the dimension of the Euclidean space R^n in which the domain Ω is located). In §§7 and 8 the principal multiplicative inequalities are given for reference. With their use results presented in this book, for equations with bounded coefficients, are quite easily generalized for the case when the coefficients of the minor terms of the equations are unbounded functions summable over Ω with certain degrees. Such generalizations were found to be important for investigating non-linear equations. Chapter I does not claim to be a complete presentation of all the questions mentioned in it. Proofs are only given for assertions that are used in the body of the book. Nevertheless, it can serve as a guide as to what a person must understand and assimilate if he or she is to get a good grasp of the modern theory of boundary value problems.

Chapter II presents equations of second-order elliptic type. In §§2, 3 and 5, Fredholm solvability of boundary value problems for them in the space $W_2^1(\Omega)$ is proved. Generalized solutions belonging to the space $W_2^1(\Omega)$ have a finite energy norm and are called "energy" solutions, for short. In §§6 and 7 the question of when these solutions belong to the space $W_2^2(\Omega)$

and satisfy the equation almost everywhere is discussed. In §8, approximate methods of solution are presented: Galerkin's method (both in its classical and modernized forms) as well as the Ritz method, and the method of least squares as its special cases. In §§4 and 7, spectral problems and the question of the convergence of Fourier series for eigenfunctions of these problems are considered. Another approximate method of solving boundary value problems—the method of finite differences—is discussed in detail in Chapter VI devoted to equations of all types.

In Chapter III initial-boundary value problems for equations of second-order parabolic type are considered. The Cauchy problem is analyzed only insofar as it can be considered as a special case of such problems. The unique solvability in the class of "energy solutions" (in the class $V_2^{1,0}(Q_T)$) is established for them. To do this under the minimally possible assumptions on the coefficients, one has to introduce and investigate generalized solutions from other functional classes.

I decided to guide the reader over such a winding path not so much from a desire to maximally lower all the requirements on the data of the problem as from an intention to illustrate the various classes of generalized solutions and the method of working with them. I could have confined myself to any of the classes I introduced and proved within them the unique solvability of the problems more concisely. Chapter III presents Galerkin's method, the functional method, the Laplace transform method, the Fourier method, and the Rothe method. As was mentioned above, the method of finite differences is presented in Chapter VI. To save space, I did not touch upon the method of continuation in a parameter. This method has two forms. One of them is presented in §7 of Chapter II with respect to elliptic equations. For equations of parabolic type it remains essentially unchanged; all that needs to be done is to connect the given equation, by means of a parameter, to the heat equation (or to some other suitable parabolic equation). The proof of this first form is based on the "second principal inequality" which is also valid for parabolic equations (see [LA 8] and §6, Chapter III, [LSU 1] for more information about this inequality). Another construction of the method of continuation in a parameter is based only on the first (energy) inequality. In this construction, the parameter connects bilinear forms corresponding to the operator of the given problem and of the auxiliary problem. This construction is also possible for both types of equations: elliptic and parabolic (see, for example, [LA 8] and [LV 1]).

Chapter IV is devoted to equations of hyperbolic type. First, these equations are considered in the entire space and the "energy inequality" is proved for them. From this follows: (1) the characteristic property of hyperbolic equations; i.e., the finiteness of the velocity of the propagation of perturbations described by this equation; and (2) the uniqueness theorem for the Cauchy problem in the class of generalized solutions from $W_2^2(\Omega \times (0, T)), \forall\Omega \subset R^n$. All these facts have been known for a long time (for equations of the general type see papers by J. Hadamard, H. Lewy, K. O.

Friedrichs, and J. Schauder; for special classes of equations they were established as far back as the nineteenth century); only they were proved for classical solutions. The transition to generalized solutions from W_2^2 does not cause any special difficulty. Thanks to (1) and (2), it is possible to consider the Cauchy problem as a special case of the initial-boundary problem with the first boundary condition. It is this and other initial-boundary problems that will be considered in the sequel; the Cauchy problem, however, will not receive separate treatment (only in Chapter VI is a finite-difference scheme constructed for it that does not involve any boundary conditions). The unique solvability in the class of generalized solutions from $W_2^1(Q_T)$ is proved for them. The solution $u(x, t)$ is obtained as the limit of Galerkin approximations $u^N(x, t)$.

Convergence of u^N to u is easily deduced with the help of the "energy inequality." On the other hand, to prove the uniqueness theorem in this class, new considerations were required which differ from those that were developed earlier to prove the uniqueness in the Cauchy problem (the Holmgren method and the proof based on the energy inequality). The proof given in the text (it was taken from the book [LA 4]) is based on an *a priori* estimate of generalized solutions from $W_2^1(Q_T)$, which differs from the energy inequality. L. Gårding named this inequality, and others similar to it, dual with respect to energy inequalities and to their corollaries. Another of the "dual" inequalities was found by us in proving the uniqueness theorem in a broader class of generalized solutions, namely, in the class $L_2(Q_T)$ (see [LA 6, 8]). As was explained in the first half of the Introduction, from such a uniqueness theorem, one can easily deduce the solvability of the problem in the class of "energy" solutions. However, to save space, I exemplified this line of investigation of boundary value problems only by the wave equation (see §6), where both "dual" inequalities are given only for zero initial and boundary conditions and zero free term (at any rate, only this case is necessary in order to prove the uniqueness theorems).

Section 4 describes how an increase in the smoothness of generalized solutions from $W_2^1(Q_T)$ is investigated, in connection with an increase in the smoothness of the data of the problem and an increase in the order of their compatibility. It is determined when such solutions belong to the "energy class" and when they have derivatives entering into the equation. Throughout the chapter I give detailed consideration to the first boundary condition; for the rest, I give the necessary references (see §5). Section 7 discusses the method of Laplace transforms and the Fourier method. For the latter, convergence in the spaces $L_2(\Omega)$, $W_2^1(\Omega)$, and $W_2^2(\Omega)$ is investigated. A more detailed analysis of these methods was given in my first book [LA 4]. Here only a small part is presented, the one pertaining to the Fourier method. When reading [LA 4], one should bear in mind that 20 years have passed since its publication, during which time less restrictive conditions on the smoothness of the domain and of the coefficients have been found which guarantee some smoothness of eigenfunctions and solutions of elliptic equations. These improvements also lead to a corresponding weakening of the conditions of Chapters II and IV of the book [LA 4] concerning the smoothness of the coefficients of the equations and of the domain.

Chapter V presents other problems that can be investigated by means of the methods set forth in Chapters II–IV. These include some boundary value problems for elliptic and parabolic equations of any order as well as for strongly elliptic, strongly parabolic, and strongly hyperbolic systems. Their special cases are the biharmonic equations and the system of equations of the theory of elasticity. Close to the elliptic case is the Navier–Stokes linearized system of equations for stationary flows of viscous incompressible fluids. It is investigated essentially in the same manner as the elliptic equation in §3 of Chapter II. Pertinent references can be found in §1. In §3, equations of the Schrödinger type (in their abstract form) are considered. It is noted that the Maxwell system of equations belongs to this type (provided all the requirements of electrodynamics have been duly taken into account). In the fourth and last section, it is shown how diffraction problems for equations of different types are embedded into problems of defining generalized solutions from class W_2^1 of ordinary boundary value problems discussed in Chapters II–IV.

The last and longest chapter of the book is devoted to the method of finite differences. It discusses the same problems as Chapters II–IV. Stress is laid on constructing converging difference schemes and proving their stability in metrics corresponding to energy metrics. In investigating the method of finite differences I do not make use of any information concerning an exact solution of the problem (not even the fact of the existence of such a solution) and, conversely, I could prove, with the help of this method, the existence theorems and carry out fairly complete qualitative investigations of these solutions and of approximate solutions converging to them. Being restricted by considerations of time and space I will not carry out these detailed qualitative investigations; as for the proofs of existence theorems, I present them rather briefly. In principle, the method of finite differences is fairly simple. It reduces all problems to the solution of finite linear algebraic systems and the analysis of these systems. It was first used for the solution of ordinary differential equations and is known in mathematics as the method of Euler broken lines. Its applications to and investigations of equations in partial derivatives did not begin until the twentieth century. In the past two decades the role of the method of finite differences has increased in connection with the emergence of high-speed computers and, at present, it has taken first place among the other approximate methods of solving boundary value problems. Chapter VI starts with an introduction explaining what considerations guided me in constructing converging difference schemes. It also gives an outline of the contents of the chapter.

The present book has little to do with textbooks on the theory of differential equations with partial derivatives including such excellent courses as Volumes II and IV of *A Course in Higher Mathematics* by V. I. Smirnov and

Lectures on Equations in Partial Derivatives by I. G. Petrovskii. They present mainly the classical methods of investigation, with respect to boundary problems, for the simplest equations. In the second volume of *Methods of Mathematical Physics* by D. Hilbert and R. Courant—a famous book, rich in content—a number of problems for equations with variable coefficients is considered. But even the second edition of this book reflects mainly what had been done by the time of its first publication, that is, by 1937).

There are several monographs ([SO 5], [LU 1], [LSU 1], [LI 1, 2], [FM 1], [KR 1], [EI, 1], [HR 1]) written in the last two decades, but they can hardly be recommended as textbooks for graduate students or as guidebooks from which mathematicians working in various fields can become acquainted with boundary problems for partial differential equations of all the principal types.

I hope that this book will fill the gap that exists in textbook literature on equations in partial derivatives. It is based on results established in the late 1940's and 1950's and contains a presentation of viewpoints and methods that are being used successfully and that continue to develop at the present time.

Basic Notation

Let us introduce a number of symbols and notations used in this book.

R^n is the n-dimensional Euclidean space.

$x = (x, \ldots, x_n)$ is a point in R^n.

Ω is a domain in R^n, in a vast majority of cases, Ω is bounded.

$S = \partial\Omega$ is the boundary of the domain Ω; $\overline{\Omega} = \Omega \cup S$,

Ω' is a bounded subdomain of Ω with $\overline{\Omega}' \subset \Omega$.

$Q_T = \{(x, t)\colon x \in \Omega, t \in (0, T)\}$ is a cylinder in R^{n+1}.

$S_T = \{(x, t)\colon x \in \partial\Omega, t \in [0, T]\}$ is the lateral surface of Q_T.

$\mathbf{n}$ is the outward (from Ω) unit normal to $\partial\Omega$.

Symbols u_{x_i} and $u_{x_i x_j}$ denote, throughout the book (except in Chapter VI), classical and generalized derivatives $\partial u/\partial x_i$ or $\partial^2 u/\partial x_i \partial x_j$.

$$u_x = (u_{x_1}, \ldots, u_{x_n}); \qquad u_{x_i}^2 = (u_{x_i})^2; \qquad u_x^2 = \sum_{i=1}^n u_{x_i}^2; \qquad |u_x| = \sqrt{u_x^2};$$

$$u_x v_x = \sum_{i=1}^n u_{x_i} v_{x_i}; \qquad u_{xx} = (u_{x_i x_j}), \qquad i, j = 1, \ldots, n; \qquad u_{x_i x_j}^2 = (u_{x_i x_j})^2;$$

$$u_{xx}^2 = \sum_{i,j=1}^n u_{x_i x_j}^2; \qquad |u_{xx}| = \sqrt{u_{xx}^2}; \qquad u_{xx} v_{xx} = \sum_{i,j=1}^n u_{x_i x_j} v_{x_i x_j}.$$

Let us list the principal function spaces appearing in the book.

$L_p(\Omega)$, $p \geq 1$ is the Banach space (i.e., the complete linear normed space) consisting of all measurable (in the sense of Lebesgue) functions on Ω

having the finite norm:

$$\|u\|_{p,\Omega} = \left(\int_\Omega |u|^p\,dx\right)^{1/p}.$$

Often the norm in $L_2(\Omega)$ will be abbreviated to $\|\cdot\|$ and the scalar product to (,). In addition:

$$\|u_x\| = \|u_x\|_{2,\Omega} \equiv \left(\int_\Omega u_x^2\,dx\right)^{1/2}, \quad \text{and} \quad \|u_{xx}\| \equiv \|u_{xx}\|_{2,\Omega} \equiv \left(\int_\Omega u_{xx}^2\,dx\right)^{1/2}.$$

The spaces $W_m^l(\Omega)$ and $\mathring{W}_m^l(\Omega)$ are defined in §5, Chapter I. In particular, the Hilbert space $W_2^1(\Omega)$ consists of the elements $L_2(\Omega)$ having square summable generalized derivatives of first order on Ω. The scalar product in it is defined by the equality

$$(u, v)_{2,\Omega}^{(1)} = \int_\Omega (uv + u_x v_x)\,dx,$$

and the norm by

$$\|u\|_{2,\Omega}^{(1)} = \sqrt{(u, v)_{2,\Omega}^{(1)}}.$$

$\dot{C}^\infty(\Omega)$ is the set of infinitely differentiable functions with compact supports lying in Ω.

$\mathring{W}_2^1(\Omega)$ is a subspace (closed) of the space $W_2^1(\Omega)$ in which the set $\dot{C}^\infty(\Omega)$ is dense (or, which is the same, the set of all smooth functions, finitary on Ω).

$W_2^2(\Omega)$ is the Hilbert space consisting of all the elements $L_2(\Omega)$ having generalized derivatives of first and second order from $L_2(\Omega)$. The scalar product in it is defined by the equality

$$(u, v)_{2,\Omega}^{(2)} = \int_\Omega (uv + u_x v_x + u_{xx} v_{xx})\,dx,$$

and the norm is denoted as follows: $\|\cdot\|_{2,\Omega}^{(2)}$.

$W_{2,0}^2(\Omega)$ is a subspace of $W_2^2(\Omega)$, in which all twice continuously differentiable functions in $\bar{\Omega}$ that are equal to zero on $\partial\Omega$ are dense. If $\partial\Omega \subset C^2$ then $W_{2,0}^2(\Omega) = W_2^2(\Omega) \cap \mathring{W}_2^1(\Omega)$.

For the cylinder Q_T the following spaces are introduced:

$W_{2,0}^1(Q_T)$ is a subspace of the space $W_2^1(Q_T)$ in which the set of smooth functions equal to zero near S_T is dense.

$\hat{W}_{2,0}^1(Q_T)$ is a subspace of $W_{2,0}^1(Q_T)$ whose elements become equal to zero when $t = T$ (in Theorem 6.3, Chapter I, it is proved that traces of the elements $W_2^1(Q_T)$ are defined on every cross-section Ω_{t_1} of the cylinder Q_T by the plane $t = t_1 \in [0, T]$ as functions from $L_2(\Omega_{t_1})$ and they change continuously with t in the norm $L_2(\Omega)$ with a change $t \in [0, T]$).

$W_2^{1,0}(Q_T) = L_2((0,T); W_2^1(\Omega))$ is the Hilbert space consisting of the elements $u(x,t)$ of the space $L_2(Q_T)$ having generalized derivatives $\partial u/\partial x_i$, $i = 1, \ldots, n$ square summable on Q_T. The scalar product and the norm are defined by the equalities

$$(u, v)_{2,Q_T}^{(1,0)} = \int_{Q_T} (uv + u_x v_x)\,dx\,dt, \qquad \|u\|_{2,Q_T}^{(1,0)} = \sqrt{(u,u)_{2,Q_T}^{(1,0)}}.$$

$\overset{\circ}{W}{}_2^{1,0}(Q_T) = L_2((0,T); \overset{\circ}{W}{}_2^1(\Omega))$ is a subspace of $W_2^{1,0}(Q_T)$ in which the set of smooth functions equal to zero near S_T is dense.

$W_2^{2,1}(Q_T)$ is the Hilbert space consisting of all the elements $L_2(Q_T)$ having generalized derivatives u_t, u_{x_i}, and $u_{x_i x_j}$ from $L_2(Q_T)$. The scalar product in it is defined by the equality

$$(u, v)_{2,Q_T}^{(2,1)} = \int_{Q_T} (uv + u_x v_x + u_t v_t + u_{xx} v_{xx})\,dx\,dt$$

and the norm is defined as follows: $\|\cdot\|_{2,Q_T}^{(2,1)}$.

$W_{2,0}^{2,1}(Q_T)$ is a subspace of $W_2^{2,1}(Q_T)$ with elements belonging to $L_2((0,T); W_{2,0}^2(\Omega))$.

$W_2^{\Delta,1}(Q_T)$ is the Hilbert space consisting of the elements $W_2^1(Q_T)$ having generalized derivatives $u_{x_i x_j}$, $i, j = 1, \ldots, n$, square summable on $Q_T' = \Omega' \times (0,T)$, where Ω' is any strictly internal subdomain of the domain Ω and having $\Delta u = \sum_{i=1}^n \partial^2 u/\partial x_i^2$ from $L_2(Q_T)$. The scalar product in it is as follows:

$$(u, v)_{2,Q_T}^{\Delta,1} = \int_{Q_T} (uv + u_x v_x + u_t v_t + \Delta u \cdot \Delta v)\,dx\,dt.$$

$W_{2,0}^{\Delta,1}(Q_T)$ is a subspace of $W_2^{\Delta,1}(Q_T)$ with elements belonging to $W_{2,0}^1(Q_T)$. Traces of the elements $u(x,t)$ from $W_{2,0}^{\Delta,1}(Q_T)$ (and even more so from $W_{2,0}^{2,1}(Q_T)$) are defined on all cross-sections Ω_t, $t \in [0,T]$, as elements of $\overset{\circ}{W}{}_2^1(\Omega)$ and depend continuously on t in the norm $W_2^1(\Omega)$. If the boundary is twice continuously differentiable (that is, $\partial\Omega \subset C^2$), then $W_{2,0}^{\Delta,1}(Q_T)$ coincides with $W_{2,0}^{2,1}(Q_T)$ (see Chapter II).

$V_2(Q_T) = W_2^{1,0}(Q_T) \cap L_\infty((0,T); L_2(\Omega))$ is the Banach space consisting of the elements $W_2^{1,0}(Q_T)$ having a finite norm

$$|u|_{Q_T} = \operatorname*{ess\,sup}_{0 \le t \le T} \|u(\cdot,t)\|_{2,\Omega} + \|u_x\|_{2,Q_T},$$

where

$$\|u_x\|_{2,Q_T} = \left(\int_{Q_T} u_x^2\,dx\,dt\right)^{1/2}.$$

$\overset{\circ}{V}_2(Q_T) = \overset{\circ}{W}{}_2^{1,0}(Q_T) \cap V_2(Q_T)$ is a subspace of $V_2(Q_T)$.

$V_2^{1,0}(Q_T) = C([0,T]; L_2(\Omega)) \cap W_2^{1,0}(Q_T)$ is a subspace of $V_2(Q_T)$, whose elements have, on cross-sections of Ω_t, traces from $L_2(\Omega)$ for all $t \in [0,T]$ continuously changing with $t \in [0,T]$ in the $L_2(\Omega)$ norm.

$\overset{\circ}{V}{}_2^{1,0}(Q_T) = V_2^{1,0}(Q_T) \cap \overset{\circ}{W}{}_2^{1,0}(Q_T)$ is a subspace of $V_2^{1,0}(Q_T)$. The smooth functions equal to zero near S_T are dense in it.

$C^l(\Omega)$ is the set of all continuous functions in Ω having derivatives up to order l inclusively that are continuous in Ω.

$C^l(\overline{\Omega})$ is the Banach space consisting of all continuous functions in $\overline{\Omega}$ having derivatives up to order l inclusively that are continuous in $\overline{\Omega}$. The norm in it is defined by the equality

$$|u|_{\Omega}^{(l)} = \sum_{|k|=0}^{l} \sum_{(k)} \sup_{x \in \Omega} \left| \frac{\partial^{|k|} u(x)}{\partial x_1^{k_1} \cdots \partial x_n^{k_n}} \right|.$$

$\overset{\circ}{C}{}^l(\Omega)$ is the set of functions from $C^l(\Omega)$ having compact supports belonging to Ω.

$\overset{\circ}{C}{}^l(\overline{\Omega})$ is the subspace of $C^l(\overline{\Omega})$ in which the set $\overset{\circ}{C}{}^l(\Omega)$ is dense.

$H^{l+\alpha}(\overline{\Omega})$ ($\alpha \in (0, 1), l = 0, 1, \ldots$) is the Banach space consisting of the elements $C^l(\overline{\Omega})$ for which the derivatives of order l satisfy the Hölder condition in $\overline{\Omega}$ with the power α. $H^{l+\alpha}(\Omega) = \cap H^{l+\alpha}(\overline{\Omega}')$ where the intersection is taken over all $\overline{\Omega}' \subset \Omega$.

Notations Used in Chapter VI

In this chapter partial derivatives are denoted by symbols $\partial/\partial_{x_i}, \partial^2/\partial_{x_i}\partial_{x_j}, \ldots$, symbols $u_{x_i}, u_{\bar{x}_i}, u_{x_i x_j}, \ldots$ are used to denote difference quotients namely:

$$u_{x_i}(x) = \frac{1}{h_i}[u(x_1, \ldots, x_i + h_i, \ldots, x_n) - u(x_1, \ldots, x_i, \ldots, x_n)]$$

is a right side difference quotient,

$$u_{\bar{x}_i}(x) = \frac{1}{h_i}[u(x_1, \ldots, x_i, \ldots, x_n) - u(x_1, \ldots, x_i - h_i, \ldots, x_n)]$$

is a left side difference quotient, where everywhere $h_i > 0$,

$$u_{x_i x_j}(x) = (u_{x_i}(x))_{x_j} = (u_{x_j}(x))_{x_i}, \text{etc.}$$

The space R^n is broken down into elementary cells (a mesh),

$$\omega_{(kh)} = \{x: k_i h_i < x_i < (k_i + 1)h_i, i = 1, \ldots, n\},$$

where k_i are integers;

$$\overline{\Omega}_h = \bigcup_{\bar{\omega}_{(kh)} \subset \overline{\Omega}} \bar{\omega}_{(kh)} \subset \overline{\Omega}.$$

The same symbol $\overline{\Omega}_h$ is used to denote the set of vertices $(kh) = (k_1 h_1, \ldots, k_n h_n)$ of our lattice belonging to the closed domain $\overline{\Omega}_h$. The boundary of the closed domain $\overline{\Omega}_h$ is denoted by S_h, and $\overline{\Omega}_h \backslash S_h = \Omega_h$ is the set of the interior points of $\overline{\Omega}_h$. The symbols S_h and Ω_h are also used to denote the set of vertices (kh) belonging to the sets S_h and Ω_h, respectively.

$\bar{\Omega}_h^*$ is the set of the vertices of the cells $\bar{\omega}_{(kh)}$ having a non-empty intersection with Ω as well as a closed domain consisting of the points of these cells $\bar{\omega}_{(kh)}$.

In accordance with the above Ω_h^* is $\bar{\Omega}_h^* \backslash S_h^*$. It is clear that $\bar{\Omega}_h \subset \bar{\Omega}_h^*$ and $\Omega_h \subset \Omega_h^*$.

Ω_h^+ is the set of vertices (kh) of the cells $\omega_{(kh)}$ belonging to Ω (that is, from each cell $\omega_{(kh)}$ one of its vertices with the coordinates (kh), is taken). It is clear that $\Omega_h \subset \Omega_h^+ \subset \bar{\Omega}_h$.

$\Omega_h^{*+} \equiv (\Omega_h^*)^+$ is defined for Ω_h^* in the same way as Ω_h^+ is for Ω_h.

For functions u_n defined on these lattices, the following norms are introduced;

$$\|u_h\|_{m,\bar{\Omega}_h} = \left(\Delta_h \sum_{\bar{\Omega}_h} |u_h|^m\right)^{1/m}, \quad \text{where} \quad \Delta_h = h_1 \cdots h_n,$$

$$\|u_{x_i}\|_{m,\bar{\omega}(kh)} = \left(\Delta_h \sum_{\omega_{(kh)}^{(i)}} |u_{x_i}|^m\right)^{1/m},$$

where the symbol $\sum_{\omega_{(kh)}^{(i)}}$ means summation over the vertices of the cell $\omega_{(kh)}$ belonging to the side $x_i = k_i h_i$.

$$\|u_x\|_{m,\bar{\omega}(kh)} = \left(\sum_{i=1}^{n} \|u_{x_i}\|_{m,\bar{\omega}(kh)}^m\right)^{1/m},$$

$$\|u_{x_i}\|_{m,\bar{\Omega}_h} = \left(\sum_{\bar{\omega}(kh) \subset \bar{\Omega}_h} \|u_{x_i}\|_{m,\bar{\omega}(kh)}^m\right)^{1/m},$$

$$\|u_x\|_{m,\bar{\Omega}_h} = \left(\sum_{i=1}^{n} \|u_{x_i}\|_{m,\bar{\Omega}_h}^m\right)^{1/m},$$

$$\|u_h\|_{m,\bar{\Omega}_h}^{(1)} = [\|u_h\|_{m,\bar{\Omega}_h}^m + \|u_x\|_{m,\bar{\Omega}_h}^m]^{1/m}.$$

The quantities $\|u_h\|_{m,\bar{\Omega}_h}$ and $\|u_h\|_{m,\bar{\Omega}_h}^{(1)}$ are called norms of the spaces $L_m(\bar{\Omega}_h)$ and $W_m^1(\bar{\Omega}_h)$ of the grid functions u_h defined on $\bar{\Omega}_h$.

For the case of the function u_h vanishing on S_h, we use other norms, equivalently to those just defined, namely:

$$\|u_x\|_{m,\Omega_h} = \left[\Delta_h \sum_{\Omega_h} |u_x|^m\right]^{1/m}$$

and

$$\|u_h\|_{m,\Omega_h}^{(1)} = [\|u_h\|_{m,\Omega_h}^m + \|u_x\|_{m,\Omega_h}^m]^{1/m},$$

where we assume that u_h is supplementarily defined by zero at the points of the lattice not belonging to $\bar{\Omega}_h$. For such functions

$$\|u_h\|_{m,\bar{\Omega}_h} = \|u_h\|_{m,\Omega_h^+} = \|u_h\|_{m,\Omega_h}.$$

The symbol $|u_x|$ means $\sqrt{u_x^2}$ and $u_x^2 = \sum_{i=1}^n u_{x_i}^2$, where $u_{x_i}^2 = (u_{x_i})^2$. Analogously $u_{x_i x_j}^2$ is $(u_{x_i x_j})^2$. In writing difference quotients $u_{x_i}, u_{x_i x_j}, \ldots$ for the grid function u_h the index "h" is omitted.

At the beginning of §3, Chapter VI, several interpolations of the grid functions u_h defined on $\overline{\Omega}_h$ are introduced, namely: $\tilde{u}_h(\cdot)$ (or $\tilde{u}_h$) is a piecewise constant function equal for $x \in \omega_{(kh)}$ to the value u_h at the vertex (kh); $u_h'(\cdot)$ (or u_h') is a continuous piecewise smooth function on $\overline{\Omega}_h$ polylinear on every $\omega_{(kh)}$ and equal to u_h at all $(kh) \in \overline{\Omega}_h$.

$u_{(m)}(\cdot)$ (or $u_{(m)}$) is a function that within each cell $\omega_{(kh)}$ is constant over x_m and on the side $x_m = k_m h_m$, u_h is interpolated by the principle of $(\ )_h'$ described in the preceding sentence.

The symbol $(\widetilde{u_{x_i}})(\cdot)$ (or $\tilde{u}_{x_i}$) denotes interpolation of the first type of the grid function u_{x_i} which is a difference quotient for the lattice function u_h.

In non-stationary problems the time variable t plays a special role. In them, the step of t in place of h_{n+1} is denoted by τ and the symbol u_h, $h = (h_1, \ldots, h_{n+1})$, is replaced by u_Δ.

It is assumed throughout the book that the coefficients a_{ij}, $i, j = 1, \ldots, n$, of equations' terms involving derivatives of second order form a symmetric matrix; that is, $a_{ij} = a_{ji}$.

CHAPTER I

Preliminary Considerations

This chapter is of an introductory character and we shall present a series of concepts and theorems from functional analysis which will be used in the sequel for studying boundary value problems for differential equations. These facts will be stated without proof.

Moreover, we introduce a number of concrete functional spaces and describe properties of these spaces of interest to us. We shall either give complete proofs of some of these theorems, or else describe the fundamental steps by which the reader can reconstruct complete proofs.

In the course of the entire book we shall use Lebesgue measure and the Lebesgue integral. The reader of this book should also be familiar with the fundamentals of real variables and functional analysis (see [AK 1], [LTS 1], [SM 1; 2], [SO 5], [RN 1].

§1. Normed Spaces and Hilbert Spaces

A set E of abstract elements is called a *real* (*complex*) *linear normed space* if:

(1) E is a linear vector space with multiplication by real (complex) numbers;

(2) To every element u of E there is a real number (called the norm of the element and denoted by $\|u\|$) satisfying the following axioms:

(a) $\|u\| \geq 0$, where $\|u\| = 0$ only for the zero element;

(b) $\|u + v\| \leq \|u\| + \|v\|$, the triangle inequality;

(c) $\|\lambda u\| = |\lambda| \cdot \|u\|$.

A natural metric can be introduced into such a space: the distance $\rho(u, v)$ between two elements u and v is defined by $\rho(u, v) = \|u - v\|$. The convergence of a sequence $\{u_n\}$ of elements of E to $u \in E$ in the norm of E (in other words, strong convergence in E) is defined by $\|u_n - u\| \to 0$ as $n \to \infty$, and in abbreviated notation by $u_n \to u$.

A collection of elements $E' \subset E$ is said to be *everywhere dense* in E if any element of E is the limit, in the norm of E, of elements of E'.

If E contains a countable, everywhere dense set of elements, then E is called *separable*. The sequence $\{u_n\}_{n=1}^{\infty}$ is called convergent (or Cauchy sequence, or fundamental) if $\|u_p - u_q\| \to 0$ when $p, q \to \infty$.

If, for every Cauchy sequence $\{u_n\}_{n=1}^{\infty}$, there is a limiting element u in E, then E is called *complete* (in this case $\|u_n - u\| \to 0$ when $n \to \infty$). A complete, linear, normed space is usually called a *space of type B* or a *Banach space*. All spaces considered below will be complete and separable.

We shall be dealing basically with a particular case of the Banach spaces, namely, the Hilbert spaces. In a real Hilbert space H we define a scalar product (u, v) for an arbitrary pair of elements u and v. It is a real number satisfying the following axioms:

(a) $(u, v) = (v, u)$;
(b) $(u_1 + u_2, v) = (u_1, v) + (u_2, v)$;
(c) $(\lambda u, v) = \lambda(u, v)$;
(d) $(u, u) \geq 0$, where $(u, u) = 0$ only for the zero element $u = 0$.

In a complex Hilbert space the scalar product (u, v) is a complex number satisfying axioms (b)–(d), together with the axiom (a′) $(u, v) = \overline{(v, u)}$ instead of axiom (a).

As the norm of an element u we take the number $\|u\| = \sqrt{(u, u)}$. In the definition of a Hilbert space we include the requirement that it be complete and separable.

For any two elements u and v in H we have the inequality of Cauchy, Bunyakovski, and Schwarz:

$$|(u, v)| \leq \|u\| \cdot \|v\|,$$

which we shall simply call Cauchy's inequality in the sequel.

In addition to convergence in norm (strong convergence) in the space H, we shall also consider weak convergence. A sequence $\{u_n\}$ is said to converge weakly in H to the element u if $(u_n - u, v) \to 0$ as $n \to \infty$ for all $v \in H$. For brevity, this will be denoted by $u_n \rightharpoonup u$. It is not difficult to understand that if the norms of the $\{u_n\}$ are uniformly bounded, then to prove the weak convergence of $\{u_n\}$ to u, it is enough to verify that $(u_n - u, v) \to 0$ as $n \to \infty$ only for some set V which is everywhere dense in H. A sequence $\{u_n\}$ cannot be weakly (much less strongly) convergent to two different elements of H. If $\{u_n\}$ converges to u in the norm of H, then it converges weakly to u. The converse is false. However, if, in addition to the weak convergence of $\{u_n\}$

to u, it is known that $\|u_n\| \to \|u\|$, then $\{u_n\}$ converges strongly to u. In the sequel we shall make frequent use of the following proposition:

Theorem 1.1. *If the sequence $\{u_n\}$ converges weakly to u in H, then*

$$\|u\| \leq \varliminf_{n\to\infty} \|u_n\| \leq \varlimsup_{n\to\infty} \|u_n\|,$$

where the right-hand side of this inequality is finite.

A Hilbert space (and we emphasize that, by definition, such spaces are complete), as well as any closed subspace of it, is complete with respect to weak convergence.†

A set M in a Banach space B is called *precompact* (or *precompact in B*) if every infinite sequence of elements of M contains a convergent subsequence. If the limits of all such subsequences belong to M, then M is called *compact* (or *compact in itself*). In a similar way we introduce in a Hilbert space H the notions of weak precompactness and weak compactness. We have the following criterion of weak compactness in H:

Theorem 1.2. *A set M of H is weakly precompact if and only if it is bounded.*

We mention two examples of real spaces B and H. The totality of all real-valued measurable functions $u(x)$, defined on a domain Ω of Euclidean space R^n with a finite integral

$$\|u\|_{p,\Omega} = \left(\int_\Omega |u(x)|^p \, dx \right)^{1/p} \tag{1.1}$$

with arbitrary fixed $p \geq 1$, forms a (complete) separable Banach space if its norm is defined by (1.1). This space is usually denoted by $L_p(\Omega)$. Strictly speaking, it must be understood that an element of $L_p(\Omega)$ is not any function $u(x)$ with the properties indicated, but rather the class of functions which are equivalent to it on Ω (that is, those functions which coincide with it almost everywhere on Ω). Nevertheless, for the sake of brevity we shall speak of the elements of $L_p(\Omega)$ as functions defined on Ω.

As examples of everywhere dense sets in $L_p(\Omega)$ we can take:

(a) all infinitely differentiable functions, or all polynomials, or even only polynomials with rational coefficients;
(b) the set $\dot{C}^\infty(\Omega)$ of all infinitely differentiable functions with compact supports belonging to Ω.

† This is also true for non-linear convex sets, but it is not true for all closed sets. For example, the set $S_R = \{u: \|u\| = R\}$ is closed but not weakly closed.

The space $L_2(\Omega)$ becomes a real Hilbert space if we introduce a scalar product by means of the equality

$$(u, v) = \int_\Omega u(x)v(x)\,dx.$$

Throughout most of the book we shall deal with the real spaces B and H. The exception consists of §§3, 4, 5, and 7 of Chapter II and of §2 of the present chapter in which we use complex Hilbert spaces, including the complex space $L_2(\Omega)$. The elements of this last space are the complex-valued functions $u(x) = u_1(x) + iu_2(x)$ with the scalar product defined as

$$(u, v) = \int_\Omega u(x)\overline{v(x)}\,dx.$$

We mention a number of algebraic and functional inequalities which we shall use frequently in the course of the entire book.

Cauchy's inequality:

$$\left|\sum_{i,j=1}^{n} a_{ij}\xi_i\eta_j\right| \le \sqrt{\sum_{i,j=1}^{n} a_{ij}\xi_i\xi_j}\,\sqrt{\sum_{i,j=1}^{n} a_{ij}\eta_i\eta_j}, \tag{1.2}$$

which is valid for any non-negative quadratic form $a_{ij}\xi_i\xi_j$ with $a_{ij} = a_{ji}$ and for arbitrary real $\xi_1, \ldots, \xi_n, \eta_1, \ldots, \eta_n$.

"Cauchy's inequality with ε":

$$|ab| \le \frac{\varepsilon}{2}|a|^2 + \frac{1}{2\varepsilon}|b|^2, \tag{1.3}$$

which holds for all $\varepsilon > 0$ and for arbitrary a and b and its generalization—Young's inequality:

$$|ab| \le \frac{1}{p}|\varepsilon a|^p + \frac{p-1}{p}\left|\frac{b}{\varepsilon}\right|^{p/(p-1)} \quad \text{for all} \quad p > 1. \tag{1.3'}$$

From the functional inequalities we need inequalities which are concrete versions of the triangle inequality and Cauchy's inequality.

For the space $L_2(\Omega)$ these take the form

$$\left(\int_\Omega (u+v)^2\,dx\right)^{1/2} \le \left(\int_\Omega u^2\,dx\right)^{1/2} + \left(\int_\Omega v^2\,dx\right)^{1/2} \tag{1.4$_1$}$$

and

$$\left|\int_\Omega uv\,dx\right| \le \left(\int_\Omega u^2\,dx\right)^{1/2}\left(\int_\Omega v^2\,dx\right)^{1/2}. \tag{1.4$_2$}$$

For the space $\mathbf{L}_2(\Omega)$ consisting of the vector functions $\mathbf{u} = (u_1, \ldots, u_N)$ with $u_i \in L_2(\Omega)$, Cauchy's inequality takes the form

$$\left| \int_\Omega \sum_{i=1}^N u_i v_i \, dx \right| \leq \left(\int_\Omega \sum_{i=1}^N u_i^2 \, dx \right)^{1/2} \left(\int_\Omega \sum_{i=1}^N v_i^2 \, dx \right)^{1/2}. \tag{1.5}$$

The left-hand side of (1.5) is the modulus of the scalar product of **u** and **v**, while the right-hand side is the product of the norms of **u** and **v**. As a generalization of (1.4_2) we have Hölder's inequality,

$$\left| \int_\Omega uv \, dx \right| \leq \left(\int_\Omega |u|^p \, dx \right)^{1/p} \left(\int_\Omega |v|^{p'} \, dx \right)^{1/p'}, \tag{1.6}$$

which holds for any $u \in L_p(\Omega)$, $v \in L_{p'}(\Omega)$ and for all $p \geq 1$ (p' will always denote the exponent conjugate to p, i.e., $p' = p/(p-1)$). For $p = 1$ we have $p' = \infty$, and by $\|v\|_{p',\Omega}$ it is necessary to take $\operatorname{ess\,sup}_\Omega |v|$. The inequality

$$\left| \sum_{i=1}^N a_i b_i \right| \leq \left(\sum_{i=1}^N |a_i|^p \right)^{1/p} \left(\sum_{i=1}^N |b_i|^{p'} \right)^{1/p'} \tag{1.7}$$

is the discrete analogue of (1.6). A generalization of (1.4_1) gives the triangle inequality for elements of $L_p(\Omega)$:

$$\|u + v\|_{p,\Omega} \leq \|u\|_{p,\Omega} + \|v\|_{p,\Omega} \qquad (p \geq 1). \tag{1.8}$$

It is also true that

$$\left| \int_\Omega \sum_{i=1}^N u_i v_i \, dx \right| \leq \left(\int_\Omega \sum_{i=1}^N |u_i|^p \, dx \right)^{1/p} \left(\int_\Omega \sum_{i=1}^N |v_i|^{p'} \, dx \right)^{1/p'} \qquad (p \geq 1). \tag{1.9}$$

§2. Some Properties of Linear Functionals and Bounded Linear Operators in Hilbert Space

A linear functional l on H (complex or real) is a linear, continuous, numerical function $l(u)$ which is defined for all $u \in H$. Linearity of l (or distributivity) means that, for arbitrary elements u_1 and u_2 of H and for arbitrary numbers λ and μ,

$$l(\lambda u_1 + \mu u_2) = \lambda l(u_1) + \mu l(u_2). \tag{2.1_1}$$

Continuity of $l(u)$ means that $l(u_n) \to l(u)$ whenever $u_n \to u$. It has been shown that if $l(u)$ satisfies (2.1_1), then continuity is equivalent to the boundedness of $l(u)$ on the surface of the unit sphere $S_1 \equiv \{u : \|u\| = 1\}$, or, similarly,

$$|l(u)| \leq c\|u\| \tag{2.1_2}$$

for all $u \in H$.

The theorem of F. Riesz asserts that a linear functional l on H may be written in the form of a scalar product

$$l(u) = (u, v),$$

where the element v is uniquely defined by $l(u)$. The quantity $\|v\|$ is called the *norm* $\|l\|$ of the linear functional l. It is clear that $\|l\| = \sup_{u \in H}(|l(u)|/\|u\|)$ is the smallest of all possible constants c for which (2.1_2) holds. Let us pass now to the linear operator on H. An operator A, defined on some set $\mathscr{D}(A)$ of H, assigns to each element $u \in \mathscr{D}(A)$ a certain element $v \in H$; this is usually written $v = Au$ or $v = A(u)$. If the equality

$$A(\lambda u_1 + \mu u_2) = \lambda A(u_1) + \mu A(u_2)$$

holds on $\mathscr{D}(A)$, then we say that A is linear (where it is assumed that $\mathscr{D}(A)$ is a linear set). If, in addition, there exists a constant c such that, for all $u \in \mathscr{D}(A)$,

$$\|Au\| \leq c\|u\|, \tag{2.2}$$

then A is called a bounded operator on $\mathscr{D}(A)$. Such an operator may be extended in a continuous way to the closure $\overline{\mathscr{D}(A)}$ in H (which will be a closed subspace of H), in which case (2.2) will hold for all $u \in \overline{\mathscr{D}(A)}$. Such an operator can be extended (in different ways if $\overline{\mathscr{D}(A)} \neq H$) to all H and still have (2.2) hold. We shall encounter various bounded operators defined on all H. The smallest c for which (2.2) holds for all $u \in H$ is called the norm of the operator A, so that

$$\|A\| = \sup_{u \in H} \frac{\|Au\|}{\|u\|}.$$

We shall be interested in two classes of bounded linear operators. One of them is the class of *self-adjoint* operators: an operator A is called *self-adjoint* if, for all $u, v \in H$,

$$(Au, v) = (u, Av). \tag{2.3}$$

The spectrum of such an operator A is real and lies in the interval $[-\|A\|, \|A\|]$. The other class is that of completely continuous operators. An operator A is called *completely continuous* if it takes any bounded set into a precompact set. The spectrum of such an operator consists of the point zero, together with an at most countable set of eigenvalues whose only possible point of accumulation is the point zero. Each of these eigenvalues, except perhaps the point zero, is of finite multiplicity. In view of this, the eigenvalues may be enumerated in the order of decreasing modulus, $|\lambda_1| \geq |\lambda_2| \geq \ldots$, with only a finite number of the λ_m on any circle of the type $|\lambda| = |\lambda_k|$. The point zero can be an eigenvalue of infinite multiplicity.

If the operator A is self-adjoint and completely continuous, then its spectrum is real and discrete with a single possible point of accumulation at zero. All eigenvalues, except perhaps zero, are of finite multiplicity, and it is possible to arrange them in order of decreasing modulus: $|\lambda_1| \geq |\lambda_2| \geq \ldots$

where $\lambda_k \to 0$ if $k \to \infty$ (here each eigenvalue, except zero, is repeated according to multiplicity). The corresponding eigenelements $\{u_k\}$ (i.e., the solutions of the equations $Au_k = \lambda_k u_k$) can be chosen in such a way that they are mutually orthogonal and normalized. The closed subspace $\mathscr{L}$ spanned by them coincides with H if $\lambda = 0$ is not a point of the discrete spectrum (i.e., if the equation $Au = 0$ has only the trivial solution $u = 0$ in H). Otherwise the orthogonal complement of $\mathscr{L}$ in H will be the eigensubspace of H corresponding to $\lambda = 0$. Let us denote the set $H \ominus \mathscr{L} = N$, and let $\{v_k\}$ be an orthogonal basis of N. Then any element $u \in H$ may be represented in the form of the sum of two series

$$u = \sum_{k=1} (u, u_k)u_k + \sum_{k=1} (u, v_k)v_k,$$

each of which may contain a finite or infinite number of terms. We have

$$\|u\|^2 = \sum_{k=1} (u, u_k)^2 + \sum_{k=1} (u, v_k)^2,$$

$$Au = \sum_{k=1} \lambda_k(u, u_k)u_k \quad \text{and} \quad \|Au\|^2 = \sum_{k=1} \lambda_k^2(u, u_k)^2.$$

Let us return now to the general completely continuous operator A and formulate some well-known results related to solving equations of the form

$$u - \lambda Au = v, \tag{2.4}$$

where v is a given element of a complex Hilbert space H and λ is a complex parameter.

For these equations we have the Fredholm solvability property, that is, the three Fredholm theorems hold for equations (2.4):

(1) If the homogeneous equation (2.4), i.e., the equation

$$u - \lambda Au = 0, \tag{2.5}$$

has only the trivial solution, then (2.4) may be solved uniquely for arbitrary $v \in H$. (In other words, this theorem asserts that the existence theorem follows from the uniqueness theorem.)

(2) The homogeneous equation (2.5) can have non-trivial (i.e., non-zero) solutions for not more than a countable number of values $\{\lambda_k\}$, each of which is of finite multiplicity. The set $\{\lambda_k\}$ cannot have a point of condensation λ_0 with $|\lambda_0| < \infty$. These exceptional values $\{\lambda_k\}$ are called characteristic numbers for A. For the adjoint operator A^* the characteristic numbers are $\{\bar{\lambda}_k\}$, that is, the equation

$$u - \lambda A^*u = 0 \tag{2.6}$$

has a non-trivial solution only for $\lambda = \bar{\lambda}_k$, and the multiplicity of $\bar{\lambda}_k$ for A^* is the same as the multiplicity of λ_k for A.

(3) Equation (2.4) with λ equal to any one of the characteristic values λ_k may be solved for those v, and only for those v, which are orthogonal to all solutions of (2.6) corresponding to $\lambda = \bar{\lambda}_k$.

If these orthogonality conditions are fulfilled, then (2.4) has infinitely many solutions. All of them can be written in the form $u = u_0 + \sum_{j=m}^{m+p} c_j u_j$, where u_0 is any solution of (2.4) with $\lambda = \lambda_k$, the c_j are arbitrary constants, and the u_j, $j = m, \ldots, m+p$, are all linearly independent solutions of (2.5) for $\lambda = \lambda_k$.

Such solvability takes place, for example, for linear algebraic systems in which u and v are vectors with n components and A is a square matrix with n^2 entries. The same thing is true of integral operators having a kernel which does not misbehave too badly.

In Chapter II we shall show that such solvability also takes place for the basic boundary value problems for elliptic equations with bounded coefficients in a bounded domain. This is not obvious, for in these problems one has to deal with unbounded operators, but they can be reduced to equations of the type (2.4) with completely continuous operators A.

We also recall a well-known fact related to the solvability of (2.4) with an arbitrary bounded operator A, namely, that such an equation may be solved uniquely for all $v \in H$ with λ such that $|\lambda| < 1/\|A\|$, and that its solution may be represented in the form of a series $u = v + \lambda A v + \lambda^2 A^2 v + \ldots$ which converges in H and also $\|u\| \leq \|v\|/(1 - |\lambda|\,\|A\|)$.

In this section we have formulated various theorems about bounded operators in a complex Hilbert space H, which are also valid in real spaces H. However, in studying the spectra of non-symmetric operators A acting on a real space H we encounter in a natural way a larger complex space, for the spectrum of such operators A may be complex.

§3. Unbounded Operators

Let us recall certain facts about unbounded linear operators A on H. Such operators are not defined for all elements u of H. The set on which A is defined is called the domain of definition of A and is denoted by $\mathscr{D}(A)$. This set is linear and, for all $u, v \in \mathscr{D}(A)$ and for all complex numbers λ, μ, satisfies

$$A(\lambda u + \mu v) = \lambda A u + \mu A v.$$

In contrast to the case of the bounded operators, a constant c does not exist in the case of an unbounded operator A for which (2.2) holds for all u in $\mathscr{D}(A)$. We shall consider only the case that $\mathscr{D}(A)$ is dense in H. The set of values of A, i.e. the range of A, will be denoted by $\mathscr{R}(A)$, so that $A(\mathscr{D}(A)) = \mathscr{R}(A) \subset H$.

We shall be interested in unbounded operators arising from differential expressions. To every such expression correspond various operators defined by indicating their domain of definition. As an example, we consider the differential expression $\mathscr{L}u(x) = d^2u(x)/dx^2$ on the interval $x \in [0, 1]$, taking as H the real functional space $L_2(0, 1)$. We can associate with $\mathscr{L}$ the operator A in this space which is defined on all infinitely differentiable functions with support in $(0, 1)$. For $u(x)$ in $\mathscr{D}(A)$ the operator A is calculated by $Au = \mathscr{L}u = d^2u(x)/dx^2$. It is easy to see that A is unbounded on $\mathscr{D}(A)$. With the expression $\mathscr{L}u$ we can associate another unbounded operator $\tilde{A}$ whose domain of definition is the set of all infinitely differentiable functions on $[0, 1]$. On these functions, $\tilde{A}$ is calculated in the same way as A on $\mathscr{D}(A)$, namely, $\tilde{A}u = d^2u(x)/dx^2$. There is a natural ordering between A and $\tilde{A}$: $\mathscr{D}(A) \subset \mathscr{D}(\tilde{A})$ and $Au = \tilde{A}u$ for $u \in \mathscr{D}(A)$. In a situation like this we say that the operator $\tilde{A}$ is an *extension* of the operator A. It is clear that the domain of definition for $\mathscr{L}$ can be chosen in an infinite number of ways, and each time we are led, generally speaking, to another unbounded operator with other properties. The operators A and $\tilde{A}$ described above have fundamentally different properties. For example, A satisfies the relationship

$$(Au, v) = \int_0^1 \frac{d^2u}{dx^2} v \, dx = (u, Av), \tag{3.1}$$

which is easily verified by integration by parts, where $u(x)$ and $v(x)$ in (3.1) are arbitrary elements of $\mathscr{D}(A)$. For the operator $\tilde{A}$ this property does not hold, for

$$(Au, v) = (u, Av) + \frac{du}{dx} v \bigg|_{x=0}^{x=1} - u \frac{dv}{dx} \bigg|_{x=0}^{x=1}, \tag{3.2}$$

and the sum of the last two terms on the right-hand side of (3.2) is not zero for all $u(x)$ and $v(x)$ in $\mathscr{D}(\tilde{A})$. The property (3.1) guarantees the symmetry of A, but the operator $\tilde{A}$ is not symmetric. As is well known in general operator theory, symmetric operators possess a large number of nice properties. The theory of symmetric operators has been well developed and can be used for studying specific classes of differential operators. One of the most important concepts is that of a self-adjoint operator.

An operator A is called *self-adjoint* if it is symmetric, that is, if

$$(Au, v) = (u, Av) \quad \text{for all} \quad u, v \in \mathscr{D}(A), \tag{3.3}$$

and if the identity

$$(Au, v) = (u, w), \tag{3.4}$$

where v and w are fixed and u is an arbitrary element of $\mathscr{D}(A)$, implies that $v \in \mathscr{D}(A)$ and $w = Av$. In other words, A is self-adjoint if its adjoint operator A^* has the same domain of definition $\mathscr{D}(A)$ and $A = A^*$ on $\mathscr{D}(A)$. The identity (3.4) defines the domain of definition of A^* and its values on the domain, namely, those v for which there exists w satisfying (3.4) for all $u \in \mathscr{D}(A)$ constitute $\mathscr{D}(A^*)$ with A^*v set equal to w. In the majority of cases of operators A arising from differential expressions it is easy to verify the validity (or non-validity) of (3.3) by means of integration by parts. It is considerably more difficult to describe the domain $\mathscr{D}(A^*)$ for these operators and to determine whether it coincides with $\mathscr{D}(A)$ or not. For differential expressions containing partial derivatives, this is usually carried out in some "roundabout" way, most often with the use of inverse operators which turn out to be bounded. In point of fact, these "roundabout" methods are used for studying elliptic differential equations under one set of boundary conditions or another. For bounded operators A defined on all H, the self-adjointness is a consequence of their symmetry.

Let us return to the example of the differential operator $\mathscr{L}u = d^2u/dx^2$. We have already determined above that the corresponding operator A is symmetric. It is not difficult to see, however, that it is not self-adjoint. Actually, (3.3) holds for A not only for u and v in $\mathscr{D}(A)$, but also, for example, for $u \in \mathscr{D}(A)$ and $v \in \mathscr{D}(\tilde{A})$. This shows that $\mathscr{D}(A^*)$ is larger than $\mathscr{D}(A)$. There arises the natural question: Can A be extended so as to be self-adjoint? It turns out that it can be done, and, indeed, in an infinite number of ways. The general theory of operators gives the first step of such an extension; this is the procedure of the closure of an operator. The procedure consists of the following. Let A (where A may be non-symmetric) be defined on a dense set $\mathscr{D}(A)$ of the space H. We adjoin to $\mathscr{D}(A)$ all elements u which are limits of those sequences $\{u_n\}$ of $\mathscr{D}(A)$ for which $\{Au_n\}$ converges to some element v. We define $v = \bar{A}u$, where the symbol $\bar{A}$ denotes the closure of A. The set $\mathscr{D}(A)$, supplemented by all such elements u, constitutes $\mathscr{D}(\bar{A})$, the domain of definition of $\bar{A}$. However, this procedure does not always lead to a linear operator $\bar{A}$, for the definition requires that it be defined in a single-valued way on $\mathscr{D}(\bar{A})$, which is equivalent to the requirement that it have the value 0 on the zero element. Actually, in the procedure described above, we can encounter the case where u_n converges to $u = 0$ and Au_n converges to $v \neq 0$. If this happens, then, by what was said above, we should set $\bar{A}0$ equal to $v \neq 0$, and, by the same token, obtain an operator $\bar{A}$ which is not linear. Examples show that the situation described is possible for certain unbounded operators, and consequently an arbitrary unbounded operator does not have a closure.

It is not difficult to prove that a necessary and sufficient condition that an operator A have a closure is that the domain of definition of the conjugate operator A^* be dense in H. This criterion is easy to verify for operators defined by differential expressions with coefficients that are "not too bad," so that such operators have a closure.

In particular, the operator A that we defined above by $\mathscr{L}u = d^2u/dx^2$ has a closure. But then the question arises: How do we find the closure $\bar{A}$ for A, that is, which functions are to be adjoined to $\mathscr{D}(A)$ and how do we calculate the operator $\bar{A}$ for them? It turns out that it is not a simple thing to answer this question, especially in the case of differential expressions with partial derivatives. The answer requires the extension of the notion of derivative and the introduction of what are called "generalized derivatives." We shall go into this notion in more detail in the next section in view of its cardinal importance for all problems which will be investigated in this book.

We shall formulate here only the final result relating to the operator $\mathscr{L}u = d^2u/dx^2$; the closure of A is achieved by adjoining to $\mathscr{D}(A)$ all continuous differentiable functions $u(x)$ vanishing at $x = 0$ and $x = 1$ for which the first derivatives du/dx are absolutely continuous and vanish at the same end-points. Moreover, the first derivative of du/dx, which exists almost everywhere, should be in $L_2(0, 1)$. We shall retain the earlier notation d^2u/dx^2 for these derivatives, but we should keep in mind that they are defined only as elements of $L_2(0, 1)$. We shall denote by $\mathring{W}_2^2(0, 1)$ this class of functions $\mathscr{D}(\bar{A})$ and say of its elements that they have generalized derivatives up to the second order inclusive. Nothing above indicates that its elements vanish at the end-points of the interval, along with its first derivative. The operator $\bar{A}$ is defined on $\mathscr{D}(\bar{A}) = \mathring{W}_2^2(0, 1)$, as the operator calculating the "generalized" derivative of second order.

It is easy to understand that the closure procedure for a symmetric operator A leads to a symmetric operator $\bar{A}$ (recall that we originally agreed that $\mathscr{D}(A)$ was dense in H, so that such operators would have closures). However, this extension can turn out to be "insufficient," that is, $\bar{A}$ need not be self-adjoint on $\mathscr{D}(\bar{A})$. This is the case for the operator $\bar{A}(u) = Lu = d^2u/dx^2$ on $\mathring{W}_2^2(0, 1)$. In the general theory of symmetric operators, abstract schemes have been developed for further extensions of an operator that preserve its symmetry, including the construction of self-adjoint extensions. However, their realization for partial differential operators is not at all transparent, for it turns out that the "defect subspace" on which one must extend $\mathscr{D}(\bar{A})$ is infinite dimensional. In view of this, we prefer not to use these general constructions of operator theory for investigating problems associated with such operators. Thus we shall not resort to them in our book, but we shall select other methods which have been developed in the theory of boundary value problems. For differential (symmetric) operators in one unknown the general theory makes it possible to describe simply and effectively all possible self-adjoint extensions, but we shall not review this here, for such operators are not the subject matter of this book (their theory having been studied in detail in a number of monographs). We cite only a few such extensions for the operator $\bar{A}u = d^2u/dx^2$ to $\mathscr{D}(\bar{A}) = \mathring{W}_2^2(0, 1)$. The first of these has the form $\hat{A}u = d^2u/dx^2$, where $\mathscr{D}(\hat{A})$ is the set of all functions of $W_2^2(0, 1)$ vanishing at the end-points of the interval. By $W_2^2(0, 1)$ we mean here the totality of all functions which have generalized derivatives up to second order inclusive and are square-summable on $(0, 1)$ (the elements $u(x) \in W_2^2(0, 1)$ are continuously differentiable functions with absolutely continuous derivatives $du(x)/dx$ on $[0, 1]$; these latter have derivatives almost everywhere as elements of $L_2(0, 1)$). Such an extension $\hat{A}$ is connected with the first boundary value problem for $\mathscr{L}u$: the problem of determining the solution $u(x)$ of the equation

$$\mathscr{L}u \equiv \frac{d^2u}{dx^2} = f(x) \tag{3.5}$$

with the boundary conditions

$$u(0) = 0, \qquad u(1) = 0. \tag{3.6}$$

It turns out that this problem has a unique solution $u(x)$ in $\mathscr{D}(\hat{A})$ for arbitrary f in $L_2(0, 1)$. It is equivalent to the unique solubility of the equation $\hat{A}u = f$ for all $f \in L_2(0, 1)$. Thus, the extension $\hat{A}$ of our original operator A corresponds to the problem (3.5), (3.6). If we had not extended the operator A, then we would not have had such a nice theorem on solvability, for the equation $Au = f$ does not have a solution in $\mathscr{D}(A)$ for arbitrary f in $L_2(0, 1)$.

Other boundary value problems for the equation (3.5) require other extensions of A. For example, if we add to (3.5) the conditions

$$u(0) = 0, \qquad \left.\frac{du}{dx}\right|_{x=1} = 0, \tag{3.7}$$

then we must extend A as the operator of repeated generalized differentiation of functions of $W_2^2(0, 1)$ satisfying (3.7). Such an extension $\hat{\hat{A}}$ is also self-adjoint and leads to the equation $\hat{\hat{A}}u = f$, which is solvable in $\mathscr{D}(\hat{\hat{A}})$ for all $f \in L_2(0, 1)$.

In this section we have reviewed those concepts in the theory of unbounded operators that will be used in the sequel. We shall not use any special theorems on the solvability of some classes of equations with unbounded operators, nor methods of extending unbounded operators.

We now go over to a description of concrete function spaces which will be useful in what follows.

§4. Generalized Derivatives and Averages

In §3 we have encountered the need to introduce generalized differentiation. This can be done in different but equivalent ways.† We shall choose one of them which, for us, seems to be the most convenient for the purposes of this book.

It is known that for two arbitrary functions $u(x)$ and $v(x)$, which are infinitely differentiable in the domain $\Omega \subset R^n$ with $v(x)$ vanishing in a boundary strip (i.e., $v \in \dot{C}^\infty(\Omega)$), we have

$$\int_\Omega \left[u \frac{\partial^k v}{\partial x_1^{k_1} \ldots \partial x_n^{k_n}} + (-1)^{k+1} v \frac{\partial^k u}{\partial x_1^{k_1} \ldots \partial x_n^{k_n}} \right] dx = 0,$$

† We remark that not all generalizations of classical differentiation proposed at the beginning of the century are equivalent, and we have chosen that which answers the needs of functional analysis and the theory of boundary value problems. But this generalization can be introduced in different ways, which, however, lead to the same extension.

which is obtained by a k-times integration by parts. This is the basic analytic relationship in the theory of boundary value problems and we shall preserve it for the generalized derivatives to be introduced below. Namely, we shall call a function $\omega_{k_1\dots k_n}$, which is summable on every strictly interior bounded sub-domain Ω' of Ω,† the generalized derivative of the form

$$\frac{\partial^k}{\partial x_1^{k_1}\dots\partial x_n^{k_n}}$$

of the function $u(x)$, summable on the same $\bar{\Omega}' \subset \Omega$, if for every $v \in \dot{C}^\infty(\Omega)$, we have

$$\int_\Omega \left[u\frac{\partial^k v}{\partial x_1^{k_1}\dots\partial x_n^{k_n}} + (-1)^{k+1} v\omega_{k_1\dots k_n}\right] dx = 0. \tag{4.1}$$

We shall denote the function $\omega_{k_1\dots k_n}$ as $\partial^k u/\partial x_1^{k_1}\dots\partial x_n^{k_n}$. This will not evoke misunderstanding for if $u \in C^k(\bar{\Omega})$ then it satisfies (4.1) with $\omega_{k_1\dots k_n} = \partial^k u/\partial x_1^{k_1}\dots\partial x_n^{k_n}$. It is clear that the concept we have introduced is an extension of the concept of the (classical) continuous partial derivative of the form $\partial^k u/\partial x_1^{k_1}\dots\partial x_n^{k_n}$. Generalized derivatives preserve many (but not all!) properties of classical derivatives. Thus, for example, if u_1 and u_2 have generalized derivatives of the form $\partial^k/\partial x_1^{k_1}\dots\partial x_n^{k_n}$ in Ω, the sum $c_1u_1 + c_2u_2$ has the same generalized derivative in Ω, and

$$\frac{\partial^k(c_1u_1 + c_2u_2)}{\partial x_1^{k_1}\dots\partial x_n^{k_n}} = c_1\frac{\partial^k u_1}{\partial x_1^{k_1}\dots\partial x_n^{k_n}} + c_2\frac{\partial^k u_2}{\partial x_1^{k_1}\dots\partial x_n^{k_n}}.$$

If v is a generalized derivative of u of the form $\partial^l/\partial x_1^{l_1}\dots\partial x_n^{l_n}$, and ω is a generalized derivative of v of the form $\partial^k/\partial x_1^{k_1}\dots\partial x_n^{k_n}$ in the domain Ω, then ω is a generalized derivative of u of the form $\partial^{k+l}/\partial x_1^{k_1+l_1}\dots\partial x_n^{k_n+l_n}$.

It is obvious from the definition of the generalized derivative $\partial^k/\partial x_1^{k_1}\dots\partial x_n^{k_n}$ that it is independent of the order of differentiation. In order to preserve the formula

$$\frac{\partial(u_1u_2)}{\partial x_i} = u_1\frac{\partial u_2}{\partial x_i} + u_2\frac{\partial u_1}{\partial x_i}$$

for the derivative of a product, we must require, for example, that u_k and $\partial u_k/\partial x_i$, $k = 1, 2$, be square-summable on every $\bar{\Omega}' \subset \Omega$. Similar requirements must be imposed upon u_k and their generalized derivatives for calculating general derivatives of any order for u_1u_2.

However, generalized derivatives do not preserve all the properties of classical derivatives, and working with them requires certain precautions. Thus, for example, the existence of a generalized derivative of $u(x)$ of the form $\partial^k/\partial x_1^{k_1}\dots\partial x_n^{k_n}$ does not imply the existence of subordinate derivatives of lower order. However, if $u(x)$ has generalized derivatives of k-th order

† For brevity, this will be written as $\bar{\Omega}' \subset \Omega$.

of all forms and if the p-th power ($p \geq 1$) of the modulus of u and of these derivatives are summable on every $\overline{\Omega}' \subset \Omega$, then u possesses all generalized derivatives of order less than k which belong to $L_p(\Omega')$ (and even to $L_q(\Omega')$ for a certain $q = q(n, p, k) > p$) (cf. [SO 5], [SM 2]).

Another important difference between generalized derivatives and classical derivatives is that the former are defined up to a set of measure zero, and are "tied" to the domain Ω, in other words, they have an "integral character"; this is obvious from the very definition of generalized derivative. We must consider each of them as an element of $L_p(\Omega)$ (or of $L_p(\Omega')$ for every $\overline{\Omega}' \subset \Omega$) for some $p \geq 1$. They are tied to the domain Ω in this way: If u has the generalized derivative $\partial^k u/\partial x_1^{k_1} \ldots \partial x_n^{k_n}$ in Ω then u has the same generalized derivative in every $\Omega' \subset \Omega$. However, if Ω is split into three parts, two disjoint subdomains Ω_1, Ω_2 and a set Γ separating Ω_1 and Ω_2, then the existence of generalized derivatives $\partial^k u/\partial x_1^{k_1} \ldots \partial x_n^{k_n}$ of u in each of Ω_1 and Ω_2 does not, in general, imply the existence of the generalized derivative $\partial^k u/\partial x_1^{k_1} \ldots \partial x_n^{k_n}$ of u in all of Ω. We can see this from the example of a function $u(x)$, defined on the interval $x \in [0, 1]$ and infinitely differentiable there except at the point $x = \frac{1}{2}$ where it has a discontinuity of the first kind, i.e., a simple jump. Such a function does not even have a generalized derivative of first order on $(0, 1)$, despite the fact that it has a classical derivative everywhere except at $x = \frac{1}{2}$. This can be verified directly, starting from the definition. On an arbitrary interval not containing $x = \frac{1}{2}$, the function $u(x)$ has generalized derivatives of any order.

However, this example should not give the impression that, for the existence of a generalized derivative, it is necessary that the function $u(x)$ be continuous. First of all, as we remarked above, a function $u(x)$ having some generalized derivative in Ω can be changed arbitrarily on a set of measure zero. Second, it is easy to give examples of functions having generalized derivatives which cannot be redefined on a set of measure zero so as to make them continuous in Ω. Thus, the function $u(x) = 1/|x|^{\varepsilon}, 0 < \varepsilon < 1$, has generalized derivative of first order $\partial u/\partial x_i = -\varepsilon x_i/|x|^{2+\varepsilon}$, in the circle $|x| = \sqrt{x_1^2 + x_2^2} \leq 1$; for this, it is necessary to verify that the corresponding identity of type (4.1) holds. Such a function cannot be made continuous (or even bounded) by redefining it on a set of measure zero.

Questions as to what sort of smoothness can be attained for a function $u(x)$ having these or other generalized derivatives, summable on Ω to these or other powers, by redefining it on a set of measure zero, form the subject of investigation of the following section; results of this type go by the name of embedding theorems. The answer depends on the dimension of the space x, the order of the generalized derivatives, and the power to which they are summable.

In this section we shall describe a number of the simpler properties of generalized derivatives. We shall begin with the connection between generalized differentiability and absolute continuity of a function. If u is a function of $x \in [0, l]$ and is absolutely continuous on $[0, l]$ then, as is well known

([SM 2], [SV 1]), $u(x)$ has a derivative du/dx almost everywhere in the ordinary sense, which is summable on $[0, l]$, and the function may be uniquely recovered from its derivative via the formula of Newton and Leibniz:

$$u(x) = u(x_1) + \int_{x_1}^{x} \frac{du(\tau)}{d\tau}\, d\tau, \qquad x_1, x \in [0, l]. \tag{4.2}$$

It is not hard to show that du/dx is the generalized derivative of $u(x)$ on $[0, l]$. The following (converse) assertion is valid: If $u(x)$ is in $L_1(0, l)$ and has a generalized derivative du/dx on $(0, l)$ belonging to $L_1(0, l)$, then $u(x)$ is equivalent to an absolutely continuous function on $[0, l]$ (which we shall also denote by $u(x)$), for which (4.2) holds, where we take the generalized derivative du/dx in lieu of du/dx.

Let us agree to formulate this fact more concisely by saying that if $u \in W_1^1(0, l)$, then $u(x)$ is absolutely continuous on $[0, l]$.

Thus, here and in what follows, in speaking of these and other properties of a function $u(x)$ in $L_p(\Omega)$, we shall mean the existence of at least one function, equivalent to $u(x)$ on Ω, for which this property holds. With this identification of all functions that are equivalent (on Ω) we have the accepted designation which uses the same symbol $u(x)$ for a concrete representative, as well as for the whole class of functions which are equivalent to u on Ω. In particular, all subsequent assertions in this section and in §5 on embedding theorems should be understood in this light.

Let us consider now a function $u(x)$ of several variables $x = (x_1, \ldots, x_n)$. Let $u(x)$ be in $L_1(\Omega)$ and have generalized derivative $\partial u/dx_1 \in L_1(\Omega)$. Then, obviously, it is absolutely continuous in x_1 for almost all values of $x_1' \equiv (x_2, \ldots, x_n)$. More precisely, if Ω_1' is the orthogonal projection of Ω onto the hyperplane $x_1 = 0$, then, for almost all values of x_1' in Ω_1', $u(x_1, x_1')$ as a function of x_1 is absolutely continuous on the closure of any interval in Ω.

If Ω contains a cylinder $\tilde{\Omega}$ of the form $\tilde{\Omega} = (x : x_1 \in (l_1, l_2); x_1' \in \tilde{\Omega}_1'\}$, where $\tilde{\Omega}_1'$ is some measurable set in Ω_1', then, for almost all x_1' in $\tilde{\Omega}_1'$ we have that

$$u(x_1, x_1') = u(x_1^0, x_1') + \int_{x_1^0}^{x_1} \frac{\partial u(\tau, x_1')}{\partial \tau}\, d\tau \tag{4.3}$$

for all x_1^0 and $x_1 \in (l_1, l_2)$. This result can be integrated with respect to $x_1' \in \tilde{\Omega}_1'$, and the result expressed in the form

$$\int_{\tilde{\Omega}_1'} u(x_1, x_1')\, dx_1' = \int_{\tilde{\Omega}_1'} u(x_1^0, x_1')\, dx_1' + \int_{x_1^0}^{x_1} \int_{\tilde{\Omega}_1'} \frac{\partial u(\tau, x_1')}{\partial \tau}\, dx_1'\, d\tau \tag{4.4}$$

where we have applied Fubini's theorem.

Everything that we have said applies not only to x_1 but also to an arbitrary coordinate direction. We assume now that the function $u(x)$ is in $L_1(\Omega)$ and possesses all generalized derivatives $\partial u/\partial x_i$ $(i = 1, \ldots, n)$ of first order and that these derivatives are also in $L_1(\Omega)$. We shall assume, further, that we can make a non-degenerate change of variables from x to y in such a way that the $y = y(x)$ are continuous in $\bar{\Omega}$ with bounded generalized derivatives $\partial y_i(x)/\partial x_k$ and with Jacobian $|\partial y(x)/\partial x| \geq c > 0$ in $\bar{\Omega}$, and such that the inverse functions $x = x(y)$ have the same properties as $y(x)$. Then the function $\tilde{u}(y) = u(x(y))$ will be in $L_1(\tilde{\Omega})$, where $\tilde{\Omega}$ is the range of y, and will have generalized derivatives $\partial\tilde{u}/\partial y_k$ in $L_1(\tilde{\Omega})$ which may be expressed in terms of the generalized derivatives $\partial u/\partial x_i$ by means of the usual formula:

$$\frac{\partial \tilde{u}(y)}{\partial y_k} = \left.\frac{\partial u(x)}{\partial x_i}\right|_{x=x(y)} \cdot \frac{\partial x_i(y)}{\partial y_k}.$$

An operation of averaging with some kernel is very useful for our purposes, more precisely, with a family of kernels $\theta(x, y, \rho)$ depending on a parameter ρ (continuous or discrete) which tends to zero. It is usually assumed that $\theta(x, y, \rho) = 0$ for $|x - y| \geq c\rho$; we shall also assume that this condition is fulfilled. In addition to this, it is essential that we have a normalization condition on θ:

$$\int_{R^n} \theta(x, y, \rho)\, dy = \int_{|x-y| \leq c\rho} \theta(x, y, \rho)\, dy = 1. \tag{4.5}$$

By the average—or the mean—of an arbitrary function $u(x)$, which we shall assume at the beginning to be defined for all $x \in R^n$, with kernel $\theta(x, y, \rho)$, we shall mean the function

$$u_\rho(x) = \int_{R^n} \theta(x, y, \rho) u(y)\, dy. \tag{4.6}$$

These means or averages u_ρ converge to u as $\rho \to 0$ for wide classes of kernels θ and functions u, and we shall make this statement more precise farther along. The purpose of the averaging operation is to obtain families of functions which possess the necessary smoothness and which approximate the averaged function. As far as we know, the first operation of this type was used extensively by V. A. Steklov for investigating a number of questions in the theory of differential equations and in the theory of series. He used a kernel of the type

$$\theta(x, y, \rho) = \frac{1}{\varkappa_n \rho^n}\, \theta\left(\frac{x - y}{\rho}\right) \tag{4.7}$$

with a concrete function $\theta(x)$ having a compact support. His averages have the form:

$$u_h(x) = \frac{1}{h^n} \int_{x_1 - h/2}^{x_1 + h/2} \cdots \int_{x_n - h/2}^{x_n + h/2} u(y)\, dy. \tag{4.8}$$

If $u(x)$ is locally summable, then the function $u_h(x)$ will be continuous and, as $h \to 0$, will converge to $u(x)$ almost everywhere and in the norms of $L_1(\Omega')$. Moreover, $u_h(x)$ has first-order generalized derivatives with respect to x_i.

More generally, if $u(x) \in L_p(\Omega)$, $p \geq 1$, and if $u(x)$ is taken to be 0 outside Ω, then the functions $u_h(x)$ are continuous in $\bar{\Omega}$ with first-order generalized derivatives in $L_p(\Omega)$. The averages $u_h(x)$ converge to $u(x)$ almost everywhere in Ω and in the norm of $L_p(\Omega)$. It is clear that we cannot get better convergence for an arbitrary element $u(x)$ of $L_p(\Omega)$.

We shall frequently use averages with sufficiently smooth—or even infinitely differentiable—kernels, in particular, with a kernel of the type (4.7), where

$$\theta(x) = \begin{cases} \exp\left\{\dfrac{|x|^2}{|x|^2 - 1}\right\} & \text{for } |x| \leq 1, \\ 0 & \text{for } |x| \geq 1, \end{cases}$$

with

$$\varkappa_n = \int_{|x|<1} \exp\left\{\frac{|x|^2}{|x|^2 - 1}\right\} dx.$$

For such kernels the averages will be sufficiently smooth (or even infinitely differentiable). The character of their convergence to the averaged function depends on the properties of the function, in particular, the same character is true for these averages as we mentioned above for the Steklov averages.

For the averages u_ρ with sufficiently smooth kernels of the type (4.7) we have the following useful property: If $u(x)$ has the generalized derivative $\partial^k u/\partial x_1^{k_1} \ldots \partial x_n^{k_n}$, then $\partial^k u_\rho/\partial x_1^{k_1} \ldots \partial x_n^{k_n} = (\partial^k u/\partial x_1^{k_1} \ldots \partial x_n^{k_n})_\rho$, that is, the operations of differentiation and averaging are interchangeable. From this and from the convergence of $u_\rho(x)$ to $u(x)$ mentioned above it follows that if $u(x)$ and $\partial^k u/\partial x_1^{k_1} \ldots \partial x_n^{k_n}$ are in $L_p(\Omega)$, then for every $\bar{\Omega}' \subset \Omega$ the averages and their derivatives of the form $\partial^k/\partial x_1^{k_1} \ldots \partial x_n^{k_n}$ converge to u and its derivatives of the same form in the norm of $L_p(\Omega')$ as $\rho \to 0$ (and almost everywhere in Ω). In formulating this result we assumed that $u(x)$ was defined only on Ω. Thus the averages $u_\rho(x)$ are defined, not on all $x \in \Omega$, but only on those $x \in \Omega$ whose distance from $\partial\Omega$ is not less than the radius of the support of the kernel $\theta(x, y, \rho)$. In what follows we shall use only the Steklov means and means of the form

$$u_\rho(x) = \frac{1}{\varkappa_n \rho^n} \int_{|x-y| \leq \rho} \theta\left(\frac{|x-y|}{\rho}\right) u(y)\, dy, \tag{4.9}$$

where $\theta(t)$ is a non-negative, infinitely differentiable function of $t \geq 0$, vanishing for $t \geq 1$, and normalized by $\int_0^1 \theta(t) t^{n-1}\, dt = 1$, and where $\varkappa_n$ is the surface area of the unit sphere. The averages $u_\rho(x)$ are defined only on the domain $\Omega_\rho \subset \Omega$ whose distance from $\partial\Omega$ is ρ, while the function $u(x)$ being averaged is defined on Ω. In order to define $u_\rho(x)$ for $x \in \Omega - \Omega_\rho$ we must somehow extend $u(x)$ to the strip of width ρ surrounding Ω. We did this above for functions $u(x) \in L_p(\Omega)$ by setting $u(x) = 0$ outside Ω. Such an extension again gave us a function (which we denote also by $u(x)$), which is p-th power summable on all $\tilde{\Omega} \supset \Omega$, that is, with such an extension we do not go outside the space L_p. This allowed us to obtain, for $u(x) \in L_p(\Omega)$, a smooth approximation $u_\rho(x)$ which is defined on Ω and which converges to $u(x)$ in the norm of $L_p(\Omega)$. In general, there is no better convergence for $u(x)$ in $L_p(\Omega)$.

However, if $u(x)$ has some nicer properties, for example, if it is continuous on $\bar{\Omega}$, then it would be desirable to have a sequence of smooth functions which would approximate $u(x)$ uniformly on $\bar{\Omega}$ (i.e., in the norm of $C(\bar{\Omega})$). For this, $u(x)$ would have to be extended outside $\bar{\Omega}$ in such a way that it remained continuous; then the averages calculated by means of this extension of $u(x)$ would yield the desired approximation. But if the extension of $u(x)$ from $\bar{\Omega}$ does not have this property, then the uniformity of the approximation is lost at the points of discontinuity of $u(x)$.

The situation is similar in the case that $u(x)$ possesses some kind of derivatives in Ω. If we wish to construct, by using the averaging operation, a sequence of smooth functions u_ρ approximating u in such a way that their derivatives approximate the corresponding derivatives of $u(x)$ in $L_p(\Omega)$, we must first extend $u(x)$ to some neighborhood of Ω in a way that preserves its smoothness properties. Thus, for example, if $u(x)$ and $\partial^k u/\partial x_1^{k_1} \ldots \partial x_n^{k_n}$ are in $L_p(\Omega)$, then $u(x)$ must be extended to some domain $\tilde{\Omega} \supset \bar{\Omega}$ in such a way that both the extension and its derivative of the form indicated are in $L_p(\tilde{\Omega})$. If this is done, then the averages u_ρ which are formed by (4.9) (or by (4.6), along with θ of the form (4.7)), will be smooth functions for all sufficiently small ρ, which converge, together with

$$\partial^k u_\rho/\partial x_1^{k_1} \ldots \partial x_n^{k_n} = (\partial^k u/\partial x_1^{k_1} \ldots \partial x_n^{k_n})_\rho$$

to $u(x)$ and $\partial^k u/\partial x_1^{k_1} \ldots \partial x_n^{k_n}$, respectively, in the norm of $L_p(\Omega)$.

We end this section by citing a useful and simple criterion for the existence of generalized partial derivatives of a function $u(x)$.

Theorem 4.1. *Let $u(x)$ be a summable function on Ω. If $u(x)$ can be approximated by a sequence of functions $u_s(x)$ $(s = 1, 2, \ldots)$ which are k-times continuously differentiable in Ω in the sense that, for every function $v(x) \in \dot{C}^\infty(\Omega)$,*

$$\lim_{s \to \infty} \int_\Omega (u_s - u)v\, dx = 0,$$

and if, in addition $\|u_s\|_{p,\Omega}$ and $\|\partial^k u_s/\partial x_1^{k_1} \ldots \partial x_n^{k_n}\|_{p,\Omega} \le c$, then the function u has generalized derivative $\partial^k u/\partial x_1^{k_1} \ldots \partial x_n^{k_n}$ and $\|u\|_{p,\Omega}$ and $\|\partial^k u/\partial x_1^{k_1} \ldots \partial x_n^{k_n}\|_{p,\Omega} \le c$, $p \ge 1$.

The conclusion is still valid if the functions $u_s(x)$ are in $L_p(\Omega)$ and have generalized derivatives of the same kind, rather than continuous derivatives.

The results cited in this section are relatively easy to prove from the definition of generalized derivatives and from certain basic facts in real-variable theory about measurable and absolutely continuous functions. Proofs may be found, for example, in [SO 5] and [SM 2].

§5. Definition of the Spaces $W^l_m(\Omega)$ and $\mathring{W}^l_m(\Omega)$

We shall say that a function $u(x)$, continuous in $\bar{\Omega}$, has a continuous derivative of some kind in $\bar{\Omega}$ if this derivative, which is defined in the usual way for points in Ω, can be extended by continuity to $\bar{\Omega}$.† Let us consider the set $C^l(\bar{\Omega})$ of all functions $u(x)$ which have in $\bar{\Omega}$ continuous derivatives with respect to $x_1, \ldots, x_n$ up to order l inclusive, and let us introduce into $C^l(\bar{\Omega})$ the norm

$$\|u\|^{(l)}_{m,\Omega} = \left\{ \int_\Omega \sum_{k=0}^{l} \sum_{(k)} \left| \frac{\partial^k u}{\partial x_1^{k_1} \ldots \partial x_n^{k_n}} \right|^m dx \right\}^{1/m}, \tag{5.1}$$

where $m \geq 1$, and where $\sum_{(k)}$ indicates the summation over all possible derivatives of order k. The closure of this set with the norm (5.1) will be called the *space* $\tilde{W}^l_m(\Omega)$, and the functions $u(x)$ in this closure will be called its elements.

Theorem 5.1. *The space $\tilde{W}^l_m(\Omega)$ is a separable space of type B (and complete because of its construction). All elements of $\tilde{W}^l_m(\Omega)$ have generalized derivatives in Ω, up to order l inclusive, which are m-th power summable on Ω.*

$\tilde{W}^l_2(\Omega)$ is a Hilbert space with scalar product

$$(u, v)^{(l)}_{2,\Omega} = \int_\Omega \sum_{k=0}^{l} \sum_{(k)} \frac{\partial^k u}{\partial x_1^{k_1} \ldots \partial x_n^{k_n}} \frac{\partial^k v}{\partial x_1^{k_1} \ldots \partial x_n^{k_n}} dx. \tag{5.2}$$

so that Theorems 1.1 and 1.2 of §1 are valid for $\tilde{W}^{(l)}_2(\Omega)$, as well as the results in §§2–3.

We denote by $W^l_m(\Omega)$ *the set of all elements in* $L_m(\Omega)$ *having generalized derivatives up to order l inclusive in* $L_m(\Omega)$. *If we introduce a norm by means of* (5.1), *we obtain a (complete) separable space of type B.*‡ In general this space is wider than $\tilde{W}^l_m(\Omega)$. However, for a domain Ω with "not too bad a boundary," $W^l_m(\Omega)$ coincides with $\tilde{W}^l_m(\Omega)$. This happens, for example, for a star-shaped domain Ω; a domain is called star-shaped if the equation of $\partial\Omega$ can be written with the help of a positive continuous function $f(\theta)$ of the point θ of the unit

† The expression "a continuous function $v(x)$ can be extended by continuity to $\bar{\Omega}$" must be made precise. By this, we mean the following: $v(x)$ can be extended to $\partial\Omega$ in such a way that $v(x)$ is a single-valued continuous function on the closed set $\bar{\Omega}$ in R^n, where the distance between two points of $\bar{\Omega}$ is taken as the ordinary distance between two points in R^n. But it is possible to take the assertion in another sense in which v is not extended to $\partial\Omega$ as a single-valued function. The distance between two points x and x' of Ω is defined in this case as the minimum of the lengths of piecewise smooth curves joining x and x' and lying in Ω. If $\{x_m\}$ $(m = 1, 2, \ldots)$, $x_m \in \Omega$, is a Cauchy sequence in the sense of this definition and if x' is its limit point, then $\{v(x_m)\}$ should converge to some number denoted by $v(x')$. For example, if Ω is a cube K minus a part Γ of some hyperplane, these two interpretations lead to different notions of continuity on $\bar{\Omega} = \overline{K - \Gamma}$. Nevertheless, what we state below will be valid for both interpretations of the continuity of $u(x)$ and its derivatives in the closure of Ω.

‡ For a proof of the separability of $W^l_m(\Omega)$, see [SM 2].

sphere in the form $x = x^0 + f(\theta)\mathbf{n}_\theta$, where $\mathbf{n}_\theta$ is the unit vector corresponding to the point θ. In fact, let $u(x) \in W_m^l(\Omega)$. We form the functions $u^\lambda(x) = u(x|\lambda)$, $1 \le \lambda \le 1 + \varepsilon$, where x^0 is taken to be the origin. It is easy to see that $u^\lambda(x) \in W_m^l(\Omega_\lambda)$, where Ω_λ is the domain obtained from Ω by dilation by a factor λ, and that $u^\lambda(x)$ converges to $u(x)$ in the norm of $W_m^l(\Omega)$ as $\lambda \to 1$. For all $u^\lambda(x)$ we form the averages $u_\rho^\lambda(x)$ of the type (4.9) with $\rho < d(\lambda - 1)$, where d is the distance from the point 0 to $\partial\Omega$. It follows from what was said in §4 that $u_\rho^\lambda(x)$ converges to $u^\lambda(x)$ in the norm of $W_m^l(\Omega)$ as $\rho \to 0$. It is clear that we can choose $\rho = \rho(\lambda)$ in such a way that as $\lambda \to 1$ the functions $u_{\rho(\lambda)}^\lambda(x)$, which are infinitely differentiable in $\overline{\Omega}$, approximate $u(x)$ in the norm of $W_m^l(\Omega)$. But this means that $u(x) \in \tilde{W}_m^l(\Omega)$, so that $W_m^l(\Omega) = \tilde{W}_m^l(\Omega)$.

The coincidence of $W_m^l(\Omega)$ with $\tilde{W}_m^l(\Omega)$ also holds for domains Ω which are C^l-diffeomorphic to a star-shaped domain; a domain Ω is said to be C^l-diffeomorphic to a domain $\hat{\Omega}$ if the functions $y = y(x)$ effecting this diffeomorphism belong to $C^l(\overline{\Omega})$. The inverse functions $x = x(y)$ should also belong to $C^l(\overline{\hat{\Omega}})$.

In the sequel we shall find the following simple fact useful, and we state it in the form of a remark for easy reference.

Remark 5.1. Let the function $u(x)$ be in $L_m(\Omega)$ and have generalized derivative u_{x_1} in $L_p(\Omega)$ in a domain of "socle type," i.e., in

$$\Omega = \{x: -\delta < x_1 < f(x_1'),\ x_1' = (x_2, \ldots, x_n) \in D\},$$

where D is an arbitrary domain in the hyperplane $x_1 = 0$, δ is any positive number, and $f(x_1')$ is a continuous function on $\overline{D}$ satisfying the inequality $f(x_1') \ge \delta$. Then $u(x)$ can be approximated by smooth functions $u_k(x)$ in $\overline{\Omega}$ in such a way that $u_k \to u$ in the norm of $L_m(\Omega)$ and $u_{kx_1}(x) \to u_{x_1}(x)$ in the norm of $L_p(\Omega)$. The functions $u_k(x)$ are constructed in roughly the same way as the functions for star-shaped domains above. Namely, we set $u^\lambda(x_1, x_1') = u(x_1/\lambda, x_1')$ for $x_1' \in D$ and $u^\lambda(x_1, x_1') = 0$ for $x_1' \notin D$, where $\lambda \in [1, 1 + \varepsilon]$. The function $u^\lambda(x)$, $\lambda > 1$, is defined in the domain

$$\tilde{\Omega}_\lambda = \{x: -\lambda\delta < x_1 < \lambda f(x_1'),\ x_1' \in D\} \cup \{x: -\infty < x_1 < \infty,\ x_1' \notin D\}$$

containing $\overline{\Omega}$, where $u^\lambda \in L_m(\tilde{\Omega}_\lambda)$ and $u_{x_1}^\lambda \in L_p(\tilde{\Omega}_\lambda)$. The averages of $u^\lambda(x)$ of the form (4.9) are defined for $\rho < \delta(\lambda - 1)$ and are smooth functions in $\overline{\Omega}$. It is easy to verify that, as $\lambda \to 1$ and for an appropriate choice of $\rho = \rho(\lambda) \to 0$, they converge to u in the way mentioned above.

For the proof of one of the central theorems of the next section (Theorem 6.2) we must extend the elements of $W_m^1(\Omega)$ with bounded Ω to a larger domain $\tilde{\Omega} \supset \overline{\Omega}$ in such a way that they become elements of $W_m^1(\tilde{\Omega})$ and satisfy the inequalities

$$\|u\|_{m,\tilde{\Omega}} \le c_m(\Omega, \tilde{\Omega})\|u\|_{m,\Omega} \tag{5.3}$$

and

$$\|u_x\|_{m,\tilde{\Omega}} \leq c'_m(\Omega, \tilde{\Omega})\|u\|_{m,\Omega}^{(1)}, \tag{5.4}$$

where the constants do not depend on $u(x) \in W_m^1(\Omega)$. (Here and below, if we use one construction for the extension of all elements of $W_m^l(\Omega)$, then we shall usually use the same symbol $u(x)$ for the continuation as we do for the original element $u(x)$. Furthermore, in speaking of extensions, we shall agree that unless we say something to the contrary, each such extension will be assumed to have the properties indicated.) If it is possible to find one such extension $u(x)$ on $\tilde{\Omega} \supset \bar{\Omega}$, then it is easy to construct a lot of others, among them those for which the extended elements vanish in the vicinity of $\partial\tilde{\Omega}$ and outside $\tilde{\Omega}$. Let the starting extension of $u(x)$ on $\tilde{\Omega}$ be denoted by $\tilde{u}(x)$. Then take some smooth (or piecewise smooth) function $\zeta(x)$ which is 1 on Ω and 0 near $\partial\tilde{\Omega}$ and outside $\tilde{\Omega}$. Then the function $u(x)$, which is equal to $\tilde{u}(x)\zeta(x)$ for $x \in \tilde{\Omega}$ and 0 outside $\tilde{\Omega}$, will be the desired extension of u to all of R^n. For the purposes of this book, it will be enough to have one extension of this type.

Let us indicate a number of domains Ω for which the necessary extension is possible and which will be used in the sequel. This is easy to do for the sphere K_R if we set the function $u(x)$ at the point $x_0 \notin K_R$ equal to the function $u(x)$ at the point $\tilde{x}_0$, where $\tilde{x}_0$ is obtained from x_0 by reflection in the sphere through reciprocal radii, i.e., $\tilde{x}_0$ is the conjugate point of x_0. It is clear that for $u(x) \in C^1(\bar{K}_R)$ this extension yields continuous and piecewise smooth functions, and that these functions belong to $W_m^1(\tilde{\Omega})$ for any bounded domain $\tilde{\Omega} \supset \bar{K}_R$ and satisfy (5.3) and (5.4). An arbitrary element $u(x)$ of $W_m^1(\Omega)$ with $\Omega = K_R$ may be approximated in the norm of $W_m^1(\Omega)$ by smooth functions $u_k(x)$ in $\bar{\Omega}$. If we extend each of the $u_k(x)$ in the way indicated, and if we apply (5.3) and (5.4) to $u_k - u_l$, we see that the function $u(x)$, which is the limit of the $u_k(x)$ in the norm of $W_m^1(\tilde{\Omega})$, will be the desired extension of $u(x)$ from Ω to $\tilde{\Omega}$. This extension can be constructed for $u(x)$ in the same way as for smooth functions.

Another example of when such an extension is possible may be obtained from the first by means of a homeomorphism $y = y(x)$ where the functions of $y = y(x)$ possess bounded first-order generalized derivatives. As we remarked in §4, a homeomorphism $y = y(x)$ which maps the domain Ω_x of x onto the range Ω_y of y such that $y = y(x)$ and its inverse $x = x(y)$ have bounded generalized derivatives of first order (such a transformation will be called *regular*) establishes a one-to-one correspondence between elements $u(x)$ of the space $W_m^1(\Omega_x)$ and elements $\hat{u}(y) = u(x(y))$ of the space $W_m^1(\Omega_y)$, where the norms $\|u\|_{m,\Omega_x}^{(1)}$ and $\|\hat{u}\|_{m,\Omega_y}^{(1)}$ are estimated by each other with constants independent of u. Because of this, if the sphere $K_{R_1} = \{y: |y| < R_1\}$ is transformed by a "regular" mapping $x = x(y)$ into the range $\tilde{\Omega}$ of x, then the space $W_m^1(K_R)$ of functions $\hat{u}(y)$, where $K_R = \{y: |y| < R < R_1\}$, gives rise to the space $W_m^1(\Omega)$ of functions $u(x) = \hat{u}(y(x))$, where $\bar{\Omega} \subset \tilde{\Omega}$, and the elements $u(x)$ of this space admit an extension to $\tilde{\Omega}$ which satisfies the requirements (5.3) and (5.4).

Since the sphere K_R can be mapped in a regular way onto a parallelopiped Π_l with some larger sphere K_{R_1} being mapped regularly onto some neighborhood $\tilde{\Pi}_l$ of Π_l, a repetition of our argument shows that it is possible to obtain an extension from Π_l. Moreover, the desired extension for Π_l can be done in another way: by means of reflections in faces so that a smooth function on $\bar{\Pi}_l$ may be extended to a continuous and piecewise smooth function on all R^n.

If Ω admits an extension of elements of $W_m^1(\Omega)$, then $W_m^1(\Omega) = \tilde{W}_m^1(\Omega)$. (It is also true for $W_m^l(\Omega)$ with arbitrary l if inequalities of the form (5.3) and (5.4) hold for the extensions for all the norms $\|\cdot\|_{m,\Omega}^{(k)}$, $k = 0, 1, \ldots, l$.) In fact, let us take an arbitrary element $u(x) \in W_m^1(\Omega)$ and its extension $u(x)$ to an element of $W_m^1(\tilde{\Omega})$ and average the latter in accordance with (4.9). As a result, we obtain smooth functions $u_\rho(x)$ which tend to $u(x)$ in the norm of $W_m^1(\Omega)$ as $\rho \to 0$.

In [MY 2] there is given a wide class of domains Ω which admit such extensions of elements of $W_m^l(\Omega)$. These are the so-called "strongly Lipschitz domains" (cf. p. 74, the lemma of Calderon and Zygmund); this class contains, for example, domains with smooth boundaries.

We introduce subspaces of $W_m^l(\Omega)$ which play an important role in the study of boundary value problems. They are obtained as the closure of the set $\dot{C}^{(\infty)}(\Omega)$ in the norm of $W_m^l(\Omega)$ and are denoted by the symbol $\overset{\circ}{W}{}_m^l(\Omega)$.† For $l \geq 1$, $\overset{\circ}{W}{}_m^l(\Omega)$ is a proper subspace of $W_m^l(\Omega)$. If we use the averaging operation, it is easy to verify that the functions of $\dot{C}^l(\Omega)$, and even those of $\dot{W}_m^l(\Omega)$, are contained in $\overset{\circ}{W}{}_m^l(\Omega)$. Moreover, it will be useful to keep in mind that if a function $u(x) \in \overset{\circ}{W}{}_m^l(\Omega)$ is extended so that it is zero for $x \in \tilde{\Omega}$, then we obtain a function in $\overset{\circ}{W}{}_m^l(\tilde{\Omega})$ for all $\tilde{\Omega} \supset \Omega$. As will be made clear below, the elements of $\overset{\circ}{W}{}_m^l(\Omega)$ vanish on the boundary of Ω, along with their derivatives up to order $l - 1$ inclusive.

We point out an important and useful property of weak convergence in $W_2^l(\Omega)$ (and also in any one of its closed subspaces). Weak convergence in $W_2^l(\Omega)$ of some subsequence $\{u_m\}$ $(m = 1, 2, \ldots)$, as in an arbitrary Hilbert space, is defined as the convergence of the projections $\{(u_m, \eta)_{2,\Omega}^{(l)}\}$ for an arbitrary element η of the space. The limiting element u has as its projection $(u, \eta)_{2,\Omega}^{(l)}$ the limits of the numerical sequences $\{(u_m, \eta)_{2,\Omega}^{(l)}\}$. The necessary and sufficient condition for the weak precompactness of the sequence $\{u_m\}$ $(m = 1, 2, \ldots)$ in $W_2^l(\Omega)$ is the uniform boundedness of the norms $\|u_m\|_{2,\Omega}^{(l)}$ (see Theorem 1.2).

We shall show that weak convergence of $\{u_m\}$ in $W_2^l(\Omega)$ (as well as in any one of its closed subspaces) is equivalent to weak convergence in $L_2(\Omega)$ of the functions $\{u_m\}$ and all of their generalized derivatives $\{D^k u_m\}$ up to order l inclusive to the limit function u and its derivatives $D^k u$, respectively. Obviously, the weak convergence in $L_2(\Omega)$ of $\{D^k u_m\}$ to $D^k u$ $(k = 0, \ldots, l)$

† A dot over $C^l(\Omega)$ or over $W_m^l(\Omega)$ means that from these sets we choose only those functions having compact support in Ω.

implies that $\{u_m\}$ converges weakly in $W_2^l(\Omega)$. The converse is proved as follows. The weak convergence of $\{u_m\}$ to u in $W_2^l(\Omega)$ implies the uniform boundedness of the norms $\|D^k u_m\|_{2,\Omega}$ $(k = 0, \ldots, l;\ m = 1, 2, \ldots)$, and this guarantees the weak precompactness in $L_2(\Omega)$ of the sequences $\{D^k u_m\}$ over the same indices. If the functions $v^{(k)}$ $(k = 0, \ldots, l)$ are weak limits in $L_2(\Omega)$ for some subsequences $\{D^k u_{m_j}\}$ $(j = 1, 2, \ldots)$ selected from $\{D^k u_m\}$, then by the properties of generalized derivatives, each $v^{(k)}$ is the derivative $D^k v$ of the function v which is the weak limit in $L_2(\Omega)$ of the subsequence $\{u_{m_j}\}$; consequently, v is the weak limit in $W_2^l(\Omega)$ for the subsequence $\{u_{m_j}\}_{j=1}^{\infty}$. But, by hypothesis, u is the weak limit of $\{u_m\}$ in $W_2^l(\Omega)$, and by virtue of the uniqueness of a weak limit, $v = u$. What we have said implies that each of the sequences $\{D^k u_m\}$ $(m = 1, 2, \ldots)$ converges weakly in $L_2(\Omega)$ to the corresponding $D^k u$.

Finally, we recall that every closed subspace of a Hilbert space is closed with respect to weak convergence. In particular, the weak convergence in $W_2^l(\Omega)$ of a sequence $\{u_m\}$ $(m = 1, 2, \ldots)$ of elements of $\mathring{W}_2^l(\Omega)$ implies that its limit is in $\mathring{W}_2^l(\Omega)$.

§6. The Spaces $\mathring{W}_2^1(\Omega)$ and $W_2^1(\Omega)$ and Their Basic Properties

The Hilbert space $\mathring{W}_2^1(\Omega)$ plays a fundamental role in studying the first boundary value problem for second-order equations of various types, the space $W_2^1(\Omega)$ in studying other boundary value problems of classical type. The scalar product in $\mathring{W}_2^1(\Omega)$ and in $W_2^1(\Omega)$ was defined by the equality

$$(u, v)_{2,\Omega}^{(1)} = \int_\Omega (uv + u_x v_x)\, dx.\dagger \tag{6.1}$$

We shall assume that Ω is a bounded domain in R^n. We shall show that in $\mathring{W}_2^1(\Omega)$ we can introduce a new scalar product

$$[u, v] = \int_\Omega u_x v_x\, dx, \tag{6.2}$$

giving rise to a norm equivalent to the original. For this we must establish, for all $u(x) \in \mathring{W}_2^1(\Omega)$, the inequality

$$\int_\Omega u^2\, dx \le c_\Omega^2 \int_\Omega u_x^2\, dx \tag{6.3}$$

† Here and in what follows $u_x v_x = \sum_{k=1}^n u_{x_k} v_{x_k}$, $u_x^2 = \sum_{i=1}^n u_{x_i}^2$ and by repeating the indices, the summation is assumed to be from 1 to n, unless the contrary is stipulated.

(the Poincaré–Friedrichs inequality), whose c_Ω is a constant depending only on Ω. It is enough to prove (6.3) only for $u(x)$ in $\dot{C}^\infty(\Omega)$, for (6.3) can then be obtained from this for all $u(x) \in \overset{\circ}{W}{}_2^1(\Omega)$ by "simple closure in the norm of $W_2^1(\Omega)$". Let us explain what we mean by this. Let $u(x)$ be any element of $\overset{\circ}{W}{}_2^1(\Omega)$, and let us approximate it in the norm of $W_2^1(\Omega)$ by functions $\{u^{(m)}\}$ ($m = 1, 2, \ldots$) of $\dot{C}^\infty(\Omega)$. Suppose that (6.3) has been proved for $\{u^{(m)}\}$. Then, if we take (6.3) for $u^{(m)}$ and go to the limit as $m \to \infty$ in the norm of $W_2^1(\Omega)$, we obtain (6.3) for $u(x)$. This procedure is referred to in brief as the "closure of (6.3) in the norm of $W_2^1(\Omega)$". We shall frequently use it in proving inequalities of the form

$$\|u\|_{B_1} \le c\|u\|_{B_2} \tag{6.4}$$

for all $u \in B_2$, where B_1, B_2 are two Banach spaces. As a first step, we shall prove (6.4) for some set $\mathfrak{M}$ of elements (usually smooth functions) that is dense in B_2 and then take the closure in the norm of B_2. But now we return to the proof of (6.3) for $u(x) \in \dot{C}^\infty(\Omega)$. We enclose Ω in some parallelopiped Π and assume, without loss of generality, that $\Pi = \{x : 0 < x_i < l_i\}$. As will be obvious from the proof, Π is the most advantageous of all to choose, so that one of its edges will have the smallest possible length. Let this be l_1. We represent $u(x)$ in the form

$$u(x_1, x_1') = \int_0^{x_1} \frac{\partial u(y_1, x_1')}{\partial y_1} dy_1, \tag{6.5}$$

where $x_1' = (x_2, \ldots, x_n) \in \Pi_1 = \{x_1' : 0 < x_i < l_i; i = 2, \ldots, n\}$, and where we assume that $u(x) = 0$ for $x \notin \Omega$. If we square both sides of (6.5), integrate over Π, and estimate the right-hand side by Cauchy's inequality, we have

$$\begin{aligned}\int_\Pi u^2(x)\,dx &= \int_0^{l_1} dx_1 \int_{\Pi_1} \left(\int_0^{x_1} \frac{\partial u(y_1, x_1')}{\partial y_1} dy_1 \right)^2 dx_1' \\ &\le \int_0^{l_1} dx_1 \int_{\Pi_1} \left[x_1 \int_0^{l_1} \left(\frac{\partial u}{\partial y_1} \right)^2 dy_1 \right] dx_1' = \frac{l_1^2}{2} \int_\Pi \left(\frac{\partial u}{\partial x_1} \right)^2 dx,\end{aligned}$$

from which it is clear that (6.3) holds with $C_\Omega^2 = l_1^2/2$. From this follows the equivalence of the norms $[u, u]^{1/2}$ and $\|u\|_{2,\Omega}^{(1)}$ corresponding to (6.2) and (6.1).

In the inequality (6.3) the constant c_Ω can also be taken in the form $c^2|\Omega|^{2/n}$, where $|\Omega| = \operatorname{mes} \Omega$ and c is an absolute constant depending only on n, i.e., for all $u \in \overset{\circ}{W}{}_2^1(\Omega)$,

$$\|u\|_{2,\Omega} \le c|\Omega|^{1/n}\|u_x\|_{2,\Omega}. \tag{6.6}$$

The proof of this fact is a bit more complicated, and since we can "make do" without it, we defer it to §7 where we deal with various pretty results which are not used in the basic text of the book.

We proceed now to the proof of the most important theorem about the space $\overset{\circ}{W}{}_2^1(\Omega)$.

Theorem 6.1 (Theorem of F. Rellich). *A bounded set in $\overset{\circ}{W}{}_2^1(\Omega)$ is precompact in $L_2(\Omega)$.*†

This assertion is usually formulated in this way: *$\overset{\circ}{W}{}_2^1(\Omega)$ is embedded compactly in $L_2(\Omega)$.* There are several methods of proof, and we give one of them.

We extend all the elements of $\overset{\circ}{W}{}_2^1(\Omega)$ by setting them equal to zero outside Ω and we consider them in the parallelopiped $\Pi = \{x : 0 < x_i < l_i\}$, where $\Omega \subset \Pi$. As we remarked at the end of §5, we obtain, for such extensions, elements of $\overset{\circ}{W}{}_2^1(\Pi)$ whose norms $\|\cdot\|_{2,\Pi}$ and $\|\cdot\|_{2,\Pi}^{(1)}$ coincide with $\|\cdot\|_{2,\Omega}$ and $\|\cdot\|_{2,\Omega}^{(1)}$, respectively. Let us decompose Π into elementary parallelopipeds ω_i with edges l_k/N $(k = 1, \ldots, n)$ and with faces parallel to the coordinate planes. Poincaré's inequality holds for an arbitrary function $u(x)$ in $W_2^1(\omega_i)$:

$$\int_{\omega_i} u^2\,dx \le \frac{1}{|\omega_i|}\left(\int_{\omega_i} u\,dx\right)^2 + \frac{n}{2}\int_{\omega_i}\sum_{k=1}^{n}\left(\frac{l_k}{N}\right)^2 u_{x_k}^2\,dx \tag{6.7}$$

(we shall carry out the proof of this inequality below). From this it follows that

$$\int_\Omega u^2\,dx = \int_\Pi u^2\,dx \le \prod_{i=1}^{N^n}\frac{1}{|\omega_i|}\left(\int_{\omega_i} u\,dx\right)^2 + \frac{n}{2N^2}\int_\Omega\sum_{k=1}^{n} l_k^2 u_{x_k}^2\,dx. \tag{6.8}$$

Let $\|u^{(m)}\|_{2,\Omega}^{(1)} \le c$. By Theorem 1.2 the set $\{u^{(m)}\}$ is weakly precompact in $L_2(\Omega)$. We may assume, without loss of generality, that the sequence $\{u^{(m)}\}$ is weakly convergent in $L_2(\Omega)$. Then, for any $u^{(p)}$ and $u^{(q)}$, (6.8) gives the estimate

$$\|u^{(p)} - u^{(q)}\|_{2,\Omega}^2 = \sum_{i=1}^{N^n}\frac{1}{|\omega_i|}\left[\int_{\omega_i}(u^{(p)} - u^{(q)})\,dx\right]^2 + \frac{n}{2N^2}\sum_{k=1}^{n} l_k^2\|u_{x_k}^{(p)} - u_{x_k}^{(q)}\|_{2,\Omega}^2. \tag{6.9}$$

The last term on the right-hand side of (6.9) can at once be made arbitrarily small for all p and q by choosing $|\omega_i|$ small (i.e., for large N), and the first term tends to zero as $p, q \to \infty$ for a fixed partition of Π because of the weak convergence of $\{u^{(m)}\}$ in $L_2(\Pi)$. By the same token $\|u^{(p)} - u^{(q)}\|_{2,\Omega} \to 0$ as $p, q \to 0$, that is, $\{u^{(m)}\}$ converges in $L_2(\Pi)$, and the theorem is proved.

The central formula in the theory of boundary value problems is valid for the elements of $\overset{\circ}{W}{}_2^1(\Omega)$, the formula for integration by parts; for these elements it has a particularly simple form and is valid for any domain Ω. For all $u(x) \in \overset{\circ}{W}{}_2^1(\Omega)$ and all $v(x) \in W_2^1(\Omega)$, it is

$$\int_\Omega u_{x_i} v\,dx = -\int_\Omega u v_{x_i}\,dx, \qquad i = 1, \ldots, n. \tag{6.10}$$

† Once again we recall that we are assuming in this section that Ω is a bounded domain.

The somewhat more general formula holds,

$$\int_\Omega \frac{\partial w}{\partial x_i}\, dx = 0 \tag{6.11}$$

for an arbitrary function w in $L_1(\Omega)$ having generalized derivative $\partial w/\partial x_i$ in $L_1(\Omega)$ which can be approximated by functions $\{w^{(m)}\}$ $(m = 1, 2, \ldots)$ from $\dot{C}^\infty(\Omega)$ in the following sense: $w^{(m)} \to w$ in $L_1(\Omega)$ and $\partial w^{(m)}/\partial x_i \to \partial w/dx_i$ in $L_1(\Omega)$. Actually, (6.11) is true for $w^{(m)}$, so that if we take the limit as $m \to \infty$, we obtain (6.11) for w. If $w \in L_1(\Omega)$ vanishes near $\partial\Omega$, and has derivative $\partial w/\partial x_i$ in $L_1(\Omega)$, then it can be approximated by functions $w^{(m)}$ in $\dot{C}^\infty(\Omega)$ (as the $w^{(m)}$ we can take the averages of w of the type (4.9)), so that (6.11) holds for w. If u and v are in $L_2(\Omega)$ and have generalized derivatives $\partial u/\partial x_i$ and $\partial v/\partial x_i$ in $L_2(\Omega)$ (here i is fixed), and if u can be approximated by functions $u^{(m)}$ in $\dot{C}^\infty(\Omega)$ so that $u^{(m)}$ and $\partial u^{(m)}/\partial x_i$ converge to u and $\partial u/\partial x_i$ in the norm of $L_2(\Omega)$, then (6.10) holds for u and v for the given value of i. Actually, the functions $w^{(m)} = u^{(m)}v$ are in $L_1(\Omega)$, vanish near $\partial\Omega$, and have generalized derivatives $\partial w^{(m)}/\partial x_i = u^{(m)}\partial v/\partial x_i + \partial u^{(m)}/\partial x_i v$ in $L_1(\Omega)$, so that (6.11) holds for the $w^{(m)}$. If we take the limit as $m \to \infty$ in this formula, which we may, we obtain (6.11) for w, or, what is the same thing, (6.10) for u and v and for that value of i. Since every u in $\mathring{W}_2^1(\Omega)$ can be approximated in the norm of $W_2^1(\Omega)$ by functions of $\dot{C}^\infty(\Omega)$, it follows that (6.10) holds for u, and for all v in $W_2^1(\Omega)$ under all $i = 1, \ldots, n$.

Remark 6.1. As we saw in §4, functions w in $L_1(\Omega)$, with w_{x_i} in $L_1(\Omega)$ for almost all lines parallel to the x_i-axis and intersecting Ω, are absolutely continuous in x_i on each interval lying on these lines and belonging to Ω. Moreover, they can be extended by continuity to the end-points of these intervals. If the function w is such that this extension causes w to vanish at all end-points of these intervals, then (6.11) is valid for w. In the second part of this section we shall define the "traces of the elements of $W_2^1(\Omega)$ on $\partial\Omega$" by a method which is different from that which has just been described. This will be done for $\partial\Omega$ possessing a certain smoothness, where the traces will be found as functions defined in a single-valued way on $\partial\Omega$ (more precisely, by elements of $L_2(\partial\Omega)$). The method mentioned here for extending w to $\partial\Omega$ (which is suitable for an arbitrary Ω) leads, in general, to a multi-valued function w on $\partial\Omega$, which is due to the fact that we can approach points of the boundary along various paths.

We prove now the inequality (6.7), or, what is the same thing, the inequality

$$\int_{\Pi_l} u^2\, dx \le \frac{1}{|\Pi_l|}\left(\int_{\Pi_l} u\, dx\right)^2 + \frac{n}{2}\int_{\Pi_l} \sum_{k=1}^n l_k^2 u_{x_k}^2\, dx, \tag{6.7'}$$

where $\Pi_l = \{x: 0 < x_i < l_i\}$. It is enough to verify it only for smooth functions $u(x)$, for these functions are dense in $W_2^1(\Pi_l)$. Thus, let $u(x) \in C^1(\overline{\Pi}_l)$.

Let us introduce new coordinates in (6.7′) by $y_i = x_i/l_i$ ($i = 1, \ldots, n$). After multiplying by $(l_1, \ldots, l_n)^{-1}$, we obtain an equivalent inequality

$$\int_{\Pi_1} \tilde{u}^2\,dy \le \left(\int_{\Pi_1} \tilde{u}\,dy\right)^2 + \frac{n}{2}\int_{\Pi_1} \tilde{u}_y^2\,dy \tag{6.7″}$$

for the function $\tilde{u}(y) = u(l_1 y_1, \ldots, l_n y_n)$ in the cube $\Pi_1 = \{y: 0 < y_i < 1\}$. For the proof of (6.7″) we take any two points $y = (y_1, \ldots, y_n)$, $y' = (y_1', \ldots, y_n')$ and the chain of points $y^{(1)} = (y_1', y_2, \ldots, y_n)$, $y^{(2)} = (y_1', y_2', \ldots, y_n), \ldots, y^{(n)} = y'$. By the Newton–Leibniz theorem,

$$\begin{aligned}\tilde{u}(y') - \tilde{u}(y) = &\int_y^{y^{(1)}} \tilde{u}_{\tau_1}(\tau_1, y_2, \ldots, y_n)\,d\tau_1 \\ &+ \int_{y^{(1)}}^{y^{(2)}} \tilde{u}_{\tau_2}(y_1', \tau_2, y_3, \ldots, y_n)\,d\tau_2 + \cdots \\ &+ \int_{y^{(n-1)}}^{y^{(n)}} \tilde{u}_{\tau_n}(y_1', y_2', \ldots, \tau_n)\,d\tau_n.\end{aligned} \tag{6.7‴}$$

We square both sides of this equality and apply Cauchy's inequality to the right-hand side to obtain

$$\tilde{u}^2(y') - 2\tilde{u}(y')\tilde{u}(y) + \tilde{u}^2(y) \le n\left(\int_0^1 \tilde{u}_{\tau_1}^2\,d\tau_1 + \cdots + \int_0^1 \tilde{u}_{\tau_n}^2\,d\tau_n\right). \tag{6.12}$$

If we integrate (6.12) over $y \in \Pi_1$ and $y' \in \Pi_1$, we have

$$2\int_{\Pi_l} \tilde{u}^2(y)\,dy - 2\left(\int_{\Pi_l} \tilde{u}(y)\,dy\right)^2 \le n\int_{\Pi_l} \sum_{k=1}^n \tilde{u}_{y_k}^2\,dy,$$

which coincides with (6.7″).

Let us turn now to the spaces $W_2^1(\Omega)$ for various domains Ω. We begin with the simplest case, when Ω is the parallelepiped $\Pi_l = \{x: 0 < x_i < l_i\}$. Here we have compactness of the embedding of $W_2^1(\Omega)$ into $L_2(\Omega)$. This is clear from the proof of Theorem 6.1, where we used the fact that the $u^{(m)}$ were in $\mathring{W}_2^1(\Omega)$ only in the beginning of the proof to show that we could extend $u^{(m)}(x)$ to Π_l in such a way that these functions still have first-order generalized derivatives and have uniformly bounded norms $\|\cdot\|_{2,\Pi_l}^{(1)}$. In the present case the functions $u^{(m)}(x)$ have these properties by hypothesis, and we can apply directly to them all further arguments from the proof of Theorem 6.1.

Now let the domain admit an extension of $W_2^1(\Omega)$ to some domain $\tilde{\Omega} \supset \bar{\Omega}$ in the sense described at the end of §5 (cf. (5.3) and (5.4)). Then the same type of extension is possible on a parallelepiped $\Pi_l \supset \bar{\Omega}$, along with the inequalities (5.3) and (5.4) with $m = 2$ and $\tilde{\Omega} = \Pi_l$. From this and from the compactness of the embedding of $W_2^1(\Pi_l)$ into $L_2(\Pi_l)$ it follows that the embedding of $W_2^1(\Omega)$ into $L_2(\Omega)$ is compact. Moreover, it is easy to see that this property still holds for domains Ω such that $\bar{\Omega}$ can be represented in the form $\bigcup_{i=1}^{N} \bar{\Omega}_i$, where the Ω_i are subdomains of Ω admitting an extension of $W_2^1(\Omega_i)$.† Thus we have proved

Theorem 6.2. *If $\bar{\Omega}$ can be represented in the form $\bigcup_{i=1}^{N} \bar{\Omega}_i$, where Ω_i is a subdomain of Ω admitting an extension of $W_2^1(\Omega_i)$, then any bounded set in $W_2^1(\Omega)$ is precompact in $L_2(\Omega)$.*‡

We turn now to the question of the traces of the elements $u(x)$ of $W_2^1(\Omega)$ on surfaces of dimension $n-1$. Let us take first a smooth function $u(x)$ in $W_2^1(\Omega)$ and a set Γ which is a domain on some hyperplane. For the sake of convenience suppose that Γ is the domain of $x_1' = (x_2, \ldots, x_n)$ lying on the plane $x_1 = 0$, and suppose that the cylinder $Q_\delta \equiv Q_\delta(\Gamma) = \{x : 0 < x_1 < \delta, x_1' \in \Gamma\}$ also belongs to Ω. By the Newton–Leibniz formula for any $u \in C(\bar{Q}_\delta)$ with continuous derivative $\partial u / \partial x_1$ in $\bar{Q}_\delta$, we have

$$u(x_1, x_1') - u(0, x_1') = \int_0^{x_1} \frac{\partial u(\tau, x_1')}{\partial \tau}\, d\tau. \tag{6.13}$$

If we square both sides of (6.13), integrate over Γ, and apply Cauchy's inequality to the result, we have

$$\|u(x_1, \cdot) - u(0, \cdot)\|_{2,\Gamma}^2 = \int_\Gamma \left(\int_0^{x_1} \frac{\partial u}{\partial \tau}\, d\tau \right)^2 dx_1' \le x_1 \int_0^{x_1} \int_\Gamma \left(\frac{\partial u}{\partial x_1} \right)^2 dx. \tag{6.14}$$

We derive another consequence of (6.13). We first transpose $u(x_1, x_1')$ to the right-hand side of (6.13), square both sides, and then integrate the result over $Q_\delta(\Gamma)$. We then divide by δ and bound the right-hand side as follows:

$$\begin{aligned} \int_\Gamma u^2(0, x_1')\, dx_1' &= \frac{1}{\delta} \int_{Q_\delta(\Gamma)} \left[-u(x) + \int_0^{x_1} \frac{\partial u}{\partial \tau}\, d\tau \right]^2 dx \\ &\le \frac{2}{\delta} \int_{Q_\delta(\Gamma)} u^2\, dx + \frac{2}{\delta} \int_\Gamma dx_1' \int_0^\delta \left(\int_0^{x_1} \frac{\partial u}{\partial \tau}\, d\tau \right)^2 dx_1 \\ &\le \frac{2}{\delta} \int_{Q_\delta(\Gamma)} u^2\, dx + \frac{2}{\delta} \int_\Gamma dx_1' \int_0^\delta x_1 \int_0^\delta \left(\frac{\partial u}{\partial \tau} \right)^2 d\tau\, dx_1 \\ &= \frac{2}{\delta} \int_{Q_\delta(\Gamma)} u^2(x)\, dx + \delta \int_{Q_\delta(\Gamma)} u_{x_1}^2\, dx. \end{aligned} \tag{6.15}$$

The inequalities (6.14) and (6.15), which we derived for "nice" functions $u(x)$, are still valid for arbitrary $u \in L_2(Q_\delta)$ having $u_{x_1} \in L_2(Q_\delta)$. For such $u(x)$ we can construct (Remark 5.1) a sequence of smooth functions $\{u^{(m)}(x)\}$

† $\bar{\Omega}_i$ and $\bar{\Omega}_j$ can have common points.

‡ The hypotheses of Theorem 6.2 are satisfied, for example, for domains with smooth boundaries.

which converges to $u(x)$ in $L_2(Q_\delta)$ and such that $\{u_{x_1}^{(m)}\}$ also converges to u_{x_1} in $L_2(Q_\delta)$. From this and from (6.15) it follows that $\{u^{(m)}(0, \cdot)\}$ converges in $L_2(\Gamma)$. It is natural to consider the function which is defined on Γ as the limit of the $u^{(m)}(0, \cdot)$ in $L_2(\Gamma)$ to be the trace of $u(x)$ on Γ. It is clear from (6.14) that $u(x)$ has such traces on all cross-sections of Q_δ by the planes $x_1 = x_1^0$, $x_1^0 \in [0, \delta]$. If we write out (6.14) and (6.15) for $u^{(m)}$ and take the limit as $m \to \infty$, we see that these relations still hold for the limit function $u(x)$ (and, moreover, for all $u \in W_2^1(\Omega)$). By (6.14) the traces of $u(x_1, x_1')$ on the cross-sections of Q_δ by the planes indicated above are the elements of $L_2(\Gamma)$ which depend continuously on the parameter $x_1 \in [0, \delta]$.

Thus we have proved the following theorem.

Theorem 6.3. *For any function $u \in L_2(Q_\delta)$ with $u_{x_1} \in L_2(Q_\delta)$ there exists a trace, as an element of $L_2(\Gamma)$, on any cross-section of Q_δ by the plane $x_1 = x_1^0$, $x_1^0 \in [0, \delta]$, and the trace depends continuously on $x_1^0 \in [0, \delta]$ in the norm of $L_2(\Gamma)$. The relations* (6.14) *and* (6.15) *hold for $u(x)$.*

For brevity, this assertion is often referred to as a bounded embedding of $W_2^1(\Omega_\delta)$ into $L_2(\Gamma)$.

Let us explain how one must think of Theorem 6.3 and similar "embedding theorems" to be formulated below. Let $u(x)$ be an arbitrary element satisfying the hypotheses of the theorem. Then, according to Theorem 6.3, there exists a representative of this element (i.e., a function equivalent to $u(x)$ on $\bar{Q}_\delta$) for which the assertions of the theorem are true. A method of finding such a representative for $u(x)$ is clear from the proof: one must take some kind of sequence $u^{(m)}$ $(m = 1, 2, \ldots)$ from $C^1(\bar{Q}_\delta)$ converging to $u(x)$ in the manner indicated above, and assign, as the trace of $u(x)$ on every cross-section of $\bar{Q}_\delta$ by the plane $x_1 = x_1^0$, the limit of the sequence $\{u^{(m)}(x_1^0, x_1')\}$ in the norm of $L_2(\Gamma)$.

Remark 6.2. It is not hard to see that the trace of an element $u(x)$ of $W_2^1(\Omega)$, defined on Γ as an element of $L_2(\Gamma)$, does not depend on the choice of the sequence of smooth functions $\{u^{(m)}\}$ that is used to approximate $u(x)$. This trace varies continuously as an element of $L_2(\Gamma)$ under a translation of Γ, not only in the direction x_1, but also in any other direction, under the condition only that the translated cross-section does not go out of Ω.

We shall prove now that $W_2^1(\Omega)$ is embedded in $L_2(\Gamma)$ not only in a bounded way, but also compactly.

Theorem 6.4. *Let the domain Ω contain a cylinder $Q_\delta(\Gamma)$ of the kind described in Theorem* 6.3, *and let the conclusion of Theorem* 6.2 *hold for $Q_\delta(\Gamma)$, or for some other domain $\Omega' \subseteq \Omega$ containing this cylinder. Then, from any sequence $\{u^{(m)}\}$ which is bounded in the norm of $W_2^1(\Omega)$ we can select a subsequence which converges in $L_2(\Gamma)$ uniformly with respect to $x_1^0 \in [0, \delta]$.*

In fact, let $\|u^{(m)}\|_{2,\Omega}^{(1)} \le c$. By Theorem 6.2 $\{u^{(m)}\}$ is precompact in $L_2(Q_\delta(\Gamma))$. We assume, without loss of generality, that the sequence $\{u^{(m)}\}$ converges to $u(x)$ in $L_2(Q_\delta(\Gamma))$. Let us consider (6.15) for the differences $u^{(p)} - u^{(q)}$ and for $\delta_1 \in (0, \delta]$. The right-hand side of (6.15) can be made arbitrarily small for sufficiently large p and q, for if we first choose δ_1 in (6.15) small enough so that $\delta_1 \|u_{x_1}^{(p)} - u_{x_1}^{(q)}\|_{2,\Omega_\delta(\Gamma)}^2$ is less than $\varepsilon/2$ for all p and q, we can then find an N_ε such that for $p, q \ge N_\varepsilon$, the first term will be less than $\varepsilon/2$. By the same token, $\|u^{(p)} - u^{(q)}\|_{2,\Gamma}^2$ will be less than ε for $p, q \ge N_\varepsilon$.

The same bound for $\|u^{(p)}(x_1, \cdot) - u^{(q)}(x_1, \cdot)\|_{2,\Gamma}$ holds for all x_1 in $[0, \delta]$ if we realize that we can replace $\int_\Gamma u^2(0, x_1')\,dx_1'$ in the left-hand side of (6.15) by $\int_\Gamma u^2(x_1, x_1')\,dx_1'$ for all $x_1 \in [0, \delta]$.

If $Q_\delta = \{x : x_1 \in (0, \delta), x_1' \in \Gamma\}$, let us denote by $\overset{\circ}{W}{}_{2,0}^1(Q_\delta)$ the closed subspace of $W_2^1(Q_\delta)$ with a dense set of infinitely differentiable functions whose support does not intersect the lateral surface $S_\delta = \{x : x_1 \in (0, \delta), x_1' \in \partial\Gamma\}$ of the cylinder Q_δ. (As will be clear from what follows, $\overset{\circ}{W}{}_{2,0}^1(Q_\delta)$ consists, roughly speaking, of elements of $W_2^1(Q_\delta)$ which vanish on S_δ.) If for all $u \in \overset{\circ}{W}{}_{2,0}^1(Q_\delta)$ we effect an extension by setting $u(x)$ equal to zero for $x_1 \in (0, \delta)$, $x_1' \notin \Gamma$, then we obtain an element of $\overset{\circ}{W}{}_{2,0}^1(\tilde{Q}_\delta)$, $\tilde{Q}_\delta = \{x : x_1 \in (0, \delta), x_1' \in \tilde{\Gamma}\}$ for all $\tilde{\Gamma} \supset \Gamma$, where its norm in $\overset{\circ}{W}{}_{2,0}^1(\tilde{Q}_\delta)$ is $\|u\|_{2,Q_\delta}^{(1)}$. From Theorems 6.2 and 6.4 we have

Corollary 6.1. *Let the elements of the sequence $\{u^{(m)}\}$ be in $\overset{\circ}{W}{}_{2,0}^1(Q_\delta)$, and let their norms $\|u^{(m)}\|_{2,Q_\delta}^{(1)}$ be uniformly bounded. Then we can select a subsequence from $\{u^{(m)}\}$ which converges strongly in $L_2(Q_\delta)$ and strongly in $L_2(\Gamma)$, uniformly with respect to $x_1 \in [0, \delta]$ (with no restrictions on the domains Γ except that it be bounded).*

All the results that we establish for planar pieces Γ can be carried over to the case where Γ is a domain on a smooth hypersurface, including a smooth part of the boundary of Ω. Let Γ be projected onto a hyperplane (without restricting generality, we may assume that this is the plane $x_1 = 0$) in the form of a domain $\Gamma^{(1)}$ so that Γ has the explicit equation $x_1 = f(x_1'), x_1' \in \Gamma^{(1)}$, $f \in C^1(\bar{\Gamma}^{(1)})$, and let the "curvilinear cylinder" $Q_\delta(\Gamma) = \{x : f(x_1') < x_1 < f(x_1') + \delta, x_1' \in \Gamma^{(1)}\}$ lie in Ω. We introduce into $Q_\delta(\Gamma)$ the new coordinates $y_1 = x_1 - f(x_1')$, $y_k = x_k$, $k = 2, \ldots, n$. The function $\tilde{u}(y) = u(x(y))$ will be an element of $W_2^1(\tilde{Q}_\delta)$, where $\tilde{Q}_\delta = \{y : 0 < y_1 < \delta, x_1' \in \Gamma^{(1)}\}$, so that it will satisfy (6.14) and (6.15) in terms of the coordinates y. If we return to the old coordinates, these inequalities yield

$$\int_\Gamma [u(x + le_1) - u(x)]^2\,ds \le cl \int_{Q_l(\Gamma)} u_x^2\,dx, \qquad 0 \le l \le \delta, \quad (6.16)$$

and

$$\|u\|_{2,\Gamma}^2 \le c\left[\frac{1}{\delta}\|u\|_{2,Q_\delta(\Gamma)}^2 + \delta\|u_x\|_{2,Q_\delta(\Gamma)}^2\right], \quad (6.17)$$

in which $\mathbf{e}_1 = (1, 0, \dots, 0)$ and the constant c is defined by the first derivatives of $f(x_1')$. In (6.16) and (6.17) a displacement of Γ, and of the arguments of $u(x)$, are taken in the direction of the coordinate axes x_1. This can be done with the same success along other paths which are not tangent to Γ if, instead of x, we introduce other non-degenerate coordinates y with bounded Jacobians $|\partial y/\partial x|$ and $|\partial x/\partial y|$. For example, we may assume in (6.16) that $\mathbf{e}_1$ is a unit normal vector to Γ at the point x and that $Q_l(\Gamma)$ is a curvilinear cylinder formed by segments of these normals, of length l, emanating from the points of Γ.

If $u \in W_2^1(\Omega)$ and if its boundary $\partial\Omega$ (or a part of it, say Γ) is a smooth hypersurface, then, if we keep in mind the availability of inequalities of the type (6.16) and (6.17) for $u(x)$, we shall say that its boundary values (that is, its trace) on $\partial\Omega$ (on Γ) yield an element of $L_2(\partial\Omega)$ [$L_2(\Gamma)$] and that they are "assumed in mean square". This happens if Ω has piecewise smooth boundary $\partial\Omega$, that is, if it is possible to cover $\partial\Omega$ by a finite number of pieces $\bar{\Gamma}_i$ such that inequalities of the type (6.16) and (6.17) hold on these pieces (these sufficient conditions on the Γ_i were formulated above). In particular, we have for such Ω the inequalities:

$$\int_{\partial\Omega} (u(x - l\mathbf{n}) - u(x))^2 \, ds \le cl \int_{\Omega_l} u_x^2 \, dx, \qquad 0 \le l \le \delta, \tag{6.18}$$

$$\|u\|_{2,\partial\Omega}^2 \le c\left[\frac{1}{\delta}\|u\|_{2,\Omega_\delta}^2 + \delta\|u_x\|_{2,\Omega_\delta}^2\right], \tag{6.19}$$

where Ω_δ is the set of points of Ω whose distance from $\partial\Omega$ does not exceed δ (such a set is frequently called a boundary strip of width δ), where δ is a sufficiently small number, and where $\mathbf{n}$ is the unit outer normal to $\partial\Omega$. Thus we have established the following generalization of Theorem 6.3.

Theorem 6.5. *For the elements $u(x)$ of $W_2^1(\Omega)$ the traces are defined on domains Γ of smooth hypersurfaces lying in Ω as elements of $L_2(\Gamma)$, and they depend continuously (as elements of $L_2(\Gamma)$) on a displacement of Γ.*† *For these traces we have inequalities of the form* (6.16) *and* (6.17). *If $\partial\Omega$ (or a part of it, Γ) is a smooth (or piecewise smooth) surface, then the traces of $u(x)$ are defined on it as elements of $L_2(\partial\Omega)$ [or of $L_2(\Gamma)$], and the inequalities* (6.18) *and* (6.19) [(6.16) *and* (6.17)] *hold for these traces.*

Theorem 6.4 can also be generalized to curvilinear pieces, and, in particular, we have

Theorem 6.6. *If Ω has a smooth boundary $\partial\Omega$, then a bounded set in $W_2^1(\Omega)$ is precompact in $L_2(\partial\Omega)$. This conclusion about compactness also holds for domains Ω with piecewise smooth boundaries $\partial\Omega = \bigcup_{i=1}^n \Gamma_i$ if for each of the*

† We have in mind a displacement described in the proof of (6.16) or by an inequality of type (6.18).

pieces Γ_i we can construct a cylinder of the type $Q_\delta(\Gamma_i) \subset \Omega$ for which (6.16) *and* (6.17) *hold, and for which the embedding of $W_2^1(\Omega)$ into $L_2(Q_\delta(\Gamma_i))$ is compact.*

Remark 6.3. For the elements $u(x)$ of $\mathring{W}_2^1(\Omega)$ the smoothness of the boundary of Ω plays no role in the derivation of the inequalities (6.16) and (6.17), for each of these functions can be extended to zero outside Ω, and considered as an element of $\mathring{W}_2^1(K)$, where K is a sphere enclosing Ω. Because of this, the set of elements $\{u^{(m)}\}$ of $\mathring{W}_2^1(\Omega)$ which have been extended to zero outside Ω will be compact in $L_2(\Gamma)$, where Γ is now the intersection of some smooth hypersurface with the sphere K. It is frequently said of the elements of $\mathring{W}_2^1(\Omega)$ that they assume the boundary value zero. This must be understood in this way: If Γ is a smooth piece of the boundary for which $Q_\delta(\Gamma) \subset \Omega$, then, for $u(x) \in \dot{C}^\infty(\Omega)$, the inequality (6.18) becomes

$$\int_\Gamma u^2(x - l\mathbf{n})\, ds \leq cl \int_{Q_l(\Gamma)} u_x^2\, dx, \qquad 0 \leq l \leq \delta. \tag{6.20}$$

If we take the closure in the norm of $W_2^1(Q_l(\Gamma))$ then (6.20) remains valid for all $u(x)$ in $\mathring{W}_2^1(\Omega)$. From this it is clear that $\int_\Gamma u^2(x - l\mathbf{n})\, ds \to 0$ as $l \to 0$.

There is a natural interest in the converse question: Let $u(x)$ belong to $W_2^1(\Omega)$ and assume the boundary value zero in mean square; does $u(x)$ then belong to $\mathring{W}_2^1(\Omega)$? The answer is affirmative, for example, for domains Ω with smooth boundary and even for strongly Lipschitz domains Ω. We shall not prove this, however, for the basic material of this book has been developed in such a way that, for its formulation and proof, it is not necessary to carry out similar laborious considerations (for these details, see [SO 5]).

We return now to the formula for integration by parts,

$$\int_\Omega \frac{\partial u}{\partial x_i} v\, dx = -\int_\Omega u \frac{\partial v}{\partial x_i}\, dx + \int_{\partial\Omega} uv \cos(\mathbf{n}, \mathbf{x}_i)\, ds. \tag{6.21}$$

Here (as everywhere else in the book) $\mathbf{n}$ is the unit exterior normal to $\partial\Omega$ and ds is the infinitesimal element of area of the surface $\partial\Omega$. It is well known that (6.21) holds for smooth surfaces $\partial\Omega$ and for functions u and v that are continuous on $\bar{\Omega}$ and have continuous derivatives u_{x_i} and v_{x_i} on $\bar{\Omega}$. From this and from the properties of traces of elements of $W_2^1(\Omega)$ proved in this section, it follows that (6.21) still holds (for all $i = 1, 2, \ldots, \mathbf{n}$) for functions $u(x)$ and $v(x)$ in $W_2^1(\Omega)$ if $\partial\Omega$ is a smooth surface. (To verify this, it is necessary to approximate u and v by smooth functions $\{u^{(m)}\}$ and $\{v^{(m)}\}$ in $\bar{\Omega}$, then write down (6.21) for $u^{(m)}$ and $v^{(m)}$, and pass to the limit as $m \to \infty$. We remark that $\tilde{W}_2^1(\Omega) = W_2^1(\Omega)$ for smooth $\partial\Omega$.) The formula (6.21) holds for smooth functions u and v of this type in domains Ω for which $\bar{\Omega} = \bar{\Omega}_1 \cup \bar{\Omega}_2$, where the domains Ω_i have smooth boundary; Ω_1 and Ω_2 can even intersect. For the proof of this it is necessary to verify that (6.21) holds for the intersection $\Omega_1 \cap \Omega_2$ and then write the integral $\int_\Omega \ldots$ in the form $\int_{\Omega_1} + \int_{\Omega_2} - \int_{\Omega_1 \cap \Omega_2}$; to each of these integrals it is necessary to apply (6.21) and then take into consideration the fact that the integrals taken over those parts of the boundaries of Ω_1 and Ω_2 not belonging to $\partial\Omega$ cancel out. Thus (6.21) is true for any Ω which can be covered not only by two, but any finite number of domains $\Omega_i \subset \Omega$ with smooth boundaries. For such domains (6.21) holds not only for smooth u and v but also for all u and v in $W_2^1(\Omega)$, a fact which can again be verified by approximating u and v in the norms of $W_2^1(\Omega_i)$ by smooth functions. Instead of domains Ω_i with smooth boundaries we may take various types of polyhedra.

We shall need (6.21) again for cylindrical domains $\Omega = \{x: 0 < x_1 < l_1;\ x_1' \in \Gamma\}$, where Γ is an $(n-1)$-dimensional domain on the hyperplane $x_1 = 0$, and where u and v are in $L_2(\Omega)$ with generalized derivatives u_{x_1} and v_{x_1} in $L_2(\Omega)$ (here we take (6.21) with $i = 1$). The relation (6.21) holds for such Ω with $i = 1$ for u and v in $C^1(\overline{\Omega})$. We can approximate functions u and v in $L_2(\Omega)$, with u_{x_1} and v_{x_1} in $L_2(\Omega)$, by smooth functions $u^{(m)}$ and $v^{(m)}$ in such a way that $u^{(m)}$, $v^{(m)}$, $u_{x_1}^{(m)}$, and $v_{x_1}^{(m)}$ converge, respectively, in $L_2(\Omega)$ to u, v, u_{x_1}, and v_{x_1} (apropos of this, see Remark 5.1). We must then take (6.21) with $i = 1$ for $u^{(m)}$ and $v^{(m)}$ and let m tend to ∞. By the above considerations, the limit process can be justified for each term, so that we have (6.21) for u and v, namely,

$$\int_\Omega \frac{\partial u}{\partial x_1} v\, dx = -\int_\Omega u \frac{\partial v}{\partial x_1} dx + \int_\Gamma uv\, dx_1' \Big|_{x_1=0}^{x_1=l_1}. \tag{6.22}$$

We shall not describe more complicated situations for which (6.21) holds, but fall back on the special literature (see [MY 2]).

We introduce one more formula which will prove useful later on, namely,

$$\int_{\partial\Omega} |u|\, ds \le c \int_\Omega (|u_x| + |u|)\, dx, \tag{6.23}$$

This formula is valid for all $u(x) \in W_1^1(\Omega)$ and for domains Ω with piecewise smooth boundary. For the proof we must cover $\partial\Omega$ by a finite number of closed domains Γ_i and on each of them construct a curvilinear cylinder $\Omega_\delta(\Gamma)$ of the same type as in the inequality (6.17); then, for $Q_\delta(\Gamma)$ we must verify the validity of an inequality of the type (6.17) with norm L_1 instead of L_2; more precisely,

$$\|u\|_{1,\Gamma_i} \le c\left[\frac{1}{\delta}\|u\|_{1,Q_\delta(\Gamma_i)} + \|u_x\|_{1,Q_\delta(\Gamma_i)}\right],$$

which can be derived in the same way as (6.17) by using the Newton–Leibniz formula (see (6.13)). If we sum these inequalities over all i, we obtain (6.23).

Let us apply (6.23) to the function $u(x) = v^2(x)$, where $v(x)$ is in $W_2^1(\Omega)$ (it is easy to see that u is in $W_1^1(\Omega)$) and extend it in the following way:

$$\int_{\partial\Omega} v^2\,ds \le c_1 \int_\Omega (|v|\cdot|v_x| + v^2)\,dx$$

$$\le c_1 \int_\Omega \left[\frac{\varepsilon}{c_1} v_x^2 + \left(\frac{c_1}{4\varepsilon} + 1\right) v^2\right] dx \equiv \int_\Omega (\varepsilon v_x^2 + c_\varepsilon v^2)\,dx. \quad (6.24)$$

Here we have used (1.3) with an arbitrary $\varepsilon > 0$.

§7. Multiplicative Inequalities for Elements of $\mathring{W}_m^1(\Omega)$ and $W_m^1(\Omega)$

The contents of this section are not used in the basic text of this book and are cited as reference material for a more complete characterization of the spaces $W_m^1(\Omega)$. We begin with the simplest of these, the space $\mathring{W}_m^1(\Omega)$, $m \ge 1$. Here we have the following theorem.

Theorem 7.1. *For all $u \in \mathring{W}_m^1(\Omega)$, $m \ge 1$, where Ω is an arbitrary (not necessarily bounded†) domain in R^n, $n \ge 2$, and for all $r \ge 1$ the multiplicative inequality*

$$\|u\|_{p,\Omega} \le \beta \|u_x\|_{m,\Omega}^{\alpha} \|u\|_{r,\Omega}^{1-\alpha} \quad (7.1)$$

holds, where

$$\alpha = \left(\frac{1}{r} - \frac{1}{p}\right)\left(\frac{1}{r} - \frac{1}{\bar{m}}\right)^{-1}, \qquad \bar{m} = \frac{nm}{n-m}, \quad (7.2)$$

and where

(1) *for $m < n$, p is an arbitrary number in $[r, \bar{m}]$ if $r \le \bar{m}$ or in $[\bar{m}, r]$ if $r \ge \bar{m}$, and $\beta = ([(n-1)/n]\bar{m})^\alpha$; as p varies between r and $\bar{m}$, α varies between 0 and 1, inclusive of the end-points; if $r = p = \bar{m}$, then α can be taken arbitrarily in $[0,1]$;*

(2) *for $m \ge n$, p is an arbitrary number in $[r, \infty)$ and*

$$\beta = \max\left\{\frac{n-1}{n} p; 1 + \frac{m-1}{m} r\right\}^\alpha.$$

As p varies from r to ∞, α varies from 0 to $nm/[nm + r(m-n)]$, except for the right-hand end-point;

(3) *if $m > n$, then (7.1) holds for $p = \infty$ also with some $\beta < \infty$, where $\alpha = nm/[nm + r(m-n)]$.*

† Remember that for any domain Ω the space $\mathring{W}_m^1(\Omega)$ is defined as the closure in the norm of $W_m^1(\Omega)$ of the set $\dot{C}^\infty(\Omega)$ of infinitely differentiable functions with compact support lying in Ω.

The proof of Theorem 7.1 may be found, for example, in [LU 1] (second edition). Various special cases of it were established independently by E. Gagliardo [2] and the author [14]. Other methods of proving inequalities of this type were given later by L. Nirenberg [3] and K. K. Golovkin [1; 2]. (See Supplement to Chapter I.)

Theorem 7.1 is meaningful if it is known that $\|u\|_{r,\Omega} < \infty$. The important feature of inequality (7.1) is that the constant β does not depend on Ω. The fact that the constant β does not vary under the similarity transformation $x \to \lambda x$ ($\lambda > 0$) makes it easy to verify directly: both sides of (7.1) are multiplied by $\lambda^{n/p}$ under this transformation (this fact follows from relation (7.2)). Moreover, as is the case with all embedding theorems, (7.1) possesses the property of homogeneity: it does not change if we multiply $u(x)$ by a constant. By the expression $\|u\|_{\infty,\Omega}$ we mean $\operatorname{ess\,sup}_{\Omega}|u(x)|$ which coincides with $\max|u(x)|$ for continuous functions on $\bar{\Omega}$. In case (3), when $m > n$, elements of $\mathring{W}^1_m(\Omega)$ are equivalent to continuous functions on $\bar{\Omega}$. Actually, all $u(x)$ in $\mathring{W}^1_2(\Omega)$ can be approximated, in the norm of $W^1_2(\Omega)$, by continuous functions $u^{(q)}(x)$ on $\bar{\Omega}$ (for example, by averages). If we apply (7.1) with $r = m$ to the differences $u^{(q)}(x) - u^{(s)}(x)$, we can verify that they converge uniformly on Ω, so that the limit function for the $u^{(q)}(x)$ will be continuous.

Remark 7.1. As examples show, the maximal exponent $p = \bar{m}$ for $r \leq \bar{m}$, which is assured by part (1) of Theorem 7.1, is precise in the sense that it cannot be increased for all $u(x)$ in $\mathring{W}^1_2(\Omega)$. With this in mind, we say that this result is precise (or sharp). Other versions of (1) are precise. In the case $m = n$ the result is also precise: we cannot take $p = \infty$ in (7.1). However, for $m = n$ and $m > n$ stronger facts than those in (2) and (3) have been established. For example, if $m = n$ for the elements $u(x)$, the integral

$$\int_\Omega \exp\left(c_1 \frac{|u(x)|}{\|u_x\|_{n,\Omega}}\right)^{n/(n-1)} dx$$

can be majorized by $c_2|\Omega|$ with some c_k depending only on n (see [TR 1]). The elements $u(x)$ of $\mathring{W}^1_m(\Omega)$ for $m > n$ are Hölder-continuous on $\bar{\Omega}$ with exponent $\alpha = (m - n)/m$.

For the case $n = 1$ we can also write down multiplicative inequalities, but this is not usually necessary, for sufficiently complete information is available from the formula of Newton and Leibniz. Namely, we write $u(x)$ in the form

$$u(x) = \int_0^x \frac{du}{d\tau}\, d\tau, \qquad u(0) = 0, \qquad x \in [0, l]. \tag{7.3}$$

From (7.3) we have, for example, the boundedness of $u(x)$:

$$\max_{x\in[0,l]} |u(x)| \leq \int_0^l \left|\frac{du}{d\tau}\right| d\tau \tag{7.4$_1$}$$

for $m = 1$ and the Hölder continuity:

$$|u(x) - u(x')| \leq \left| \int_{x'}^{x} \frac{du}{d\tau}\, d\tau \right| \leq |x - x'|^{1/m'} \left(\int_{x'}^{x} \left| \frac{du}{d\tau} \right|^m d\tau \right)^{1/m} \qquad (7.4_2)$$

for $m > 1$. The compactness of an arbitrary bounded set of elements of $\mathring{W}_m^1((0, l))$, $m > 1$, in $C([0, l])$ for $l < \infty$ (Arzela's theorem) follows from (7.4_1) and (7.4_2).

Theorem 7.1 and Theorem 6.1 imply the compactness of the embedding of $\mathring{W}_m^1(\Omega)$ into $L_{\tilde{p}}(\Omega)$ with any $\tilde{p}$ less than the maximum possible in (7.1) if Ω is bounded. To prove this, it is necessary to distinguish two cases: (1) $m \geq 2$, and (2) $m \in [1, 2)$. Let $\|u^{(q)}\|_{m,\Omega}^{(1)} \leq c$, $q = 1, 2, \ldots$, and $m \geq 2$. By Rellich's theorem, $\{u^{(q)}\}$ is precompact in $L_2(\Omega)$. We select a subsequence $\{u^{(q_k)}\}$ of $\{u^{(q)}\}$ which converges in $L_2(\Omega)$ and apply (7.1) with $r = 2$ to the difference $u^{(q_k)} - u^{(q_{k+l})}$, taking $p = \tilde{p}$ less than $\bar{m}$ for $m < n$, p an arbitrary finite number for $m = n$, and $p = \infty$ for $m > n$. For such r and $\tilde{p}$ the exponent $\alpha < 1$, so that (7.1) implies that $\|u^{(q_k)} - u^{(q_{k+l})}\|_{\tilde{p},\Omega} \to 0$ as $k, l \to \infty$. By the same token the precompactness of $\{u^{(q)}\}$ is proved.

In the case $m \in [1, 2)$ and $n \geq 2$ we must argue somewhat differently: First we must show that the embedding of $\mathring{W}_m^1(\Omega)$ into $L_m(\Omega)$ is compact, and then, by using (7.1) with $r = m$, verify that the embedding of $\mathring{W}_m^1(\Omega)$ into $L_{\tilde{p}}(\Omega)$ is compact for all $\tilde{p} < \bar{m}$. The proof of the compactness of the embedding of $\mathring{W}_m^1(\Omega)$ into $L_m(\Omega)$ is similar to the proof given above of Theorem 6.1, except that, instead of (6.7′), we must bring in the inequality

$$\int_{\Pi_l} |u|^m\, dx \leq \frac{c}{|\Pi_l|^{m-1}} \left| \int_{\Pi_l} u\, dx \right|^m + c \int_{\Pi_l} \sum_{k=1}^{n} l_k^m |u_{x_k}|^m\, dx,$$

where $\Pi_l = \{x : 0 < x_i < l_i\}$. It can be proved in this way: Instead of x_i we introduce the variables $y_i = x_i l_i^{-1}$ so that Π_l is mapped into $\Pi_1 = \{y : 0 < y_i < 1\}$ and $u(x) = \tilde{u}(y)$. We take y and y' in Π_1 and represent $|\tilde{u}(y')|$ in the form

$$|\tilde{u}(y')| = \left| \int_{\Pi_1} [\tilde{u}(y') - \tilde{u}(y)]\, dy + \int_{\Pi_1} \tilde{u}(y)\, dy \right|.$$

If we raise both sides of this equality to the m-th power and integrate the result over $y' \in \Pi_1$, we can bound the right-hand side by the inequality $|a + b|^m \leq 2^{m-1}(|a|^m + |b|^m)$ and use the representation (6.7‴) for $\tilde{u}(y') - \tilde{u}(y)$, given on p. 27. The first term on the right-hand side can be bounded in accordance with the inequality

$$\left| \int_{\Omega} \sum_{i=1}^{N} f_i(x)\, dx \right|^m \leq N^{m-1} |\Omega|^{m-1} \int_{\Omega} \sum_{i=1}^{N} |f_i|^m\, dx$$

and the observation that none of the intervals of integration exceeds 1. This gives us the desired majorant for $\int_{\Pi_1} |\tilde{u}(y')|^m\, dy'$.

We summarize the results obtained as follows.

Theorem 7.2. *A bounded set in $\overset{\circ}{W}{}^1_m(\Omega)$, $m \geq 1$, is precompact in $L_{\tilde{p}}(\Omega)$ with $\tilde{p} < \bar{m}$ for $m \leq n$ and with arbitrary finite $\tilde{p}$ for $m = n$. For $m > n$ it is precompact in $C(\bar{\Omega})$. Here the domain Ω is assumed to be bounded.*

Let us go over to the spaces $W^1_m(\Omega)$. First we consider those domains Ω for which the elements $u(x)$ of $W^1_m(\Omega)$ can be extended to a larger domain $\tilde{\Omega} \supset \bar{\Omega}$ such that $u(x)$ will be an element of $\overset{\circ}{W}{}^1_m(\tilde{\Omega})$ and

$$\|u\|_{r,\tilde{\Omega}} \leq c_r(\Omega, \tilde{\Omega})\|u\|_{r,\Omega}, \tag{7.5}$$

and

$$\|u_x\|_{m,\tilde{\Omega}} \leq c'_m(\Omega, \tilde{\Omega})\|u\|^{(1)}_{m,\Omega} \tag{7.6}$$

for all r such that $\|u\|_{r,\Omega}$ is finite. The constants c_r and c'_m in these inequalities should not depend on $u(x)$. The domains Ω, indicated at the end of §5, for which an extension is possible with $r = m$ in (7.5) admit an extension for all r, where the method of extension is the same as that for $r = m$. If we apply Theorem 7.1 to a $u(x)$ in $\overset{\circ}{W}{}^1_m(\tilde{\Omega})$ and use (7.5) and (7.6), we obtain, for the restriction of $u(x)$ to Ω, the inequality

$$\|u\|_{p,\Omega} \leq \beta c_{p,r,m}(\Omega)(\|u\|^{(1)}_{m,\Omega})^{\alpha}\|u\|^{1-\alpha}_{r,\Omega}\dagger \tag{7.7}$$

with the same parameters as in (7.1). If $\bar{\Omega}$ can be represented in the form $\bigcup_{i=1}^{N} \bar{\Omega}_i$ (the domains Ω_i, $i = 1, \ldots, N$, can have a non-empty intersection), where Ω_i admits an extension of elements of $W^1_m(\Omega_i)$, then the inequality (7.7) is also valid for Ω. Actually, if we write down (7.7) for each of the Ω_i, raise them to the p-th power, add then up over i from 1 to N, and then take the p-th root of the inequality obtained, this will give us an inequality with $\|u\|_{p,\Omega}$ on the left-hand side. The right-hand side will be increased if in each term we replace the integral over Ω_i by the same integral over Ω. As a result, we obtain (7.7) with a certain constant $c_{p,r,m}(\Omega)$.

In exactly the same way, we can establish Theorem 7.2 for $W^1_m(\Omega)$ with the same Ω as in (7.7). Thus we have proved

Theorem 7.3. *Let $\bar{\Omega} = \bigcup_{i=1}^{N} \bar{\Omega}_i$ (here the Ω_i, $i = 1, \ldots, N$, can intersect), where the Ω_i are domains admitting such extension of elements of $W^1_m(\Omega)$ that (7.5) and (7.6) are satisfied. Then, for all $u(x) \in W^1_m(\Omega)$, inequality (7.7) holds with the same parameters m, r, p, and α as in Theorem 7.1. The conclusions of Theorem 7.2 concerning the compactness of the embedding into $L_{\tilde{p}}(\Omega)$ for bounded Ω still hold for the space $W^1_m(\Omega)$.*

† If we assume in (7.6) and (7.7) that the norm $\|u\|^{(1)}_{m,\Omega}$ is defined as

$$\left[|\Omega|^{-m/n}\int_{\Omega} |u|^m\,dx + \int_{\Omega} |u_x|^m\,dx\right]^{1/m},$$

then the constants in these inequalities are invariant under a similarity transformation of the independent variables x.

We conclude this section by deriving the inequality (6.6) from (7.1). If we take $m = 2$, $r = 1$, and $p = 2$ in (7.1), we can bound the last factor in (7.1) by Cauchy's inequality as follows:

$$\|u\|_{2,\Omega} \leq \beta \|u_x\|_{2,\Omega}^{n/(n+2)} \|u\|_{1,\Omega}^{2/(n+2)} \leq \beta \|u_x\|_{2,\Omega}^{n/(n+2)} |\Omega|^{1/(n+2)} \|u\|_{2,\Omega}^{2/(n+2)}.$$

If we cancel $\|u\|_{2,\Omega}^{2/(n+2)}$ from both sides and raise the result to the power $(n + 2)/n$, we obtain (6.6):

$$\|u\|_{2,\Omega} \leq \beta^{(n+2)/n} |\Omega|^{1/n} \|u_x\|_{2,\Omega}. \tag{7.8}$$

§8. Embedding Theorems for the Spaces $\mathring{W}^l_m(\Omega)$ and $W^l_m(\Omega)$

From Theorem 7.1 we can obtain various multiplicative inequalities for the elements $u \in \mathring{W}^l_m(\Omega)$, which give the bounds of the norms of an arbitrary derivative $D^k u$ with $k < l$ by means of the product of the norms $D^l u$ and $D^p u$ for arbitrary $p \leq k$, where these norms are chosen in various L_{p_i}. We shall derive the most commonly used inequalities and indicate the highest powers to which the $D^k u$ are summable. The derivatives $D^{l-1}u$ may be considered as elements of $\mathring{W}^1_m(\Omega)$, and we can apply (7.1) to them. From the inequality (7.1) with $r = m$ and $\alpha = 1$ (if $m < n$) we conclude that $D^{l-1}u \in L_{m_1}(\Omega)$, where m_1 is determined by the equation $1/m_1 = 1/m - 1/n$. Consequently, $D^{l-2}u \in \mathring{W}^1_{m_1}(\Omega)$. If $m_1 < n$, then again by applying to $D^{l-2}u$ the inequality (7.1) with $m = m_1$, $r = m_1$, and $\alpha = 1$, we can verify that $D^{l-2} \in L_{m_2}(\Omega)$ where $1/m_2 = 1/m_1 - 1/n = 1/m - 2/n$. If we continue this argument, we arrive either at the inequality

$$\|u\|_{p,\Omega} \leq c \|D^l u\|_{m,\Omega} \tag{8.1}$$

where $1/p = 1/m_l = 1/m - l/n$, that is, where $p = nm/(n - lm)$ if $lm < n$, or at (8.1) with arbitrary $p < \infty$ if $lm = n$, or at the inequality

$$\max_{\Omega} |u| \leq c \|D^l u\|_{m,\Omega}, \tag{8.2}$$

if $lm > n$. The constants c in (8.1) and (8.2) are defined by n, p, l, m; they depend also on $|\Omega|$ when $lm = n$ or $lm > n$. We shall not cite further consequences, but go over to the spaces $W^l_m(\Omega)$ with bounded Ω. The following inequalities for the elements of $W^l_m(\Omega)$

$$\|u\|_{p,\Omega} \leq c(\Omega) \|u\|_{m,\Omega}^{(l)} \tag{8.3}$$

were proved, where the parameters are the same as in (8.1), and

$$\max_{\Omega} |u| \leq c(\Omega) \|u\|_{m,\Omega}^{(l)} \tag{8.4}$$

for $lm > n$. The assumptions about Ω are the same as in Theorem 7.3 and the $c(\Omega)$ depend on Ω in a complicated way; they also depend on n, p, l, m.

Special cases of (8.3) and (8.4) have been encountered in the works of various authors (e.g., in the works of A. Poincaré, J. Leray and others). In general form these inequalities were proved by S. L. Sobolev ([3], [5]) and were formulated as theorems "on the bounded embedding of the spaces $W^l_m(\Omega)$ into $L_p(\Omega)$ or into $C(\bar{\Omega})$"; Sobolev assumed that $m > 1$. His student V. I. Kondrashov proved the compactness of the embedding of $W^l_m(\Omega)$ into $L_{\tilde{p}}(\Omega)$ for all $\tilde{p} < p = nm/(n - lm)$ for $lm < n$, and for all $\tilde{p} < \infty$, for $lm = n$; for $lm > n$, $W^l_m(\Omega)$ can be embedded compactly into $C(\bar{\Omega})$. This fact can be derived from Theorem 7.3 on the compactness of the embedding of $W^1_m(\Omega)$ into $L_p(\Omega)$ if we apply it to the same elements $u(x)$ of $W^l_m(\Omega)$ and their derivatives $D^k u$, $k < l$, as was done for the proof of (8.1).

In addition to (8.3) and (8.4), Sobolev established more general and detailed facts about a bounded embedding of $W^l_m(\Omega)$ into $L_q(\Gamma_s)$, where Γ_s is a piece of a sufficiently smooth manifold (more precisely, $\Gamma_s \in C^l$) of dimension s lying in Ω. The parameters n, l, m, q, and s are related by the following conditions:

$$m > 1, \qquad lm < n, \qquad s > n - lm, \qquad q \leq \frac{ms}{n - lm},\dagger \tag{8.5}$$

and, for $lm = n$, the number $q < \infty$.

This proposition asserts that, for $u(x)$ in $W^l_m(\Omega)$, we can define traces on Γ_s and

$$\|u\|_{q, \Gamma_s} \leq c(\Gamma_s, \Omega)\|u\|^{(l)}_{m,\Omega}. \tag{8.6}$$

These traces vary continuously (as elements of $L_q(\Gamma_s)$) under a continuous displacement of Γ_s. In the case $lm > n$, as we indicated above, $W^l_m(\Omega)$ can be embedded into $C(\bar{\Omega})$, so that traces are defined for its elements at all points of $\bar{\Omega}$: "manifolds Γ_s of dimension zero".

For the exponents q less than $q = ms/(n - lm)$ we have a compact embedding of $W^l_m(\Omega)$ into $L_{\tilde{q}}(\Gamma_s)$. There are several different ways of proving the theorems of Sobolev that we have enumerated. Sobolev himself started from a certain integral representation for the elements $u(x)$ of $W^l_m(\Omega)$ which expressed $u(x)$ in the form of a sum of polynomials of degree $l - 1$ and of an integral in which only the derivatives $D^l u$ appeared. This representation was applicable for domains Ω which were star-shaped relative to some sphere $\bar{K}_R \subset \Omega$ (we remark that $\mathring{W}^l_m(\Omega) = W^l_m(\Omega)$ for such domains). All assertions were proved first for such Ω, and from this, conclusions were drawn about their validity for unions of domains of this kind.

We point out one more question relating to the spaces $W^l_m(\Omega)$ and considered by Sobolev. This is the question of equivalent norms for the spaces $W^l_m(\Omega)$. We touched on this in §6 for the spaces $\mathring{W}^1_2(\Omega)$ and $W^1_2(\Omega)$

† The case of equality was studied by V. P. Ilin.

by proving the inequalities of Poincaré, (6.3) and (6.7′). The first of these implies the equivalence of the original norm in $\mathring{W}_2^1(\Omega)$ and the norm $\|u_x\|_{2,\Omega}$. The second guarantees the equivalence in $W_2^1(\Omega)$ of the norm $\|u\|_{2,\Omega}^{(1)}$ and the norm defined by the equality

$$|u| \equiv \left[\left(\int_\Omega u\,dx\right)^2 + \int_\Omega u_x^2\,dx\right]^{1/2}. \tag{8.7}$$

This fact is valid not only for parallelopipeds, as was shown in §6, but also for a wide class of domains. In the spaces $\mathring{W}_m^l(\Omega)$, just as in the space $\mathring{W}_2^1(\Omega)$, the norm can be defined in an especially simple and convenient way:

$$\left(\int_\Omega \sum_{(l)} |D^l u|^m\,dx\right)^{1/m} \approx \|u\|_{m,\Omega}^{(l)} \tag{8.8}$$

where $\sum_{(l)}$ denotes the summation over all derivatives of order l. Moreover, this is valid for subspaces of $W_m^l(\Omega)$ wider than $\mathring{W}_m^l(\Omega)$, namely for those which are obtained as the closure in the norm of $W_m^l(\Omega)$ of a set of elements in $W_m^l(\Omega)$ which vanish in the neighborhood of some part Γ_{n-1} of $\partial\Omega$, which is the closure of a domain on $\partial\Omega$. In his book [5], Sobolev indicates a wide class of norms which are equivalent to the norm $\|u\|_{m,\Omega}^{(l)}$ for the spaces $W_m^l(\Omega)$.

Finally we indicate a problem related to the spaces $W_m^1(\Omega)$ and their use in the theory of boundary value problems. This is the problem of "bounded extension" of functions $u(s)$ from the boundary $\partial\Omega$ of a domain Ω to the whole domain Ω. It is the converse of the question of traces of the elements of $W_m^1(\Omega)$ on the boundary $\partial\Omega$. As we indicated above (see (8.6)), the traces of elements of $W_m^1(\Omega)$ for $m > 1$ and $m < n$ are elements of $L_q(\partial\Omega)$, $q = (n-1)m/(n-m)$. Examples show that this power q cannot be increased for all $W_m^1(\Omega)$. The question arises: Is this characterization of traces "precise," that is, is it possible for an arbitrary function $u(s)$, defined on $\partial\Omega$ and belonging to $L_q(\partial\Omega)$, $q = (n-1)m/(n-m)$, to be extended to Ω in such a way that it is an element $u(x)$ of $W_m^1(\Omega)$ and such that the inequality

$$\|u\|_{m,\Omega}^{(1)} \le c\|u\|_{q,\partial\Omega}$$

holds for a constant c that is independent of u?

The answer appears to be negative. For a solution of this problem a finer characterization of traces of elements of $W_m^1(\Omega)$ is necessary. It is expressed in terms of spaces of functions having generalized derivatives of fractional order. In order to avoid a study of these spaces and of a solution of the question posed about the extension of functions from $\partial\Omega$, we have chosen the following course in this book: Instead of assigning, for example, the boundary values in the Dirichlet problem for the unknown function $u(x)$ in the usual form,

$$u|_{\partial\Omega} = \varphi(s),$$

with an indication of the smoothness properties of $\varphi(s)$ on $\partial\Omega$, we require that $u(x) - \varphi(x)$ be in $\mathring{W}_2^1(\Omega)$, where $\varphi(x)$ is a prescribed element of $W_2^1(\Omega)$.

This form of prescribing the boundary condition allows us to separate the problem of extending $\varphi(s)$ to Ω from the problem, which interests us here, of the solvability of boundary value problems. The transition from results on boundary value problems obtained in this form to a reformulation in the ordinary form of prescribing boundary conditions is then straightforward: one may simply refer to the corresponding results on extensions.

Supplements and Problems

1. Inequalities with constants not depending on the domain Ω are rather useful for the study of boundary value problems (in particular, elliptic problems) in unbounded domains.

Such inequalities were first proved by Poincaré. They are

$$\int_{\Omega} \frac{u^2(x)}{|x-y|^2}\,dx \le \frac{4}{(n-2)^2}\int_{\Omega} u_x^2(x)\,dx, \qquad n > 2, \tag{1}$$

where Ω is any domain in R^n (including the case when $\Omega = R^n$), $u(\cdot)$ is an element of $\dot{C}^\infty(\Omega)$, and $y \in R^n$. If $n = 2$ then the inequality

$$\int_{\Omega} \frac{u^2(x)}{|x-y|^2(\ln|x-y|)^2}\,dx \le 4\int_{\Omega} u_x^2(x)\,dx \tag{2}$$

is true for any domain $\Omega \subset R^2$, $y \in \Omega$, and $u \in \dot{C}^\infty(\Omega)$.

Prove (1) and (2). Extend for this purpose $u(x)$ by zero outside Ω (if $\Omega \ne R^n$), then introduce spherical coordinates (r, θ) with center y and represent $|x-y|^{-2}\,dx$ as

$$r^{n-3}\,dr\,d\theta = \frac{1}{n-2}\frac{\partial}{\partial r} r^{n-2}\,dr\,d\theta \quad \text{if } n > 2,$$

and

$$|x-y|^{-2}(\ln|x-y|)^{-2}\,dx \quad \text{as} \quad r^{-1}(\ln r)^{-2}\,dr\,d\theta$$

$$= -\frac{\partial}{\partial r}(\ln r)^{-1}\,dr\,d\theta \quad \text{if } n = 2.$$

After that integrate by parts passing the derivative $\partial/\partial r$ onto u^2 and make use of Cauchy's inequality.

The inequalities

$$\int_{\Omega} \frac{|u(x)|^p}{|x-y|^l}\,dx \le \left|\frac{p}{n-l}\right|^p \int_{\Omega} \frac{|u_x(x)|^p}{|x-y|^{l-p}}\,dx \tag{1'}$$

are proved in the same way for all $p > 0, l \neq n$ and $u \in \dot{C}^\infty(\Omega)$. Consider also the case $l = n$. The inequalities (1), (2) enable us to understand properties of elements of Hilbert spaces obtained by closure of the set $\dot{C}^\infty(\Omega)$ in the norm corresponding to the inner product

$$[u, v] = \int_\Omega u_x v_x \, dx. \tag{3}$$

Elucidate it and pay attention to the case $\Omega = R^2$.

Inequalities (1) and (2) served as a prototype for obtaining more general inequalities "with weights" (see V. P. Il'in [2] and others).

2. I have found some other inequalities with constants independent of Ω while studying boundary value problems for Navier–Stokes equations (see [LA 14] and [LA 15]). The first such inequalities were

$$\|u\|_{4,\Omega}^4 \le \|u\|_{2,\Omega}^2 \|u_{x_1}\|_{2,\Omega} \|u_{x_2}\|_{2,\Omega} \le \tfrac{1}{2} \|u\|_{2,\Omega}^2 \|u_x\|_{2,\Omega}^2 \tag{4}$$

for every $u \in \mathring{W}_2^1(\Omega)$ and $\Omega \subseteq R^2$,

$$\|u\|_{4,\Omega}^4 \le \|u\|_{2,\Omega} \prod_{i=1}^{3} \|u_{x_i}\|_{2,\Omega} \le 3^{-3/2} \|u\|_{2,\Omega} \|u_x\|_{2,\Omega}^3 \tag{5}$$

and

$$\|u\|_{6,\Omega}^3 \le \frac{9}{2} \prod_{i=1}^{3} \|u_{x_i}\|_{2,\Omega} \le \frac{\sqrt{3}}{2} \|u_x\|_{2,\Omega}^3 \tag{6}$$

for every $u \in \mathring{W}_2^1(\Omega)$ and $\Omega \subseteq R^3$. It is sufficient to check these only for smooth functions u with compact supports and all integrals may be written as integrals over R^n.

Inequality (4) was proved as follows:

$$\begin{aligned}
\int_{R^2} u^4(x_1, x_2)\, dx_1\, dx_2 &\le \int_{R^1} \max_{x_2 \in R^1} u^2(x_1, x_2)\, dx_1 \int_{R^1} \max_{x_1 \in R^1} u^2(x_1, x_2)\, dx_2 \\
&\le \int_{R^2} |u u_{x_2}|\, dx_1\, dx_2 \int_{R^2} |u u_{x_1}|\, dx_1\, dx_2 \\
&\le \|u\|_{2,R^2}^2 \|u_{x_1}\|_{2,R^2} \|u_{x_2}\|_{2,R^2}.
\end{aligned}$$

Inequalities (5) and (6) were obtained using (4) and the estimates

$$\int_{R^1} |\phi_1(x)\phi_2(x)|\, dx_k \le \tfrac{1}{2} \int_{R^1} |\phi_{1x_k}|\, dx_1 \int_{R^1} |\phi_2|\, dx_k.$$

(In my publications the constants in (4)–(6) are overstated owing to the fact that instead of $|\phi_1(x)| \le \frac{1}{2} \int_{R^1} |\phi_{1x_k}|\, dx_k$ I used the same inequality but with constant 1.) By means of a similar approach E. Gagliardo [2], K. K. Golovkin

[1, 2] and L. Nirenberg [3] proved multiplicative inequalities for all dimensions n. Let us give here the proof of the inequality

$$\|u\|_{n/(n-1),\Omega} \le \frac{1}{2}\prod_{i=1}^{n}\|u_{x_i}\|_{1,\Omega}^{1/n}, \qquad n \ge 1, \tag{7}$$

from Nirenberg's paper [3]. Many other inequalities are deduced from it by applying (7) to functions $u = v^\alpha$ with different $\alpha > 0$ (in §7, Chapter I we presented such inequalities for elements of $\overset{\circ}{W}{}_m^1(\Omega)$). As above let $u(\cdot)$ be a smooth function equaling zero outside Ω (the norm $\|u\|_{\infty,\Omega}$ is defined as $\operatorname{ess\,sup}_{x\in\Omega}|u(x)|$ if $n = 1$). The inequality (7) follows in the case $n = 1$ from

$$u(x_1) = \int_{-\infty}^{x_1} u_\tau(\tau)\,d\tau = \int_{\infty}^{x_1} u_\tau(\tau)\,d\tau,$$

used in the proof of (4). Assuming that (7) is true for some $n - 1 \ge 0$ let us prove that it is true for n (hence all $n \ge 2$). With the help of this inductive assumption and the Hölder inequality we obtain

$$\begin{aligned}
&\int_{R^n} |u(x_1 \ldots x_n)|^{n/(n-1)}\,dx_1 \ldots dx_n \\
&\quad\le \int_{R^1} dx_n\left[\left(\int_{R^{n-1}} |u|\,dx_1 \ldots dx_{n-1}\right)^{1/(n-1)} \cdot \right. \\
&\left.\left(\int_{R^{n-1}} |u|^{(n-1)/(n-2)}\,dx_1 \ldots dx_{n-1}\right)^{(n-2)/(n-1)}\right] \\
&\quad\le \left(\max_{x_n\in R^1}\int_{R^{n-1}} |u|\,dx_1 \ldots dx_{n-1}\right)^{1/(n-1)} \cdot \\
&\int_{R^1}\frac{1}{2}\prod_{i=1}^{n-1}\|u_{x_i}\|_{1,R^{n-1}}^{1/(n-1)}\,dx_n \\
&\quad\le \left(\frac{1}{2}\right)^{n/(n-1)}\prod_{i=1}^{n}\|u_{x_i}\|_{1,R^n}^{1/(n-1)},
\end{aligned}$$

i.e., (7) is true for all $n \ge 1$.

Multiplicative inequalities, especially (4)–(6), have had many applications in the study of non-linear equations of mathematical physics. It would be interesting to find the smallest constants in these and other multiplicative inequalities.

CHAPTER II

Equations of Elliptic Type

§1. Posing of Boundary Value Problems. Description of the Basic Material of the Chapter

In this section we shall consider linear second-order equations

$$\mathscr{L}u = \sum_{i,j=1}^{n} \frac{\partial}{\partial x_i}(a_{ij}(x)u_{x_j} + a_i(x)u(x))$$

$$+ \sum_{i=1}^{n} b_i(x)u_{x_i} + a(x)u = f(x) + \sum_{i=1}^{n} \frac{\partial f_i(x)}{\partial x_i}, \qquad a_{ij}(x) = a_{ji}(x), \quad (1.1)$$

with real coefficients which satisfy the condition of uniform ellipticity in a bounded domain Ω of Euclidean space R^n. By uniform ellipticity of (1.1) in Ω we mean that the $a_{ij}(x)$ satisfy the inequalities

$$\nu\xi^2 \le a_{ij}(x)\xi_i\xi_j \le \mu\xi^2, \qquad \xi^2 = \sum_{i=1}^{n} \xi_i^2, \quad (1.2)$$

for some positive constants ν and μ for all $x \in \overline{\Omega}$ and for any real parameters $\xi_1, \ldots, \xi_n$. The left-hand side of (1.2) expresses the requirement of ellipticity, while the right-hand side expresses the boundedness of the coefficients $a_{ij}(x)$. We shall also assume that the other coefficients in (1.1)—a_i, b_i, a—are bounded functions in Ω, although the results brought out below remain valid under more general assumptions, namely, that these coefficients belong to $L_{p_k}(\Omega)$ for some p_k depending on n; for more detailed results, see Ladyzhenskaya and Ural'čeva [1]. All functions considered in this book are Lebesgue-measurable functions; this property will be assumed everywhere and will not be stipulated explicitly in the sequel. In many sections the functions a_{ij}, a_i, and f_i need not have derivatives (even generalized derivatives). How one must interpret (1.1) in this case is explained in the next section. In those cases where a_{ij}, a_i, and f_i have generalized derivatives, (1.1) can be written in the traditional form:

$$\mathscr{L}u = a_{ij}u_{x_i x_j} + \tilde{a}_i u_{x_i} + \tilde{a}u = \mathscr{F}. \tag{1.1'}$$

We shall consider the following three boundary value problems for (1.1) (or for (1.1')): (1) the Dirichlet problem (or first boundary value problem), which consists of finding a function $u(x)$ satisfying (1.1) (or (1.1')) in Ω and the boundary condition

$$u|_S = \varphi(s) \tag{1.3}$$

on the boundary S of Ω; (2) the Neumann problem (or second boundary value problem) in which we seek a solution $u(x)$ of (1.1) (or (1.1')) satisfying the boundary condition

$$\left.\frac{\partial u}{\partial N}\right|_S = \varphi(s), \tag{1.4}$$

where $\partial u/\partial N \equiv a_{ij}u_{x_j}\cos(\mathbf{n}, \mathbf{x_i})$ with $\mathbf{n}$ the unit exterior normal to S; and (3) the third boundary value problem, in which the boundary condition assumes the form

$$\frac{\partial u}{\partial N} + \sigma(s)u|_S = \varphi(s). \tag{1.5}$$

In all of these problems the function $\varphi(s)$, as well as Ω, σ, f, f_i and the coefficients, is assumed to be known. Only the function $u(x)$ is to be determined. All three of the problems cited can be reduced to problems with homogeneous boundary conditions, that is, to problems in which $\varphi(s) \equiv 0$. In fact, if we introduce the new unknown function $v(x) = u(x) - \Phi(x)$ in place of $u(x)$, where $\Phi(x)$ is an arbitrary function satisfying only the boundary condition selected (i.e., (1.3), (1.4), or (1.5)), then the original problem reduces to the same problem for the function $v(x)$, but with a homogeneous boundary condition. The equation for $v(x)$

$$\mathscr{L}v = \mathscr{F} + \frac{\partial \mathscr{F}_i}{\partial x_i} \tag{1.6}$$

differs from (1.1) only in the free terms (the right-hand side), namely in (1.6),

$$\mathscr{F} = f - b_i\Phi_{x_i} - a\Phi, \qquad \mathscr{F}_i = f_i - a_{ij}\Phi_{x_j} - a_i\Phi. \tag{1.7}$$

We shall assume everywhere that such a transformation of the problem has already taken place, i.e., that the unknown function $u(x)$ satisfies a homogeneous boundary condition. From results on these problems and from known results on the extension of functions from the boundary of the domain to the whole domain (see [MY 2]) follow the corresponding conclusions on the solvability of these problems for non-homogeneous boundary conditions.

In §§2, 3, and 5 we shall investigate the solvability of the three boundary value problems in $W_2^1(\Omega)$. It is in the space $W_2^1(\Omega)$ that we can, with comparative simplicity, reduce these problems to equations with completely continuous operators and prove their Fredholm solvability.

Concerning these generalized solutions $u(x)$ we know only that they are elements of $W_2^1(\Omega)$. Examples show that, under these minimal hypotheses on the functions given in the problems, and the Ω for which the solvability is proved, these solutions actually do not have second-order derivatives (or even generalized derivatives) and they satisfy the conditions of the problem in some generalized sense. It will turn out later that, under a minor improvement of the coefficients and free terms of (1.1), and also under a certain increase in the smoothness of the boundary S, all generalized solutions $u(x)$ in $W_2^1(\Omega)$ will belong to $W_2^2(\Omega)$ and will satisfy (1.1) in the ordinary sense for almost all $x \in \Omega$. This is done in §§6 and 7 for the first boundary value problem. Analogous results hold for the other boundary value problems. Moreover, it can be shown that, irrespective of the boundary condition, any generalized solution $u(x)$ of (1.1) in $W_2^1(\Omega)$ belongs, in fact, to $W_2^2(\Omega')$, $\bar{\Omega}' \subset \Omega$, if the coefficients and free terms of (1.1) are not "too bad," i.e., if they satisfy the conditions given in §6. All these properties of generalized solutions of (1.1) are derived from the so-called second fundamental inequality (see §6) and its consequences. By using a generalization of this inequality, we could show a further increase in the smoothness of all solutions $u(x)$ of (1.1) under a corresponding increase in the smoothness of the functions appearing in (1.1). We shall not do this in view of the comparative unwieldiness of the corresponding analysis, especially near the boundary. Let us point out only that this question has been studied extensively and that best-possible results have been obtained in terms of the spaces $W_p^l(\Omega)$ and the Hölder spaces $H^{l+\alpha}(\bar{\Omega})$ (cf. [LU 1]).

There is a certain natural ordering among the generalized solutions of boundary value problems in various functional classes: If $u(x)$ is a generalized solution of a problem in a class M, then it is a generalized solution of this problem in any larger class $M' \supset M$.

In §8 we shall present certain approximation methods of finding the solutions of boundary value problems, namely, the Galerkin method in its original and modern forms. It is shown that the Ritz method and the method of least squares are special cases of it. In Chapter 6 we shall give another method of finding approximate solutions of boundary value problems, the method of finite differences (§7), and we shall explain its connection with the method of Galerkin (§12). In §§4 and 7 we shall prove theorems on the series expansion in terms of eigen functions of symmetric elliptic operators (for more complete results, see [LA 4]).

§2. Generalized Solutions in $W_2^1(\Omega)$. The First (Energy) Inequality

In this section, and in the following sections, we shall investigate the solvability of the first boundary value problem, that is, the Dirichlet problem

$$\mathscr{L}u \equiv \frac{\partial}{\partial x_i}(a_{ij}u_{x_j} + a_i u) + b_i u_{x_i} + au = f + \frac{\partial f_i}{\partial x_i}, \tag{2.1}$$

$$u|_S = 0 \tag{2.2}$$

in the space $W_2^1(\Omega)$. We shall assume that the equation (2.1) is elliptic and that its coefficients are bounded measurable functions, i.e.,

$$\nu\xi^2 \le a_{ij}\xi_i\xi_j \le \mu\xi^2, \qquad \nu, \mu = \text{const} > 0, \quad a_{ij} = a_{ji}, \tag{2.3}$$

and

$$\sqrt{\sum_{i=1}^n a_i^2}, \quad \sqrt{\sum_{i=1}^n b_i^2} \le \mu_1, \qquad \sqrt{\sum_i (a_i - b_i)^2} \le \mu_2, \qquad \mu_3 \le a(x) \le \mu_4. \tag{2.4}$$

The functions f and f_i, which represent the free term in (2.1), will be assumed to be square-summable in Ω, i.e.,

$$\|f\|_{2,\Omega} < \infty, \qquad \|\mathbf{f}\|_{2,\Omega} \equiv \left\| \sqrt{\sum_{i=1}^n f_i^2} \right\|_{2,\Omega} < \infty. \tag{2.5}$$

We shall call a function $u(x) \in W_2^1(\Omega)$ a *generalized solution in* $W_2^1(\Omega)$ *of the equation* (2.1) if it satisfies the integral identity

$$\mathscr{L}(u, \eta) \equiv \int_\Omega (a_{ij}u_{x_i}\eta_{x_j} + a_i u\eta_{x_i} - b_i u_{x_i}\eta - au\eta)\,dx$$
$$= \int_\Omega (-f\eta + f_i\eta_{x_i})\,dx \tag{2.6}$$

for all $\eta(x) \in \dot{C}^\infty(\Omega)$.

It is easy to see that this definition is meaningful in the present situation, for all the integrals appearing in (2.6) are finite.

The identity (2.6) can be obtained formally from the identity

$$-\int_\Omega \left(\mathscr{L}u - f - \frac{\partial f_i}{\partial x_i}\right)\eta\,dx = 0, \qquad \eta \in \dot{C}^\infty(\Omega) \tag{2.6'}$$

by means of a single integration by parts involving the terms $-(a_{ij}u_{x_j} + a_i u)_{x_i}\eta$ and $(f_i)_{x_i}\eta$. If the coefficients a_{ij}, a_i have bounded first-order generalized derivatives, if the f_i have generalized derivatives $\partial f_i/\partial x_i$ in $L_2(\Omega)$, and if u belongs to $W_2^1(\Omega) \cap W_2^2(\Omega')$ for all $\overline{\Omega}' \subset \Omega$ and satisfies (2.1) almost everywhere in Ω, then (2.6') follows from (2.1) and (2.6) follows from (2.6').

Under these conditions $u(x)$ is then the generalized solution in $W_2^1(\Omega)$ of the equation. The converse is also true: Under the same conditions on a_{ij}, a_i, and f_i, any generalized solution of (2.1) in $W_2^1(\Omega)$ which belongs to $W_2^2(\Omega')$ for all $\overline{\Omega}' \subset \Omega$ satisfies (2.1) for almost all x in Ω. In fact, (2.6′) follows from (2.6) for all $\eta \in \dot{C}^\infty(\Omega)$, and (2.1) follows from (2.6′) for almost all x in Ω, since $\mathscr{L}u - f - \partial f_i/\partial x_i \in L_2(\Omega')$ for all $\overline{\Omega}' \in \Omega$, and $\dot{C}^\infty(\Omega')$ is dense in $L_2(\Omega')$. Because of the direct and converse assertions, one can say that, for differentiable a_{ij}, a_i, and f_i, equation (2.1) and the identity (2.6) carry the same information about $u(x)$. However, the identity (2.6) is meaningful when a_{ij}, a_i, and f_i are not differentiable, and it is known only that $u(x)$ belongs to $W_2^1(\Omega)$. Consequently, our definition is actually an extension of the conventional notion of a solution of (2.1). We shall show that such an extension of the notion of solution is possible for all the fundamental boundary value problems for (2.1), that is, we shall show that, for such an extension, we shall still have the fundamental properties of these problems: their Fredholm solvability.

Let us determine the *generalized solution in* $W_2^1(\Omega)$ *of the problem* (2.1), (2.2) (or simply a generalized solution), meaning a function $u(x)$ in $\overset{\circ}{W}{}_2^1(\Omega)$ which satisfies (2.6) for any $\eta \in \overset{\circ}{W}{}_2^1(\Omega)$.

We obtain for such solutions the first fundamental inequality. For this we consider the quadratic form $\mathscr{L}(u, u)$. By (2.3) and (2.4), and by Cauchy's inequality, we have, for all $\varepsilon_1 > 0$,

$$\begin{aligned}\mathscr{L}(u, u) &= \int_\Omega [a_{ij}u_{x_i}u_{x_j} + (a_i - b_i)u_{x_i}u - au^2]\,dx \\ &\geq \int_\Omega [\nu u_x^2 - \mu_4 u^2]\,dx - \mu_2\|u\| \cdot \|u_x\| \\ &\geq (\nu - \varepsilon_1)\|u_x\|^2 - \left(\mu_4 + \frac{\mu_2^2}{4\varepsilon_1}\right)\|u\|^2 \quad \text{for all} \quad \varepsilon_1 > 0.\end{aligned} \tag{2.7}$$

Here and below the symbol $\|u\|$ is used to denote the norm in $L_2(\Omega)$, and (u, v) the scalar product in $L_2(\Omega)$. If we take $\varepsilon_1 = \nu/2$ in (2.7), we have

$$\mathscr{L}(u, u) \geq \frac{\nu}{2}\|u_x\|^2 - \left(\mu_4 + \frac{\mu_2^2}{2\nu}\right)\|u\|^2. \tag{2.8}$$

Because of the inequality (6.3) in Chapter I, the right-hand side of (2.7) is not less than the expression $[(\nu - \varepsilon_1)c_\Omega^{-2} - \mu_4 - \mu_2^2/(4\varepsilon_1)]\|u\|^2$ taken over all $\varepsilon_1 \in (0, \nu]$, so that

$$\mathscr{L}(u, u) \geq \delta_1\|u\|^2, \tag{2.9}$$

where

$$\delta_1 = \max_{0 < \varepsilon_1 \leq \nu}\left[(\nu - \varepsilon_1)c_\Omega^{-2} - \mu_4 - \frac{\mu_2^2}{4\varepsilon_1}\right]. \tag{2.10}$$

If $\delta_1 > 0$, then, by using (2.8) and (2.9), we can bound $\|u_x\|^2$ in terms of $\mathscr{L}(u, u)$, namely,

$$\frac{\nu}{2}\|u_x\|^2 \le \mathscr{L}(u, u)\left[1 + \delta_1^{-1}\max\left\{0; \mu_4 + \frac{\mu_2^2}{2\nu}\right\}\right]$$

or, it is the same

$$\delta_2\|u_x\|^2 \le \mathscr{L}(u, u), \tag{2.11}$$

where

$$\delta_2 = \frac{\nu}{2}\left[1 + \delta_1^{-1}\max\left\{0; \mu_4 + \frac{\mu_2^2}{2\nu}\right\}\right]^{-1} > 0. \tag{2.12}$$

Let $u(x)$ be a generalized solution in $W_2^1(\Omega)$ of the problem (2.1), (2.2). Then it follows from (2.6) that we have for $u(x)$ the inequality

$$\begin{aligned}\mathscr{L}(u, u) &= -(f, u) + (f_i, u_{x_i}) \le \|f\| \cdot \|u\| + \|\mathbf{f}\|\,\|u_x\| \\ &\le \varepsilon_2\|u\|^2 + \frac{1}{4\varepsilon_2}\|f\|^2 + \varepsilon_3\|u_x\|^2 + \frac{1}{4\varepsilon_3}\|\mathbf{f}\|^2, \end{aligned} \tag{2.13}$$

for all $\varepsilon_2, \varepsilon_3 > 0$, which, along with (2.7), leads to the *first fundamental inequality*, the *energy inequality*,

$$\begin{aligned}(\nu - \varepsilon_1 - \varepsilon_3)\|u_x\|^2 &\le \frac{1}{4\varepsilon_2}\|f\|^2 \\ &\quad + \frac{1}{4\varepsilon_3}\|\mathbf{f}\|^2 + \left(\mu_4 + \frac{\mu_2^2}{4\varepsilon_1} + \varepsilon_2\right)\|u\|^2 \end{aligned} \tag{2.14}$$

for all $\varepsilon_i > 0$, $i = 1, 2, 3$. For $\varepsilon_1 + \varepsilon_3 < \nu$ it gives a bound on $\|u_x\|$ in terms of $\|u\|$, $\|f\|$, and $\|\mathbf{f}\|$. In particular, for $\varepsilon_1 = \varepsilon_3 = \nu/4$, (2.14) assumes the form

$$\|u_x\|^2 \le \frac{1}{2\nu\varepsilon_2}\|f\|^2 + \frac{2}{\nu^2}\|\mathbf{f}\|^2 + \frac{2}{\nu}\left(\mu_4 + \frac{\mu_2^2}{\nu} + \varepsilon_2\right)\|u\|^2 \tag{2.15}$$

for all $\varepsilon_2 > 0$. The remarkable property of the inequalities (2.14) and (2.15) lies in the fact that they allow us to find a bound for the "big norm" $\|u_x\|$ for the solution $u(x)$ of the problem (2.1), (2.2) in terms of the "subordinate norm" $\|u\|$ and the known quantities in the problem. Of special interest are those cases where it is possible to discard the term with $\|u\|^2$. This is possible, for example, if $\delta_1 > 0$. In fact, in this case we have the inequalities (2.9) and (2.11) with $\delta_i > 0$, $i = 1, 2$. If we apply (2.13) and (6.3) of Chapter I to (2.11), we have

$$\delta_2\|u_x\|^2 \le \mathscr{L}(u, u) \le (\varepsilon_2 c_\Omega^2 + \varepsilon_3)\|u_x\|^2 + \frac{1}{4\varepsilon_2}\|f\|^2 + \frac{1}{4\varepsilon_3}\|\mathbf{f}\|^2, \tag{2.16}$$

so that, by setting $\varepsilon_3 = \varepsilon_2 c_\Omega^2 = \delta_2/4$, we obtain the desired *a priori* bound

$$\|u_x\|^2 \le \frac{2}{\delta_2^2}\left[c_\Omega^2\|f\|^2 + \|\mathbf{f}\|^2\right]. \tag{2.17}$$

From this it is clear that if $f = \mathbf{f} = 0$, the solution $u(x)$ must be zero, and consequently the problem (2.1), (2.2) cannot have more than one generalized solution in $W_2^1(\Omega)$ whenever $\delta_1 > 0$. In fact, if u' and u'' are two generalized solutions, it follows from the linearity of the problem (2.1), (2.2) that the difference $u = u' - u''$ is a generalized solution of the same problem with $f = \mathbf{f} = 0$ for which we have the bound (2.17) with zero on the right-hand side. From this and from $u \in \overset{\circ}{W}{}_2^1(\Omega)$ it follows that u' coincides with u''. Thus we have proved the following uniqueness theorem:

Theorem 2.1. *The problem* (2.1), (2.2) *cannot have more than one generalized solution in* $W_2^1(\Omega)$ *if* (2.3)–(2.5) *are satisfied and if* $\delta_1 > 0$.

The condition $\delta_1 > 0$ will be fulfilled for (2.1) with coefficients satisfying the inequalities (2.2)–(2.5) if the constant c_Ω is "sufficiently small" (we have this if $|\Omega|$ is small—see (6.6) of Chapter I), or if the upper bound μ_4 on the coefficient $a(x)$ is a negative number which is "sufficiently large" in modulus. The latter condition is automatically satisfied for the equation

$$\mathscr{L}u - \lambda u = f + \frac{\partial f_i}{\partial x_i}$$

for "sufficiently large" numbers λ. The expressions "sufficiently small" and "sufficiently large" are made more precise by the condition $\delta_1 > 0$, where δ_1 is defined by (2.10).†

§3. Solvability of the Dirichlet Problem in the Space $W_2^1(\Omega)$. Three Theorems of Fredholm

We shall show that the problem (2.1), (2.2) is Fredholm-solvable in the space $W_2^1(\Omega)$. For this we introduce into $\overset{\circ}{W}{}_2^1(\Omega)$ a new scalar product

$$[u, v] = \int_\Omega a_{ij} u_{x_i} v_{x_j} \, dx. \tag{3.1}$$

By (2.3) and the inequality (6.3) of Chapter I the norm $\|u\|_1 \equiv \sqrt{[u, u]}$ is equivalent to the norm $\|u_x\|$ and the original norm $\|u\|_{2,\Omega}^{(1)}$ of the space $\overset{\circ}{W}{}_2^1(\Omega)$. We write (2.6) in the form

$$[u, \eta] + l(u, \eta) = -(f, \eta) + (f_i, \eta_{x_i}), \tag{3.2}$$

where

$$l(u, \eta) \equiv \int_\Omega (a_i u \eta_{x_i} - b_i u_{x_i} \eta - a u \eta) \, dx. \tag{3.3}$$

† It is easy to see that the uniqueness of the generalized solution also takes place for the case when $a_i - b_i$ $(i = 1, \ldots, n)$ have bounded first derivatives and $\frac{1}{2} \sum_{i=1}^n (a_i - b_i)_{x_i} + a \leq 0$.

By the assumption (2.4),

$$|l(u,\eta)| \le \mu_1\|u\|\cdot\|\eta_x\| + \mu_1\|u_x\|\cdot\|\eta\| + \max(|\mu_3|;|\mu_4|)\|u\|\,\|\eta\| \le c\|u\|_1\cdot\|\eta\|_1, \tag{3.4}$$

that is, for any fixed element $u \in \mathring{W}_2^1(\Omega)$, $l(u,\eta)$ is a linear functional on η in the space $\mathring{W}_2^1(\Omega)$. By the theorem of F. Riesz (see §2, Chapter I), we can represent $l(u,\eta)$ uniquely in the form of a scalar product

$$l(u,\eta) = [Au,\eta], \tag{3.5}$$

for all $\eta \in \mathring{W}_2^1(\Omega)$, where A is a bounded linear operator in $\mathring{W}_2^1(\Omega)$ with norm not exceeding the c in (3.4). The sum $-(f,\eta) + (f_i,\eta_{x_i})$ also defined a linear functional in $\mathring{W}_2^1(\Omega)$ on η, and, by the Riesz theorem, there exists a unique element $F \in \mathring{W}_2^1(\Omega)$ for which

$$-(f,\eta) + (f_i,\eta_{x_i}) = [F,\eta] \tag{3.6}$$

for all $\eta \in \mathring{W}_2^1(\Omega)$. Because of (3.5) and (3.6), the identity (3.2) is equivalent to

$$[u,\eta] + [Au,\eta] = [F,\eta]. \tag{3.7}$$

Since (3.7) holds for all $\eta \in \mathring{W}_2^1(\Omega)$, (3.7) is equivalent to the operator equation

$$u + Au = F \tag{3.8}$$

in the space $\mathring{W}_2^1(\Omega)$. We shall show that A is a completely continuous operator in $\mathring{W}_2^1(\Omega)$. For this we shall verify that any weakly convergent sequence $\{u_k\}$ $(k = 1, 2, \ldots)$ in $\mathring{W}_2^1(\Omega)$ is transformed by A into a strongly convergent sequence $\{Au_k\}$. Because the operator A is bounded, the sequence $\{Au_k\}$ converges weakly to Au, where $u(x)$ is the weak limit of $\{u_k\}$. Moreover, because of the compactness of the embedding operator of $\mathring{W}_2^1(\Omega)$ into $L_2(\Omega)$ (Theorem 6.1, Chapter I), the sequences $\{u_k\}$ and $\{Au_k\}$ converge strongly in $L_2(\Omega)$ to u and Au, respectively. If we use the definition (3.5) of the operator A and the inequalities (3.4), we obtain the bounds

$$\begin{aligned}[A(u_k - u_m), A(u_k - u_m)] &= l(u_k - u_m, A(u_k - y_m)) \\ &\le \mu_1\|u_k - u_m\|\cdot\|(Au_k)_x - (Au_m)_x\| \\ &\quad + \mu_1\|u_{kx} - u_{mx}\|\cdot\|Au_k - Au_m\| \\ &\quad + \max(|\mu_3|;|\mu_4|)\|u_k - u_m\|\,\|Au_k - Au_m\|.\end{aligned}$$

From this it is clear that the right-hand side tends to zero as $k, m \to \infty$, so that $\{Au_k\}$ is actually a strongly convergent sequence in $\mathring{W}_2^1(\Omega)$. This proves the complete continuity of A. Hence the first theorem of Fredholm holds for the equation (3.8): The solvability of (3.8) for all $F \in \mathring{W}_2^1(\Omega)$ is a consequence of the uniqueness theorem for (3.8). Because (3.8) is equivalent to the identity (2.6) for all $\eta \in \mathring{W}_2^1(\Omega)$ with $u(x)$ in $\mathring{W}_2^1(\Omega)$, we can formulate this theorem of Fredholm as follows:

Theorem 3.1 (First Theorem of Fredholm). *If the problem* (2.1), (2.2) *has no more than one generalized solution in* $W_2^1(\Omega)$, *then it is solvable in* $W_2^1(\Omega)$ *for any* f *and* $\mathbf{f}$ *in* $L_2(\Omega)$.

Sufficient conditions for uniqueness, and hence solvability, of the problem (2.1), (2.2) are given by Theorem 2.1. However, these are far from containing all possible cases. In order to clear up this question completely, we consider, not the equation (2.1) taken separately, but rather the family of equations

$$\mathscr{L}u = \lambda u + f + \frac{\partial f_i}{\partial x_i} \tag{3.9}$$

with a complex parameter λ. As before, the coefficients of $\mathscr{L}$ are taken to be real, but the solution of (3.9) will be, in general, a complex-valued function $u(x) = u'(x) + iu''(x)$. Because of this, we introduce the complex spaces $L_2(\Omega)$ and $\mathring{W}_2^1(\Omega)$, retaining in them the same notation as in the real spaces. The elements of these spaces are complex-valued functions of the variable $x \in \Omega$, and the scalar products are defined by

$$(u, v) = \int_\Omega u\bar{v}\,dx, \qquad (u, v)_{2,\Omega}^{(1)} = \int_\Omega (u\bar{v} + u_x \bar{v}_x)\,dx$$

respectively.

In accordance with what we said above, we define a *generalized solution of the problem* (3.9), (2.2) *in* $W_2^1(\Omega)$ as an element of $\mathring{W}_2^1(\Omega)$ which satisfies the identity

$$\begin{aligned}\mathscr{L}(u, \bar{\eta}) &\equiv \int_\Omega (a_{ij}u_{x_i}\bar{\eta}_{x_j} + a_i u\bar{\eta}_{x_i} - b_i u_{x_i}\bar{\eta} - au\bar{\eta})\,dx \\ &= -\lambda \int_\Omega u\bar{\eta}\,dx + \int_\Omega (-f\bar{\eta} + f_i \bar{\eta}_{x_i})\,dx \end{aligned} \tag{3.10}$$

for all $\eta \in \mathring{W}_2^1(\Omega)$.

To find the solution we transform (3.10) to an equation similar to (3.8). For this we introduce a new scalar product into $\mathring{W}_2^1(\Omega)$,

$$[u, v] = \int_\Omega a_{ij}u_{x_i}\bar{v}_{x_j}\,dx,$$

which gives rise, as in the real case, to the norm $\|u\|_1 = \sqrt{[u, u]}$ in $\mathring{W}_2^1(\Omega)$ which is equivalent to the old one, $\|u\|_{2,\Omega}^{(1)} = \sqrt{(u, u)_{2,\Omega}^{(1)}}$. Furthermore, if we argue in the same way as we did above in the real case, we arrive at the functional equation

$$u + Au = \lambda Bu + F \tag{3.11}$$

in the space $\mathring{W}_2^1(\Omega)$. Here the operators A and B are defined on all of $\mathring{W}_2^1(\Omega)$ by their bilinear forms:

$$[Au, \eta] = \int_\Omega (a_i u \bar{\eta}_{x_i} - b_i u_{x_i} \bar{\eta} - au\bar{\eta})\, dx, \tag{3.12}$$

and

$$[Bu, \eta] = -\int_\Omega u\bar{\eta}\, dx, \tag{3.13}$$

and the element F by the identity

$$[F, \eta] = \int_\Omega (-f\bar{\eta} + f_i \bar{\eta}_{x_i})\, dx. \tag{3.14}$$

The relations (3.12)–(3.14) are satisfied by all $\eta \in \mathring{W}_2^1(\Omega)$. In the same way as before, it can be proved that the operators A and B are linear and completely continuous. In addition, the operator B is symmetric (and consequently self-adjoint), and negative, that is, $[Bu, \eta] = [u, B\eta]$ for all u and η in $\mathring{W}_2^1(\Omega)$, and $[Bu, u] < 0$ for $u \not\equiv 0$. Both of these properties follow directly from the identity (3.13) which defined the operator B. Because of this, B has an inverse on its range $R(B)$ of values, but this inverse is unbounded. Let us write (3.11) in the form

$$u + Au - \lambda_0 Bu = (\lambda - \lambda_0)Bu + F \tag{3.15}$$

and verify that, for sufficiently large real λ_0, the operator $(E + A - \lambda_0 B) \equiv \mathscr{D}$ has a bounded inverse. Let us denote $\mathscr{D}v \equiv w$. According to (3.12)–(3.14) this equality is equivalent to (3.10) with $u = v$, $\lambda = \lambda_0$, and $[w, \eta]$ instead of the last integral. From this identity with $\eta = v$, along with (2.8) (for complex v the inequality (2.8) is readily seen to take the form

$$\operatorname{Re} \mathscr{L}(v, \bar{v}) \geq \frac{\nu}{2} \|v_x\|^2 - \left(\mu_4 + \frac{\mu_2^2}{2\nu}\right) \|v\|^2,$$

where $\|v\|^2 = \int_\Omega |v|^2\, dx$ and $\|v_x\|^2 = \int_\Omega |v_x|^2\, dx$), we derive the inequality

$$\begin{aligned} \operatorname{Re}[w, v] &= \operatorname{Re}[(E + A - \lambda_0 B)v, v] \\ &= \operatorname{Re} \mathscr{L}(v, \bar{v}) + \lambda_0 \|v\|^2 \geq \frac{\nu}{2} \|v_x\|^2 + \left(\lambda_0 - \mu_4 - \frac{\mu_2^2}{2\nu}\right) \|v\|^2. \end{aligned} \tag{3.16}$$

For $\lambda_0 \geq \mu_4 + \mu_2^2/2\nu$ it follows from (3.16) that

$$\|v\|_1 \leq c\|w\|_1 = c\|\mathscr{D}v\|_1, \tag{3.17}$$

that is, the operator $\mathscr{D}$ actually has a bounded inverse defined on all $\mathring{W}_2^1(\Omega)$ for such a λ_0 (see Theorem 3.1). Let us choose $\lambda_0 = \mu_4 + \mu_2^2/2\nu$ and rewrite (3.15) in the form

$$u = (\lambda - \lambda_0)\mathscr{D}^{-1}Bu + \mathscr{D}^{-1}F. \tag{3.18}$$

The operator $\mathscr{D}^{-1}B$, as the product of a bounded operator and a completely continuous operator, is completely continuous. As a result of this, we have the three theorems of Fredholm for the equation (3.18). The first of these asserts that the existence theorem follows from the uniqueness theorem for any free term. Since (3.18) is equivalent to (3.10) for all $\eta \in \overset{\circ}{W}{}_2^1(\Omega)$, this theorem guarantees the existence of a generalized solution $u(x)$ in $W_2^1(\Omega)$ of the problem (3.9), (2.2) for all f and $\mathbf{f}$ in $L_2(\Omega)$ whenever it is known that this problem cannot have two different solutions in $W_2^1(\Omega)$. We go over to the next theorem of Fredholm, which asserts that the uniqueness theorem for equation (3.18) fails only for a certain set of values λ which is at most countable and which has a single possible point of accumulation at infinity. These exceptional values of λ are called *spectral values* for the problem (3.9), (2.2). We shall denote them by $\{\lambda_k\}$ $(k = 1, 2, \ldots)$ and let $|\lambda_1| \leq |\lambda_2| \leq \cdots$. Each such λ corresponds to at least one non-trivial solution $u(x)$ (i.e., not identically zero) of the homogeneous problem

$$u = (\lambda - \lambda_0)\mathscr{D}^{-1}Bu \tag{3.19}$$

or, what is the same thing, to a $u(x) \in \overset{\circ}{W}{}_2^1(\Omega)$ satisfying the identity

$$\mathscr{L}(u, \bar{\eta}) = -\lambda(u, \eta) \tag{3.20}$$

for all $\eta \in \overset{\circ}{W}{}_2^1(\Omega)$. The second theorem of Fredholm also asserts that each spectral value λ_k is of finite multiplicity† and that the complex conjugate $\bar{\lambda}_k$ of λ_k is a spectral value of the equation adjoint to (3.19), that is, the equation‡

$$v = (\lambda - \lambda_0)B\mathscr{D}^{*-1}v, \tag{3.21}$$

where the multiplicity of $\bar{\lambda}_k$ for (3.21) is the same as the multiplicity of λ_k for (3.19). The relation (3.21) yields for $w = \mathscr{D}^{*-1}v$ the equation

$$\mathscr{D}^*w = (\lambda - \lambda_0)Bw, \tag{3.22}$$

which is equivalent, because of (3.12)–(3.14) and the definition of [,], to the identity

$$\begin{aligned} \mathscr{L}^*(w, \bar{\eta}) &\equiv \int_\Omega (a_{ij}w_{x_i}\bar{\eta}_{x_j} + a_i w_{x_i}\bar{\eta} - b_i w\bar{\eta}_{x_i} - aw\bar{\eta})\,dx \\ &= -\lambda \int_\Omega w\bar{\eta}\,dx \end{aligned} \tag{3.23}$$

for all $n \in \overset{\circ}{W}{}_2^1(\Omega)$. The identity (3.20) expresses the fact that $u(x)$ is a *generalized eigenfunction* in $W_2^1(\Omega)$ for the problem

$$\mathscr{L}u = \lambda u, \qquad u|_S = 0, \tag{3.24}$$

† The multiplicity of λ_k is the dimension of the eigensubspace corresponding to λ_k; in other words, the number of linearly independent solutions of (3.19) for $\lambda = \lambda_k$.

‡ We recall that $B^* = B$ and that $(\mathscr{D}^{-1})^* = (\mathscr{D}^*)^{-1}$, so that $(\mathscr{D}^{-1}B)^* = B\mathscr{D}^{*-1}$.

and that λ is its corresponding spectral value or eigenvalue. The identity (3.23) defines an eigenfunction $w(x) \in W_2^1(\Omega)$ of the problem

$$\mathscr{L}^*w \equiv \frac{\partial}{\partial x_i}(a_{ij}w_{x_j} - b_i w) - a_i w_{x_i} + aw = \lambda w, \qquad w|_S = 0, \quad (3.25)$$

which is the adjoint of (3.24). The second theorem of Fredholm for (3.19) guarantees that the problem (3.24) has non-trivial solutions $u(x)$ in $W_2^1(\Omega)$ only for the values $\{\lambda_k\}$ $(k = 1, 2, \ldots)$, and that each λ_k is of finite multiplicity; similarly, the problem (3.25) has non-trivial solutions w in $W_2^1(\Omega)$ only for the values $\bar{\lambda}_k$ $(k = 1, 2, \ldots)$, and each $\bar{\lambda}_k$ is of the same multiplicity as λ_k. Moreover, because the coefficients of $\mathscr{L}$ are real, it is easy to see that if λ_k is a spectral value for the problem (3.24) and $u_k(x)$ is one of the corresponding eigenfunctions, then $\bar{\lambda}_k$ and $\bar{u}_k(x)$ are solutions of the same problem (3.24), so that, in this situation, the spectral set $\{\lambda_k\}$ and $\{\bar{\lambda}_k\}$ of the problems (3.24) and (3.25) coincide.

Let us go on now to the third theorem of Fredholm, which gives us necessary and sufficient conditions for the solvability of (3.18) for the spectral values λ. That is, if $\lambda = \lambda_k$, then the problem (3.18) is solvable if and only if the free terms $\mathscr{D}^{-1}F$ are orthogonal to all the solutions v_k of the problem (3.21) corresponding to $\lambda = \bar{\lambda}_k$, i.e., if $\mathscr{D}^{-1}F$ satisfies the conditions

$$[\mathscr{D}^{-1}F, v_k] = 0. \quad (3.26)$$

Let us show that (3.26) is equivalent to

$$\int_\Omega (-f\bar{w}_k + f_i \bar{w}_{kx_i})\, dx = 0, \quad (3.27)$$

where $w_k = \mathscr{D}^{*-1}v_k$ is any generalized solution of (3.25) with $\lambda = \bar{\lambda}_k$; for $\mathbf{f} \equiv 0$, (3.27) expresses the fact that $f(x)$ is orthogonal in $L_2(\Omega)$ to all solutions of (3.25) with $\lambda = \bar{\lambda}_k$.

Actually, because of (3.14)

$$0 = [\mathscr{D}^{-1}F, v_k] = [F, \mathscr{D}^{*-1}v_k] = [F, w_k] = \int_\Omega (-f\bar{w}_k + f_i \bar{w}_{kx_i})\, dx,$$

that is, (3.27) is equivalent to (3.26). Thus, we have proved all three of the Fredholm theorems, which we state as the following theorem.

Theorem 3.2. *The problem (3.9), (2.2) is uniquely solvable in the space $W_2^1(\Omega)$ for any f and $\mathbf{f}$ in $L_2(\Omega)$ for all $\lambda = \lambda' + i\lambda''$, except for a set $\{\lambda_k\}$ $(k = 1, 2, \ldots)$ of values λ which is at most countable and which forms the spectrum of the problem (3.9), (2.2). Each λ_k is of finite multiplicity, and the only possible limit point of the set $\{\lambda_k\}$ is the point $\lambda = \infty$. Among the numbers $\{\lambda_k\}$ are the numbers $\{\bar{\lambda}_k\}$ and their multiplicities coincide. The same set $\{\lambda_k\}$ forms the spectrum of the problem (3.25) which is the adjoint of the homogeneous problem (3.9), (2.2), i.e., the problem (3.24). The multiplicities of λ_k and $\bar{\lambda}_k$ coincide* *for the problems* (3.24) *and* (3.25). *In order that the problem* (3.9), (2.2) *be solvable for* $\lambda = \lambda_k$, *it is necessary and sufficient that* f *and* f_i *satisfy the conditions* (3.27), *where* w_k *is any one of the generalized solutions of* (3.25) *with* $\lambda = \bar{\lambda}_k$; *the solution of the problem* (3.9), (2.2) *is not unique in this case. Its general solution is the sum of some particular solution and* $\sum_{m=1}^{N_k} c_m v_k^{(m)}(x)$, *where the* c_m *are arbitrary constants and* $v_k^{(m)}(x)$ *is an eigenfunction of the problem* (3.24) *corresponding to* $\lambda = \lambda_k$.

If λ is not in the spectrum of the problem (3.9), (2.2), then (3.18) can be solved in $\mathring{W}_2^1(\Omega)$ for all $\mathscr{D}^{-1}F$, and the operator $\mathscr{E} \equiv E - (\lambda - \lambda_0)\mathscr{D}^{-1}B$ has a bounded inverse in $\mathring{W}_2^1(\Omega)$, according to the theory of linear equations of the form (3.18) with a completely continuous operator $\mathscr{D}^{-1}B$. Consequently,

$$\|u\|_1 = \|\mathscr{E}^{-1}\mathscr{D}^{-1}F\|_1 \leq c_\lambda \|\mathscr{D}^{-1}F\|_1, \tag{3.28}$$

where c_λ is the norm of $\mathscr{E}^{-1}$ as an operator in $\mathring{W}_2^1(\Omega)$. The operator $\mathscr{D}$ also has a bounded inverse in $\mathring{W}_2^1(\Omega)$, where for the norm of $\mathscr{D}^{-1}$ we have the known majorant c, which may be expressed explicitly in terms of known quantities (cf. (3.17)). Thus,

$$\|u\|_1 \leq c c_\lambda \|F\|_1. \tag{3.29}$$

Finally, if we recall the connection of F with f and f_i (cf. (3.14)), we can bound $\|F\|_1$ in terms of $\|f\|$ and $\|\mathbf{f}\|$,

$$\|F\|_1 \leq c_2(\|f\| + \|\mathbf{f}\|), \tag{3.30}$$

where c_2 depends only of Ω. A bound for the solution $u(x)$ of the problem (3.9), (2.2) given by the free terms of (3.9) follows from (3.29) and (3.30):

$$\|u\|_1 \leq c c_\lambda c_2(\|f\| + \|\mathbf{f}\|). \tag{3.31}$$

In this inequality, we do not actually know the constant c_λ. The fact that it is finite is guaranteed by the theory of equations with completely continuous operators that we mentioned above. But to express it explicitly in terms of some properties of quantities that we know and in terms of λ is impossible in the general case. It depends on how close λ is to the spectrum $\sum_\lambda \equiv \{\lambda_k\}$ ($k = 1, 2, \ldots$) of the problem (3.9), (2.2). The calculation of the eigenvalues λ_k is also an extremely difficult problem. As a result, there is some interest in various conclusions concerning the location of the spectrum. One such assertion can be proved easily, namely, the spectrum $\sum_\lambda$ lies in a parabola of the form $-\lambda' = c + (\lambda'')^2$ in the plane of $\lambda = \lambda' + i\lambda''$. The constant c can be calculated from the constants ν, μ, and μ_i in conditions (2.3), (2.4). To prove this, we must set $\eta = u$ in (3.10) and from this relation obtain a bound for $\|u\|_1$ in terms of $\|f\|$ and $\|\mathbf{f}\|$, assuming that λ lies outside the above parabola. The existence of such a bound indicates that, for λ outside the parabola, the uniqueness theorem holds, that is, that such values λ do not belong to $\sum_\lambda$. Significantly more difficult considerations are required for obtaining the asymptotic behavior of the eigenvalues λ_k as a function of k.

This was done by T. Carleman (see [SM 1]). In particular, he showed that $\sum_\lambda$ consists of infinitely many eigenvalues.

Remark 3.1. As we recall in §1, all the results which we prove in this chapter carry over to the equations (2.1) with unbounded coefficients a_i, b_i, a which are summable to certain powers. Thus, for example, the conclusions of Theorems 3.1 and 3.2 are valid if a_i and b_i belong to $L_{q_i}(\Omega)$ with $q_i > n$, $a \in L_q(\Omega)$ with $q > n/2$, and $f \in L_q(\Omega)$ with $q > 1$ for $n = 2$ and with $q \geq 2n/(n+2)$ for $n \geq 3$ (cf. [LU 1]).

Remark 3.2. In every chapter, if we do not indicate otherwise; we assume that Ω is a bounded domain in R^n. However, some of our results can be extended easily to the case of unbounded domains. For example, the problem (3.9), (2.2) is solvable in $W_2^1(\Omega)$ for any domain Ω and for any f and f_i in $L_2(\Omega)$ provided that λ lies outside a parabola of the form $-\lambda' = c + (\lambda'')^2$. The constant c is easy to compute from the constants ν, μ, and μ_i in the conditions (2.3), (2.4). For such λ, we can establish at first the *a priori* bound

$$\|u\|_{2,\Omega}^{(1)} \leq c(\|f\| + \|\mathbf{f}\|), \tag{3.32}$$

where the constant c is independent of Ω. The solvability can be proved in this way: We can choose a sequence of bounded, expanding domains Ω_m ($m = 1, 2, \ldots$) which exhaust the domain Ω. The *a priori* bound (3.32) holds for any solution $u_m(x)$ of the problem (3.9), (2.2) in the domain Ω_m, for all m. The existence of such solutions u_m in $\mathring{W}_2^1(\Omega_m)$ is guaranteed by Theorem 3.1. We extend $u_m(x)$ to the whole domain by taking it to be zero outside Ω_m, and we retain the same notation $u_m(x)$ for this extension.

All the $u_m(x)$ will be elements of $\mathring{W}_2^1(\Omega)$, and by (3.32), they will be of uniformly bounded norm $\|u_m\|_{2,\Omega}^{(1)}$. We choose from $\{u_m\}$ a subsequence $\{u_{m_k}\}$ ($k = 1, 2, \ldots$) which, together with the derivatives $\{\partial u_{m_k}/\partial x_i\}$, converges weakly in $L_2(\Omega)$ to some function $u(x)$ with derivatives $\partial u/\partial x_i$. This function will be the desired generalized solution in $W_2^1(\Omega)$ of the problem (3.9), (2.2) in Ω. Actually, the functions $u_m(x)$ satisfy integral identities of the form (3.10) in the domains Ω_m for all $\eta \in \mathring{W}_2^1(\Omega_m)$. These identities can be rewritten in the form

$$\mathscr{L}(u_m, \bar{\eta}) = -\lambda(u_m, \eta) + \int_\Omega (-f\bar{\eta} + f_i\bar{\eta}_{x_i})\,dx, \tag{3.33}$$

where we assume that all integrations are taken over the domain Ω and that u_m and η are zero outside Ω_m. In (3.33) we fix some $\eta(x) \in \mathring{W}_2^1(\Omega)$ which is zero outside Ω_m, and then we take the limit in the subsequence $\{m_k\}$ that we chose above, assuming that $m_k \geq m$. In the limit we obtain (3.10) for u with the function $\eta(x)$ that we took. But these $\eta(x)$ form a dense set in $\mathring{W}_2^1(\Omega)$. Hence $u(x)$ will satisfy the identity (3.10) for all $\eta \in \mathring{W}_2^1(\Omega)$. By the same token, it follows that $u(x)$ is a generalized solution of the problem (3.9), (2.2) in Ω. By the uniqueness, which is guaranteed by the bound (3.32), the whole sequence $\{u_m\}$ will converge in this way to the solution $u(x)$.

§4. Expansion in Eigenfunctions of Symmetric Operators

Let us consider symmetric elliptic operators $\mathscr{L}$, that is, operators for which $\mathscr{L}^* = \mathscr{L}$. If we compare (2.1) and (3.25), we see that, in this case, $a_i = -b_i$ and

$$\mathscr{L}u = \frac{\partial}{\partial x_i}(a_{ij}u_{x_j} + a_i u) - a_i u_{x_i} + au. \tag{4.1}$$

We shall show that for such $\mathscr{L}$ the spectrum $\{\lambda_k\}$ and the eigenfunctions $\{u_k(x)\}$ are real, that $\lambda_k \to -\infty$ as $k \to \infty$, and that $\{u_k(x)\}$ forms a basis in the real spaces $L_2(\Omega)$ and $\mathring{W}_2^1(\Omega)$. In order to do this, we introduce a new scalar product into the complex space $\mathring{W}_2^1(\Omega)$,

$$\begin{aligned}[u, v] &= \int_\Omega (a_{ij}u_{x_i}\bar{v}_{x_j} + a_i(u_{x_i}\bar{v} + u\bar{v}_{x_i}) + (\lambda_0 - a)u\bar{v})\,dx \\ &\equiv \mathscr{L}(u, \bar{v}) + \lambda_0(u, v),\end{aligned} \tag{4.2}$$

where λ_0 is "sufficiently large" in the sense that

$$\mu_1^2 < \nu(\lambda_0 - \mu_4) \tag{4.3}$$

(cf. (2.3) and (2.4)). If this condition is fulfilled, then

$$\begin{aligned}[u, u] &\geq \nu\|u_x\|^2 - 2\mu_1\|u_x\| \cdot \|u\| + (\lambda_0 - \mu_4)\|u\|^2 \\ &\geq \nu_1(\|u_x\|^2 + \|u\|^2), \qquad \nu_1 = \text{const} > 0,\end{aligned} \tag{4.4}$$

which, together with (2.3) and (2.4), shows that the norm $\|u\|_1 = \sqrt{[u, u]}$ is equivalent to the old norm $\|u\|_{2,\Omega}^{(1)}$ in $\mathring{W}_2^1(\Omega)$. The scalar product (4.2) differs from the one introduced in §3, but we shall use the same notation for it, $[\ ,\ ]$, as we did in §3.

Let us consider the spectral problem

$$\mathscr{L}u = \lambda u, \qquad u|_S = 0. \tag{4.5}$$

Its *generalized solutions in* $W_2^1(\Omega)$ are elements of $\mathring{W}_2^1(\Omega)$ satisfying the identity $\mathscr{L}(u, \bar{\eta}) = -\lambda(u, \eta)$, or, what is the same thing, the identity

$$[u, \eta] = (\lambda_0 - \lambda)(u, \eta) \tag{4.6}$$

for all $\eta \in \mathring{W}_2^1(\Omega)$. In analogy with (3.13) we introduce the operator B by the identity

$$[Bu, \eta] = -(u, \eta), \tag{4.7}$$

for all $\eta \in \mathring{W}_2^1(\Omega)$. As in §3 above, it can be proved that B is completely continuous and symmetric (and consequently self-adjoint), as well as negative. Its inverse B^{-1} is defined on $\mathscr{R}(B)$ (we remark that it differs from the operator B in §3, for the scalar product (4.2) is different from the scalar product [,] in §3). By (4.7) the identity (4.6) is equivalent to the equation

$$u = (\lambda - \lambda_0)Bu, \tag{4.8}$$

and (4.8) to the equation

$$Bu = \mu u, \tag{4.9}$$

where $\mu = (\lambda - \lambda_0)^{-1}$. It follows from the general theory of self-adjoint, completely continuous operators that the spectrum of B is real and negative, and that the eigenvalues μ_k ($k = 1, 2, \ldots$) can be enumerated in the order of decreasing moduli, where each is counted according to its multiplicity.† The sequence $\{\mu_k\}$ can have only $\mu = 0$ as its limit point. The functions $\{u_k(x)\}$ corresponding to $\{\mu_k\}$ can be assumed to be real and mutually orthogonal, so that‡

$$[u_k, u_l] = 0, \qquad k \neq l. \tag{4.10}$$

For $\mu = 0$, (4.9) has only the zero solution, $u \equiv 0$, for it is clear from (4.7) that $Bu = 0$ implies that $u = 0$. From what has been said, it follows that $\{u_k\}$ constitutes a basis of $\mathring{W}_2^1(\Omega)$ (cf. §2, Chapter I). Since $\mathring{W}_2^1(\Omega)$ is an infinite-dimensional space, there are infinitely many elements in the basis $\{u_k\}$, and since each eigenvalue is of finite multiplicity, there are infinitely many distinct eigenvalues, and $\lim_{k\to\infty} \mu_k = 0$. Furthermore, any element $F \in \mathring{W}_2^1(\Omega)$ can be expanded into a Fourier series in terms of the basis elements $\{u_k\}$, that is, into the series

$$F(x) = \sum_{k=1}^{\infty} \frac{[F, u_k]}{[u_k, u_k]} u_k(x), \tag{4.11}$$

which converges in the space $\mathring{W}_2^1(\Omega)$. We recall that convergence in $\mathring{W}_2^1(\Omega)$ means the convergence of (4.11) in $L_2(\Omega)$, as well as of series obtained from it by (a single) term-by-term differentiation with respect to any x_i ($i = 1, \ldots, n$). Such convergence of (4.11) is quite remarkable, especially as series obtained by term-by-term differentiation with respect to x_i are not, in general, orthogonal in $L_2(\Omega)$. The series (4.11) itself is orthogonal in $L_2(\Omega)$ for it follows from (4.7), (4.9), and (4.10) that

$$\mu_k[u_k, u_l] = [Bu_k, u_l] = -(u_k, u_l) = 0, \qquad k \neq l. \tag{4.12}$$

If we normalize the eigenfunctions $u_k(x)$ so that $\|u_k\| = 1$, then

$$(u_k, u_l) = \delta_k^l = -(\lambda_k - \lambda_0)^{-1}[u_k, u_l], \tag{4.13}$$

† That is, an m-tuple spectral value appears m times in the sequence $\{\mu_k\}$.

‡ Functions u_k and u_l corresponding to different eigenvalues are orthogonal by the general theory, while functions corresponding to the same eigenvalue must be orthogonalized (which is possible because any linear combinations of them are eigenfunctions for the same eigenvalue λ).

where $\lambda_k = \lambda_0 + \mu_k^{-1}$ are the eigenvalues of the problem (4.5), and the series (4.11) can be written in the form

$$F(x) = \sum_{k=1}^{\infty} (F, u_k)u_k(x). \tag{4.14}$$

Since $\mathring{W}_2^1(\Omega)$ is dense in $L_2(\Omega)$, and since $\{u_k\}$, being a basis of $\mathring{W}_2^1(\Omega)$, is orthogonal in $L_2(\Omega)$, the expansion (4.14) holds not only for $F \in \mathring{W}_2^1(\Omega)$ but also for any element F in $L_2(\Omega)$, where the series (4.14) converges to F in the norm of $L_2(\Omega)$. Let us summarize in the form of a theorem the results obtained in this section.

Theorem 4.1. *Suppose that an operator $\mathscr{L}$ of the form (4.1), defined in a bounded domain Ω, satisfies (2.3) and (2.4). Then the spectral problem (4.5) for $\mathscr{L}$ in $\mathring{W}_2^1(\Omega)$ has a countable set of solutions, $\lambda = \lambda_k, u = u_k(x), k = 1, 2, \ldots$. The eigenvalues λ_k, with the possible exception of the first few, are negative and $\lambda_k \to -\infty$ as $k \to \infty$. The eigenfunctions $\{u_k(x)\}$ form a basis in $L_2(\Omega)$ and in $\mathring{W}_2^1(\Omega)$; this basis can be orthonormalized in $L_2(\Omega)$ and is orthogonal in the sense of the scalar product (4.2) (cf. (4.13)).*

Later, in §7, we shall show that the eigenfunction expansions in $\{u_k\}$ catch delicately a further refinement of the differential properties of the expanded function $F(x)$, by converging just in that function space $W_2^l(\Omega)$ (more precisely, in a certain subspace of it) containing $F(x)$ and $u_k(x)$ $(k = 1, 2, \ldots)$.

From the general theory of self-adjoint, completely continuous operators B it is known that the eigenfunctions and eigenvalues of the problem (4.9) can be obtained as solutions of the following variational problems. The first and least (left) eigenvalue μ_1 (we recall that μ_1 and all other μ_k are negative) can be found as the infimum of $[Bu, u]$ among all elements of $\mathring{W}_2^1(\Omega)$ satisfying the condition $[u, u] = 1$, or, what is the same thing, as the infimum of $J(u) \equiv [Bu, u]/[u, u]$ over all of $\mathring{W}_2^1(\Omega)$. The eigenfunction $u_1(x)$ realizes this infimum. The spaces $L_2(\Omega)$ and $\mathring{W}_2^1(\Omega)$ here and below can be taken to be real. The next eigenvalue μ_2 is the infimum of $J(u)$ over all functions in $\mathring{W}_2^1(\Omega)$ satisfying the condition $[u, u_1] = 0$. This infimum is realized by one or more eigenfunctions; we choose one of these functions as a solution, say $u_2(x)$. The next value μ_3 (which can coincide numerically with μ_2) is the infimum of $J(u)$ taken over the set of all elements of $\mathring{W}_2^1(\Omega)$ which are orthogonal to $u_1(x)$ and $u_2(x)$ relative to [,]. If we continue the process in this way, we can find all the μ_k and $u_k(x)$ $(k = 1, 2, \ldots)$.

R. Courant established the so-called minimax principle of determining $\{\mu_k\}$ and $\{u_k(x)\}$ which allows us to find μ_m and $u_m(x)$ without first determining μ_k and $u_k(x)$ for $k = 1, 2, \ldots, m - 1$ (cf. [CHI, Bd 1]). He proved that

$$\mu_m = \sup_{\mathfrak{M}_{m-1}} \inf_{\substack{[u,v]=0 \\ v \in \mathfrak{M}_{m-1}}} \frac{[Bu, u]}{[u, u]}, \tag{4.15}$$

where $\mathfrak{M}_{m-1}$ is an arbitrary $(m-1)$-dimensional subspace of $\mathring{W}_2^1(\Omega)$. If we recall the definition (4.7) of the operator B, we see that $J(u) = -(u, u)/[u, u]$, so that the function $u_1(x)$ realizing $\inf J(u)$ in $\mathring{W}_2^1(\Omega)$ also realizes $\inf[u, u]/(u, u) = -1/\mu_1$ in $\mathring{W}_2^1(\Omega)$. Since $[u, u] = \mathscr{L}(u, u) + \lambda_0(u, u)$, $u_1(x)$ also yields $\inf \mathscr{L}(u, u) \equiv \inf \int_\Omega [a_{ij}u_{x_i}u_{x_j} + 2a_iu_{x_i}u - au^2]\,dx$ under the condition $\int_\Omega u^2\,dx = 1$. (As we said above, the spaces $L_2(\Omega)$ and $\mathring{W}_2^1(\Omega)$, and the scalar products in them, can be chosen to be real. This is possible, since it is known in advance that μ_k and $u_k(x)$ are real.) It is not hard to show that the $\inf \mathscr{L}(u, u)$ in $\mathring{W}_2^1(\Omega)$ with $\|u\| = 1$ is equal to $-1/\mu_1 - \lambda_0$, i.e., $-\lambda_1$, so that $u_1(x)$ and λ_1 satisfy

$$\inf \mathscr{L}(u, u) = \mathscr{L}(u_1, u_1) = -\lambda_1, \tag{4.16}$$

where the infimum is taken over all real $u \in \mathring{W}_2^1(\Omega)$ with $\|u\| = 1$.

By what we said above, the function $u_2(x)$ is to be found among the elements $u \in \mathring{W}_2^1(\Omega)$ satisfying $[u_1, u] = 0$. But by (4.6), $[u_1, u] = (\lambda_0 - \lambda_1)(u_1, u)$, where $\lambda_0 - \lambda_1 \neq 0$, so that the requirement $[u_1, u] = 0$ is equivalent to $(u_1, u) = 0$, and u_2, λ_2 can be found as solutions of the problem

$$\inf \mathscr{L}(u, u) = \mathscr{L}(u_2, u_2) = -\lambda_2, \tag{4.17}$$

where the infimum is taken over all $u \in \mathring{W}_2^1(\Omega)$ with $\|u\| = 1$ and $(u_1, u) = 0$. In this way we find successively all λ_k and $u_k(x)$ $(k = 1, 2, \ldots)$. The principle (4.15) for λ_m and $u_m(x)$ yields the following: $-\lambda_m = \sup \inf \mathscr{L}(u, u)$, where the infimum is taken over all $u \in \mathring{W}_2^1(\Omega)$ with $\|u\| = 1$ and $(u, v) = 0$, and v ranges over some $(m-1)$-dimensional subspace $\mathfrak{M}_{m-1}$ of $L_2(\Omega)$; then the supremum is sought over all subspaces $\mathfrak{M}_{m-1}$. Thus, $\sup \inf \mathscr{L}(u, u)$ is realized by the eigenfunction u_m.

§5. The Second and Third Boundary Value Problems

Let us consider the problem of finding the solutions of the equation

$$\mathscr{L}u \equiv \frac{\partial}{\partial x_i}(a_{ij}u_{x_j}) + b_iu_{x_i} + au = \lambda u + f, \tag{5.1}$$

which satisfy the boundary condition

$$\frac{\partial u}{\partial N} + \sigma u|_S = 0, \tag{5.2}$$

where $\sigma = \sigma(s)$ is a given function on S, $\partial u/\partial N = a_{ij}x_{x_j}\cos(\mathbf{n}, \mathbf{x}_i)$ is the so-called co-normal derivative, and $\mathbf{n}$ is the outer normal to S. Let the conditions (2.3)–(2.5) be fulfilled, let $|\sigma(s)| \leq \mu_5$, and let S be piecewise smooth. More precisely, S should be such that the elements of $W_2^1(\Omega)$ have traces on S in $L_2(S)$ for which the formula on integration by parts holds and for which the embeddings of $W_2^1(\Omega)$ into $L_2(\Omega)$ and $L_2(S)$ are compact; sufficient conditions were given in §6, Chapter I. First, let us explain what is natural to understand as a generalized solution of the problem (5.1), (5.2) in $W_2^1(\Omega)$, where λ and $u(x)$ are assumed to be complex-valued. It is clear that an arbitrary element $u(x)$ of $W_2^1(\Omega)$ need not satisfy either the equation or the boundary condition in the form (5.1) and (5.2) (for it need not have second-order generalized derivatives in Ω, and its first-order derivatives need not be defined on S). Hence both equations (5.1) and (5.2) must be "embedded" into an integral identity, and to do this, at least formally, we multiply (5.1) by an arbitrary function $-\bar{\eta}(x)$ in $W_2^1(\Omega)$ and integrate over Ω. After this, we integrate the first term by parts and take (5.2) into account. This leads us to the identity

$$\int_\Omega (a_{ij}u_{x_j}\bar{\eta}_{x_i} - b_i u_{x_i}\bar{\eta} + (\lambda - a)u\bar{\eta})\,dx + \int_S \sigma u\bar{\eta}\,ds = -\int_\Omega f\bar{\eta}\,dx, \tag{5.3}$$

which holds for all $\eta \in W_2^1(\Omega)$. The relation (5.3) is meaningful for all $u(x)$ and $\eta(x)$ in $W_2^1(\Omega)$, for according to the results concerning such functions presented in §6, Chapter I, we can say that they assume on S their boundary values $u(x)$ and $\eta(s)$, where these values are elements of $L_2(S)$. If the coefficients a_{ij} have bounded first derivatives, and if $u(x)$ belongs to $W_2^2(\Omega)$ and satisfies equations (5.1), (5.2), then $u(x)$ satisfies (5.3) for all $\eta \in W_2^1(\Omega)$. Conversely, the identity (5.3) implies both (5.1) and (5.2); and to show that we must change (5.3) into the form

$$-\int_\Omega \mathscr{L}u\cdot\bar{\eta}\,dx + \int_S \left(\frac{\partial u}{\partial N} + \sigma u\right)\bar{\eta}\,ds = -\int_\Omega f\bar{\eta}\,dx$$

and take advantage of a sufficient arbitrariness in the choice of $\eta(x)$.

A function $u(x)$ in $W_2^1(\Omega)$ satisfying (5.3) for all $\eta \in W_2^1(\Omega)$ will be called a *generalized solution in* $W_2^1(\Omega)$ *of the problem* (5.1), (5.2). The role of the fundamental functional space in this problem will be played by $W_2^1(\Omega)$, and not by $\mathring{W}_2^1(\Omega)$, as in the Dirichlet problem. We introduce into $W_2^1(\Omega)$ a new scalar product

$$[u, v] = \int_\Omega (a_{ij}u_{x_i}\bar{v}_{x_j} + u\bar{v})\,dx. \tag{5.4}$$

The identity (5.3) can be represented in the form

$$[u, \eta] + [Au, \eta] - \lambda[Bu, \eta] + [Cu, \eta] = [F, \eta], \tag{5.5}$$

where the operators A, B, and C are defined by their bilinear forms:

$$[Au, \eta] = \int_\Omega (-b_i u_{x_i}\bar{\eta} - (a + 1)u\bar{\eta})\,dx,$$

$$[Bu, \eta] = -\int_\Omega u\bar{\eta}\,dx, \qquad [Cu, \eta] = \int_S \sigma u\bar{\eta}\,ds,$$

and

$$[F, \eta] = -\int_\Omega f\bar{\eta}\,dx.$$

Just as in §3, it can be proved that A, B, and C are completely continuous on $W_2^1(\Omega)$, where the complete continuity of C is a consequence of the compactness of the embedding of $W_2^1(\Omega)$ into $L_2(S)$. For the latter we require a certain regularity of S.

The identity (5.3) is equivalent to the equation

$$u + Au - \lambda Bu + Cu = F. \tag{5.6}$$

The investigation of (5.6) and of the problem (5.1), (5.2) can be carried out in a way that is totally analogous to what we did in §3 for the Dirichlet problem, and leads to the three theorems of Fredholm (i.e., to Theorem 3.2) for the problem (5.1), (5.2). We shall not rewrite all these arguments and conclusions, but shall remark only that the problem adjoint to (5.1), (5.2) has the form

$$\left.\begin{aligned} \mathscr{L}^*v &\equiv \frac{\partial}{\partial x_i}(a_{ij}v_{x_j}) - (b_i v)_{x_i} + av = \lambda v + \tilde{f}, \\ \frac{\partial v}{\partial N} &+ (\sigma - b_i \cos(\mathbf{n}, \mathbf{x}_i))v|_S = 0, \end{aligned}\right\} \tag{5.7}$$

which can be seen from the identity

$$\int_\Omega \bar{v}\mathscr{L}u\,dx = \int_\Omega u\mathscr{L}^*\bar{v}\,dx + \int_S \left(\frac{\partial u}{\partial N}\bar{v} - u\frac{\partial \bar{v}}{\partial N} + b_i\cos(\mathbf{n}, \mathbf{x}_i)u\bar{v}\right)ds. \tag{5.8}$$

This last identity is valid for any u and v in $W_2^2(\Omega)$ if, besides (2.3)–(2.5), the derivatives $\partial a_{ij}/\partial x_j$ and $\partial b_i/\partial x_i$ exist and are bounded.

A *generalized solution v in $W_2^1(\Omega)$ of the problem* (5.7) is defined to be an element of $W_2^1(\Omega)$ satisfying the identity

$$\int_\Omega (a_{ij}v_{x_j}\bar{\eta}_{x_i} - b_i v\bar{\eta}_{x_i} + (\lambda - a)v\bar{\eta})\,dx + \int_S \sigma v\bar{\eta}\,ds = -\int_\Omega \tilde{f}\bar{\eta}\,dx$$

for all $\eta \in W_2^1(\Omega)$.

If $b_i \equiv 0$ then the problem (5.7) coincides with (5.1), (5.2). In this case the eigenvalues λ_k and the eigenfunctions u_k of the problem

$$\mathscr{L}u = \lambda u, \qquad \left.\frac{\partial u}{\partial N} + \sigma u\right|_S = 0 \tag{5.9}$$

are real. Conclusions now hold for λ_k and $u_k(x)$ which are analogous to those proved in §4 for the first boundary condition. Thus the system of all eigenfunctions $\{u_k\}$ forms a basis for the spaces $L_2(\Omega)$ and $W_2^1(\Omega)$, where we can choose the $u_k(x)$ in such a way that

$$(u_k, u_l) = \delta_k^l, \tag{5.10}$$

and

$$\int_\Omega (a_{ij} u_{kx_j} u_{lx_i} - a u_k u_l)\, dx + \int_S \sigma u_k u_l\, ds = -\lambda_k \delta_k^l. \tag{5.11}$$

Any function $u(x)$ in $L_2(\Omega)$ can be expanded into a Fourier series

$$u(x) = \sum_{k=1}^{\infty} (u, u_k) u_k(x), \tag{5.12}$$

which converges in the norm of $L_2(\Omega)$. If $u(x) \in W_2^1(\Omega)$ then (5.12) converges to $u(x)$ in the norm of $W_2^1(\Omega)$; in other words, the series (5.12) and the series obtained from it by term-by-term differentiation with respect to any x_i converge in $L_2(\Omega)$ to u and $\partial u/\partial x_i$, respectively.

Remark 5.1. Problems for equation (5.1) with mixed boundary conditions of the form

$$u\Big|_{S_1} = 0, \qquad \frac{\partial u}{\partial N} + \sigma u\Big|_{S_2} = 0, \qquad S_1 \bigcup S_2 = S \tag{5.13}$$

can be considered in a way similar to the problem (5.1), (5.2). Their *generalized solutions in* $W_2^1(\Omega)$ are defined as elements of $W_{2,0}^1(\Omega)$ satisfying

$$\int_\Omega (a_{ij} u_{x_i} \bar{\eta}_{x_j} - b_i u_{x_i} \bar{\eta} + (\lambda - a) u \bar{\eta})\, dx + \int_{S_2} \sigma u \bar{\eta}\, ds = -\int_\Omega f \bar{\eta}\, dx \tag{5.14}$$

for all $\eta \in W_{2,0}^1(\Omega)$. Here $W_{2,0}^1(\Omega)$ is a subspace of $W_2^1(\Omega)$ in which the set of all functions of $C^1(\bar{\Omega})$ vanishing near S_1 is dense. For $u(x) \in W_{2,0}^1(\Omega)$ we have

$$\int_\Omega u^2\, dx \le c_{\Omega, S_1} \int_\Omega u_x^2\, dx \tag{5.15}$$

with the constant c_{Ω, S_1} depending only on Ω and S_1 (for this "area" of the piece S_1 of S should be positive).

§6. The Second Fundamental Inequality for Elliptic Operators

In the previous sections we explained what the situation is with the solvability of the basic boundary value problems for the equations (2.1) in the space $W_2^1(\Omega)$, where Ω is a bounded domain. In this and the following sections we shall show that all the generalized solutions in $W_2^1(\Omega)$ of the Dirichlet problem will be in $W_2^2(\Omega)$ whenever the known functions and the boundary S possess a certain smoothness—more than that required in §§2–5. This will also be true of the other boundary value problems. We shall assume that the coefficients, besides satisfying (2.3) and (2.4), have bounded derivatives $\partial a_{ij}/\partial x_k$ and $\partial a_i/\partial x_i$ in Ω and that f and $\partial f_i/\partial x_i$ are in $L_2(\Omega)$. Because of this it is sufficient to consider, instead of (2.1), an equation of the form

$$\mathscr{L}u \equiv \frac{\partial}{\partial x_i}(a_{ij}u_{x_j}) + a_i u_{x_i} + au = f, \tag{6.1}$$

where we have rewritten f in place of $f + \partial f_i/\partial x_i$ and $a_i u_{x_i} + au$ in place of $\partial(a_i u)/\partial x_i + b_i u_{x_i} + au$. Thus, suppose that (2.3) and (2.4) are fulfilled for an $\mathscr{L}$ of the form (6.1) and that

$$\left|\frac{\partial a_{ij}}{\partial x_k}\right| \leq \mu_5, \tag{6.2}$$

and

$$\|f\| < \infty. \tag{6.3}$$

Let us assume that the boundary S is twice continuously differentiable (in short, $S \in C^2$). We shall show that, when these conditions concerning $\mathscr{L}$ and S are fulfilled, then an arbitrary function $u(x)$ in $C^2(\bar{\Omega})$ which vanishes on S satisfies the inequality

$$\|u_{xx}\|^2 \leq \frac{2}{\nu^2}\|\mathscr{L}u\|^2 + c(\|u\|_{2,\Omega}^{(1)})^2 \tag{6.4}$$

with a constant c defined by S and the constants in (2.3), (2.4), and (6.2)†. In particular, if $\mathscr{L} = \Delta$ in a convex domain Ω, (6.4) takes the form

$$\|u_{xx}\| \leq \|\Delta u\|. \tag{6.5}$$

The inequalities (6.4) and (6.5) are rather remarkable and unexpected, as can be seen from (6.5): it turns out that the sum $\Delta u = \sum_{i=1}^n \partial^2 u/\partial x_i^2$ bounds each term separately if the norm is taken to be the norm of $L_2(\Omega)$, and this, for any function $u(x)$ in $C^2(\bar{\Omega})$, vanishes on S. If we take any point $x_0 \in \Omega$, then it is clear that $|\partial^2 u/\partial x_i^2|$ cannot be bounded at x_0 in terms of $|\Delta u|$. If (6.4) is true for all $u \in C^2(\bar{\Omega})$ and $u|_s = 0$, then (6.4) is true for *the closure of this set in the norm of* $W_2^2(\Omega)$, *which we denote by* $W_{2,0}^2(\Omega)$. Actually, for each $u(x) \in W_{2,0}^2(\Omega)$ there is a sequence $\{u_m\}$, $u_m \in C^2(\bar{\Omega})$, $u_m|_s = 0$, which converges to $u(x)$ in the norm of $W_2^2(\Omega)$. For each of the u_m the inequality (6.4) holds, where c is independent of m. If we take the limit as $m \to \infty$, we obtain (6.4) for $u(x)$; this procedure is called, in brief, "the closure of the inequality (6.4) in the norm of $W_2^2(\Omega)$."

As is obvious from (2.15), the inequality

$$\|u_x\|^2 \leq \frac{1}{2\nu\varepsilon_2}\|\mathscr{L}u\|^2 + c_{\varepsilon_2}\|u\|^2 \tag{6.6}$$

† The coefficient $2/\nu^2$ can be replaced by $(1 + \varepsilon)/\nu^2$ for all $\varepsilon > 0$, but then c must be replaced by the corresponding constant c_ε.

is valid for all $u(x) \in \overset{\circ}{W}{}_2^2(\Omega)$, where $c_{\varepsilon_2} = (2/\nu)|\mu_4 + \mu_2^2/\nu + \varepsilon_2|$ and all $\varepsilon_2 > 0$, for $u(x)$ is a generalized solution in $\overset{\circ}{W}{}_2^1(\Omega)$ of the problem $\mathscr{L}u = f$, $u|_s = 0$ with $f(x)$ equal to $\mathscr{L}u(x) \in L_2(\Omega)$. Let us add the sum $\|u_x\|^2 + \|u\|^2$ to both sides of (6.4), and, on the right-hand side of the inequality obtained, replace $\|u_x\|^2$ by its majorant from (6.6) and then reduce similar terms. Then we take square roots of both sides of the resulting inequality and apply the inequality $\sqrt{a+b} \leq \sqrt{a} + \sqrt{b}$ to the right-hand side. For $\varepsilon_2 = \nu(c+1)/4$, this gives us the *second fundamental inequality* for elliptic operators,

$$\|u\|_{2,\Omega}^{(2)} \leq \frac{2}{\nu}\|\mathscr{L}u\| + c_2\|u\|, \tag{6.7}$$

where $c_2 = \sqrt{(c+1)(c_1+1)}$, and $c_1 = c_{\varepsilon_2}$ for $\varepsilon_2 = \nu(c+1)/4$; this inequality is valid for all $u \in \overset{\circ}{W}{}_2^2(\Omega)$. In those cases where $\mathscr{L}(u, u) \geq \delta_1\|u\|^2$, $\delta_1 = \text{const.} > 0$, the last term $\|u\|$ in (6.7) can be bounded in terms of $\|\mathscr{L}u\|$, and, instead of (6.7), we have the inequality

$$\|u\|_{2,\Omega}^{(2)} \leq c_3\|\mathscr{L}u\|, \tag{6.8}$$

where $c_3 = 2/\nu + c_2/\delta_1$. In fact,

$$|\mathscr{L}(u, u)| = |(\mathscr{L}u, u)| \leq \|\mathscr{L}u\| \cdot \|u\|,$$

hence $\delta_1\|u\| \leq \|\mathscr{L}u\|$ follows from our assumption, and (6.7) implies (6.8). The inequality (6.8) holds, for example, if

$$\max_{\varepsilon_1 \in (0,\nu]}\left[(\nu - \varepsilon_1)c_\Omega^{-2} - \left(\mu_4 + \frac{\mu_2^2}{4\varepsilon_1}\right)\right] \equiv \delta_1 > 0 \tag{6.9}$$

(cf. (2.9), (2.10)). In the general case, (6.8) holds if and only if $\lambda = 0$ is not in the spectrum of $\mathscr{L}$, for only in this case does the inequality $\|u\| \leq c\|\mathscr{L}u\|$ hold (cf. (3.31)), and consequently (6.8). For solutions of (6.1) vanishing on S, the inequalities (6.7) and (6.8) give an *a priori* bound for the norm $\|u\|_{2,\Omega}^{(2)}$ in terms of $\|\mathscr{L}u\| = \|f\|$ and $\|u\|$.

We go over to the proof of (6.4) for $u \in C^2(\overline{\Omega})$, $u|_s = 0$. For this we consider the integral $\int_\Omega (\mathscr{L}u)^2\,dx$ and bound it from below in the following way:

$$\begin{aligned}\int_\Omega (\mathscr{L}u)^2\,dx &= \int_\Omega \left[(a_{ij}u_{x_ix_j})^2 + 2a_{ij}u_{x_ix_j}\left(\frac{\partial a_{ij}}{\partial x_i}u_{x_j} + a_iu_{x_i} + au\right)\right.\\ &\quad \left. + \left(\frac{\partial a_{ij}}{\partial x_i}u_{x_j} + a_iu_{x_i} + au\right)^2\right]dx \geq (1-\varepsilon)\int_\Omega (a_{ij}u_{x_ix_j})^2\,dx\\ &\quad + \left(1 - \frac{1}{\varepsilon}\right)\int_\Omega\left(\frac{\partial a_{ij}}{\partial x_i}u_{x_j} + a_iu_{x_i} + au\right)^2 dx\\ &\geq (1-\varepsilon)\int_\Omega (a_{ij}u_{x_ix_j})^2\,dx - c_1\left(-1 + \frac{1}{\varepsilon}\right)\int_\Omega (u_x^2 + u^2)\,dx\end{aligned} \tag{6.10}$$

for all $\varepsilon \in (0, 1)$. By means of a double integration by parts we transform $\int_\Omega (a_{ij}u_{x_ix_j})^2\,dx$ as follows:

$$\int_\Omega a_{ij}u_{x_ix_j}a_{kl}u_{x_kx_l}\,dx$$

$$= \int_\Omega \left[a_{ij}u_{x_ix_k}a_{kl}u_{x_jx_l} - \frac{\partial}{\partial x_j}(a_{ij}a_{kl})u_{x_i}u_{x_kx_l} + \frac{\partial}{\partial x_k}(a_{ij}a_{kl})u_{x_i}u_{x_jx_l}\right]dx + \int_S I(s)\,ds\dagger \geq \int_\Omega I_1(x)\,dx$$

$$+ \int_S I(s)\,ds - c_2\int_\Omega \left(\varepsilon_1 u_{xx}^2 + \frac{1}{\varepsilon_1}u_x^2\right)dx \tag{6.11}$$

for all $\varepsilon_1 \in (0, 1)$. Here the constant c_2, as well as c_1, is defined by the constants from the conditions (2.3), (2.4), (6.2), and

$$I_1(x) \equiv a_{ij}(x)a_{kl}(x)u_{x_ix_k}u_{x_jx_l},$$

$$I(s) \equiv a_{ij}a_{kl}u_{x_i}[u_{x_kx_l}\cos(\mathbf{n}, \mathbf{x}_j) - u_{x_jx_l}\cos(\mathbf{n}, \mathbf{x}_k)]|_{x\in S}.$$

It is not hard to see that

$$I_1(x) \geq \nu^2 u_{xx}^2(x). \tag{6.12}$$

In fact, we fix an arbitrary point $x^0 \in \Omega$ and introduce new Cartesian coordinates into some neighborhood of x^0, $y_k = \alpha_{kl}(x_l - x_l^0)$. We choose an orthogonal matrix (α_{kl}) in such a way that it reduces the quadratic form $a_{ij}(x^0)\xi_i\xi_j$ to diagonal form, i.e., so that $a_{ij}(x^0)\alpha_{ki}\alpha_{lj} = \lambda_k(x^0)\delta_k^l$, where $\lambda_k(x)$ $(k = 1, \ldots, n)$ are the eigenvalues of the form $a_{ij}(x)\xi_i\xi_j$. Then, by using (2.3), we obtain

$$I_1(x^0) = \sum_{s,t=1}^{n} \lambda_s(x^0)\lambda_t(x^0)u_{y_sy_t}^2(x^0) \geq \nu^2 u_{yy}^2(x^0).$$

But $u_{yy}^2(x^0) = u_{xx}^2(x^0)$, so that (6.12) is valid. From (6.10), (6.11), and (6.12) we deduce the inequality

$$\int_\Omega (\mathscr{L}u)^2\,dx \geq (1 - \varepsilon)\left[\nu^2\|u_{xx}\|^2 + \int_S I\,ds - c_2\varepsilon_1\|u_{xx}\|^2 - \frac{c_2}{\varepsilon_1}\|u_x\|^2\right] - c_1\left(\frac{1}{\varepsilon} - 1\right)(\|u\|_{2,\Omega}^{(1)})^2,$$

† For the first integration by parts, $u(x)$ should have derivatives of third order, but our equality, obtained as a result of a double integration by parts, requires of $u(x)$ only second-order derivatives, and it is valid for all $u(x)$ in $C^2(\bar{\Omega})$. In order to show this, we must extend $u(x)$ from Ω to a larger domain Ω_1 in such a way that $u(x) \in C^2(\Omega_1)$ and to take averages $u_\rho(x)$ with a smooth kernel. The equality of interest to us for u_ρ is valid, and we then go to the limit as $\rho \to 0$. As a result, we recapture the same equality for the function $u(x)$.

from which it follows that

$$(1-\varepsilon)(v^2-\varepsilon_1 c_2)\int_\Omega u_{xx}^2\,dx \le \int_\Omega (\mathscr{L}u)^2\,dx - (1-\varepsilon)\int_S I(s)\,ds + \left[c_1\left(\frac{1}{\varepsilon}-1\right)+\frac{c_2}{\varepsilon_1}(1-\varepsilon)\right](\|u\|_{2,\Omega}^{(1)})^2. \tag{6.13}$$

If $\varepsilon = \frac{1}{7}$ and $\varepsilon_1 = v^2/(8c_2)$, (6.13) assumes the form

$$\frac{3v^2}{4}\int_\Omega u_{xx}^2\,dx \le \int_\Omega (\mathscr{L}u)^2\,dx - \frac{6}{7}\int_S I(s)\,ds + c_3(\|u\|_{2,\Omega}^{(1)})^2. \tag{6.14}$$

The inequality (6.14) is of little use in itself, for the right-hand side contains the integral over S which involves the second derivative of $u(x)$. However, it turns out (and this is the "nugget" in the whole proof) that if we use the boundary condition $u|_s = 0$, then $I(s)$ reduces to a form containing only the first-order derivatives of $u(x)$. In order to prove this, we consider an arbitrary point x^0 of S and introduce local Cartesian coordinates $y_1, \ldots, y_n$, $y_k = c_{kl}(x_l - x_l^0)$, into a neighborhood of x^0 in such a way that the y_n-axis is directed along the outer normal $\mathbf{n}$ to S at x^0 and such that the matrix (c_{kl}) is orthogonal. Let $y_n = \omega(y_1, \ldots, y_{n-1})$ be the equation of S in the neighborhood of the origin $y = (0, \ldots, 0)$. By hypothesis, $\omega(y_1, \ldots, y_{n-1}) \in C^2$. From the orthogonality of the matrix (c_{kl}) we have $x_l - x_l^0 = c_{kl} y_k$ $(l = 1, \ldots, n)$, whence $\cos(\mathbf{n}, \mathbf{x}_l) = c_{nl}$ $(l = 1, \ldots, n)$. Let us consider the expression $I(s)$ at the point x^0 and change it over to the y-coordinates:

$$\begin{aligned} I(x^0) &= a_{ij}a_{kl}c_{mi}\frac{\partial u}{\partial y_m}c_{pk}c_{ql}\frac{\partial^2 u}{\partial y_p \partial y_q}c_{nj} \\ &\quad - a_{ij}a_{kl}c_{mi}\frac{\partial u}{\partial y_m}c_{pj}c_{ql}\frac{\partial^2 u}{\partial y_p\,\partial y_q}c_{nk} \\ &\equiv (b_{nm}b_{pq} - b_{mp}b_{qn})\frac{\partial u}{\partial y_m}\frac{\partial^2 u}{\partial y_p\,\partial y_q}, \end{aligned} \tag{6.15}$$

where $b_{pq} = a_{kl}c_{pk}c_{ql}$ $(p, q = 1, \ldots, n)$. Now let us use the boundary condition $u|_s = 0$. Near the point x^0 with coordinates $y_1 = \cdots = y_n = 0$, this condition has the form

$$u(y_1, \ldots, y_{n-1}, \omega(y_1, \ldots, y_{n-1})) = 0,$$

where it is satisfied identically for $y_1, \ldots, y_{n-1}$ near 0. We shall differentiate this identity with respect to y_i and y_j $(i, j = 1, \ldots, n-1)$, keeping in mind the fact that, at x^0, $\omega_{y_i} = 0$ $(i = 1, \ldots, n-1)$. At x_0 this gives

$$\frac{\partial u}{\partial y_i} = 0, \qquad \frac{\partial^2 u}{\partial y_i\,\partial y_j} = -\frac{\partial u}{\partial y_n}\frac{\partial^2 \omega}{\partial y_i\,\partial y_j} \equiv -\frac{\partial u}{\partial n}\frac{\partial^2 \omega}{\partial y_i\,\partial y_j}, \qquad i, j = 1, \ldots, n-1. \tag{6.16}$$

Because $u_{y_i}(x^0) = 0$ $(i = 1, \ldots, n-1)$, $I(s)$ assumes at x^0 the form

$$I(x^0) = (b_{nn}b_{pq} - b_{np}b_{qn})\left(\frac{\partial u}{\partial n}\right)^2 \frac{\partial^2 u}{\partial y_p\,\partial y_q}. \tag{6.17}$$

For $p = n$ and arbitrary q, and also for $q = n$ and arbitrary p, the terms in parentheses in (6.17) cancel out, a fact which with (6.16) leads to

$$I(x^0) = -\sum_{p,q=1}^{n-1} (b_{nn}b_{pq} - b_{np}b_{qn})\left(\frac{\partial u}{\partial n}\right)^2 \frac{\partial^2 \omega}{\partial y_p\,\partial y_q}. \tag{6.18}$$

Let us assume that the coordinates $y_1, \ldots, y_{n-1}$ in the tangent plane to S at x^0 have been chosen in such a way that all the mixed derivatives $\partial^2\omega/\partial y_p\,\partial y_q$ $(p, q = 1, \ldots, n-1)$ vanish at x^0 (this can always be done by an orthogonal transformation of $y_1, \ldots, y_{n-1}$). Then

$$I(x^0) = -\sum_{p=1}^{n-1} (b_{nn}b_{pp} - b_{np}^2)\left(\frac{\partial u}{\partial n}\right)^2 \frac{\partial^2 \omega}{\partial y_p^2}. \tag{6.19}$$

Let

$$\left.\frac{\partial^2 \omega}{\partial y_p^2}\right|_{x^0} \le K, \qquad p = 1, \ldots, n-1, \tag{6.20}$$

where $K \ge 0$; then

$$-I(x^0) \le \mu^2(n-1)K\left(\frac{\partial u}{\partial n}\right)^2 \tag{6.21}$$

for the matrix (b_{ij}), being related to the matrix (a_{ij}) by a similarity transformation induced by the orthogonal matrix (c_{ij}), is itself positive definite and satisfies (2.3); from this it follows that $0 < b_{nn}b_{pp} - b_{np}^2 \le \mu^2$. For convex domains Ω we have that $\partial^2\omega/\partial y_p^2 \le 0$, so that we can take K to be zero. Thus, for any Ω, it follows from (6.14) and (6.21) that

$$\frac{3v^2}{4}\int_\Omega u_{xx}^2\,dx \le \int_\Omega (\mathscr{L}u)^2\,dx$$

$$+\tfrac{6}{7}\mu^2(n-1)K\int_S \left(\frac{\partial u}{\partial n}\right)^2 ds + c_3(\|u\|_{2,\Omega}^{(1)})^2, \tag{6.22}$$

and for convex Ω that

$$\frac{3v^2}{4}\int_\Omega u_{xx}^2\,dx \le \int_\Omega (\mathscr{L}u)^2\,dx + c_3(\|u\|_{2,\Omega}^{(1)})^2. \tag{6.23}$$

On the basis of (1.3) in Chapter I, the integral $\int_S (\partial u/\partial n)^2\, ds$ can be bounded as follows for all $\varepsilon > 0$:

$$\int_S \left(\frac{\partial u}{\partial n}\right)^2 ds = \int_S u_x^2\, ds \le c_4 \int_\Omega (u_x^2 + |\nabla u_x^2|)\, dx$$

$$\le c_4 \int_\Omega (u_x^2 + 2|u_x||u_{xx}|)\, dx$$

$$\le c_4 \int_\Omega \left[\varepsilon u_{xx}^2 + \left(1 + \frac{1}{\varepsilon}\right) u_x^2\right] dx. \tag{6.24}$$

From (6.23) and (6.24) with $\varepsilon = (\nu^2/4)[\frac{6}{7}\mu^2(n-1)c_4 K]^{-1}$ we have

$$\frac{\nu^2}{2} \int_\Omega u_{xx}^2\, dx \le \int_\Omega (\mathscr{L}u)^2\, dx + c_5(\|u\|_{2,\Omega}^{(1)})^2, \tag{6.25}$$

which is the inequality (6.4).

The validity of (6.5) for convex domains follows from (6.11) which, for $\mathscr{L} = \Delta$, has the form

$$\int_\Omega (\Delta u)^2\, dx = \int_\Omega u_{xx}^2\, dx + \int_S I(s)\, ds, \tag{6.26}$$

where $I(s) = -(\partial u/\partial n)^2 \sum_{p=1}^{n-1} \partial^2\omega/\partial y_p^2$, and from the fact that $\partial^2\omega/\partial y_p^2 \le 0$ for convex domains, i.e., $I(s) \ge 0$. Moreover, for $u \in C^2(\overline{\Omega})$ and $u|_s = 0$, and taking into account (6.3) of Chapter I, we have

$$\int_\Omega u_x^2\, dx = -\int_\Omega u\, \Delta u\, dx \le \|u\| \cdot \|\Delta u\| \le c_\Omega \|\Delta u\| \|u_x\|,$$

which yields

$$\|u\|_{2,\Omega}^{(1)} \le c_\Omega \sqrt{1 + c_\Omega^2}\, \|\Delta u\|. \tag{6.27}$$

From (6.26) and (6.27) for convex Ω we obtain

$$\|u\|_{2,\Omega}^{(2)} \le [1 + c_\Omega^2(1 + c_\Omega^2)]^{1/2} \|\Delta u\|, \tag{6.28}$$

and, for arbitrary Ω with S in C^2, it follows from (6.25) and (6.27) that

$$\|u\|_{2,\Omega}^{(2)} \le c\|\Delta u\|, \tag{6.29}$$

where the constant c is defined not only by the rough characteristics of Ω (like the constant c_Ω) but also by the smoothness properties of S.

Remark 6.1. It is obvious from the derivation of (6.7) that the hypothesis that S be of class C^2 can be weakened. To preserve (6.7), it is enough to know that the boundary S is piecewise smooth and that (6.20) is fulfilled for all points of S except those lying on the "ridges" of S (such points form a set on S of $(n-1)$-dimensional measure zero and thus do not affect the value of the integral $\int_S I(x)\, ds$). Such conditions are fulfilled, for example, by polyhedra. Another example of a domain Ω for which (6.7) holds is an arbitrary convex domain. For such surfaces we can take K to be 0 in (6.20).

Remark 6.2. The inequality (6.7) was proved by the author in the papers [2–4], and it is the simplest special case of the inequalities established in those papers. Analogous inequalities were proved there not only for the first boundary condition, but also for any boundary condition of the type

$$\frac{\partial u}{\partial l} + \sigma u|_s = 0,$$

where $\partial/\partial l$ indicates differentiation in any direction not tangent to S. Independently, and by another method, R. Caccioppoli [1] had proved (6.7) (under the condition $u|_s = 0$). As it subsequently turned out, (6.7) for $n = 2$ and with Ω a disc, had been in the work of S. N. Bernstein ([1], [2]). In this case the inequality (6.7) for the operator $\mathscr{L}u$ can be derived from the same inequality for the Laplace operator.

§7. Solvability of the Dirichlet Problem in $W_2^2(\Omega)$

The object of this section is to prove that if the conditions (2.3), (2.4), (6.2), and (6.3) are fulfilled for equation (6.1) and if the boundary S is "sufficiently nice", then any generalized solution in $W_2^1(\Omega)$ of (6.1), (2.2) is an element of $\overset{\circ}{W}{}_{2,0}^2(\Omega)$. We shall assume that $\mathscr{L}$ and S satisfy the same hypotheses used in the proof of the second fundamental inequality in §6. Then if λ_0 is sufficiently large and if $\mathscr{L}_1 \equiv \mathscr{L} - \lambda_0 E$, we have

$$\mathscr{L}_1(u, u) \geq \delta_1 \|u\|^2, \qquad \delta_1 = \text{const} > 0, \tag{7.1}$$

for all $u \in W_{2,0}^2(\Omega)$. As we showed in §6, (7.1) and (6.7) imply (6.8), that is

$$\|u\|_{2,\Omega}^{(2)} \leq c\|\mathscr{L}_1 u\|, \tag{7.2}$$

where u is an arbitrary element in $W_{2,0}^2(\Omega)$. We shall prove the following assertion.

Theorem 7.1. *Let $\mathscr{L}_1$ and $\mathscr{L}_0$ have the same form as $\mathscr{L}$ and satisfy* (2.3), (2.4), (6.2), *and* (7.1), *and let the boundary S satisfy the hypotheses under which the second fundamental inequality is held. In addition, let the problem*

$$\mathscr{L}_0 u = f, \qquad u|_S = 0 \tag{7.3}$$

have solutions $u(x)$ in $W_{2,0}^2(\Omega)$ for some set $\mathfrak{M}$ of elements $f(x)$ which is dense in $L_2(\Omega)$. Then the problems

$$\mathscr{L}_\tau u = f, \qquad u|_S = 0, \qquad \text{where} \qquad \mathscr{L}_\tau = \mathscr{L}_0 + \tau(\mathscr{L}_1 - \mathscr{L}_0) \tag{7.4}$$

for all $\tau \in [0, 1]$ are uniquely solvable in $W_{2,0}^2(\Omega)$ for all $f \in L_2(\Omega)$.

It follows from the hypotheses of the theorem that (7.1) and (7.2) are valid for $\mathscr{L}_0$, that is,

$$\mathscr{L}_0(u, u) \geq \delta_1 \|u\|^2, \qquad \delta_1 > 0, \tag{7.5}$$

and

$$\|u\|_{2,\Omega}^{(2)} \leq c\|\mathscr{L}_0 u\| \tag{7.6}$$

for all $u \in W_{2,0}^2(\Omega)$. By (7.6), the problem (7.3) is solvable in $W_{2,0}^2(\Omega)$ for all $f \in L_2(\Omega)$. In fact, for f in $\mathfrak{M}$ the solvability is given by one of the hypotheses of the theorem, and the uniqueness follows from (7.6). If $f \in L_2(\Omega)$ but $f \notin \mathfrak{M}$, then we take a sequence f_m $(m = 1, 2, \ldots)$ from $\mathfrak{M}$ which converges to f in the norm of $L_2(\Omega)$. For each of the f_m there exists a solution $u_m \in W_{2,0}^2(\Omega)$ of (7.3) with $f = f_m$. By the linearity of the problem, the difference $u_k - u_m$ is a solution of (7.3) with $f = f_k - f_m$. The inequality (7.6) holds for this difference, i.e.,

$$\|u_k - u_m\|_{2,\Omega}^{(2)} \leq c\|f_k - f_m\|,$$

from which it follows that u_k converges in $W_2^2(\Omega)$ to some element $u \in W_{2,0}^2(\Omega)$. Since the coefficients of $\mathscr{L}_0$ are bounded, the functions $\mathscr{L}_0 u_k$ converge in $L_2(\Omega)$ to $\mathscr{L}_0 u$, that is, $\mathscr{L}_0 u = f$. Thus we have shown that, for all f in $L_2(\Omega)$, the problem (7.3) has a solution in $W_{2,0}^2(\Omega)$. Its uniqueness in $W_{2,0}^2(\Omega)$ follows from (7.6). At the same time we have proved that the operator $\mathscr{L}_0$ establishes a one-to-one correspondence between the spaces $W_{2,0}^2(\Omega)$ and $L_2(\Omega)$.

We consider now the family of operators

$$\mathscr{L}_\tau = \mathscr{L}_0 + \tau(\mathscr{L}_1 - \mathscr{L}_0), \qquad \tau \in [0, 1].$$

Obviously, $\mathscr{L}_\tau$ coincides with $\mathscr{L}_0$ for $\tau = 0$ and with $\mathscr{L}_1$ for $\tau = 1$. We shall show that, for all τ in $[0, 1]$, $\mathscr{L}_\tau$ establishes a one-to-one correspondence between $W_{2,0}^2(\Omega)$ and $L_2(\Omega)$. Since the operator $\mathscr{L}_0$ has this property, the problem

$$\mathscr{L}_\tau u = f, \qquad u|_S = 0 \tag{7.7}$$

is equivalent to the problem

$$[E + \tau\mathscr{L}_0^{-1}(\mathscr{L}_1 - \mathscr{L}_0)]u = \mathscr{L}_0^{-1}f \tag{7.8}$$

in the space $W_{2,0}^2(\Omega)$. The operator $\mathscr{L}_0^{-1}(\mathscr{L}_1 - \mathscr{L}_0)$ is bounded in $W_{2,0}^2(\Omega)$, for the boundedness of the coefficients of $\mathscr{L}$ and $\mathscr{L}_0$, together with (7.6), gives us

$$\|\mathscr{L}_0^{-1}(\mathscr{L}_1 - \mathscr{L}_0)u\|_{2,\Omega}^{(2)} \leq c\|(\mathscr{L}_1 - \mathscr{L}_0)u\| \leq c_1\|u\|_{2,\Omega}^{(2)}, \tag{7.9}$$

that is, the norm $\|\mathscr{L}_0^{-1}(\mathscr{L}_1 - \mathscr{L}_0)\|^{(2)}$ in $W_{2,0}^2(\Omega)$ does not exceed c_1. Hence (7.8) is solvable for all $\tau < 1/c_1$ (cf. §2, Chapter I), that is, for $\tau < 1/c_1$ the operators $\mathscr{L}_\tau$ establish a one-to-one correspondence between $W_{2,0}^2(\Omega)$ and

$L_2(\Omega)$. If $1/c_1 \leq 1$, we take $\tau_1 = 1/(2c_1)$ and apply the operator $\mathscr{L}_{\tau_1}^{-1}$ to (7.7). Because $\mathscr{L}_\tau = \mathscr{L}_{\tau_1} + (\tau - \tau_1)(\mathscr{L}_1 - \mathscr{L}_0)$, this gives us the equation

$$[E + (\tau - \tau_1)\mathscr{L}_{\tau_1}^{-1}(\mathscr{L}_1 - \mathscr{L}_0)]u = \mathscr{L}_{\tau_1}^{-1}f, \tag{7.10}$$

which is equivalent to (7.7). In order to investigate the solvability of (7.10), we shall bound the norm of the operator $\mathscr{L}_{\tau_1}^{-1}(\mathscr{L}_1 - \mathscr{L}_0)$ in $\overset{\circ}{W}{}_{2,0}^2(\Omega)$. For this we remark that from (7.1) and (7.5) we have the inequality

$$\mathscr{L}_\tau(u, u) = (1 - \tau)\mathscr{L}_0(u, u) + \tau\mathscr{L}_1(u, u) \geq \delta_1 \|u\|^2, \tag{7.11}$$

and from the conditions (2.3), (2.4), and (6.2) for $\mathscr{L}_0$ and $\mathscr{L}_1$ we have the fulfillment of these conditions with the same constants for all $\mathscr{L}_\tau$, $\tau \in [0, 1]$. Because of this, we have, for all $u \in \overset{\circ}{W}{}_{2,0}^2(\Omega)$ and for all operators $\mathscr{L}_\tau$ with $\tau \in [0, 1]$, the inequality (7.6), that is,

$$\|u\|_{2,\Omega}^{(2)} \leq c\|\mathscr{L}_\tau u\| \tag{7.12}$$

with the same constant c as in (7.6). The bound for the norm, $\|\mathscr{L}_\tau^{-1}(\mathscr{L}_1 - \mathscr{L}_0)\|^{(2)} \leq c_1$, follows from (7.11) and (7.12), provided that $\mathscr{L}_\tau^{-1}$ exists. Returning to (7.10), we conclude that (7.10) is solvable for $\tau - \tau_1 < 1/c_1$, in particular for $\tau = 2\tau_1$. So we have proved the existence of the inverse of $\mathscr{L}_{2\tau_1}$. If we continue this process, then after a finite number of steps we can prove the existence of $\mathscr{L}_\tau^{-1}$ for all $\tau \in [0, 1]$. Thus we have proved Theorem 7.1.

To apply the theorem, we must have the solvability in $\overset{\circ}{W}{}_{2,0}^2(\Omega)$ of the problem (7.3) for some operator $\mathscr{L}_0$ which possesses the properties demanded by Theorem 7.1. If Ω is the sphere K_ρ, or the spherical shell $K_{\rho,\rho_1} = \{x: \rho < |x| < \rho_1\}$, or a parallelopiped Π, then we can take $\mathscr{L}_0$ to be the Laplace operator. In fact, the complete system of eigenfunctions $\{u_k(x)\}$ of the Laplace operator for the first boundary condition is known for these domains (as well as for many others), and the $u_k(x)$ are infinitely differentiable in $\bar{\Omega}$. Because of this fact, the solution of the problem

$$\Delta u = \sum_{k=1}^{N} c_k u_k(x), \quad u|_S = 0$$

for any numbers c_k and for all $N \geq 1$ is

$$u = \sum_{k=1}^{N} \frac{c_k}{\lambda_k} u_k(x) \in \overset{\circ}{W}{}_{2,0}^2(\Omega),$$

where $\Delta u_k = \lambda_k u_k$ and the sums $\sum_{k=1}^{N} c_k u_k(x)$ are dense in $L_2(\Omega)$. All the remaining hypotheses of Theorem 7.1 are obviously satisfied, if we only take suitable constants as the ν and μ_i for $\mathscr{L}_0$ and $\mathscr{L}_1$. Consequently, the role of $\mathscr{L}_0$ in the regions described above can be played by Δ. A similar argument holds for domains which can be transformed into one of the indicated domains by means of a non-degenerate change of variables† $y = y(x)$ with $y(x) \in C^2(\bar{\Omega})$.

† That is, the function $y = y(x)$ should give a diffeomorphic mapping of $\bar{\Omega}$ onto $\bar{\tilde{\Omega}}$, $y(x) \in C^2(\bar{\Omega})$, and the Jacobians $\partial(y)/\partial(x)$ and $\partial(x)/\partial(y)$ should be strictly positive.

In fact, if we express the equation $\mathscr{L}u - \lambda_0 u = f$ in terms of y, we obtain the equation $\tilde{\mathscr{L}}u - \lambda_0 u = f$, where

$$\tilde{\mathscr{L}}u \equiv \frac{\partial}{\partial y_i}(b_{ij}u_{y_j}) + b_i u_{y_i} + bu,$$

$$b_{ij} = a_{kl}\frac{\partial y_i}{\partial x_k}\frac{\partial y_j}{\partial x_l}$$

$$b_i = a_k\frac{\partial y_i}{\partial x_k} - a_{ij}\frac{\partial y_i}{\partial x_j}\frac{\partial}{\partial y_k}\left(\frac{\partial y_k}{\partial x_j}\right),$$

and $b = a$ in the range $\tilde{\Omega}$ of y. The coefficients of $\tilde{\mathscr{L}}$ satisfy the conditions of the form (2.3), (2.4), (6.2). Thus, for sufficiently large λ_0 we shall have the inequalities (7.1) and (7.2) for $\tilde{\mathscr{L}}_1 \equiv \tilde{\mathscr{L}} - \lambda_0 E$ (with other constants in general), so that Theorem 7.1 will hold for $\tilde{\mathscr{L}}_1$. As $\tilde{\mathscr{L}}_0$ we can take the operator $\sum_{i=1}^n \partial^2/\partial y_i^2 - \lambda_0 E$. Then Theorem 7.1 guarantees the unique solvability in $W_{2,0}^2(\tilde{\Omega})$ of the problem

$$(\tilde{\mathscr{L}} - \lambda_0 E)u = f, \quad u|_{\partial\tilde{\Omega}} = 0. \tag{7.13}$$

If we return to the variable x, we have shown that the problem

$$(\mathscr{L} - \lambda_0 E)u = f, \qquad u|_S = 0 \tag{7.14}$$

is uniquely solvable in $W_{2,0}^2(\Omega)$. Thus, we have proved

Theorem 7.2. *If the coefficients of $\mathscr{L}$ in* (6.1) *satisfy* (2.3), (2.4) *and* (6.2), *if $f \in L_2(\Omega)$, and if Ω is a sphere, a spherical shell, a parallelopiped, or a domain that can be transformed into one of these domains by a regular mapping $y = y(x) \in C^2(\bar{\Omega})$, then* (7.14) *is uniquely solvable in $W_{2,0}^2(\Omega)$ for sufficiently large λ_0.*

Let us take now an arbitrary generalized solution $u(x)$ in $W_2^1(\Omega)$ of the problem

$$(\mathscr{L} - \lambda E)u = f, \qquad u|_S = 0 \tag{7.15}$$

with $f \in L_2(\Omega)$. It can be considered as a generalized solution in $W_2^1(\Omega)$ of the problem (7.14) with free term $f + (\lambda - \lambda_0)u \in L_2(\Omega)$. By Theorems 7.2 and 2.1 this problem is solvable in $W_{2,0}^2(\Omega)$, and the uniqueness theorem holds for it in the class $W_2^1(\Omega)$. Hence the $u(x)$ taken by us will belong to $W_{2,0}^2(\Omega)$. Thus we have proved the following theorem.

Theorem 7.3. *Under the hypotheses of Theorem 7.2 for $\mathscr{L}$, f, and Ω, any generalized solution in $W_2^1(\Omega)$ of the problem* (7.15) *is an element of $W_{2,0}^2(\Omega)$.*

It follows from this theorem and from the results of §3 on the Fredholm solvability of the problem

$$\mathscr{L}u \equiv \lambda u + f, \qquad u|_S = 0 \tag{7.16}$$

in $W_2^1(\Omega)$ that if the hypotheses of Theorem 7.3 are fulfilled, this problem is Fredholm solvable in $\overset{\circ}{W}{}_{2}^{2}(\Omega)$. Its spectrum $\{\lambda_k\}$ $(k = 1, 2, \ldots)$ does not depend on the space in which we consider the problem. If $\lambda \neq \lambda_k$ $(k = 1, 2, \ldots)$, then the operator $\mathscr{L} - \lambda E$ has a bounded inverse, which, under the hypotheses of Theorem 7.2, guarantees the existence of the bound

$$\|u\|_{2,\Omega}^{(2)} \leq c_\lambda \|(\mathscr{L} - \lambda E)u\|. \tag{7.17}$$

In the general case we cannot write the constant c_λ explicitly in terms of the coefficients of $\mathscr{L} - \lambda E$ and of S, as we were able to do for (6.9) in §6, but the existence of c_λ is guaranteed by the theorems of Fredholm.

Remark 7.1. Theorem 7.3 shows that increasing the smoothness of the coefficients of $\mathscr{L}$ and of f and S guarantees an increase in the smoothness of all the generalized solutions in $W_2^1(\Omega)$ of (7.15).† It can be shown that this improvement of the properties of the solutions has a local character, for if f and the coefficients of $\mathscr{L}$ satisfy the hypotheses of Theorem 7.2 only in some subdomain Ω_1 of Ω, then every generalized solution $u(x) \in W_2^1(\Omega)$ of (7.15) will be an element of $W_2^2(\Omega_1')$ for all $\overline{\Omega}_1' \subset \Omega_1$. If Ω_1 abuts the boundary of Ω along a piece $S_1 \subset S$, and if $\mathscr{L}$, f, and Ω_1 satisfy the hypotheses of Theorem 7.2, then every generalized solution $u \in W_2^1(\Omega)$ will be in $W_2^2(\tilde{\Omega}_1)$ for every $\tilde{\Omega}_1 \subset \Omega_1$ which is at a positive distance from that part of the boundary of Ω_1 not belonging to S. It follows from these results that Theorems 7.2 and 7.3 hold for a wider class of domains Ω, namely, for domains which can be represented as the union $\bigcup_{i=1}^{N} \Omega_i$ of domains Ω_i for each of which there is an ε-neighborhood $\Omega_i^\varepsilon \supset \Omega_i$ such that the intersection $\Omega_i^\varepsilon \cap \Omega$ satisfies the requirements of Theorem 7.2. *In particular, this condition is satisfied by domains with twice continuously differentiable boundaries.*

On the other hand, examples show that Theorems 7.2 and 7.3 are not true for domains whose boundaries have edges with interior face angle greater than π. Thus, for the planar domain $\Omega = \{x: x = \rho e^{i\theta},\ 0 < \rho < 1,\ 0 < \theta < \pi + \varepsilon\}$ with $\varepsilon > 0$, the problem $\Delta u = f$, $u|_s = 0$, $f \in L_2(\Omega)$ has generalized solutions in $W_2^1(\Omega)$ not belonging to $W_2^2(\Omega)$, and this despite the fact that the second fundamental inequality holds for such a domain.

The local character of increasing the smoothness of generalized solutions of (7.15) depends on the fact that, in addition to bounds of the form (6.4), there are "local bounds" of the form

$$\int_\Omega u_{xx}^2 \zeta^4\, dx \leq \frac{2}{\nu^2} \int_\Omega (\mathscr{L}u)^2 \zeta^4\, dx + c \int_\Omega u^2(\zeta^4 + \zeta_x^4 + \zeta^2 \zeta_{xx}^2)\, dx, \tag{*}$$

which are valid for all $u(x) \in W_2^2(\Omega)$ and for every function $\zeta(x) \in C^2(\overline{\Omega})$ with $\zeta|_s = 0$; there are analogous bounds for $u(x) \in C^2(\overline{\Omega}_1)$ in domains

† We recall that, by Remark 3.1 in §3 of Chapter II, the results of §§2–3 are valid for $f \in L_{2N/(N+2)}(\Omega)$, where $N = n$ for $n > 2$ and $N = 2 + \varepsilon$, $\varepsilon > 0$ for $n = 2$.

Ω_1 which abut some smooth piece S_1 of $\partial\Omega$. In the latter case $u(x)$ should vanish on S_1, but on the other hand, the support of $\zeta(x)$ can intersect the boundary of Ω_1 in such a way that this intersection belongs to S_1. In the first case the smoothness of the boundary is not necessary. These inequalities are derived from considerations of the integral $\int_\Omega (\mathscr{L}u)^2\zeta^4\,dx$, similar to those carried out above.

Proofs of the facts stated here and references to investigations of the Dirichlet problem in planar domains whose boundaries have interior angles greater than π are given in [LU 1].

Remark 7.2. Theorems 7.2 and 7.3, as well as the inequality (7.17) remain valid for complex λ, $u(x)$, and $f(x)$.

If the operator $\mathscr{L}$ is symmetric, that is, if $\mathscr{L}$ is of the form (4.1), or, what is the same thing, of the form

$$\mathscr{L}u = \frac{\partial}{\partial x_i}(a_{ij}u_{x_j}) + au, \tag{7.18}$$

and if the hypotheses of Theorem 7.2 are satisfied for $\mathscr{L}$ and Ω, then the eigenfunctions of the problem

$$\mathscr{L}u = \lambda u, \qquad u|_S = 0 \tag{7.19}$$

are in $W^2_{2,0}(\Omega)$. Moreover, any function f in $W^2_{2,0}(\Omega)$ can be expanded into a Fourier series in terms of these eigenfunctions $u_k(x)$,

$$f(x) = \sum_{k=1}^{\infty} (f, u_k)u_k(x), \tag{7.20}$$

which converge to $f(x)$ in the norm of $W^2_2(\Omega)$ (in other words, the series (7.20) and series which are obtained from it by term-by-term differentiation with respect to x_i once or twice, converge in $L_2(\Omega)$)†. In fact, the convergence of (7.20) to f in the norm of $W^1_2(\Omega)$ has been proved in §4. All that remains is to prove its convergence in the norm of $W^2_2(\Omega)$. Because of (6.8), which holds for the operator $\mathscr{L} - \lambda_0 E$ with $\lambda_0 \geq \max_x a(x)$, the norm $\|\cdot\|^{(2)}_{2,\Omega}$ in $W^2_{2,0}(\Omega)$ is equivalent to the norm $\|u\|_2 \equiv \|\mathscr{L}u - \lambda_0 u\|$. The norm $\|u\|_2$ corresponds to the scalar product $\{u, v\}_2 \equiv (\mathscr{L}u - \lambda_0 u, \mathscr{L}v - \lambda_0 v)$. The eigenfunctions u_k of the problem (7.19) are orthogonal with respect to this scalar product, for

$$\{u_k, u_l\} = (\lambda_k - \lambda_0)(\lambda_l - \lambda_0)(u_k, u_l) = (\lambda_k - \lambda_0)^2\delta_k^l,$$

so that

$$\left\|\sum_{k=1}^{\infty} (f, u_k)u_k\right\|_2^2 = \sum_{k=1}^{\infty} (f, u_k)^2(\lambda_k - \lambda_0)^2. \tag{7.21}$$

† We remark that series resulting from the differentiation of (7.20) are not orthogonal in $L_2(\Omega)$!

To prove the convergence of (7.20) in the norm of $W^2_{2,0}(\Omega)$, it is enough to show the convergence of the numerical series (7.21), and for this we write (f, u_k) in the form

$$(f, u_k) = \frac{1}{\lambda_k}(f, \mathscr{L}u_k) = \frac{1}{\lambda_k}(\mathscr{L}f, u_k) \equiv \frac{\alpha_k}{\lambda_k}, \tag{7.22}$$

if $\lambda_k \neq 0$; this can be done since f and u_k are in $W^2_{2,0}(\Omega)$. The series $\sum_{k=1}^{\infty} \alpha_k^2$ converges and is equal to $\|\mathscr{L}f\|^2$. But then the series (7.21) converges if we take into consideration the fact that only a finite number of eigenfunctions correspond to the eigenvalue $\lambda = 0$. Thus, we have proved the following theorem.

Theorem 7.4. *If Ω and the coefficients of $\mathscr{L}$ satisfy the hypotheses of Theorem 7.2, then any function $f \in W^2_{2,0}(\Omega)$ can be expanded into the series (7.20) which converges to f in the norm of $W^2_{2,0}(\Omega)$.*

§8. Approximate Methods of Solving Boundary Value Problems

1. Galerkin's Method. The solutions of boundary value problems for elliptic equations can be obtained as limits of approximate solutions calculated by Galerkin's method.

The classical form of Galerkin's method for the problem (2.1), (2.2) is the following. In the space $\overset{\circ}{W}{}^1_2(\Omega)$ we choose some fundamental system $\{\varphi_k(x)\}$ of linearly independent functions $\varphi_1, \varphi_2, \ldots$ (that is, the functions φ_k $(k = 1, \ldots, N)$ are linearly independent for all $N < \infty$), and we seek an approximate solution u^N in the form $u^N = \sum_{k=1}^{N} c_k^N \varphi_k(x)$, where the constants c_k^N are defined by the conditions

$$(\mathscr{L}u^N, \varphi_l) = \left(f + \frac{\partial f_i}{\partial x_i}, \varphi_l\right), \qquad l = 1, \ldots, N. \tag{8.1}$$

The equalities (8.1) consists of a system of N linear algebraic equations in the c_k^N $(k = 1, \ldots, N)$. If the uniqueness theorem holds for the problem (2.1), (2.2), then the system (8.1) proves to be solvable for sufficiently large N and its solutions u^N tend to the solution of the problem (2.1), (2.2) in the norm of $W^1_2(\Omega)$. Convergence in a stronger norm (in particular, convergence of the remainder $\mathscr{L}u^N - f - \partial f_i/\partial x_i$ to zero in $L_2(\Omega)$) is not possible, generally speaking, even under the hypotheses of §7 (see, for example, S. G. Michlin [3]). We shall justify Galerkin's method under the assumption that

$$\mathscr{L}(u, u) \geq \nu_1(\|u\|^{(1)}_{2,\Omega})^2, \qquad \nu_1 = \text{const} > 0, \tag{8.2}$$

which is more stringent than the assumption concerning the uniqueness of the solution of the problem. In this case the arguments are quite simple since the systems for the c_k^N are uniquely solvable for all N. Thus, suppose that the assumptions (2.3)–(2.5) and (8.2) are fulfilled. The domain Ω can be arbitrary, even unbounded. The desired generalized solution $u(x)$ of the problem (2.1), (2.2) should satisfy (2.6), i.e.,

$$\mathscr{L}(u, \eta) = -(f, \eta) + (f_i, \eta_{x_i}) \tag{8.3}$$

for all $\eta \in \mathring{W}_2^1(\Omega)$. We look for approximate solutions in the form $u^N = \sum_{k=1}^N c_k^N \varphi_k(x)$, where $\{\varphi_k(x)\}$ is a fundamental system in $\mathring{W}_2^1(\Omega)$. We define the coefficients c_k^N as solutions of the linear algebraic system

$$\mathscr{L}(u^N, \varphi_l) = -(f, \varphi_l) + (f_i, \varphi_{lx_i}), \qquad l = 1, \ldots, N. \tag{8.4}$$

The relations (8.4) coincide with (8.1) if $\varphi_k \in W_{2,0}^2(\Omega)$, $\partial f_i/\partial x_i \in L_2(\Omega)$, and the coefficients of $\mathscr{L}$ satisfy the conditions of §6. The system (8.4) is solvable for any f and f_i because the uniqueness theorem holds for it. In fact, if u^N were a solution of the homogeneous system (8.4), then, multiplying each equation by the corresponding c_l^N and adding them up over l from 1 to N, we would arrive at the result $\mathscr{L}(u^N, u^N) = 0$, and it would follow from (8.2) that $u^N \equiv 0$, that is, all the c_k^N are zero for $k = 1, \ldots, N$. Hence the system (8.4) defines u^N uniquely. Let us multiply each equation of (8.4) by c_l^N and sum over l from $l = 1$ to N. This gives us

$$\mathscr{L}(u^N, u^N) = -(f, u^N) + (f_i, u_{x_i}^N),$$

which, coupled with (8.2), implies that

$$\nu_1(\|u^N\|_{2,\Omega}^{(1)})^2 \leq \|f\| \|u^N\| + \|\mathbf{f}\| \|u_x^N\|,$$

and this guarantees the bound

$$\nu_1 \|u^N\|_{2,\Omega}^{(1)} \leq (\|f\| + \|\mathbf{f}\|). \tag{8.5}$$

Let us denote by H_N the subspace of $\mathring{W}_2^1(\Omega)$ consisting of the elements $\sum_{k=1}^N d_k \varphi_k(x)$ with arbitrary d_k. Because of (8.4), u^N satisfies the identity

$$\mathscr{L}(u^N, \eta) = -(f, \eta) + (f_i, \eta_{x_i}) \tag{8.6}$$

for all $\eta \in H_N$. The uniform boundedness of the norms $\|u^N\|_{2,\Omega}^{(1)}$ in the Hilbert space $\mathring{W}_2^1(\Omega)$ makes it possible to choose a subsequence u^{N_k} $(k = 1, 2, \ldots)$ which converges weakly to some element $u \in \mathring{W}_2^1(\Omega)$.† It is not difficult to show that u is a solution of the problem (2.1), (2.2). In fact, we can take

† By what was proved in §5, Chapter I, the weak convergence in $\mathring{W}_2^1(\Omega)$ of $\{u^{N_k}\}$ to u shows that $\{u^{N_k}\}$ and $\{u_{x_i}^{N_k}\}$ converge weakly in $L_2(\Omega)$ to u and u_{x_i}, respectively. It follows from the theorem of Rellich that $\{u^{N_k}\}$ converges strongly to $u(x)$ in $L_2(\Omega)$.

the limit in (8.6) over the selected subsequence N_k $(k = k_0, k_0 + 1, \ldots)$ for a fixed η in $H_{N_{k0}}$ and obtain, in the limit

$$\mathscr{L}(u, \eta) = -(f, \eta) + (f_i, \eta_{x_i}) \tag{8.7}$$

for all $\eta \in H_{N_{k0}}$. The index k_0 here is arbitrary, so that (8.7) holds for all η in $\bigcup_{N=1}^{\infty} H_N$, that is, for all η of the form $\sum_{k=1}^{N} d_k \varphi_k(x)$ with any N. But these η are dense in $\mathring{W}_2^1(\Omega)$ since, by hypothesis, $\{\varphi_k\}$ is a fundamental system in $\mathring{W}_2^1(\Omega)$. Hence the function $u(x)$ will satisfy (8.7) for all $\eta \in \mathring{W}_2^1(\Omega)$ and consequently will be a generalized solution of the problem (2.1), (2.2). If we choose from $\{u^N\}$ $(N = 1, 2, \ldots)$ any other subsequence which converges in the same sense as $\{u^{N_k}\}$ we obtain again as limit function a generalized solution of the problem (2.1), (2.2), and since it is unique, this shows that the set $\{u^N\}$ $(N = 1, 2, \ldots)$ has a unique limit point $u(x)$ (in the sense of the weak topology in $\mathring{W}_2^1(\Omega)$). Because of the weak compactness of the set $\{u^N\}$ in $\mathring{W}_2^1(\Omega)$, this leads to the conclusion that the whole sequence u^N $(N = 1, 2, \ldots)$ converges weakly in $\mathring{W}_2^1(\Omega)$ to $u(x)$.

It can be shown that u^N converges strongly to u in $\mathring{W}_2^1(\Omega)$ as $N \to \infty$. For this we choose a sequence u_N $(N = 1, 2, \ldots)$ such that $u_N \in H_N$ and $\|u - u_N\|_{2,\Omega}^{(1)} \to 0$ as $N \to \infty$, and then subtract (8.6) from (8.7), replacing η by $u_N - u^N$. The result can be written

$$\mathscr{L}(u - u^N, u - u^N) = \mathscr{L}(u - u^N, u - u_N). \tag{8.8}$$

Since u^N converges weakly to u in $\mathring{W}_2^1(\Omega)$ and u_N converges strongly to u in $\mathring{W}_2^1(\Omega)$, the right-hand side of (8.8) tends to zero as $N \to \infty$. This together with (8.2), guarantees the strong convergence of u^N to u in $\mathring{W}_2^1(\Omega)$. Let us formulate what we have proved in the form of a theorem.

Theorem 8.1. *If the conditions* (2.3)–(2.5) *and* (8.2) *are satisfied, then the Galerkin approximations u^N can be calculated uniquely from the system* (8.4) *for all N; the bound* (8.5) *holds for these approximations and, as $N \to \infty$, they converge in the norm of $W_2^1(\Omega)$ to the solution of the problem* (2.1), (2.2). *The domain Ω can be unbounded.*

Let us assume that the hypotheses of Theorem 7.2 are satisfied for $\mathscr{L}, f$, and Ω, and that the operator $\mathscr{L}$ of the type (6.1) establishes a one-to-one correspondence between $W_{2,0}^2(\Omega)$ and $L_2(\Omega)$, so that

$$\|u\|_{2,\Omega}^{(2)} \leq c\|\mathscr{L}u\| \tag{8.9}$$

for all $u \in W_{2,0}^2(\Omega)$. In this case, we can change the procedure of calculating the approximate solutions u^N so that u^N will converge to a solution of the problem in the norm of $W_2^2(\Omega)$ and, by the same token, the remainder in the equation, i.e. $\mathscr{L}u^N - f$, will converge to zero in the norm of $L_2(\Omega)$. Namely, we take a fundamental system $\{\psi_k(x)\}$ in $W_{2,0}^2(\Omega)$. Then the functions $\{\varphi_k = \mathscr{L}\psi_k\}$ form a fundamental system in $L_2(\Omega)$. In fact, if f is any element of $L_2(\Omega)$, then $\mathscr{L}^{-1}f \equiv v$, as a solution of the problem $\mathscr{L}v = f$, $v|_S = 0$, exists and is in $W_{2,0}^2(\Omega)$. By hypothesis, v is a limit in $W_{2,0}^2(\Omega)$ of elements of the form $v_N = \sum_{k=1}^N b_k^N \psi_k(x)$, but then $\mathscr{L}v_N = \sum_{k=1}^N b_k^N \varphi_k(x)$ is in $L_2(\Omega)$ and converges in $L_2(\Omega)$ to $\mathscr{L}v = f$, that is, f is in the closure in $L_2(\Omega)$ of the finite-dimensional subspaces spanned by the functions $\{\varphi_k\}$. We shall look for approximate solutions of the problem (6.1), (2.2) of the form $u^N = \sum_{k=1}^N c_k^N \psi_k(x)$ from the system of equations

$$(\mathscr{L}u^N, \mathscr{L}\psi_l) = (f, \mathscr{L}\psi_l), \qquad l = 1, \ldots, N. \tag{8.10}$$

This system is uniquely solvable for each N, and its solutions satisfy

$$\|\mathscr{L}u^N\|^2 = (f, \mathscr{L}u^N) \leq \|f\| \, \|\mathscr{L}u^N\|, \tag{8.11}$$

which, along with (8.9), gives us the bound

$$\|u^N\|_{2,\Omega}^{(2)} \leq c\|f\|. \tag{8.11'}$$

We can obtain (8.11) by multiplying each equality in (8.10) by c_l^N and then summing all these equations over l from 1 to N. Similarly, we can obtain from (8.10) the identity

$$(\mathscr{L}u^N, \mathscr{L}\eta) = (f, \mathscr{L}\eta) \tag{8.12}$$

for all $\eta \in \mathscr{H}_N$, where $\mathscr{H}_N$ is the linear span of the elements $\psi_1, \ldots, \psi_N$. The weak convergence of u^N in $W_2^2(\Omega)$ to a solution of the problem can be proved on the basis of (8.11′) in the same way that it was done above in $\overset{\circ}{W}{}_2^1(\Omega)$ on the basis of the bound (8.5). For the proof of the strong convergence of u^N to u in $W_2^2(\Omega)$ we must subtract the identity (8.12) for u^N from the identity (8.12) for u, taking $\eta = u_N - u^N$, where u_N $(N = 1, 2, \ldots)$ is a sequence approximating u in the norm of $W_2^2(\Omega)$ and such that $u_N \in \mathscr{H}_N$. This gives us

$$(\mathscr{L}(u - u^N), \mathscr{L}(u_N - u^N)) = 0,$$

from which it follows that

$$\begin{aligned}(\mathscr{L}(u - u^N), \mathscr{L}(u - u^N)) &= (\mathscr{L}(u - u^N), \mathscr{L}(u - u_N)) \\ &\leq \|\mathscr{L}(u - u^N)\| \, \|\mathscr{L}(u - u_N)\|.\end{aligned}$$

From this inequality, from (8.9), and from the convergence of u_N to u in the norm of $W_2^2(\Omega)$ we conclude that u^N converges to u in the norm of $W_2^2(\Omega)$. Thus we have the theorem:

Theorem 8.2. *If f, $\mathscr{L}$, and Ω satisfy the hypotheses of Theorem 7.2 and the condition* (8.9), *then the approximations $u^N = \sum_{k=1}^N c_k^N \psi_k(x)$ are uniquely determined by* (8.10) *and converge to the solution $u(x)$ of the problem* (2.1), (2.2) *in the norm of $W_2^2(\Omega)$. Any fundamental system in $W_{2,0}^2(\Omega)$ can be taken as $\{\psi_k(x)\}$.*

The convergence of the approximations u^N to the exact solution in the norm of $W_2^2(\Omega)$, and, consequently, the convergence of the remainders $\mathscr{L}u^N - f$ to zero in the norm of $L_2(\Omega)$ can also be achieved by the following modernization of Galerkin's method. Let (8.2) be satisfied by $\mathscr{L}$. We take any fundamental system $\{\psi_k(x)\}$ in $W_{2,0}^2(\Omega)$ and an arbitrary elliptic operator

$$\mathscr{M}u = (b_{ij}(x)u_{x_j})_{x_i} + b_i(x)u_{x_i} + b(x)u \qquad (8.13)$$

whose coefficients satisfy the conditions

$$v_2\xi^2 \le b_{ij}\xi_i\xi_j \le \mu_2\xi^2, \quad v_2 > 0; \qquad \left|\frac{\partial b_{ij}}{\partial x_k}\right|, \quad \sqrt{\sum_i b_i^2}, \quad |b| \le \mu_2. \quad (8.14)$$

We shall look for approximate solutions of the form $u^N = \sum_{k=1}^N c_k^N \psi_k(x)$ from the system of equations

$$(\mathscr{L}u^N, \mathscr{M}\psi_l - \lambda_0\psi_l) = (f, \mathscr{M}\psi_l - \lambda_0\psi_l), \qquad l = 1, \dots, N. \quad (8.15)$$

The number λ_0 should be sufficiently large (just how large will be indicated below). To prove the unique solvability of the system (8.15) and the uniform boundedness of the norms $\|u^N\|_{2,\Omega}^{(2)}$, we multiply each of the equations of (8.15) by the respective c_l^N and add them up from 1 to N. As a result we obtain

$$(\mathscr{L}u^N, \mathscr{M}u^N - \lambda_0 u^N) = (f, \mathscr{M}u^N - \lambda_0 u^N). \quad (8.16)$$

For all $u \in W_{2,0}^2(\Omega)$ and for any elliptic operators $\mathscr{L}$ and $\mathscr{M}$ satisfying only conditions of the type (8.14) the inequality

$$(\mathscr{L}u, \mathscr{M}u) \ge v_3\|u_{xx}\|^2 - \mu_3(\|u\|_{2,\Omega}^{(1)})^2, \quad v_3 > 0, \quad (8.17)$$

is valid.

This inequality can be proved in the same way as in the case $\mathscr{L} = \mathscr{M}$ in §6 above (i.e., we begin with the integral $\int_\Omega \mathscr{L}u \cdot \mathscr{M}u \, dx$ and transform it in accordance with (6.10); then we develop transformations and estimates similar to (6.11)–(6.25). In this connection we must use the fact that it is possible to reduce simultaneously two positive quadratic forms to diagonal form; see the Supplements no. 13). From (8.17) and (8.2) it follows that

$$(\mathscr{L}u, \mathscr{M}u - \lambda_0 u) \ge v_3\|u_{xx}\|^2 + (\lambda_0 v_1 - \mu_3)(\|u\|_{2,\Omega}^{(1)})^2. \quad (8.18)$$

We choose λ_0 in such a way that $\lambda_0 v_1 - \mu_3 > 0$ (the equality can be achieved, $\lambda_0 v_1 - \mu_3 = 0$). Then, for all $u \in W_{2,0}^2(\Omega)$ we have the bound

$$(\mathscr{L}u, \mathscr{M}u - \lambda_0 u) \ge v_4(\|u\|_{2,\Omega}^{(2)})^2, \quad v_4 > 0, \quad (8.19)$$

which, together with (8.16), easily implies the bound

$$\|u^N\|_{2,\Omega}^{(2)} \le c\|f\| \quad (8.20)$$

with a constant c which is independent of N. It guarantees the unique solvability of the algebraic system (8.15) and the weak convergence in $L_2(\Omega)$ of u^N, $u_{x_i}^N$ and $u_{x_ix_j}^N$ to u, u_{x_i} and $u_{x_ix_j}$, respectively. An additional argument, similar to that carried through above for the case $\mathscr{M} = \mathscr{L}$, allows us to prove the strong convergence of u^N to u in the norm of $W_{2,0}^2(\Omega)$.

The modernizations (8.10) and (8.16) and some other modernizations of Galerkin's method which lead to even better convergence of u^N to u have been proposed in the author's paper [11].

Remark 8.1. In the case when the $\{\psi_l\}$ are eigenfunctions of the problem $\mathscr{M}\psi = \lambda\psi$, $\psi|_s = 0$, the system (8.15) is equivalent to (8.4) (in which the φ_l have been replaced by the ψ_l). Thus, for this choice of the fundamental system $\{\psi_l\}$, the original form of Galerkin's method coincides with the modernized form and leads to solutions u^N which converge to u in the norm of $W_2^2(\Omega)$.

Some generalizations of Galerkin's method were proposed by G. I. Petrov [1].

Remark 8.2. We shall return again to Galerkin's method in §12, Chapter VI, and establish its connections with the method of finite differences and other projection methods.

We presented Galerkin's method in connection with the first boundary condition. If, instead of this condition, the desired solution is subject to the condition (5.2), then we must introduce some changes in the scheme described above. We recall that a generalized solution $u(x)$ in $W_2^1(\Omega)$ of the problem

$$\mathscr{L}u \equiv \frac{\partial}{\partial x_i}(a_{ij}u_{x_j}) + b_i u_{x_i} + au = f, \qquad \left.\frac{\partial u}{\partial N} + \sigma u\right|_S = 0 \tag{8.21}$$

is defined as an element of $W_2^1(\Omega)$ satisfying the integral identity

$$\mathscr{L}(u, \eta) + \int_S \sigma u\eta\, ds = -(f, \eta) \tag{8.22}$$

for all $\eta \in W_2^1(\Omega)$. Here

$$\mathscr{L}(u, \eta) = \int_\Omega (a_{ij}u_{x_j}\eta_{x_i} - b_i u_{x_i}\eta - au\eta)\, dx.$$

We take any fundamental system $\{\varphi_k(x)\}$ in $W_2^1(\Omega)$ and look for approximate solutions of the form $\sum_{k=1}^N c_k^N \varphi_k(x)$ from the equations

$$\mathscr{L}(u^N, \psi_l) + \int_S \sigma u^N \varphi_l\, ds = -(f, \varphi_l), \qquad l = 1, \ldots, N. \tag{8.23}$$

These form a system of N linear algebraic equations in N unknowns c_k^N $(k = 1, \ldots, N)$. The unique solvability and the uniform bounds for the norms $\|u^N\|_{2,\Omega}^{(1)}$ can be derived from the relations

$$\mathscr{L}(u^N, u^N) + \int_S \sigma(u^N)^2\, ds = -(f, u^N), \tag{8.24}$$

which are the sum of the equations (8.23) multiplied by the corresponding c_l^N. If $\mathscr{L}$, σ, and Ω are such that, for all $u \in W_2^1(\Omega)$,

$$\mathscr{L}(u, u) + \int_S \sigma u^2 \, ds \geq \nu_1(\|u\|_{2,\Omega}^{(1)})^2, \qquad \nu_1 > 0, \tag{8.25}$$

then it follows from (8.24) that

$$\|u^N\|_{2,\Omega}^{(1)} \leq \frac{1}{\nu_1} \|f\|.$$

If we argue in the same way that we did in the beginning of the section, we can conclude that u^N converges to u in the norm of $W_2^1(\Omega)$. We shall not speak here of modernizations of Galerkin's method for the problem (8.21), which guarantees the convergence of u^N to u in the norm of $W_2^2(\Omega)$ since we have not derived the second fundamental inequality for boundary conditions of the third kind.

2. The Method of Ritz. For the symmetric operators

$$\mathscr{L}u \equiv \frac{\partial}{\partial x_i}(a_{ij}u_{x_j} + a_i u) - a_i u_{x_i} + au \tag{8.26}$$

the integral identity

$$\mathscr{L}(u, \eta) = \int_\Omega (a_{ij}u_{x_j}\eta_{x_i} + a_i u \eta_{x_i} + a_i u_{x_i}\eta - au\eta) \, dx$$
$$= \int_\Omega (-f\eta + f_i \eta_{x_i}) \, dx, \tag{8.27}$$

for all $\eta \in \mathring{W}_2^1(\Omega)$, which is the basis of the definition of a generalized solution in $W_2^1(\Omega)$ of the problem (2.1), (2.2), is nothing else than the vanishing of the first variation of the quadratic functional

$$J(u) = \int_\Omega (a_{ij}u_{x_i}u_{x_j} + 2a_i u u_{x_i} - au^2 + 2fu - 2f_i u_{x_i}) \, dx,$$

considered over all elements of $\mathring{W}_2^1(\Omega)$. The function η in (8.27) plays the role of the admissible variation of u. If the function $u(x)$ yields the smallest value for the functional $J(u)$ over $\mathring{W}_2^1(\Omega)$, then, as is well known, $\delta J(u; \delta u) = 0$ for all $\delta u \in \mathring{W}_2^1(\Omega)$, that is, u satisfies (8.27) for all $\eta \in \mathring{W}_2^1(\Omega)$. The Ritz method is based upon this idea and allows us to construct a minimizing sequence u^N ($N = 1, 2, \ldots$) for the functional $J(u)$, i.e., a sequence on which $J(u^N)$ tends to inf $J(u)$. Under certain assumptions about $\mathscr{L}$, f, and f_i the functions u^N converge in the norm of $W_2^1(\Omega)$ to a function $u(x)$ which realizes inf $J(u)$.

Let a_{ij}, a_i, and a be bounded functions, and let

$$\mathscr{L}(u, u) \geq \nu_1(\|u\|_{2,\Omega}^{(1)})^2, \qquad \nu_1 > 0, \tag{8.28}$$

for any $u \in \mathring{W}_2^1(\Omega)$. It is easy to see that, under these conditions, the problem

$$\mathscr{L}u = f + \frac{\partial f_i}{\partial x_i}, \qquad u|_S = 0 \tag{8.29}$$

cannot have two distinct generalized solutions in $W_2^1(\Omega)$, and that the functional $J(u)$ is bounded from below on $\mathring{W}_2^1(\Omega)$, since, for all $u \in \mathring{W}_2^1(\Omega)$,

$$J(u) \geq \nu_1(\|u\|_{2,\Omega}^{(1)})^2 - 2\|f\|\,\|u\| - 2\|\mathbf{f}\|\,\|u_x\| \geq -\frac{1}{\nu_1}(\|f\|^2 + \|\mathbf{f}\|^2). \tag{8.30}$$

In view of (8.30), $\inf J(u) > -\infty$ and there exists a minimizing sequence u^N $(N = 1, 2, \ldots)$, which can be constructed as follows. We take a fundamental sequence $\{\varphi_k(x)\}$ in $\mathring{W}_2^1(\Omega)$ and its corresponding sequence of expanding spaces H_N $(N = 1, 2, \ldots)$, where H_N is the linear span of $\varphi_1, \ldots, \varphi_N$. The functional $J(u)$ on H_N is a quadratic form in the coefficients of an arbitrary element $u = \sum_{k=1}^N d_k\varphi_k$ in H_N. This form, considered as a function of second degree in the variables $d_1, \ldots, d_N$ in Euclidean space R^N, is bounded from below because of (8.30) and hence assumes its smallest value at some point $d_1^0, \ldots, d_N^0$. At this point the equalities $\partial/\partial d_l\, J(\sum_{k=1}^N d_k\varphi_k) = 0$ $(l = 1, \ldots, N)$ should be satisfied, which, as is not hard to see, can be written as follows:

$$\mathscr{L}(u^N, \varphi_l) = -(f, \varphi_l) + (f_i, \varphi_{lx_i}), \qquad l = 1, \ldots, N, \tag{8.31}$$

where $u^N = \sum_{k=1}^N d_k^0\varphi_k$ is a function realizing $\inf J(u)$ on H_N. The system (8.31), serving to define $d_1^0, \ldots, d_N^0$, coincides with the relations (8.4) which define the Galerkin approximations.

This allows us to bring in Theorem 8.1, which guarantees the convergence of u^N to the solution u of the problem (8.29). We must remark that the method of Ritz was formulated much earlier than the method of Galerkin, so that we may say that the Galerkin method is a generalization of the Ritz method to the case of non-symmetric operators $\mathscr{L}$.

3. The Method of Least Squares. Let us consider the problem

$$\mathscr{L}u = f, \qquad u|_S = 0, \tag{8.32}$$

under the hypotheses of Theorem 7.2, and let $\lambda = 0$ not be in the spectrum of this problem. Then, for all $u \in W_{2,0}^2(\Omega)$, we have the inequality

$$\|u\|_{2,\Omega}^{(2)} \leq c\|\mathscr{L}u\| \tag{8.33}$$

and (8.32) is uniquely solvable in $W_{2,0}^2(\Omega)$ for all $f \in L_2(\Omega)$. According to the method of least squares, the problem (8.32) is replaced by the variational problem of finding the infimum of the functional

$$J(v) = \int_\Omega (\mathscr{L}v - f)^2\, dx \tag{8.34}$$

over $W_{2,0}^2(\Omega)$. It is clear that, at the solution $u \in W_{2,0}^2(\Omega)$ of the problem (8.32), $J(v)$ assumes its smallest value: $J(u) = 0$. It follows from (8.33) that both problems have a unique solution $u(x)$ in $W_{2,0}^2(\Omega)$. The first variation of J has the form

$$\delta J(v, \eta) = 2 \int_\Omega (\mathscr{L}v - f)\mathscr{L}\eta \, dx. \tag{8.35}$$

At the solution $u(x)$,

$$\delta J(u, \eta) = 2 \int_\Omega (\mathscr{L}u - f)\mathscr{L}\eta \, dx = 0 \tag{8.36}$$

for all $\eta \in W_{2,0}^2(\Omega)$. For the solution of the variational problem we introduce a fundamental system $\{\psi_k(x)\}$ in $W_{2,0}^2(\Omega)$ and choose finite-dimensional subspaces $\mathscr{H}_N$ ($N = 1, 2, \ldots$) having the functions $\{\psi_1, \ldots, \psi_N\}$ as basis. An approximate solution u^N is sought as an element of $\mathscr{H}_N$ on which $\inf_{v \in \mathscr{H}_N} J(v)$ is realized. It is easy to see that this problem is uniquely solvable, and that its solution $u^N = \sum_{k=1}^N c_k^N \psi_k$ is defined by the conditions

$$\delta J(u^N, \eta) = 0 \tag{8.37}$$

for all $\eta \in \mathscr{H}_N$, or, what is the same thing, by the system of linear algebraic equations

$$(\mathscr{L}u^N - f, \mathscr{L}\psi_l) = 0, \qquad l = 1, \ldots, N. \tag{8.38}$$

But this system coincides with the system (8.10), and, by the same token, the approximiate solutions u^N coincide with the approximations u^N of one of the variants of the modernized Galerkin method considered above, where it was proved that u^N converges to the solution u in the norm of $W_2^2(\Omega)$.

Supplements and Problems

1. Throughout all of the chapters of this book it is assumed that the coefficients in equations are real and $a_{ij} = a_{ji}$. If the functions a_{ij} have bounded first generalized derivatives then the condition $a_{ij} = a_{ji}$ is not essential since

$$\mathscr{L}u = (a_{ij}u_{x_j} + a_i u)_{x_i} + b_i u_{x_i} + au = (A_{ij}u_{x_j} + a_i u)_{x_i} + \tilde{b}_i u_{x_i} + au,$$

where

$$A_{ij} = \tfrac{1}{2}(a_{ij} + a_{ji}) = A_{ji} \quad \text{and} \quad \tilde{b}_i = b_i + \tfrac{1}{2}(a_{ji} - a_{ij})_{x_j}.$$

But if a_{ij} have no derivatives in Ω then the condition $a_{ij} = a_{ji}$ restricts the class of equations (1.1). Let us show how to apply the approach described in §§2, 3 to equations (1.1) when $a_{ij} \neq a_{ji}$. For this purpose we denote

$\frac{1}{2}(a_{ij}+a_{ji})$ by A_{ij} and $\frac{1}{2}(a_{ij}-a_{ji})$ by B_{ij}, and introduce in $H = \overset{\circ}{W}{}_2^1(\Omega)$ the new inner product

$$[u, v] = \int_\Omega A_{ij} u_{x_j} v_{x_i}\, dx, \tag{1}$$

and in $H \equiv W_2^1(\Omega)$ the inner product

$$[u, v] = \int_\Omega (A_{ij} u_{x_j} v_{x_i} + uv)\, dx. \tag{2}$$

We consider, for example, the Dirichlet problem (2.1), (2.2). The integral identity (2.6) can be rewritten in the form

$$[u, \eta] + [Ku, \eta] + [Cu, \eta] = [F, \eta], \tag{3}$$

where [,] is defined by (1), K and C are the operators defined by the identities

$$[Ku, \eta] = \int_\Omega B_{ij} u_{x_j} \eta_{x_i}\, dx, \qquad \forall \eta \in H, \tag{4_1}$$

and

$$[Cu, \eta] = \int_\Omega (a_i u \eta_{x_i} - b_i u_{x_i} \eta - au\eta)\, dx, \qquad \forall \eta \in H, \tag{4_2}$$

and F is defined by (3.6). Just as in §3 it is proved that F is an element of H, C is a compact linear operator and K is a skew-symmetric bounded operator in H, so that $[Ku, \eta] = -[u, K\eta]$ for all $u, \eta \in H$. Identity (3) is equivalent to the equation

$$u + Ku + Cu = F$$

in H, which is in turn equivalent to the equation

$$u + (I + K)^{-1} Cu = (I + K)^{-1} F \tag{5}$$

since the operator $I + K$ has the bounded inverse operator $(I + K)^{-1}$ defined on H. Equation (5) has the same properties as equation (3.8) and therefore the conclusions of §§3 and 5 are valid for it.

Prove that the equation

$$v + Kv = \psi$$

with a skew-symmetric bounded operator K is uniquely solvable in H for $\psi \in H$ and $\|v\| \le \|\psi\|$; i.e., the operator $(I + K)^{-1}$ exists and its norm is bounded by 1.

For this purpose use the approximations

$$v_n + K_n v_n = P_n \psi, \qquad K_n \equiv P_n K P_n,$$

where P_n is the orthogonal projection on an n-dimensional subspace of H.

2. The equation (1.1) with complex non-differentiable $a_{ij} \neq a_{ji}$ may be treated in the same way. Let us denote by A the matrix (a_{ij}) and represent it as the sum $A = A' + A''$, $A' = \frac{1}{2}(A + A^*)$, $A'' = \frac{1}{2}(A - A^*)$, where $A^* = (\bar{a}_{ji})$ is the Hermitian conjugate matrix for A (the dash above a symbol denotes complex conjugation). We suppose that for all complex $\xi = (\xi_1, \ldots, \xi_n)$ the inequalities

$$\operatorname{Re}(A\xi \cdot \xi) \equiv \operatorname{Re}(a_{ij}\xi_j\bar{\xi}_i) \geq \nu|\xi|^2 \equiv \nu \sum_{i=1}^{n} |\xi_i|^2, \qquad \nu = \text{const.} > 0, \tag{6}$$

are fulfilled. It is easy to see that $A'\xi \cdot \xi$ is real and $A''\xi \cdot \xi$ is purely imaginary and so $\operatorname{Re}(A\xi \cdot \xi) = A'\xi \cdot \xi$. If $a_{ij} = a_{ji}$ then $A\xi \cdot \xi = A\xi' \cdot \xi' + A\xi'' \cdot \xi''$, where $\xi = \xi' + \sqrt{-1}\xi''$ and ξ', ξ'' are real, and so the condition (6) is equivalent to the same inequality restricted for real ξ.

Convince yourself that the first, second, and third boundary value problems for (3.9) with complex bounded coefficients and $a_{ij} \neq a_{ji}$ satisfying the condition (6) are Fredholm solvable. As the new inner product in $W_2^1(\Omega)$ (now it is the complex space) take

$$[u, v] = \int_\Omega (A'u_x \cdot \bar{v}_x + u\bar{v})\, dx \equiv \int_\Omega (a'_{ij}u_{x_j}\bar{v}_{x_i} + u\bar{v})\, dx,$$

where a'_{ij} are the elements of A'. The integral identity defining solutions will be equivalent to the equation $u + Ku + Cu + \lambda Bu = F$, where K is the bounded skew-symmetric operator, C the compact operator, and B the positive compact operator considered in $\mathring{W}_2^1(\Omega)$ for the first boundary value problem, and in $W_2^1(\Omega)$ for the second and third boundary value problems.

3. In §8 we proved that generalized solutions of boundary value problems for (1.1) can be obtained as limits of Galerkin approximations. For this purpose we used only two properties of the bilinear form $\mathscr{L}(u, \eta)$:

$$\mathscr{L}(u, u) \geq \nu_1 \|u\|_H^2, \qquad \nu_1 = \text{const.} > 0, \tag{7}$$

and

$$|\mathscr{L}(u, \eta)| \leq \mu_1 \|u\|_H \|\eta\|_H. \tag{8}$$

These inequalities must be fulfilled for all elements u and η of the Hilbert space H in which we want to find a generalized solution. ($H = \mathring{W}_2^1(\Omega)$ for Dirichlet problem and $H = W_2^1(\Omega)$ for the other two.) No other properties of $\mathscr{L}$ and Ω were used in the standard Galerkin method (in particular, Ω could be unbounded and $a_{ij} \neq a_{ji}$). So the following theorem was actually proved.

Theorem 1. *Let us assume that in a Hilbert space H there is a bilinear form $\mathscr{L}(u, \eta)$ satisfying the inequalities (7) and (8). Then for an arbitrary linear functional F on H there is a unique element u in H for which the identity*

$$\mathscr{L}(u, \eta) = F(\eta) \tag{9}$$

is true. (Here, $F(\eta)$ is the value of F on η.)

This fact was known to some mathematicians (including myself) at the end of the 1940's and was practically proved in their studies. It was published as a separate statement in the paper [1] of Lax and Milgram and is frequently cited as the lemma of Lax–Milgram. It could be proved without any finite-dimensional approximation with the help of the Riesz representation theorem. But the proof with the aid of Galerkin approximations has an advantage of admitting generalization to the case when $\mathscr{L}(u, \eta)$ is linear in η only and H is any separable Banach space. Indeed, the following statement is true:

Theorem 2. *Let us assume that B is a separable reflexive Banach space and that on $B \times B$ there is a real scalar function $\mathscr{L}(u, \eta)$ which in linear in η and has the following properties:*

(a)

$$|\mathscr{L}(u, \eta)| \leq \psi_1(\|u\|_B)\|\eta\|_B, \qquad \forall u, \eta \in B, \tag{10}$$

where $\psi_1(\tau)$ is a continuous function of $\tau \geq 0$.

(b)

$$\mathscr{L}(u, u) \geq \psi_2(\|u\|_B), \qquad \forall u \in B, \tag{11}$$

where $\psi_2(\tau)$ is a positive continuous function of $\tau > 0$ satisfying $\tau^{-1}\psi_2(\tau) \to \infty$ when $\tau \to \infty$.

(c) *$\mathscr{L}(u, \eta)$ is hemi-continuous ("weakly" continuous) in u, i.e. the function $\mathscr{L}(u + \varepsilon v, \eta)$ is continuous in $\varepsilon \in (-1, 1)$ for every fixed $u, v, \eta \in B$.*

(d)

$$\mathscr{L}(u, u - v) - \mathscr{L}(v, u - v) \geq 0, \qquad \forall u, v \in B, \tag{12}$$

with equality occurring only when $u = v$.

Then for each $F \in B^*$ (i.e. for each linear functional F on B) there is a unique element u of B for which the identity (9) is true.

Verify that if $\mathscr{L}(u, \eta)$ is also linear in u then the conditions (a)–(d) are equivalent to (7), (8). Try to prove Theorem 2 using Galerkin approximations $u^m = \sum_{k=1}^{m} c_k^m \varphi_k$ where $\{\varphi_k\}_{k=1}^{\infty}$ is a fundamental system in B. For c_k^m you will obtain a non-linear algebraic system. Its solvability follows from the Brouwer fixed point theorem and an *a priori* estimate of $\|u^m\|_B$. The latter and the estimate of $\|u\|_B$ are derived based on (11). After that it is necessary to prove that a subsequence $\{u^{m_k}\}_{k=1}^{\infty}$ converges to an element $u \in B$ which satisfies (9). For this the properties (c) and (d) are essential. The property (d) is called the monotonicity condition. Theorem 2 was proved by Minty [1], [2] and F. Browder [4], [5] (for Theorem 2 and generalizations see [LU 1], [LI 2]). It should be noted that in the both cases (linear and non-linear) the identity (9) is equivalent to the equation

$$\tilde{\mathscr{L}}(u) = F$$

in the space B^* since for each fixed u from B, $\mathscr{L}(u, \cdot)$ is a linear functional on B. So the operator $\tilde{\mathscr{L}}$ transforms elements u from B into elements of B^*.

Theorems 1 and 2 have found many interesting applications for the solution of some boundary value problems (both linear and non-linear). Reduction of each concrete problem to one of these theorems is frequently a non-trivial task. In Chapter II we have shown how three principal boundary value problems for equation (1.1) can be reduced to equation (9) having properties (7) and (8). In Chapter V we do the same for diffraction problems and for some boundary value problems for a large class of systems including the linear systems of elasticity and hydrodynamics. Here we confine ourself to scalar equations of the second order.

4. Let us show how the boundary value problem "with an oblique derivative" for equation (5.1) can be reduced to the third boundary problem. The boundary condition in this problem has the form

$$\left.\frac{\partial u}{\partial l} + \sigma u\right|_S \equiv \left.\sum_{i=1}^{n} l_i u_{x_i} + \sigma u\right|_S = 0, \tag{13}$$

where $\mathbf{l} = (l_1, \ldots, l_n)$ is a smooth vector field on S with $|\mathbf{l}| = 1$.

Theorem 3. *If the field* $\mathbf{l}(\cdot)$ *is not tangent to* S *at any point of* S *and* S *is a smooth surface then problem* (5.1), (13) *is Fredholm solvable.*

Let us represent $\mathscr{L}u$ in the form

$$\mathscr{L}u = \frac{\partial}{\partial x_i}[(a_{ij} + b_{ij})u_{x_j}] - b_{ij}u_{x_i x_j} - \frac{\partial b_{ij}}{\partial x_i}u_{x_j} + b_i u_{x_i} + au$$

and choose the functions b_{ij} in such a way that $b_{ij}u_{x_i x_j} \equiv 0$ and the condition (13) has the form

$$(a_{ij} + b_{ij})u_{x_j} n_i + \sigma\sigma_1 u|_S = 0. \tag{14}$$

The functions

$$\begin{aligned} b_{ij} = (\mathbf{l} \cdot \mathbf{n})^{-1}(n_i l_j - n_j l_i)a_{km} n_k n_m \\ - a_{kj} n_k n_i + a_{ki} n_k n_j, \end{aligned}$$

where $\mathbf{l}(\cdot)$ and $\mathbf{n}(\cdot)$ are smooth vector functions on $\bar{\Omega}$ coinciding with $\mathbf{l}(\cdot)$ and $\mathbf{n}(\cdot)$ on S, satisfy both these requirements (these functions b_{ij} were constructed by H. Weinberger, see [FI 3]). In fact, $b_{ij} = -b_{ji}$ (consequently $b_{ij}u_{x_i x_j} = 0$) and, as it is not difficult to calculate, $(a_{ij} + b_{ij})n_i = (\mathbf{l} \cdot \mathbf{n})^{-1} \times a_{km} n_k n_m l_j$ on S. Hence the condition (14) with $\sigma_1 = (\mathbf{l} \cdot \mathbf{n})^{-1} a_{km} n_k n_m$ is equivalent to (13). So we see that (5.1), (13) is the third boundary value problem

$$\mathscr{L}u \equiv \frac{\partial}{\partial x_i}(\tilde{a}_{ij}u_{x_j}) + \tilde{b}_i u_{x_i} + au = \lambda u + f,$$

$$\frac{\partial u}{\partial N} + \sigma_2 u \equiv \tilde{a}_{ij}u_{x_j}n_i + \sigma_2 u\bigg|_S = 0, \tag{15}$$

with $\tilde{a}_{ij} = a_{ij} + b_{ij}$, $\tilde{b}_i = b_i + \partial b_{ij}/\partial x_j$ and $\sigma_2 = \sigma\sigma_1$, which is Fredholm solvable.

5. The Neumann problem

$$\Delta u = f, \qquad \frac{\partial u}{\partial n}\bigg|_S = 0, \tag{16}$$

is a particular case of the problems considered in §5. Hence for the problem

$$\Delta u = \lambda u + f, \qquad \frac{\partial u}{\partial n}\bigg|_S = 0, \tag{17}$$

three Fredholm theorems are true and the question about the solvability of (16) is reduced to the question whether $\lambda = 0$ is the eigenvalue of the appropriate homogeneous problem; and if this is the case, one should investigate the eigenfunctions corresponding to $\lambda = 0$. The latter is needed to find necessary and sufficient conditions under which the problem (16) is solvable.

Prove that $u(x) \equiv 1$ is the unique eigenfunction corresponding to $\lambda = 0$ and these conditions consist of the single requirement

$$\int_\Omega f\,dx = 0. \tag{18}$$

This result can be also proved "directly." We sketch here this proof. Let us choose as the "large" space from which we shall take f the space $\hat{L}_2(\Omega)$ consisting of all elements of $L_2(\Omega)$ orthogonal to 1 (i.e., elements f of $\hat{L}_2(\Omega)$ satisfy (18)) and as the space H (in which we shall look for solutions of (16)) the subspace $\hat{W}_2^1(\Omega) \equiv W_2^1(\Omega) \cap \hat{L}_2(\Omega)$ of $W_2^1(\Omega)$. In $\hat{W}_2^1(\Omega)$ we introduce the new inner product

$$[u, v] = \int_\Omega u_x v_x\,dx. \tag{19}$$

It is easy to prove that the problem (16) is uniquely solvable in $\hat{W}_2^1(\Omega)$ for every $f \in \hat{L}_2(\Omega)$.

In Chapter V and in the Supplement to it we shall illustrate how the second approach works in some problems of elasticity and hydrodynamics.

6. For problems in unbounded domains Ω it is sometimes useful not to restrict but to extend the space $W_2^1(\Omega)$ in order to pick up the solution. Let us consider Dirichlet problem

$$\Delta u = f, \qquad u|_S = 0, \tag{20}$$

in an exterior domain Ω (i.e. $R^n \backslash \Omega$ is bounded). The inequality (6.3) of Chapter I is not fulfilled. The spectrum of the corresponding problem

$$\Delta u = \lambda u, \qquad u|_S = 0 \tag{21}$$

fills the semi-axis $(-\infty, 0]$, and the Fredholm theorems for $\Delta u = \lambda u + f$, $u|_S = 0$, are not held. In this case the inclusion of the problem (20) in such a family is not useful. But the approach of Chapter II enables one to obtain some meaningful results if we apply it in the following way. Let us take as space H the closure of the set $C^\infty(\Omega)$ in the norm of the Dirichlet integral, i.e. in the norm corresponding to the inner product (19), and provide H with the same Hilbert structure.

It follows from (7.1), Chapter I, that the inequality

$$\|u\|_{2n/(n-2), \Omega} \leq \beta \|u_x\|_{2, \Omega}$$

is true for all $u \in \dot{C}^\infty(\Omega)$ if $n > 2$, and therefore the elements of H will belong to $L_{2n/(n-2)}(\Omega)$.

Prove that the problem (20) has a unique solution from H for any $f \in L_{2n/(n+2)}(\Omega)$ $(n > 2)$. Find the space H for $n = 2$ and determine the behavior of the generalized solutions of (20) when $|x| \to \infty$. In the framework of these spaces H it is easy to see the differences both in the statement and final results of exterior problems for Laplace's operator when $n > 2$ and $n = 2$—the differences well known from the classical Potential Theory. The vector analogs of these spaces helped to elucidate the Stokes paradox for viscous incompressible fluids (see [LA 15]).

7. Freedom given by the generalized treatment of problems made it possible to investigate equations (1.1) not only with bounded coefficients satisfying the conditions (2.3), (2.4). In Remark 3.1 we indicated conditions for a_i, b_i, a and also for f and f_i under which all Fredholm theorems are retained. Verify this.

You may also weaken the condition (2.3) of uniform ellipticity. Assume, for example, that instead of (2.3) the inequalities

$$c_1 \mu(x) \xi^2 \leq a_{ij}(x) \xi_i \xi_j \leq c_2 \mu(x) \xi^2, \qquad c_k = \text{const.} > 0, \tag{22}$$

are fulfilled for some non-negative integrable function μ. Then for investigation of the first boundary value problem it is reasonable to take as the space H the space $\mathring{W}^1_{2,\mu}(\Omega)$ with the inner product

$$(u, v)_H = \int_\Omega (\mu u_x v_x + uv)\, dx \tag{23}$$

or

$$(u, v)_H = \int_\Omega \mu u_x v_x \, dx, \tag{24}$$

and with the dense set $\dot{C}^\infty(\Omega)$ (i.e. to define $\mathring{W}^1_{2,\mu}(\Omega)$ as the closure of set $\dot{C}^\infty(\Omega)$ in the norm corresponding to (23) or (24)). In this context some questions arise:

(a) Will elements of H have zero traces on S or on a part of S?
(b) For what functions f and f_i do the integrals $\int_\Omega f\eta \, dx \equiv l(\eta)$ and $\int_\Omega f_i \eta_{x_i} \, dx \equiv l_1(\eta)$ determine linear functionals on H?
(c) For what a_i, b_i, and a does the identity (3.5) determine the completely continuous operator A?
(d) When are norms corresponding to (23) and (24) equivalent?
(e) To ensure discreteness of the spectrum we need compactness of the embedding of $\mathring{W}^1_{2,\mu}(\Omega)$ into $L_2(\Omega)$.
(f) The conditions (7) and (8) impose some restrictions on the coefficients a_i, b_i, a.

It is desirable to formulate these restrictions in a more explicit way.

The answers to these questions depend essentially on the function μ and primarily on the sets on which the function μ is equal to zero or infinity and how fast $\mu(x)$ tends to zero or infinity when x approaches these sets.

Let us consider, for example, the first boundary value problem

$$\frac{d}{dx}\left(x^\alpha \frac{du}{dx}\right) = f(x), \qquad \alpha \in (0, 2], \quad u(0) = u(1) = 0, \tag{25}$$

on the interval $\Omega = (0, 1)$. In accordance with the above let us define H as the closure of $\dot{C}^\infty(\Omega)$ in the norm corresponding to the inner product

$$(u, v)_H = \int_0^1 x^\alpha u_x(x) v_x(x) \, dx. \tag{26}$$

It is easy to see that elements u of H are continuous on $(0, 1]$, have generalized first derivatives on $(0, 1)$, and $u(1) = 0$. If $\alpha > 0$ then (25) is an elliptic equation degenerating at the point $x = 0$. Show that for $\alpha \in (0, 2]$ H is embedded into $L_2(\Omega)$, and for $\alpha \in (0, 2)$ this embedding is compact. For $\alpha \in (0, 1)$ all elements of H are equal to zero at the point $x = 0$ but for $\alpha \in [1, 2]$ it is not the case. Therefore, if we want to solve (25) with $\alpha \in [1, 2]$ for each $f \in L_2(\Omega)$, we must remove the condition $u(0) = 0$. Prove all this using the inequality

$$\int_0^1 x^{\alpha - 2} u^2(x) \, dx \le 4(\alpha - 1)^{-2} \int_0^1 x^\alpha u_x^2(x) \, dx, \tag{27}$$

true for an arbitrary $u \in \dot{C}^\infty(\Omega)$ and therefore for all $u \in H$. The inequality (27) is easily derived from

$$\int_0^1 x^{\alpha-2} u^2(x)\, dx = \frac{1}{\alpha - 1} \int_0^1 u^2\, d(x^{\alpha-1}).$$

The case $\alpha = 1$ requires some special considerations (investigate it). Such analysis of the one-dimensional problem (25) is necessary also for studying the many-dimensional problem (2.1), (2.2) in which degeneration of the form $a_{ij}(x)\xi_i\xi_j$ occurs on a part $S_1 \subset S$ or on a $(n-1)$-dimensional smooth manifold lying in $\bar{\Omega}$. The behavior of this quadratic form evaluated on the vector $\xi = (\xi_i, \ldots, \xi_n), |\xi| = 1$, directed along the normal to S_1 is responsible for the preservation of zero boundary value on S_1: that is, in the case $a_{ij}(x)\xi_i\xi_j \approx d^\alpha(x)$, where $d(x)$ is the distance from the point x to S_1, the zero boundary value on S_1 is preserved if $\alpha \in (0, 1)$ and is not preserved if $\alpha \in [1, 2]$, in full agreement with the one-dimensional case. The papers [KE 1], [OL 2], [MI 5], [VI 4] and others are devoted to elliptic equations degenerating on the boundary.

The situation is different when $\mu(x)$ becomes infinity on a subset $\Omega_1 \subset \Omega$ of positive measure. Let us first consider the simplest one-dimensional case

$$\frac{d}{dx}\left(\mu(x)\frac{du}{dx}\right) - u = f(x), \qquad u(0) = u(1) = 0, \tag{28}$$

with a measurable function μ which is equal to ∞ on the interval $(0, \delta)$ and satisfies $0 < \nu_1 \le \mu(x) \le \nu_2$, $\nu_k = \text{const.} > 0$, on the interval $[\delta, 1]$. As the inner product we take

$$(u, v)_H = \int_0^1 (\mu u_x v_x + uv)\, dx. \tag{29}$$

From $\dot{C}^\infty(\Omega)$ $(\Omega = (0, 1))$ we choose only those functions which are equal to zero on the segment $[0, \delta]$ and close this set of functions in the norm corresponding to (29). It is easy to see that in the final result we shall come to functions $u(\cdot)$ belonging to $\mathring{W}_2^1(\Omega)$ and equal to zero on $[0, \delta]$. These functions considered only on $(\delta, 1)$ are elements of $\mathring{W}_2^1((\delta, 1))$. On this interval we come to the first boundary value problem for (28) considered in Chapter II. It is uniquely solvable in $\mathring{W}_2^1((\delta, 1))$ for an arbitrary $f \in L_2((\delta, 1))$. For the case studied, our speculations seem to be awkward as we actually threw out the interval where $\mu(x) = \infty$. But they proved useful for numerical solution of the many-dimensional Dirichlet problem in domains of complicated shape. Let us first explain it for equation (28) on $(\delta, 1)$ with $u(\delta) = u(1) = 0$. Let us denote by $\mu^\varepsilon(x)$, $\varepsilon > 0$, the function which equals $\mu(x)$ for $x \in [\delta, 1]$ and ε^{-1} for $x \in [0, \delta)$ and by $f^\delta(x)$ the function which equals $f(x)$ for $x \in [\delta, 1]$ and zero for $x \in [0, \delta)$, supposing that $f \in L_2((\delta, 1))$. Equation (28) with $\mu(x)$ replaced by $\mu^\varepsilon(x)$ and $f(x)$ replaced by $f^\delta(x)$ has the unique generalized solution $u^\varepsilon \in \overset{\circ}{W}{}_2^1((0, 1))$. The inequality

$$\int_0^1 [2\mu^\varepsilon(u_x^\varepsilon)^2 + (u^\varepsilon)^2]\, dx \le \int_\delta^1 f^2\, dx \tag{30}$$

is true for u^ε. Let us now let ε tend to zero. Show that as a limit of $\{u^\varepsilon(x)\}$, $x \in (\delta, 1)$, we shall have the generalized solution of the Dirichlet problem for equation (28) on the interval $(\delta, 1)$. This method of finding solutions in the one-dimensional case is not expendient but all of the preceding is applicable to many-dimensional equations (2.1) in an arbitrary domain Ω. If Ω has a complicated form we can include it into a larger domain $\tilde{\Omega}$ of a simpler form and solve the Dirichlet problem in $\tilde{\Omega}$ assuming that in $\tilde{\Omega}\backslash\Omega$ the coefficients $a_{ij}(x)$ are equal to $\varepsilon^{-1}\delta_i^j$ and that other coefficients and f are zero. The desired solution is a limit of the solutions u^ε of such problems. V. J. Rivkind ([1]–[3]) calculated the rate of convergence of solutions u^ε to the exact solution of the Dirichlet problem in Ω and did the same for the solutions u_h^ε of some difference equations which approximate the equations of the problem.

A similar technique known as the method of "infinite barriers" was used by physicists for the investigation of the Schrödinger operators $-\Delta u + V(x)u$.

8. In this section we suppose that the coefficients a_{ij} and a_i of $\mathscr{L}$ from (6.1), in addition to (2.3) and (2.4), have bounded first derivatives in Ω and that Ω satisfies the conditions of §7. Then the differential operators $\mathscr{L}$ and $\mathscr{L}^*$ may be evaluated on any $u \in W_2^2(\Omega)$ and

$$\mathscr{L}u = \frac{\partial}{\partial x_i}(a_{ij}u_{x_j}) + a_i u_{x_i} + au$$

and

$$\mathscr{L}^*u = \frac{\partial}{\partial x_i}(a_{ij}u_{x_j}) - \frac{\partial}{\partial x_i}(a_i u) + au$$

will be elements of $L_2(\Omega)$. We can associate with $\mathscr{L}$ and $\mathscr{L}^*$ the unbounded operators $\hat{\mathscr{L}}$ and $\widehat{(\mathscr{L}^*)}$ in $L_2(\Omega)$ which are defined on $\mathscr{D}(\hat{\mathscr{L}}) = \mathscr{D}(\widehat{(\mathscr{L}^*)}) = W_{2.0}^2(\Omega)$ and calculated on $u \in W_{2.0}^2(\Omega)$ as $\mathscr{L}u$ and $\mathscr{L}^*u$ correspondingly. It is not hard to prove that $\widehat{(\mathscr{L}^*)}$ is adjoint to $\hat{\mathscr{L}}$, i.e. $\widehat{(\mathscr{L}^*)} = (\hat{\mathscr{L}})^*$. If $\mathscr{L}^* = \mathscr{L}$ (i.e. $a_i \equiv 0$) then $\hat{\mathscr{L}}$ is a self-adjoint operator, (i.e. $\hat{\mathscr{L}}^* = \hat{\mathscr{L}}, \mathscr{D}(\hat{\mathscr{L}}^*) = \mathscr{D}(\hat{\mathscr{L}})$).

For the problem

$$\mathscr{L}u = \lambda u + f, \qquad u|_S = 0 \tag{$3 1_1$}$$

we can define a generalized solution in $L_2(\Omega)$ as an element u of $L_2(\Omega)$ which satisfies the identity

$$\int_\Omega u\mathscr{L}^*\eta\, dx = \int_\Omega (\lambda u + f)\eta\, dx \tag{$3 1_2$}$$

for all $\eta \in W^2_{2,0}(\Omega)$. This definition is sensible for "rather bad" f (see [LA 11]). Note that the boundary condition $u|_S = 0$ is hidden in (31_2): if u satisfies (31_2) and u belongs to $W^1_2(\Omega)$ then $u \in \overset{\circ}{W}{}^1_2(\Omega)$. If $f \in L_2(\Omega)$ then any generalized solution of $(31)_1$ from $L_2(\Omega)$ will belong to $W^2_{2,0}(\Omega)$.

Verify these statements.

The existence of a generalized solution in $L_2(\Omega)$ of the problem (31_1) with $\lambda = 0$ may be proved if we know only that for any $v \in W^2_{2,0}(\Omega)$ we have the inequality

$$\mathcal{L}(v, v) \geq \nu_1 \|v\|^2_{2,\Omega}, \qquad \nu_1 = \text{const.} > 0 \tag{32}$$

The operator $\mathcal{L}$ may degenerate in an arbitrary way (i.e. $a_{ij}(x)\xi_i\xi_j \geq 0$). Prove this using the auxiliary family of operators $\mathcal{L}^\varepsilon = \varepsilon\Delta + \mathcal{L}$, $\varepsilon \to 0+$, and results of §2 (see also No. 10 of this supplement). A special analysis is needed to learn whether such a solution (for the degenerate $\mathcal{L}$) is unique, how smooth it is, and on what part of S the boundary condition is satisfied [FI 1; 2], [OL 3], [KN 1], etc.).

9. We considered in Chapter II spectral problems of the form (3.24) and reduced them to the problem (3.19) of the form $u = \lambda Cu$ with a completely continuous operator C. As basic spaces we took the spaces $\overset{\circ}{W}{}^1_2(\Omega)$ and $L_2(\Omega)$ and used the compactness of the embedding of $\overset{\circ}{W}{}^1_2(\Omega)$ into $L_2(\Omega)$. But there are problems in which the spectrum parameter λ enters the equation in a different way. Consider, for example,

$$\mathcal{L}u \equiv \Delta u + a(x)u = \lambda\mu(x)u, \qquad u|_S = 0, \tag{33}$$

where $\mu(x) > 0$. It is natural to associate with this problem the Hilbert space $L_{2,\mu}(\Omega)$ with the inner product

$$(u, v)_\mu = \int_\Omega \mu uv\, dx,$$

and the Hilbert space H with the inner product

$$(u, v)_H = \int_\Omega (\mu uv + u_x v_x)\, dx$$

Investigate the solvability of (33) using these spaces. For what $a(x)$, $\mu(x)$ and Ω does this problem have a discrete spectrum?

10. A great number of publications are devoted to problems with a small parameter in principal terms. One such problem is

$$\mathcal{L}^\varepsilon u \equiv \varepsilon\frac{\partial}{\partial x_i}(a_{ij}u_{x_j}) + b_i u_{x_i} + au = f, \qquad u|_S = 0. \tag{34}$$

Assume that it has the unique solution u^ε for any $\varepsilon > 0$ (e.g. the generalized solution from $\overset{\circ}{W}{}^1_2(\Omega)$). The question arises as to what function is the limit of $\{u^\varepsilon\}$ when $\varepsilon \to 0+$. It would be natural to expect that the limit is a solution u^0 of the degenerate equation

$$\mathscr{L}^0 u^0 = f, \qquad \mathscr{L}^0 = \mathscr{L}^\varepsilon|_{\varepsilon=0}. \tag{35}$$

But what is preserved from the boundary condition? It is clear that for an arbitrary function f the function u^0 cannot be zero on the whole of S. On the other hand, equation (35) in the general case has infinitely many solutions. Which of them is the limit of $\{u^\varepsilon\}$? This depends on where on S the zero boundary value is preserved. The answers to these questions are related to the position of characteristic lines of equation (35), i.e. the lines $x_i = x_i(\tau)$ determined by $dx_i/d\tau = b_i(x)$, $i = 1, \ldots, n$.

Consider the case when these lines cover Ω regularly (that is, when they could be chosen as coordinate lines of one family) and $a(x) \le a_0 < 0$.

Let us analyze a simpler problem

$$\varepsilon \Delta^2 u - \mathscr{L}u = f_1 \qquad u|_S = 0, \qquad \left.\frac{\partial u}{\partial n}\right|_S = 0, \tag{36}$$

when $\varepsilon \to 0+$. Here $\mathscr{L}u$ has the form (2.1) and satisfies the condition

$$\mathscr{L}(u, u) \ge \nu_1 \int_\Omega (u_x^2 + u^2)\, dx, \qquad \nu_1 = \text{const.} > 0.$$

The unique solvability of the problem (36) in $\mathring{W}_2^2(\Omega)$ for each $\varepsilon > 0$ and each $f \in L_2(\Omega)$ is proved similarly as for (2.1), (2.2) under condition (8.2). (For this purpose one may use Theorem 1 from No. 3 of this Supplement taking $\mathring{W}_2^2(\Omega)$ for H.) The solution u^ε satisfies the identity

$$\varepsilon \int_\Omega \Delta u^\varepsilon \cdot \Delta \eta\, dx + \mathscr{L}(u^\varepsilon, \eta) = \int_\Omega f\eta\, dx \tag{36'}$$

for all $\eta \in \mathring{W}_2^2(\Omega)$ and

$$\varepsilon \int_\Omega (\Delta u^\varepsilon)^2\, dx + \nu_1 \int_\Omega [(u_x^\varepsilon)^2 + \tfrac{1}{2}(u^\varepsilon)^2]\, dx \le \frac{1}{2\nu_1} \int_\Omega f^2\, dx.$$

Therefore

$$\left| \varepsilon \int_\Omega \Delta u^\varepsilon \cdot \Delta \eta\, dx \right| \le \varepsilon \|\Delta u^\varepsilon\|_{2,\Omega} \|\Delta \eta\|_{2,\Omega}$$

$$\le \sqrt{\frac{\varepsilon}{2\nu_1}}\, \|f\|_{2,\Omega} \|\Delta \eta\|_{2,\Omega} \to 0 \qquad \text{when} \quad \varepsilon \to 0+$$

and from $\{u^\varepsilon\}$, $\varepsilon \to 0+$, we can choose a sequence $\{u^{\varepsilon_k}\}_{k=1}^{\infty}$ which converges weakly in $\mathring{W}_2^1(\Omega)$ to some $u^0 \in \mathring{W}_2^1(\Omega)$. Taking the limit in (36′) when $k \to \infty$ we prove that u^0 satisfies the identity

$$\mathscr{L}(u^0, \eta) = \int_\Omega f\eta \, dx$$

for all $\eta \in \mathring{W}_2^2(\Omega)$. Since $\mathring{W}_2^2(\Omega)$ is dense in $\mathring{W}_2^1(\Omega)$, u^0 is a generalized solution of the problem $-\mathscr{L}u = f$, $u|_S = 0$ from the space $\mathring{W}_2^1(\Omega)$. Owing to the uniqueness of such a solution, the family $\{u^\varepsilon\}$, $\varepsilon > 0$, converges to u when $\varepsilon \to 0+$.

Such an approach for the study of problems with small parameter was suggested in my paper [10]. In this paper some other problems were considered, the problem (34) in particular. (In connection with this work see also Solonnikov [1].) The paper [1] by N. Levinson was the first one devoted to the problem (34). It was followed by the works of O. A. Oleinik [1], L. A. Lusternik and M. I. Višik [1] and others. We cannot reproduce here the vast list of publications concerning problems with a small parameter. The majority of them are devoted to linear problems. For non-linear partial differential equations, rigorous results are scarce, although they are of great interest for mechanicists, physicists, and mathematicians.

Note that for the solution of some boundary value problems it may be useful to exploit auxiliary problems of higher order with a small parameter in the principal terms; these problems must be simpler and more amenable to analysis. In the Supplements to succeeding chapters we shall demonstrate how this idea is realized. The term "vanishing viscosity" is adopted from hydrodynamics where ε is the coefficient of viscosity.

11. We were unable to include here a wide range of questions belonging to the so-called qualitative theory of elliptic equations. They were mainly borrowed from the theory of analytic and harmonic functions and concern the maximum principle, removable singularities, Harnack's theorem, and theorems of Liouville, Fragmen-Lindelöf, and others. When extending these theorems to the solution of elliptic equations the investigators tried to minimize the restrictions imposed on the coefficients and on the class of solutions. Many results of this kind are contained in the monographs [LD 1], [LU 1] and others.

Here we shall describe one rather simple technique for the proof of a Fragmen-Lindelöf theorem which is in the spirit of this book. For the sake of brevity consider the elliptic equation

$$\mathscr{L}u = \frac{\partial}{\partial x_i}(a_{ij}u_{x_j}) = 0, \qquad a_{ij} = a_{ji}, \tag{37}$$

in the unbounded domain $\Omega = \{x: x_1 > 0,\ x' = (x_2, \ldots, x_n) \in \sum(x_1)\}$, where $\sum(x_1)$ for $x_1 \geq x_1^0 > 0$ results from a bounded domain $\sum(x_1^0) \subset R^{n-1}$

by dilation with the coefficient $g(x_1) > 0$, and $\Omega' \equiv \{x: 0 < x_1 < x_1^0\} \cap \Omega$ is a bounded domain. Let us consider a solution of (37) which is zero on $\partial\Omega$ but is not identically zero in Ω and belongs to $W_2^2(\Omega_\tau)$, for all $\tau > 0$, where $\Omega_\tau = \{x \in \Omega: x_1 \in (0, \tau)\}$. Let us multiply (37) by u, integrate the result over Ω_τ, $\tau > x_1^0$, and transform it by integration by parts:

$$\int_{\Omega_\tau} a_{ij} u_{x_j} u_{x_i}\, dx = \int_{\Sigma(\tau)} \frac{\partial u}{\partial N} u\, dx', \qquad \tau > x_1^0.$$

If a_{ij} satisfy the condition (2.3) then

$$y(\tau) \equiv \int_{\Omega_\tau} u_x^2\, dx \le \mu\nu^{-1} \left(\int_{\Sigma(\tau)} u_x^2\, dx' \int_{\Sigma(\tau)} u^2\, dx' \right)^{1/2}. \tag{38}$$

Let us denote by $\lambda(\tau)$ the smallest eigenvalue of the spectral problem $-\sum_{i=2}^n \partial^2 v/\partial x_i^2 = \lambda v$, $v|_{\partial\Sigma(\tau)} = 0$ in $\sum(\tau)$. It is positive and tends to zero if $g(\tau) \to \infty$. As follows from §4, Chapter II, the inequality

$$\int_{\Sigma(\tau)} w^2(x')\, dx' \le \lambda^{-1}(\tau) \int_{\Sigma(\tau)} w_{x'}^2\, dx'$$

is true for arbitrary $w \in \overset{\circ}{W}{}_2^1(\sum(\tau))$ and for $w(x') = u(\tau, x')$ in particular. Therefore from (38) it follows that

$$y(\tau) \le \mu\nu^{-1}\lambda^{-1/2}(\tau) \int_{\Sigma(\tau)} u_x^2\, dx' = \gamma^{-1}(\tau) \frac{dy(\tau)}{d\tau}, \tag{39}$$

where $\gamma(\tau) = \mu^{-1}\nu\lambda^{1/2}(\tau)$. By integrating (39) from $\tau_1 \ge x_1^0$ to $\tau_2 > \tau_1$ we obtain

$$y(\tau_2) \ge y(\tau_1) \exp\left\{ \int_{\tau_1}^{\tau_2} \gamma(\tau)\, d\tau \right\}. \tag{40}$$

This inequality is the integral analog of the Fragmen-Lindelöf theorem for equation (37). It implies that if $u(\cdot)$ is not identically zero then $y(\tau)$ must grow "sufficiently rapidly." If, for example, $g(\tau) \le c_1$ then $\gamma(\tau) \ge c_2 > 0$ and $y(\tau)$ has to grow at least as $c_3 \exp(c_2 \tau)$.

Convince yourself that for $\mathscr{L} = \Delta$ and a cylindrical tube such method enables us to obtain the precise growth rate of $y(\tau)$ for the most slowly growing solution.

The inequality (40) may also be interpreted as the integral version of Saint-Venant's principle: the portion $y(\tau_1)$ of energy $y(\tau_2)$ decreases with the distance from τ_1 to τ_2.

A similar consideration for the non-homogeneous problem

$$\mathscr{L}u = f, \qquad u|_{\partial\Omega} = 0, \tag{41}$$

makes it possible to prove the unique solvability of (41) in the class of functions for which $y(t)$ grows slower than the exponent in (40). The function f here may be increasing. Analyze what growth rate of $\int_{\Omega_\tau} f^2\, dx$ is admissible. This depends on the behavior of $\lambda(\tau)$ when $\tau \to \infty$. Such an approach proved useful for study of some non-linear problems. (See the paper [4] of Ladyzhenskaya and Solonnikov devoted to the viscous fluids in tubes. It contains a list of publications where this approach was used. It first appeared in the works of Toupin [1] and Knoweles [1], [2] in connection with the equations of the linear theory of elasticity).

12. Solution of a boundary value problem for an elliptic equation in Ω depends on the coefficients and the free term of this equation and on the function from the boundary condition in whole domains of their definition. Thus it cannot be obtained without utilizing all this information. However if we have a solution of the equation we may investigate its regularity locally, singling out one or another part of Ω. This is especially simple to do for the equation with constant coefficients in the principal terms. We shall explain it on the example of Laplace operator. Let us assume that $u \in W_2^1(\Omega)$ and satisfies the identity

$$\int_\Omega u_{x_k} \eta_{x_k}\, dx = -\int_\Omega f\eta\, dx \tag{42}$$

for all $\eta \in \overset{\circ}{W}{}_2^1(\Omega)$. We take $\eta = [(u_{\rho x_i}\zeta^2)_{x_i}]_\rho$ where $\zeta \in \dot{C}^\infty(\Omega)$ and u_ρ is the averaging of u of the type (4.9) from Chapter I. For sufficiently small ρ the function η belongs to $\overset{\circ}{W}{}_2^1(\Omega)$. The operation of such averaging commutes with differentiation and therefore (42) can be transformed by integration by parts into

$$\int_\Omega u_{\rho x_i x_k}(u_{\rho x_i}\zeta^2)_{x_k}\, dx = -\int_\Omega f_\rho (u_{\rho x_i}\zeta^2)_{x_i}\, dx. \tag{43}$$

From (43) it is easy to obtain

$$\int_\Omega \sum_{i,k=1}^n \zeta^2 (u_{\rho x_i x_k})^2\, dx \le c_\zeta \int_\Omega (u_x^2 + f^2)\, dx, \tag{44}$$

where the constant $c_\zeta = \psi(\max_{x\in\Omega}|\zeta(x)|, \max_{x\in\Omega}|\zeta_x(x)|)$ does not depend on ρ. This permits us to conclude that u has in Ω generalized second derivatives belonging to $W_2^2(\Omega')$, $\forall \overline{\Omega}' \subset \Omega$ (see Theorem 4.1, Chapter I). Do such an analysis for the equation $a_{ij}u_{x_i x_j} + a_i u_{x_i} + au = f$ with constant coefficients a_{ij}. Regularity of a solution near $S_1 \subset S$ depends not only on the smoothness of the coefficients and f but also on that of S_1 and of the function determining the boundary condition on S_1. In subsequent items we shall describe some methods for investigation of smoothness of solutions near the boundary.

13. Prove the statements of the Remark 7.1. For the treatment near a $S_1 \subset S$ it is useful to rectify S_1, i.e. to introduce new coordinates $y = y(x)$ so that $y_n = 0$ for S_1. This will simplify the investigation of the boundary integral arising when the second inequality with a cutoff function is deduced. (See (*) from §7, Chapter II.)

Let us describe briefly the proof of (8.17) for $\mathscr{L}_0 u = a_{ij}u_{x_ix_j}$ and $\mathscr{M}_0 u = b_{ke}u_{x_kx_e}$. Integrating twice by parts we obtain the equality

$$\int_\Omega \mathscr{L}_0 u \cdot \mathscr{M}_0 u \, dx = \int_\Omega a_{ij}b_{ke}u_{x_ix_k}u_{x_jx_e}$$

$$- \int_\Omega [(a_{ij}b_{ke})_{x_j}u_{x_i}u_{x_kx_e} - (a_{ij}b_{ke})_{xk}u_{x_i}u_{x_jx_e}]\,dx + \int_S I(s)\,ds,$$

where

$$I(s) = a_{ij}b_{ke}u_{x_i}[u_{x_kx_e}\cos(\mathbf{n}, \mathbf{x}_j) - u_{x_jx_e}\cos(\mathbf{n}, \mathbf{x}_k)].$$

The first term of the right-hand part is not less than $vv_1 \int_\Omega u_{xx}^2\,dx$, where v and v_1 are the lower bounds of the forms $a_{ij}(x)\xi_i\xi_j$ and $b_{ke}(x|\xi_k\xi_e)$ on the unit sphere $\xi^2 = 1$ for all $\bar{x} \in \Omega$. This is proved in the same way as (6.12), taking into account that the both forms can be diagonalized simultaneously. The second term is also majorized in just the same way as was done in §6 for $\mathscr{M}_0 = \mathscr{L}_0$ (see (6.11)). To estimate $\int_S I(s)\,ds$ we must first calculate $I(s)$ in each point x^0 of S in the new "local" cartesian coordinates $y_k = c_{ke}(x_e - x_e^0)$ (the same ones as in (6.15)). This gives the following expression for $I(x^0)$:

$$I(x^0) = \sum_{m,p,q=1}^{n} (\tilde{a}_{nn}\tilde{b}_{pq} - \tilde{a}_{mp}\tilde{b}_{qn})u_{y_m}u_{y_py_q}.$$

Due to the boundary condition $u|_S = 0$ we have $u_{y_m} = 0$, $m = 1, \ldots, n-1$, and so the summation in m may be eliminated and the index m may be replaced by n. The terms with $p = n$ are cancelled and only two sums remain:

$$I_1(x^0) = \sum_{p=1}^{n-1} (\tilde{a}_{nn}\tilde{b}_{pn} - \tilde{a}_{np}\tilde{b}_{nn})\frac{1}{2}\frac{\partial}{\partial y_p}(u_{y_n}^2) \equiv \sum_{p=1}^{n-1} A_p \frac{\partial}{\partial y_p}(u_{y_n}^2)$$

and

$$I_2(x^0) = \sum_{p,q=1}^{n-1} (\tilde{a}_{nn}\tilde{b}_{pq} - \tilde{a}_{np}\tilde{b}_{qn})u_{y_n}u_{y_py_q},$$

where $\tilde{a}_{pq} = a_{ij}c_{pi}c_{qj}$, $\tilde{b}_{pq} = b_{ij}c_{pi}c_{qj}$. The sum $I_2(x^0)$ contains derivatives of second order only in tangential directions. As shown in (6.16) they can be expressed in terms of the derivative u_{y_n} of the first order. The term $I_1(x^0)$ however, includes derivatives which cannot be expressed in terms of first-order derivatives of u at the point x^0. But they are the sums of derivatives of $u_{y_n}^2$ in tangential directions. Therefore, if S is a surface of the class C^2 and a_{ij} and b_{ij} are smooth functions than $\int_S I_1(s)\,ds$ may be transformed into an integral over S containing only $u_{y_n}^2$. For the latter we can use the inequality (6.24). If all calculations are carried out with a cutoff function ζ, $\zeta|_{S\setminus S_1} = 0$, and coordinates $y = y(x)$ are chosen so that S_1 has the equation $y_n = 0$, then the integral

$$J_1(S_1) = \int_{S_1} \sum_{p=1}^{n-1} \zeta^4 A_p \frac{\partial}{\partial y_p}(u_{y_n}^2)\,dy_1 \cdots dy_{n-1}$$

arises instead of $\int_S I_1(s)\,dS$. Routine integration by parts gives for $J_1(S_1)$ the expression

$$J_1(S_1) = -\int_{S_1} \sum_{p=1}^{n-1} \frac{\partial}{\partial y_p}(\zeta^4 A_p) u_{y_n}^2 \, dy_1 \cdots dy_{n-1}$$

since $\zeta|_{\partial S_1} = 0$.

Prove the inequalities (*) in §7 with cutoff functions ζ for inner subdomains and for subdomains bordering S and also the inequality (8.17) for whole of Ω using our hints.

14. Prove Theorem 7.1 in a different way according to the following plan:

(a) verify that for $\mathscr{L}_1$ and $\mathscr{M} = \mathscr{L}_0 - \lambda_0 I$ the inequality (8.17) holds with $\mu_3 = 0$ if λ_0 is sufficiently large;
(b) show that in $L_2(\Omega)$ there is no orthogonal complement to the range $R(\mathscr{L}_1)$ of $\mathscr{L}_1$ on $\mathscr{D}(\mathscr{L}_1) = \overset{\circ}{W}{}_2^2(\Omega)$;
(c) prove that $R(\mathscr{L}_1)$ is closed in $L_2(\Omega)$.

Therefore $R(\mathscr{L}_1) = L_2(\Omega)$ and the problem $\mathscr{L}_1 u = f$, $u|_S = 0$ is solvable in $\overset{\circ}{W}{}_2^2(\Omega)$ for every $f \in L_2(\Omega)$.

15. Show that the regularity of solutions has a local character (see the first part of the Remark 7.1 and 12 of this Supplement).
Hints: take a ball $K_\rho \subset \Omega$ in which the coefficients a_{ij} have the properties (6.2), the radius ρ being so small that the uniqueness theorem holds for the problem

$$\mathscr{L}v = \hat{f}, \qquad v|_{\partial K_\rho} = \varphi, \qquad x \in K_\rho. \tag{45}$$

Approximate the investigated generalized solution $u \in W_2^1(\Omega)$ of the equation $\mathscr{L}u = \hat{f}$, $\hat{f} = f + \sum_{i=1}^n \partial f_i/\partial x_i$ in the norm of $W_2^1(K_\rho)$ by smooth functions u_m and solve (45) with $\varphi = u_m|_{\partial K_\rho}$. Look for the solution v_m of the latter in the form $v_m = u_m - w_m$. Show that for v_m the estimates

$$\|v_m\|_{2,K_\rho}^{(1)} \le c, \qquad \|v_m \zeta\|_{2,K_\rho}^{(2)} \le c,$$

hold with constants c independent of m. In the second inequality, $\zeta \in \dot{C}^\infty(K_\rho)$. With the help of these inequalities pass to the limit $m \to \infty$ and make sure that the limit coincides with u in K_ρ.

Make a similar analysis in a small domain touching a smooth part of the boundary S. This domain should first be mapped onto a ball by a smooth transformation $y = y(x)$. This approach is realized in §3, Chapter III of the book [LU 1]. It allows one to observe the further increase of smoothness of solutions as the smoothness of $\mathscr{L}$, f, and S increases.

16. Another method of the investigation of regularity of solutions is based on their finite-difference approximations. It was proposed in my first monograph [4]. Let us explain it in brief for the problem (2.1), (2.2) under the conditions of Theorem 7.1. First, the equations of the problem are approximated by some difference equations $\mathscr{L}_h u_h = f_h$, $u_h|_{S_h} = 0$, which form a uniquely solvable linear algebraic system. From a solution u_h of this system the continuous function $u'_h \in \overset{\circ}{W}{}_2^1(\Omega)$ is constructed and then it is proved that for some $h_k \to 0$ the sequence $\{u'_{h_k}\}_{k=1}^{\infty}$ has a weak limit u in the space $\overset{\circ}{W}{}_2^1(\Omega)$. This limit u turns out to be the solution of the problem. This is shown in §7, Chapter VI of this book. The existence of the second derivatives for u, square integrable on $\overline{\Omega}' \subset \Omega$, follows from the estimate

$$h^n \sum_{\Omega'_h} u_{hx\bar{x}}^2 \le c(\Omega'), \tag{46}$$

where the constant $c(\Omega')$ does not depend on h. The proof of (46) requires some labor. Due to (46) the limiting function u will belong to $W_2^2(\Omega')$. To prove this in a subdomain Ω' adherent to the boundary S it is necessary to introduce new coordinates $y = y(x)(y(\cdot) \in C^2(\overline{\Omega}))$ in which $\partial\Omega' \cap S$ has the equation $y_n = 0$. It is necessary to rewrite the equation $\mathscr{L}u = f$ and use the previous difference scheme in the coordinates y. The corresponding algebraic system determines uniquely the mesh function $\hat{u}_h$. In the limit the functions $\hat{u}'_h(y)$ produce the function $\hat{u} \in \overset{\circ}{W}{}_2^1(\hat{\Omega})$, where $\hat{\Omega}$ is the image of Ω. By virtue of the uniqueness theorem, $\hat{u}(y(x))$ coincides with $u(x)$. The estimate $h^n \sum_{\hat{\Omega}'_h} \hat{u}_{hy\bar{y}}^2 \le c_1(\hat{\Omega}')$, with a constant $c_1(\hat{\Omega}')$ independent of h, is proved for $\hat{u}_h$. This estimate guarantees that $\hat{u} \in W_2^2(\hat{\Omega}')$ and therefore $u \in W_2^2(\Omega')$. The main step of this method is the derivation of the estimates (46).

17. Prove the Fredholm theorems for the problem (5.1), (5.2) in the space $W_2^2(\Omega)$. To this end, prove first the second inequality (see [LA 4] in which it is done for the boundary condition $\partial u/\partial l + \sigma u|_S = 0$ where $\mathbf{l}(\cdot)$ is a smooth non-tangential vector field on S) and after that use the method of §7 including the parameter τ in the equation $\mathscr{L}u = f$ and in the boundary equation $\partial u/\partial v + \sigma u|_S = 0$.

CHAPTER III

Equations of Parabolic Type

In this chapter we shall consider second-order parabolic equations and prove the unique solvability of the initial-boundary value problem in the domains $Q_T = \{(x, t): x \in \Omega,\ t \in (0, T)\}$ for the first, second, and third boundary conditions. We shall assume that the domain Ω is bounded, although all the results, except for the representation of solutions by Fourier series, will be valid for an arbitrary unbounded domain Ω. Moreover, the methods of solution given for bounded Ω are applicable to unbounded Ω (in particular, for $\Omega = R^n$), but need minor modification which we shall point out.

Basically, we shall carry out a detailed study of the first initial-boundary value problem, and we shall prove the unique solvability of it in the three spaces $\mathring{W}_2^{1,0}(Q_T)$, $\mathring{V}_2^{1,0}(Q_T)$, and $W_{2,0}^{2,1}(Q_T)$ under various assumptions about the data of the problem. All the theorems of §§2–3 are related in a distinctive way. This permits the maximal relaxation of restrictions on the data of the problem and also demonstrates the fruitfulness of taking into consideration different classes of generalized solutions of the same problem.† However, if we place restrictions on the coefficients of the equations which are somewhat more stringent than in Theorem 3.2, then we can prove the unique solvability of all the initial boundary value problems in the space $W_2^{1,0}(Q_T)$ rather briefly, as described in the second subsection of §4.

† We remark that the assumptions concerning the boundedness of the coefficients a_i, b_i, and a of the lower-order terms can be replaced by conditions requiring them to belong to spaces of the type $L_{q_i, r_i}(Q_T)$. This relaxation of requirements is possible within the framework of the method of investigation chosen here and leads to the same results on the unique solvability. However, we shall not present generalizations of this kind in this book, since we do not care to have purely technical complications, (which can be handled by using the embedding theorems of §§7–8 of Chapter I) obscure the basic line of thought.

In §§4 and 5 and in the Supplement we give a short description of some other methods. All the material in this chapter is taken from our early papers in which not only scalar equations, but also strong parabolic systems and abstract parabolic equations, were investigated (cf. [2, 4, 6, 8, 12]). We leave aside the potential theory, and its applications to the analysis of parabolic equations, since the traditional material for equations with constant coefficients is described in many textbooks. Deeper and sharper results on the solvability of the initial-boundary value problems for equations with variable coefficients in the spaces $W_p^{2l,l}(Q_T)$, $p > 1$, $l \geq 1$, $H^{2l+2\alpha,l+\alpha}(\bar{Q}_T)$, $l \geq 1$, and others are based on significantly more complicated analytic considerations (cf. the papers of Solonnikov [2, 6, 8, 9], the monographs of A. Friedman [1], Eidel'man [1], Ladyzhenskaya *et al.* [1]).

The Cauchy problem is not studied separately; for parabolic equations it has a number of specific peculiarities connected with the fact that the initial data and the solutions can have an extremely strong growth as $|x| \to \infty$ (see, for example, [GS 1], [EI 1], [FM 1]).

We shall return to parabolic equations in Chapter VI and describe yet another method of studying and solving them: the method of finite differences.

§1. Posing Initial-Boundary Value Problems and the Cauchy Problem

As is well known, the equation

$$\mathscr{M}u \equiv \sum_{i,j=1}^{n+1} a_{ij}(x)u_{x_ix_j} + \sum_{i=1}^{n+1} a_i u_{x_i} + au = f \tag{1.1}$$

is called parabolic at the point x^0 if, in the new coordinates $y_i = \alpha_{ij}(x_j - x_j^0)$ $(i = 1, \ldots, n+1)$, in which $a_{ij}(x^0)\alpha_{ki}\alpha_{lj} = \lambda_k(x^0)\delta_k^l$, it assumes the form

$$\sum_{k=1}^{n+1} \lambda_k(x^0)u_{y_ky_k} + \sum_{k=1}^{n+1} b_k(x^0)u_{y_k} + b(x^0)u = f(x^0), \tag{1.2}$$

at the point x^0 where one of the $\lambda_k(x^0)$ (say, $\lambda_{n+1}(x^0)$) is zero while the remaining $\lambda_k(x^0)$ have the same sign, and $b_{n+1}(x^0) \neq 0$. If we divide (1.2) by $b_{n+1}(x^0)$, we have an equation of the form

$$u_{y_{n+1}} + \sum_{k=1}^{n} \mu_k(x^0)u_{y_ky_k} + \sum_{k=1}^{n} \bar{b}_k(x^0)u_{y_k} + \bar{b}u = \bar{f}. \tag{1.3}$$

If $\mu_k(x^0) < 0$ $(k = 1, \ldots, n)$, then (1.3) has what is called the "standard form"; if $\mu_k(x^0) > 0$, then, by changing the direction of the y_{n+1}-axis and multiplying (1.3) by -1, we again obtain an equation in standard form. If

(1.1) is parabolic at all points x of some domain, then we say that it is parabolic in this domain. If the coefficients of $\mathscr{M}$ are smooth functions and if (1.1) is parabolic in the domain, then in a neighborhood (generally, a small neighborhood) of an arbitrary point of the region it can be reduced by a non-degenerate change of variables to the form

$$u_{y_{n+1}} - \sum_{i,j=1}^{n} b_{ij} u_{y_i y_j} + \sum_{i=1}^{n} b_i u_{y_i} + bu = \tilde{f}, \tag{1.4}$$

where the form $\sum_{i,j=1}^{n} b_{ij}\xi_i\xi_j$ is positive definite. The variable y_{n+1} plays an exceptional role. In describing heat (and a number of other) phenomena, this variable is none other than the time. Accordingly, we shall designate it by t; the remaining variables $y_1 \ldots, y_n$ describe the position of a point in space in physical problems. We shall consider parabolic equations which have already been reduced to the form (1.4). Moreover, it will be more convenient to begin our study of parabolic equations with equations of the form

$$\mathscr{M}u \equiv u_t - \sum_{i,j=1}^{n} \frac{\partial}{\partial x_i}(a_{ij}(x,t)u_{x_j} + a_i(x,t)u)$$
$$+ \sum_{i=1}^{n} b_i(x,t)u_{x_i} + a(x,t)u = f(x,t) + \frac{\partial f_i(x,t)}{\partial x_i}. \tag{1.5}$$

For differentiable a_{ij}, a_i, and f_i, (1.5) can be transformed in an elementary way to an equation of the form (1.4), and, conversely, for differentiable b_{ij}, (1.4) can be written in the form (1.5). We begin by studying (1.5), whose coefficients and external forces f_i are, in general, non-differentiable.

A special case of (1.5) is the heat equation

$$u_t - \sum_{i=1}^{n} \frac{\partial^2 u}{\partial x_i^2} = f(x,t), \tag{1.6}$$

which describes the flow of heat in some part of space $(x) \in R^n$. The following problems are fundamental for equation (1.5):

(1) The Cauchy problem in which we seek a function $u(x, t)$ satisfying (1.5) for $x \in R^n$ and $t > 0$, and satisfying, for $t = 0$, the initial condition

$$u|_{t=0} = \varphi(x). \tag{1.7}$$

(2) The first initial-boundary value problem, in which (1.5) is assumed given for $(x, t) \in \bar{Q}_T = \bar{\Omega} \times [0, T]$, where Ω is some domain in R^n. In this problem the function u must satisfy (1.5) in Q_T, the initial condition

$$u|_{t=0} = \varphi(x), \quad x \in \Omega, \tag{1.8}$$

and, for all $t \in [0, T]$, the boundary condition

$$u|_{x\in\partial\Omega} = \psi(s,t). \tag{1.9}$$

In R^{n+1} the domain Q_T is called, naturally, a cylinder, $S_T = S \times [0, T]$ its lateral surface, and the set $\{(x, t): x \in \Omega, t = 0\}$ its lower base. In these terms, the first initial-boundary value problem consists in determining a solution of (1.5) in the cylinder Q_T which coincides with the prescribed functions φ and ψ on the lower base of Q_T and on its lateral surface S_T. We can consider the more general case when the domain Ω_t of the variable x changes with time. In this case $Q_T = \{(x, t): x \in \Omega_t, t \in [0, T]\}$ will not be a cylinder (in any case, a "right cylinder"). However, if the "lateral surface" S_T is smooth and if the "speed" of variation of the $\partial\Omega_t$ is finite, then a change of variables can transform it into the lateral surface of a right cylinder (at least for small T) and reduce the problem to one with a fixed boundary $\partial\Omega$.

(3) The second and third initial-boundary value problems differ from the first only by the condition (1.9), which, for (1.5), is replaced either by the second boundary condition

$$\left.\frac{\partial u}{\partial N}\right|_{S_T} \equiv a_{ij}u_{x_j}\cos(\mathbf{n}, \mathbf{x}_i)|_{S_T} = \chi(s, t), \tag{1.10}$$

or by the third,

$$\left.\frac{\partial u}{\partial N} + \sigma u\right|_{S_T} = \chi(s, t), \tag{1.11}$$

respectively.

We shall consider in detail the first initial-boundary value problem in a bounded domain Ω. The second and third can be considered in a similar way. We impose the assumption of boundedness on Ω only for the sake of a little convenience. It is easy to get rid of this assumption, the results remaining the same for bounded and unbounded domains.

The object of this section of the lectures is a proof of the fact that the above problems are always uniquely solvable if their data are not too bad. We shall introduce here, as we did in Chapter II, various classes of generalized solutions. The solvability can be proved most simply in the Hilbert space $W_2^{1,0}(Q_T)$, where the scalar product is defined by

$$(u, v)_{2,Q_T}^{(1,0)} \equiv \int_{Q_T} (uv + u_x v_x)\,dx\,dt. \tag{1.12}$$

The space $W_2^{1,0}(Q_T)$ consists of all elements $u \in L_2(Q_T)$ having generalized derivatives u_{x_i} in $L_2(Q_T)$. We denote the norm in $W_2^{1,0}(Q_T)$ by $\|\cdot\|_{2,Q_T}^{(1,0)}$. We denote by $\overset{\circ}{W}{}_2^{1,0}(Q_T)$ the subspace of $W_2^{1,0}(Q_T)$ which is the closure in the norm $\|\cdot\|_{2,Q_T}^{(1,0)}$ of the set of smooth functions which vanish in the vicinity of S_T. The space $\overset{\circ}{W}{}_2^{1,0}(Q_T)$ is a proper subspace of $W_2^{1,0}(Q_T)$, which is obvious from the formula of integration by parts,

$$\int_{Q_T} u_{x_i} v\,dx\,dt = -\int_{Q_T} uv_{x_i}\,dx\,dt, \tag{1.13}$$

which holds for arbitrary smooth v and for any smooth u vanishing near S_T. Actually, under closure in the norm of $W_2^{1,0}(Q_T)$, (1.13) is still valid for all $v \in W_2^{1,0}(Q_T)$ and all $u \in \overset{\circ}{W}{}_2^{1,0}(Q_T)$. If neither u nor v vanishes on S_T, then (1.13) is not true in general, so that (1.13) does not hold in all of $W_2^{1,0}(Q_T)$. This reasoning shows that the elements $u \in \overset{\circ}{W}{}_2^{1,0}(Q_T)$ vanish on S_T in a well-defined sense. If the boundary S is smooth, then the embedding theorem 6.5 of Chapter I (see Remark 6.3) guarantees that the integral $\int_{S_{T'}} u^2\, ds\, dt$ tends to zero as $S'_T \equiv \partial\Omega' \times [0, T] \to S_T$.

With parabolic equations of the form (1.5) is closely connected the "energy" norm defined by

$$|u|_{Q_T} = \operatorname*{ess\,sup}_{0 \le t \le T} \|u(\cdot, t)\|_{2,\Omega} + \|u_x\|_{2,Q_T} \tag{1.14}$$

In the next section we shall explain this by showing that the fundamental relation of the "balance of energy" implies an *a priori* bound for the norm $|u|_{Q_T}$ of the solution of initial value problems in terms of the data of the problem. Here we shall define Banach spaces whose norm is defined by (1.14). The largest of these, $V_2(Q_T)$, consists of all elements of $W_2^{1,0}(Q_T)$ having finite norm $|\cdot|_{Q_T}$. Its subspace $\overset{\circ}{V}_2(Q_T)$ consists of elements of $\overset{\circ}{W}{}_2^{1,0}(Q_T)$ having finite norm $|\cdot|_{Q_T}$.

Another subspace of $V_2(Q_T)$ is the space $V_2^{1,0}(Q_T)$ consisting of all elements $u \in V_2(Q_T)$ which are strongly continuous in t in the norm of $L_2(\Omega)$, that is, such that $\|u(\cdot, t + \Delta t) - u(\cdot, t)\|_{2,\Omega} \to 0$ as $\Delta t \to 0$, uniformly on the interval $[0, T]$. Finally, $\overset{\circ}{V}{}_2^{1,0}(Q_T)$ is the intersection of $V_2^{1,0}(Q_T)$ and $\overset{\circ}{W}{}_2^{1,0}(Q_T)$.† In the second half of §3 we shall prove that any generalized solution of (1.5) in $\overset{\circ}{W}{}_2^{1,0}(Q_T)$ (under the same hypotheses by which we established its existence at the beginning of §3) is, in fact, in $\overset{\circ}{V}{}_2^{1,0}(Q_T)$ and that the energy-balance relation holds for it. This will allow us to prove a uniqueness theorem for the first initial-boundary value problem in $W_2^{1,0}(Q_T)$ under the same hypotheses on the coefficients in (1.5) which are needed for the energy-balance relation and the existence theorem for generalized solutions in $W_2^{1,0}(Q_T)$. In order to realize our projected program, we first study in detail the heat equation.

In concluding this section, we shall prove a well-known lemma which can be used to obtain *a priori* bounds for the solutions of various non-stationary problems.

† In modern terminology:

$$W_2^{1,0}(Q_T) = L_2((0,T); W_2^1(\Omega)); \ \overset{\circ}{W}{}_2^{1,0}(Q_T) = L_2((0,T); \overset{\circ}{W}{}_2^1(\Omega));$$

$$V_2(Q_T) = L_\infty((0,T); L_2(\Omega)) \cap W_2^{1,0}(Q_T);$$

$$\overset{\circ}{V}_2(Q_T) = V_2(Q_T) \cap \overset{\circ}{W}{}_2^{1,0}(Q_T);$$

$$V_2^{1,0}(Q_T) = C([0,T]; L_2(\Omega)) \cap W_2^{1,0}(Q_T);$$

$$\overset{\circ}{V}{}_2^{1,0}(Q_T) = C([0,T]; L_2(\Omega)) \cap \overset{\circ}{W}{}_2^{1,0}(Q_T).$$

Lemma 1.1. *Let $y(t)$ be non-negative and absolutely continuous on $[0, T]$, and for almost all t in $[0, T]$ satisfy the inequality*

$$\frac{dy(t)}{dt} \leq c_1(t)y(t) + c_2(t), \tag{1.15}$$

where the $c_i(t)$ are summable, non-negative functions on $[0, T]$. Then

$$y(t) \leq \exp\left\{\int_0^t c_1(\tau)\,d\tau\right\} \cdot \left[y(0) + \int_0^t c_2(\xi) \exp\left(-\int_0^\xi c_1(\tau)\,d\tau\right) d\xi\right]$$
$$\leq \exp\left\{\int_0^t c_1(\tau)\,d\tau\right\}\left[y(0) + \int_0^t c_2(\tau)\,d\tau\right]. \tag{1.16}$$

In fact, if we multiply (1.15) by $\exp\{-\int_0^t c_1(\tau)\,d\tau\}$, we can write the result in the form

$$\frac{d}{dt}\left[y \exp\left(-\int_0^t c_1(\tau)\,d\tau\right)\right] \leq c_2(t) \exp\left(-\int_0^t c_1(\tau)\,d\tau\right)$$

and if we integrate from 0 to t, then (1.16) follows in an obvious way from the inequality which we have obtained. If $c_1(t) = c_1 = \text{const.} > 0$ and $c_2(\cdot)$ is a non-decreasing function on t then from (1.15) and (1.16) we have the following inequalities:

$$y'(t) \leq e^{c_1 t}[c_1 y(0) + c_2(t)],$$
$$y(t) \leq e^{c_1 t} y(0) + c_1^{-1} c_2(t)[e^{c_1 t} - 1]. \tag{1.17}$$

§2. First Initial-Boundary Value Problem for the Heat Equation

For the equation

$$\mathscr{M}_0 u \equiv u_t - \Delta u = f + \frac{\partial f_i}{\partial x_i} \tag{2.1}$$

let us consider the problem of finding its solution $u(x, t)$ in the bounded domain $Q_T = \Omega \times (0, T)$ which satisfies the initial condition

$$u|_{t=0} = \varphi(x) \tag{2.2}$$

and the boundary condition

$$u|_{S_T} = 0. \tag{2.3}$$

We introduce and study certain classes of generalized solutions for the problem (2.1)–(2.3). We begin with the comparatively smooth generalized solutions. For S in C^2 the generalized solutions of this class are elements of the Hilbert space $W_2^{2,1}(Q_T)$, whose scalar product is defined by

$$(u, v)_{2,Q_T}^{(2,1)} \equiv \int_{Q_T} (uv + u_t v_t + u_x v_x + u_{xx} v_{xx})\, dx\, dt.\dagger \tag{2.4}$$

More precisely, *a generalized solution of the problem* (2.1)–(2.3) *in this class* belongs to the subspace $W_{2,0}^{2,1}(Q_T) = W_2^{2,1}(Q_T) \cap \overset{\circ}{W}{}_2^{1,0}(Q_T)$. If the boundary S is not smooth, then the generalized derivatives u_{xx} for these generalized solutions will not generally be square-summable on Q_T, but the Laplace operator will be defined for them, and Δu will be in $L_2(Q_T)$.

Accordingly, we introduce the Hilbert space $W_2^{\Delta,1}(Q_T)$, whose elements $u(x, t)$ are in $L_2(Q_T)$ along with u_t and u_x, and have in Q_T generalized derivatives u_{xx} and the finite norm

$$\left(\int_{Q_T} [u^2 + u_t^2 + u_x^2 + (\Delta u)^2]\, dx\, dt \right)^{1/2}. \tag{2.5}$$

The scalar product in $W_2^{\Delta,1}(Q_T)$ is defined by

$$\int_{Q_T} (uv + u_t v_t + u_x v_x + \Delta u\, \Delta v)\, dx\, dt. \tag{2.6}$$

We shall be interested only in the subspace $W_{2,0}^{\Delta,1}(Q_T) \equiv W_2^{\Delta,1}(Q_T) \cap \overset{\circ}{W}{}_2^{1,0}(Q_T)$. Instead of (2.5) we find it more convenient to take the equivalent norm

$$\|u\|_{2,Q_T}^{(\Delta,1)} = \left(\int_{Q_T} [u_t^2 + (\Delta u)^2]\, dx\, dt \right)^{1/2} \tag{2.7}$$

and the corresponding scalar product

$$(u, v)_{2,Q_T}^{(\Delta,1)} = \int_{Q_T} (u_t v_t + \Delta u\, \Delta v)\, dx\, dt. \tag{2.8}$$

For $S \in C^2$ the space $W_{2,0}^{2,1}(Q_T)$ coincides with $W_{2,0}^{\Delta,1}(Q_T)$ (this follows from the results of §§6 and 7, Chapter II). We shall investigate the problem (2.1)–(2.3) basically for an arbitrary bounded domain Ω with non-smooth boundary S.

Let first the free term $f + \partial f_i/\partial x_i \equiv F$ be in $L_2(Q_T)$.

By *a generalized solution of the problem* (2.1)–(2.3) *in* $W_2^{\Delta,1}(Q_T)$ we naturally understand an element u of the space $W_{2,0}^{\Delta,1}(Q_T)$ satisfying (2.1) almost everywhere in Q_T and equal to $\varphi(x)$ for $t = 0$. The latter condition can be understood as $\|u(\cdot, t) - \varphi(\cdot)\|_{2,\Omega} \to 0$ as $t \to 0$. This is meaningful for elements of $W_{2,0}^{\Delta,1}(Q_T)$ since they are defined for all $t \in [0, T]$ as elements of $L_2(\Omega)$ (and even as elements of $\overset{\circ}{W}{}_2^1(\Omega)$, as we shall see below) and are continuous with respect to t in the norm of $L_2(\Omega)$ (and even in the norm of

† Here, as everywhere, $u_x v_x = \sum_{i=1}^n u_{x_i} v_{x_i}$; $u_x^2 = u_x u_x$; $u_{xx} v_{xx} = \sum_{i,j=1}^n u_{x_i x_j} v_{x_i x_j}$; $u_{xx}^2 = u_{xx} u_{xx}$.

$\mathring{W}_2^1(\Omega)$). We reformulate the problem (2.1)–(2.3) with $f + \partial f_i/\partial x_i = F$ as the problem of solving the operator equation

$$Au = \{F; \varphi\}, \tag{2.9}$$

where A is the operator which maps $u(x, t)$ to the pair of elements $\mathscr{M}_0 u(x, t)$ and $u(x, 0)$, so that

$$Au = \{\mathscr{M}_0 u; u(\cdot, 0)\}. \tag{2.10}$$

We consider A as an unbounded operator acting from the space $L_2(Q_T)$ into the Hilbert space $\mathscr{W}$ which is the direct product of $L_2(Q_T)$ and $\mathring{W}_2^1(\Omega)$. The elements of $\mathscr{W}$ are the pairs $\{f; \psi\}$ with $f \in L_2(Q_T)$ and $\psi \in \mathring{W}_2^1(\Omega)$, and the scalar product is defined by

$$(\{f'; \psi'\}, \{f''; \psi''\})_{\mathscr{W}} = \int_{Q_T} f'f''\, dx\, dt + \int_\Omega \psi'_x \psi''_x\, dx. \tag{2.11}$$

As the domain of definition $\mathscr{D}(A)$ of A we take the elements of the form $\psi(x) + \int_0^t \chi(x, \tau)\, d\tau$, where $\psi \in \mathscr{D}(\Delta)$, $\chi(\cdot, t) \in \mathscr{D}(\Delta)$ for almost all t in $[0, T]$ and $\Delta\chi \in L_2(Q_T)$. Here, by $\mathscr{D}(\Delta)$ we mean the set of generalized solutions in $W_2^1(\Omega)$ of the problem

$$\Delta u = f(x), \qquad u|_S = 0, \tag{2.12}$$

as f ranges over all of $L_2(\Omega)$. From the results of Chapter II follow these facts: If $f(x) = \hat{f}(x, t)$, $\hat{f} \in L_2(Q_T)$, then the solution $\hat{u}(x, t)$ of (2.12) is in $L_2(Q_T)$ along with $\hat{u}_x$; the derivatives $\hat{u}_{xx}$ exist and are square-summable over $Q'_T = \Omega' \times (0, T)$ for all $\overline{\Omega}' \subset \Omega$; for $\hat{u}$ and for all $v \in W_2^{1,0}(Q_T)$ we have

$$\int_{Q_T} \Delta\hat{u} \cdot v\, dx\, dt = -\int_{Q_T} \hat{u}_x v_x\, dx\, dt, \tag{2.13}$$

and the equation $\Delta\hat{u} = \sum_{i=1}^n \hat{u}_{x_i x_i} = \hat{f}$ is satisfied for almost all $(x, t) \in Q_T$.

Moreover, if $f = \int_0^t \hat{f}(x, \tau)\, d\tau$ then the solution $u(x, t)$ of (2.12) is equal to $\int_0^t \hat{u}(x, \tau)\, d\tau$ (so that

$$\Delta u(x, t) = \Delta \int_0^t \hat{u}(x, \tau)\, d\tau = \int_0^t \Delta\hat{u}(x, \tau)\, d\tau = \int_0^t \hat{f}(x, \tau)\, d\tau),$$

and $u(\cdot, t)$ will be an element of $\mathring{W}_2^1(\Omega)$ which is continuous in t (in the norm of that space).

In view of this, the elements $v(x, t) = \psi(x) + \int_0^t \chi(x, \tau)\, d\tau$ forming $\mathscr{D}(A)$ are in $\mathscr{D}(\Delta)$ for all $t \in [0, T]$, $\Delta v = \Delta\psi + \int_0^t \Delta\chi\, d\tau$, $v_x = \psi_x + \int_0^t \chi_x\, d\tau$ are elements of $C([0, T]; L_2(\Omega))$, and $v_{xt} \in L_2(Q_T)$. The operator A on $v(x, t)$ can be written as

$$Av = \left\{\chi - \Delta\psi - \int_0^t \Delta\chi\, d\tau; \psi\right\}. \tag{2.14}$$

It is easy to see that the set $\mathscr{D}(A)$ is dense in $L_2(Q_T)$.

We shall prove that A admits a closure. For this we must either show that it follows from a known theorem in the theory of unbounded operators that the operator A^* adjoint to A is defined on a dense set, or else verify directly the following assertion: If $v_m \in \mathscr{D}(A)$ $(m = 1, 2, \dots)$, if $v_m \to 0$ in the norm of $L_2(Q_T)$, and if $Av_m \equiv \{f_m, \varphi_m\} \to \{f, \varphi\}$ in the norm of $\mathscr{W}$, then $f \equiv \varphi \equiv 0$. We shall prove the latter assertion. For this we take an arbitrary, sufficiently smooth function $\eta(x, t)$ which is zero on S_T, with $\eta(x, T) = 0$, and consider the corresponding integral $\int_{Q_T} \mathscr{M}_0(v_m)\eta\, dx\, dt$. We integrate it by parts and obtain

$$\int_{Q_T} \mathscr{M}_0(v_m)\eta\, dx\, dt = \int_{Q_T} v_m \mathscr{M}_0^*(\eta)\, dx\, dt - \int_\Omega v_m \eta\, dx|^{t=0}.$$

We can take the limit as $m \to \infty$ to obtain

$$\int_{Q_T} f\eta\, dx\, dt = -\int_\Omega \varphi\eta(x, 0)\, dx$$

for any $\eta(x, t)$ with the properties indicated above. Hence, by a well-known method, we can conclude that f and φ are identically zero, so that the operator A admits a closure $\bar{A}$. In order to describe the domain of definition $\mathscr{D}(\bar{A})$ and to calculate $\bar{A}$ on the elements of $\mathscr{D}(\bar{A})$, we shall prove that for $\mathscr{M}_0$ the equality

$$\|v_x(\cdot, t)\|_{2,\Omega}^2 + \int_{Q_t} (v_t^2 + (\Delta v)^2)^2\, dx\, dt = \|v_x(\cdot, 0)\|_{2,\Omega}^2 + \int_{Q_t} (\mathscr{M}_0 v)^2\, dx\, dt \tag{2.15}$$

is true. Here $v(x, t)$ is an arbitrary element of $\mathscr{D}(A)$ and t is any number in $[0, T]$. We remark that for general parabolic operators there is an analogous equality from which we can derive a so-called "second fundamental inequality" similar to the second fundamental inequality for elliptic operators (cf. subsection 4, §4 below).

The validity of (2.15) is clear from the relation

$$\int_{Q_t} (\mathscr{M}_0 v)^2\, dx\, dt = \int_{Q_t} \left[v_t^2 + (\Delta v)^2 + \frac{\partial v_x^2}{\partial t} \right] dx\, dt$$

$$= \int_{Q_t} [v_t^2 + (\Delta v)^2]\, dx\, dt + \int_\Omega v_x^2\, dx \Big|_{t=0}^{t=t}.$$

It is clear from (2.15) that the convergence of Av_m $(v_m \in \mathscr{D}(A))$ in $\mathscr{W}$ implies the convergence of v_m in the norm of $W_2^{\Delta,1}(Q_T)$ and in the norm $\sup_{0 \le t \le T} \|\cdot\|_{2,\Omega}^{(1)}$. This proves that the elements of the domain $\mathscr{D}(\bar{A})$ of the closure $\bar{A}$ of A are "not much worse" than the elements of $\mathscr{D}(A)$, namely, they are in $W_{2,0}^{\Delta,1}(Q_T)$, they depend continuously on t in the norm of $\overset{\circ}{W}{}_2^1(\Omega)$, and they satisfy (2.15). The operator $\bar{A}$ is defined on them by the same equality (2.10), that is,

$$\bar{A}v = \{\mathscr{M}_0 v; v(\cdot, 0)\}. \tag{2.16}$$

Finally, it follows from (2.15) that the closure procedure for A reduces to the closure of the range $\mathscr{R}(A)$ in $\mathscr{W}$, so that $\overline{\mathscr{R}(A)} = \mathscr{R}(\bar{A})$, and a bounded inverse $\bar{A}^{-1}$ exists on $\mathscr{R}(\bar{A})$. If we prove that $\mathscr{R}(A)$ does not have an orthogonal complement in $\mathscr{W}$, then it will follow from what has just been said about $\bar{A}$ that $\bar{A}^{-1}$ is defined on all of $\mathscr{W}$, that is, equation (2.9), or more precisely,

$$\bar{A}u = \{F; \varphi\} \tag{2.17}$$

is uniquely solvable for all $\{F; \varphi\} \in \mathscr{W}$. By the same token, we shall have proved

Theorem 2.1. *For any bounded domain Ω the problem (2.1)–(2.3) is uniquely solvable in $W_{2,0}^{\Delta,1}(Q_T)$ if $F \equiv f + \partial f_i/\partial x_i \in L_2(Q_T)$ and $\varphi \in \mathring{W}_2^1(\Omega)$. The solution $u(x, t)$ depends continuously on t in the norm of $\mathring{W}_2^1(\Omega)$.*

Thus it remains for us to prove that $\mathscr{R}(A)$ has no orthogonal complement in $\mathscr{W}$, i.e., to show that it follows from the identity

$$\int_{Q_T} w(v_t - \Delta v)\, dx\, dt + \int_\Omega \psi_x v_x(x, 0)\, dx = 0, \tag{2.18}$$

that $w \equiv 0$ and $\psi \equiv 0$, where v is an arbitrary element of $\mathscr{D}(A)$ and $\{w; \psi\} \in \mathscr{W}$. For this we take $v(x, t)$ equal to $\int_{t_1}^t \Delta^{-1} w(x, \tau)\, d\tau$ for $t \geq t_1$ and zero for $t < t_1$, where t_1 is some number in $[0, T]$. It is easy to see that such a v belongs to $\mathscr{D}(A)$. If we substitute it into (2.18), we obtain $\int_{t_1}^T \int_\Omega \Delta v_t(v_t - \Delta v)\, dx\, dt = 0$, whence it follows that

$$-\int_{t_1}^T \int_\Omega v_{xt}^2\, dx\, dt - \frac{1}{2} \int_\Omega (\Delta v)^2\, dx \Big|_{t=t_1}^{t=T} = 0. \tag{2.19}$$

Since $\Delta v|_{t=t_1} = 0$ and t_1 is arbitrary, it follows from (2.19) that $v_{xt} \equiv 0$ in Q_T, so that $w = \Delta v_t \equiv 0$. Thus the identity (2.18) takes the form

$$\int_\Omega \psi_x v_x(x, 0)\, dx = 0 \tag{2.20}$$

for any $v(x, 0)$ in $\mathscr{D}(\Delta)$. Since $\psi \in \mathring{W}_2^1(\Omega)$ and $\mathscr{D}(\Delta)|_{t=0}$ is dense in $\mathring{W}_2^1(\Omega)$, it follows that $\psi \equiv 0$, and so $\overline{\mathscr{R}(A)} = \mathscr{W}$. This proves Theorem 2.1. Simultaneously, we have proved essentially the uniqueness theorem for the problem (2.1)–(2.3) in the *class of generalized solutions in $L_2(Q_T)$*. By such a solution we shall mean an element $u \in L_2(Q_T)$ satisfying the identity

$$\int_{Q_T} u(\eta_t + \Delta\eta)\, dx\, dt + \int_\Omega \varphi\eta(x, 0)\, dx = \int_{Q_T} (-f\eta + f_i \eta_{x_i})\, dx\, dt \tag{2.21}$$

for all $\eta \in W_{2,0}^{\Delta,1}(Q_T)$ which vanishes for $t = T$.

If u is a generalized solution in $L_2(Q_T)$ of the problem (2.1)–(2.3) with $f = f_i = \varphi \equiv 0$, i.e., if (2.21) holds for u with $f = f_i = \varphi = 0$, then it follows from this identity that $u \equiv 0$. In fact, when t is replaced by $-t$, this identity is transformed into an identity of the form (2.18) with $\psi \equiv 0$. Here u plays the role of w and η that of v, where the set of η in (2.21) is even bigger than that of v in (2.18). In view of this and of the conclusion, from (2.18), that w vanishes, it follows that $u \equiv 0$ if φ, f, and f_i are zero in (2.21). Thus we have proved

Theorem 2.2. *The problem* (2.1)–(2.3) *cannot have more than one generalized solution in* $L_2(Q_T)$.

Remark 2.1. If $S \in C^2$, then, by what we proved in §§6–7, Chapter II, it follows from the fact that Δu is in $L_2(\Omega)$ and u is in $\mathring{W}_2^1(\Omega)$ that $u \in W_{2,0}^2(\Omega)$. So, for $S \in C^2$ the space $W_{2,0}^{\Delta,1}(Q_T)$ coincides with the space $W_{2,0}^{2,1}(Q_T)$, and Theorem 2.1 guarantees the solvability of the problem (2.1)–(2.3) in $W_{2,0}^{2,1}(Q_T)$.

Remark 2.2. Any element u of $W_{2,0}^{\Delta,1}(Q_T)$ belongs to $\mathring{W}_2^1(\Omega)$ for all $t \in [0, T]$ and depends continuously on t in the norm of $\mathring{W}_2^1(\Omega)$. Moreover, $\|u_x(\cdot, t)\|_{2,\Omega}^2$ is an absolutely continuous function of t on $[0, T]$, and

$$\|u_x(\cdot, t)\|_{2,\Omega}^2 = \|u_x(\cdot, 0)\|_{2,\Omega}^2 - 2\int_0^t (u_\tau(\cdot, \tau), \Delta u(\cdot, \tau))\, d\tau.$$

Actually, these properties hold for elements u of $W_{2,0}^{\Delta,1}(Q_T)$ which have the derivatives $u_{tx} \in L_2(Q_T)$, and the totality $\mathfrak{M}$ of all such functions is dense in $W_{2,0}^{\Delta,1}(Q_T)$. These properties still hold under the closure of $\mathfrak{M}$ in the Banach norm

$$\|u\|_{W_T} \equiv \left[\sup_{0 \le t \le T} \|u_x(\cdot, t)\|_{2,\Omega}^2 + \int_0^T \int_\Omega (u_t^2 + (\Delta u)^2)\, dx\, dt\right]^{1/2}.$$

But this norm is equivalent on $\mathfrak{M}$ to the norm $\|\cdot\|_{2,Q_T}^{(\Delta,1)}$ since, for all $u \in \mathfrak{M}$ and for any smooth function $\zeta(t)$

$$\left|\|u_x(\cdot, t)\zeta(t)\|_{2,\Omega}^2 - \|u_x(\cdot, t_1)\zeta(t_1)\|_{2,\Omega}^2\right|$$

$$= \left|\int_{t_1}^t \frac{d}{d\tau}\|u_x(\cdot, \tau)\zeta(\tau)\|_{2,\Omega}^2\, d\tau\right|$$

$$= 2\left|\int_{t_1}^t \int_\Omega (-\Delta u \cdot u_\tau \zeta^2 + u_x^2 \zeta\zeta')\, dx\, dt\right|$$

$$\le c(\max|\zeta|^2 + \max|\zeta'|^2)\int_{t_1}^t \int_\Omega [u_\tau^2 + (\Delta u)^2]\, dx\, dt,$$

whence it follows that

$$\sup_{0 \le t \le T} \|u_x(\cdot, t)\|_{2,\Omega} \le c\|u\|_{2,Q_T}^{(\Delta,1)}.$$

Let us now consider the problem (2.1)–(2.3) for $\varphi \in L_2(\Omega)$, $f_i \in L_2(Q_T)$, and $f \in L_{2,1}(Q_T)$. The space $L_{q,r}(Q_T)$ consists of all elements of $L_1(Q_T)$ with the finite norm

$$\|u\|_{q,r,Q_T} = \left(\int_0^T \left(\int_\Omega |u|^q \, dx\right)^{r/q} dt\right)^{1/r}.$$

We shall show that in this case the solutions belong to the space $\mathring{V}_2^{1,0}(Q_T)$ and satisfy the energy relation (the energy-balance equation),

$$\tfrac{1}{2}\|u(\cdot, t)\|_{2,\Omega}^2 + \|u_x\|_{2,Q_t}^2 = \int_{Q_t} (fu - f_i u_{x_i}) \, dx \, dt + \tfrac{1}{2}\|u(x, 0)\|_{2,\Omega}^2 \tag{2.22}$$

for all $t \in [0, T]$. We shall call such solutions *generalized solutions of the problem* (2.1)–(2.3) *in the class* $\mathring{V}_2^{1,0}(Q_T)$, or, what is the same thing, *generalized solutions in the energy class.* These solutions satisfy the equation (2.1) and the initial condition (2.2) in the sense of the integral identity

$$\int_\Omega u(x, t)\eta(x, t) \, dx - \int_\Omega \varphi\eta(x, 0) \, dx + \int_{Q_t} (-u\eta_t + u_x\eta_x) \, dx \, dt$$

$$= \int_{Q_t} (f\eta - f_i\eta_{x_i}) \, dx \, dt, \tag{2.23}$$

which should be fulfilled for all $t \in [0, T]$ and all $\eta \in W_{2,0}^1(Q_T)$. The fact that $u \in \mathring{V}_2^{1,0}(Q_T)$ gives that the boundary condition is satisfied. The space $W_{2,0}^1(Q_T)$ consists of all elements $v \in \mathring{W}_2^{1,0}(Q_T)$ with v_t in $L_2(Q_T)$; the scalar product in $W_{2,0}^1(Q_T)$ is defined by $(u, v)_{2,Q_T}^{(1,1)} = \int_{Q_T}(uv + u_t v_t + u_x v_x) \, dx \, dt$, and the norm is denoted by $\|\cdot\|_{2,Q_T}^{(1,1)}$.

The identity (2.23) can be obtained formally by integrating by parts the relation

$$\int_{Q_t} \mathcal{M}_0 u \cdot \eta \, dx \, dt = \int_{Q_t} \left(f + \frac{\partial f_i}{\partial x_i}\right)\eta \, dx \, dt \tag{2.24}$$

and using (2.2) and the fact that η is in $W_{2,0}^1(Q_T)$. The energy equation can be obtained from the equality

$$\int_{Q_t} \mathcal{M}_0 u \cdot u \, dx \, dt = \int_{Q_t} \left(f + \frac{\partial f_i}{\partial x_i}\right) u \, dx \, dt \tag{2.25}$$

by integrating by parts and using the fact that u vanishes on S_T. However, (2.24) and (2.25) presupposes the existence of u_t and Δu in $L_2(Q_T)$, whereas (2.23) and (2.22) have meaning for any element in $\mathring{V}_2^{1,0}(Q_T)$. Consequently, the definition of a generalized solution of (2.1)–(2.3) in $\mathring{V}_2^{1,0}(Q_T)$ which we introduced is an extension of the notion of a generalized solution of this problem in $W_{2,0}^{\Delta,1}(Q_T)$, and, as is easy to see, a restriction of the notion of a generalized solution in $L_2(Q_T)$. There is a peculiarity in this definition which distinguishes it from the definitions of generalized solution given in Chapter II and in the present chapter, namely, it is not immediately obvious from the definition that if u' is a generalized solution of (2.1)–(2.3) in $\mathring{V}_2^{1,0}(Q_T)$ with $\varphi = \varphi', f = f', f_i = f'_i$, and if u'' is some other solution with $\varphi = \varphi'', f = f''$, and $f_i = f''_i$ then $u' + u''$ is a generalized solution of (2.1)–(2.3) in $\mathring{V}_2^{1,0}(Q_T)$ corresponding to $\varphi = \varphi' + \varphi'', f = f' + f''$, and $f_i = f'_i + f''_i$. The reason for this is the non-linearity of the relation (2.22), which by itself does not imply that (2.22) holds for $u' + u''$ whenever it holds for u' and u''. However, subsequent arguments, along with Theorem 2.1, will enable us to show that (2.22) holds for $u' + u''$. The work spent on the proof of this fact will be compensated for in the following section, where we shall study equations with variable coefficients and where (2.22) will be used in an essential way.

From (2.22) we can derive an *a priori* bound for the "energy norm" of the solution, that is, for the norm $|u|_{Q_T}$. To do this, we majorize the right-hand side of (2.22) by the quantity

$$\max_{0 \le \tau \le t} \|u(\cdot, \tau)\|_{2,\Omega} \|f\|_{2,1,Q_t} + \|\mathbf{f}\|_{2,Q_t}\|u_x\|_{2,Q_t} + \tfrac{1}{2}\|u(\cdot, 0)\|_{2,Q}^2,$$

where $\mathbf{f} = (f_1, \ldots, f_n)$, $\|\mathbf{f}\|_{2,Q_t} = (\int_{Q_t} \sum_{i=1} f_i^2 \, dx \, dt)^{1/2}$, and we derive from (2.22) two consequences:

$$\max_{0 \le \tau \le t} \|u\|_{2,\Omega}^2 \le 2 \max_{0 \le \tau \le t} \|u\|_{2,\Omega} \|f\|_{2,1,Q_t} + 2\|\mathbf{f}\|_{2,Q_t}\|u_x\|_{2,Q_t} + \|u(\cdot, 0)\|_{2,\Omega}^2, \tag{2.26}$$

and

$$\|u_x\|_{2,Q_t}^2 \le \max_{0 \le \tau \le t} \|u\|_{2,\Omega}\|f\|_{2,1,Q_t} + \|\mathbf{f}\|_{2,Q_t}\|u_x\|_{2,Q_t} + \tfrac{1}{2}\|u(\cdot, 0)\|_{2,\Omega}^2. \tag{2.27}$$

Let us take the square roots of both sides of (2.26) and (2.27), and add the results. Then, on the right-hand side, we replace $\max_{0 \le \tau \le t} \|u\|_{2,\Omega}$, $\|u_x\|_{2,Q_t}$ and $\|u(\cdot, 0)\|_{2,\Omega}$ by the larger quantity $|u|_{Q_t}$ and cancel $|u|_{Q_t}^{1/2}$ from the inequality obtained. If we square this inequality, we have the desired bound

$$|u|_{Q_t} \le (1 + \sqrt{2})^2(\|f\|_{2,1,Q_t} + \|\mathbf{f}\|_{2,Q_t} + \tfrac{1}{2}\|u(\cdot, 0)\|_{2,\Omega}). \tag{2.28}$$

We shall now prove the following existence theorem

Theorem 2.3. *The problem* (2.1)–(2.3) *has a generalized solution in* $\mathring{V}_2^{1,0}(Q_T)$ *for* $\varphi \in L_2(\Omega)$, $f \in L_{2,1}(Q_T)$, *and* $f_i \in L_2(Q_T)$.

The uniqueness in $\mathring{V}_2^{1,0}(Q_T)$ is a consequence of Theorem 2.2. For the proof of existence we approximate φ in the norm of $L_2(\Omega)$ by the functions φ_m $(m = 1, 2, \ldots)$ from $\mathring{W}_2^1(\Omega)$, f in the norm of $L_{2,1}(Q_T)$ by the functions f_m $(m = 1, 2, \ldots)$ from $L_2(Q_T)$ and f_i in the norm of $L_2(Q_T)$ by the functions

f_{im} $(m = 1, 2, \ldots)$ from $W_2^{1,0}(Q_T)$. By Theorem 2.1 the problem (2.1)–(2.3) has a solution u_m in $W_{2,0}^{\Delta,1}(Q_T)$ corresponding to φ_m and $F_m = f_m + \partial f_{im}/\partial x_i$. The differences $u_m - u_p$ are solutions in $W_{2,0}^{\Delta,1}(Q_T)$ of (2.1)–(2.3) corresponding to $\varphi_m - \varphi_p$ and $F_m - F_p$. Therefore they satisfy (2.22) and (2.28), that is,

$$|u_m - u_p|_{Q_t} \leq (1 + \sqrt{2})^2[\|f_m - f_p\|_{2,1,Q_t} + \|\mathbf{f}_m - \mathbf{f}_p\|_{2,Q_t} + \tfrac{1}{2}\|u_m(\cdot, 0) - u_p(\cdot, 0)\|_{2,\Omega}].$$

This shows that $\{u_m\}$ converges in the norm of $\mathring{V}_2^{1,0}(Q_T)$. Because $\mathring{V}_2^{1,0}(Q_T)$ is complete, the limit function u of the $\{u_m\}$ is in $\mathring{V}_2^{1,0}(Q_T)$; moreover, (2.22) and (2.23) hold for u, a fact which follows from writing these two relations for u_m and then taking the limit as $m \to \infty$. This proves Theorem 2.3.

Remark 2.3. It turns out that the generalized solutions which are guaranteed by Theorem 2.3 have a derivative of order $\frac{1}{2}$ with respect to t. For an explanation of this statement, as well as proof, see [LSU 1].

Remark 2.4. The operator B which maps a vector function $\{f; f_i; \varphi\}$ from $L_{2,1}(Q_T) \times L_2(Q_T) \times L_2(\Omega)$ to the generalized solution $u(x, t)$ of (2.1)–(2.3) in $\mathring{V}_2^{1,0}(Q_T)$ is linear. Indeed, if $u' = B(\{f'; f_i'; \varphi'\})$ and $u'' = B(\{f''; f_i''; \varphi''\})$, then $u = u' + u''$ is a generalized solution of (2.1)–(2.3) in $L_2(Q_T)$ corresponding to $\{f' + f''; f_i' + f_i''; \varphi' + \varphi''\}$. But, on the other hand, this latter problem has a generalized solution v in $\mathring{V}_2^{1,0}(Q_T)$ by Theorem 2.3, and, by Theorem 2.2, v must coincide with $u = u' + u''$, that is, $u = v = B(\{f' + f''; f_i' + f_i''; \varphi' + \varphi''\})$. By the same token, we have proved that the set of generalized solutions of (2.1)–(2.3) in $\mathring{V}_2^{1,0}(Q_T)$ corresponding to all possible $\{f; f_i; \varphi\}$ in $L_{2,1}(Q_T) \times L_2(Q_T) \times L_2(\Omega)$ is linear, despite the non-linearity of (2.22), which must be satisfied by such solutions. The reason for this is the fact that (2.22) follows from (2.23), and even from (2.21).

§3. First Initial-Boundary Value Problem for General Parabolic Equations

In this section we investigate the problem

$$\mathscr{M}u \equiv u_t - \frac{\partial}{\partial x_i}(a_{ij}(x, t)u_{x_j} + a_i(x, t)u) + b_i(x, t)u_{x_i} + a(x, t)u = f + \frac{\partial f_i}{\partial x_i}, \tag{3.1}$$

$$u|_{t=0} = \varphi(x), \qquad u|_{S_T} = 0 \tag{3.2}$$

for a bounded domain Ω under the conditions $a_{ij} = a_{ji}$,

$$\sqrt{\sum_{i=1}^{n} a_i^2}, \qquad \sqrt{\sum_{i=1}^{n} b_i^2}, \qquad |a| \leq \mu, \tag{3.3}$$

$$\varphi \in L_2(\Omega), \qquad f \in L_{2,1}(Q_T), \qquad f_i \in L_2(Q_T), \tag{3.4}$$

and under the condition of uniform parabolicity

$$\nu\xi^2 \leq a_{ij}(x, t)\xi_i\xi_j \leq \mu\xi^2, \qquad \nu, \mu = \text{const} > 0. \tag{3.5}$$

First we shall prove that this problem has a generalized solution in $W_2^{1,0}(Q_T)$. This can be done in several ways. We shall use Galerkin's method for this; then we shall use Theorems 2.2 and 2.3 to show that each such solution is actually in $\mathring{V}_2^{1,0}(Q_T)$ and satisfies the energy-balance equation. Finally, we shall use these facts to derive the uniqueness theorem for the problem (3.1)–(3.2) in the class of generalized solutions in $W_2^{1,0}(Q_T)$.

The energy-balance equation for the problem (3.1)–(3.2) has the form

$$\tfrac{1}{2}\|u(\cdot, t)\|_{2,\Omega}^2 + \int_{Q_t} (a_{ij}u_{x_j}u_{x_i} + a_i u u_{x_i} + b_i u_{x_i} u + au^2)\, dx\, dt$$

$$= \tfrac{1}{2}\|u(\cdot, 0)\|_{2,\Omega}^2 + \int_{Q_t} (fu - f_i u_{x_i})\, dx\, dt. \tag{3.6}$$

This equation can be obtained formally by integrating by parts the equality

$$\int_{Q_t} \mathscr{M}u \cdot u\, dx\, dt = \int_{Q_t} \left(f + \frac{\partial f_i}{\partial x_i}\right) u\, dx\, dt \tag{3.7}$$

and using the boundary condition $u|_{S_T} = 0$.

We can derive a bound for $|u|_{Q_T}$ from (3.6), in much the same way as the bound (2.28) was derived from (2.22). In fact, on the basis of (3.3)–(3.5) it follows from (3.6) that

$$\tfrac{1}{2}\|u(\cdot, t)\|_{2,\Omega}^2 + \nu\|u_x\|_{2,Q_t}^2$$

$$\leq \tfrac{1}{2}\|u(\cdot, 0)\|_{2,\Omega}^2 + 2\mu\|u\|_{2,Q_t}\|u_x\|_{2,Q_t} + \mu\|u\|_{2,Q_t}^2$$

$$+ \|f\|_{2,1,Q_t} \max_{0 \leq \tau \leq t} \|u(\cdot, \tau)\|_{2,\Omega} + \|\mathbf{f}\|_{2,Q_t}\|u_x\|_{2,Q_t}$$

$$\leq \tfrac{1}{2}\|u(\cdot, 0)\|_{2,\Omega}^2 + \frac{\nu}{2}\|u_x\|_{2,Q_t}^2 + \frac{2\mu^2}{\nu}\|u\|_{2,Q_t}^2 + \mu\|u\|_{2,Q_t}^2$$

$$+ \|f\|_{2,1,Q_t} \max_{0 \leq \tau \leq t} \|u(\cdot, \tau)\| + \|\mathbf{f}\|_{2,Q_t}\|u_x\|_{2,Q_t}. \tag{3.8}$$

We combine similar terms, then multiply both sides by 2 and replace $\|u\|^2_{2,Q_t}$ by $ty^2(t)$, where $y(t) \equiv \max_{0 \le \tau \le t} \|u(\cdot, t)\|_{2,\Omega}$, and $\|u(\cdot, 0)\|^2_{2,\Omega}$ by $y(t)\|u(\cdot, 0)\|_{2,\Omega}$. This gives us the inequality

$$\|u(\cdot, t)\|^2_{2,\Omega} + \nu\|u_x\|^2_{2,Q_t} \le y(t)\|u(\cdot, 0)\|_{2,\Omega} + cty^2(t) + 2y(t)\|f\|_{2,1,Q_t} + 2\|\mathbf{f}\|_{2,Q_t}\|u_x\|_{2,Q_t} \equiv j(t), \tag{3.9}$$

where $c = 2(2\mu^2/\nu + \mu)$. From this follows the two inequalities

$$y^2(t) \le j(t), \tag{3.10}$$

and

$$\|u_x\|^2_{2,Q_t} \le \nu^{-1}j(t). \tag{3.11}$$

We take the square root of both sides of (3.10) and (3.11), add together the resulting inequalities and then majorize the right-hand side in the following way:

$$\begin{aligned} |u|_{Q_t} \equiv y(t) + \|u_x\|_{2,Q_t} &\le (1 + \nu^{-1/2})j^{1/2}(t) \\ &\le (1 + \nu^{-1/2})\sqrt{ct}|u|_{Q_t} + (1 + \nu^{-1/2})|u|^{1/2}_{Q_t} \\ &\quad \times [\|u(\cdot, 0)\|_{2,\Omega} + 2\|f\|_{2,1,Q_t} + 2\|\mathbf{f}\|_{2,Q_t}]^{1/2}. \end{aligned}$$

For

$$t < t_1 \equiv c^{-1}(1 + \nu^{-1/2})^{-2}, \qquad c = 2\left(\frac{2\mu^2}{\nu} + \mu\right), \tag{3.12}$$

we obtain the following bound for $|u|_{Q_t}$:

$$\begin{aligned} |u|_{Q_t} &\le [1 - (1 + \nu^{-1/2})\sqrt{ct}]^{-2}(1 + \nu^{-1/2})^2 \\ &\quad \times [\|u(\cdot, 0)\|_{2,\Omega} + 2\|f\|_{2,1,Q_t} + 2\|\mathbf{f}\|_{2,Q_t}]. \end{aligned} \tag{3.13}$$

We subdivide the interval $[0, t]$ into subintervals $\Delta_1 = [0, \frac{1}{2}t_1]$, $\Delta_2 = [\frac{1}{2}t_1, t_1], \ldots$ and the last Δ_N of length not exceeding $\frac{1}{2}t_1$. For each of them we have a bound of the form (3.13). If we take into account that $\|u(\cdot, t)\|_{2,\Omega} \le |u|_{Q_t}$, we derive the energy inequality

$$|u|_{Q_t} \le c(t)[\|u(\cdot, 0)\|_{2,Q} + 2\|f\|_{2,1,Q_t} + 2[\|\mathbf{f}\|_{2,Q_t}] \equiv c(t)\mathscr{F}(t), \tag{3.14}$$

which holds for any t in $[0, T]$. The function $c(t)$ is defined by t and by the constants ν and μ in (3.3) and (3.5).

We define now a *generalized solution* $u(x, t)$ *of the problem* (3.1)–(3.2) *in* $W_2^{1,0}(Q_T)$ (or, what is the same thing, in $\overset{\circ}{W}{}_2^{1,0}(Q_T)$) as an element of $\overset{\circ}{W}{}_2^{1,0}(Q_T)$ satisfying the identity

$$\begin{aligned} \mathscr{M}(u, \eta) &\equiv \int_{Q_T} (-u\eta_t + a_{ij}u_{x_j}\eta_{x_i} + a_iu\eta_{x_i} + b_iu_{x_i}\eta + au\eta)\,dx\,dt \\ &= \int_\Omega \varphi\eta(x, 0)\,dx + \int_{Q_T} (f\eta - f_i\eta_{x_i})\,dx\,dt \end{aligned} \tag{3.15}$$

for all $\eta \in W^1_{2,0}(Q_T)$ vanishing for $t = T$. It is clear that the set of such solutions is linear. For the proof of the solvability of (3.1)–(3.2) in $\overset{\circ}{W}{}^{1,0}_2(Q_T)$ we take a fundamental system $\{\varphi_k(x)\}$ in $\overset{\circ}{W}{}^1_2(\Omega)$ and assume for convenience that it has been orthonormalized in $L_2(\Omega)$. We shall look for approximate solutions $u^N(x, t)$ in the form $u^N(x, t) = \sum_{k=1}^N c_k^N(t)\varphi_k(x)$ from the system of relations†

$$(u_t^N, \varphi_l) + (a_{ij}u_{x_j}^N + a_i u^N, \varphi_{lx_i}) + (b_i u_{x_i}^N + au^N, \varphi_l)$$
$$= (f, \varphi_l) - (f_i, \varphi_{ix_i}), \qquad l = 1, \ldots, N, \tag{3.16}$$

and the equalities

$$c_l^N(0) = (\varphi, \varphi_l). \tag{3.17}$$

The relation (3.16) is simply a system of N linear ordinary differential equations in the unknowns $c_l(t) \equiv c_l^N(t)$, $(t = 1, \ldots, N)$, whose principal terms are of the form $dc_l(t)/dt$, the coefficients of $c_k(t)$ being bounded functions of t and the free terms summable functions on $(0, T)$. By a well-known theorem on the solvability of such systems, we see that (3.16) and (3.17) uniquely determine absolutely continuous functions $c_l^N(t)$ on $[0, T]$. Let us obtain bounds for the u^N which do not depend on N. To do this, let us multiply each equation of (3.16) by the appropriate c_l^N, add them up from 1 to N, and then integrate with respect to t from 0 to $t \le T$; as a result, we arrive at (3.6) for $u = u^N$. As we showed earlier, (3.6) implies (3.14) with $\mathscr{F}(t) = 2\|f\|_{2,1,Q_t} + 2\|\mathbf{f}\|_{2,Q_t} + \|u^N(\cdot, 0)\|_{2,\Omega}$. But $\|u^N(\cdot, 0)\|_{2,\Omega} \le \|\varphi\|_{2,\Omega}$, so that we have the bound

$$|u^N|_{Q_T} \le c_1 \tag{3.18}$$

with a constant c_1 independent of N. Because of (3.18), we can select a subsequence $\{u^{N_k}\}$ $(k = 1, 2, \ldots)$ from the sequence $\{u^N\}$ $(N = 1, 2, \ldots)$ which converges weakly in $L_2(Q_T)$, together with the derivatives $u_x^{N_k}$, to some element $u \in \overset{\circ}{W}{}^{1,0}_2(Q_T)$.‡ This element $u(x, t)$ is the desired generalized solution of the problem (3.1)–(3.2). Indeed, let us multiply (3.16) by an arbitrary absolutely continuous function $d_l(t)$ with $dd_l/dt \in L_2(0, T)$, $d_l(T) = 0$, add up the equations thus obtained from 1 to N, and then integrate the result from 0 to T. If we then integrate the first term by parts with respect to t, we obtain an identity

$$\mathscr{M}(u^N, \Phi) = \int_\Omega u^N\Phi|_{t=0}\, dx + \int_{Q_T} (f\Phi - f_i\Phi_{x_i})\, dx\, dt \tag{3.19}$$

† These relations are nothing more than the equalities

$$\left(\mathscr{M}u^N - f - \frac{\partial f_i}{\partial x_i}, \varphi_l\right) = 0, \qquad l = 1, \ldots, N,$$

transformed to a form corresponding to our choice of space.

‡ As a result of subsequent arguments, we shall see that the entire sequence $\{u^N\}$ converges to u.

of the form (3.15) in which $\Phi = \sum_{l=1}^{N} d_l(t)\varphi_l(x)$. Let us denote by $\mathfrak{M}_N$ the set of all such functions Φ with $d_l(t)$ having the properties indicated above. The totality $\bigcup_{p=1}^{\infty} \mathfrak{M}_p$ is dense in the subspace $\hat{W}^1_{2,0}(Q_T$ of $W^1_{2,0}(Q_T)$ consisting of all those elements of $W^1_{2,0}(Q_T)$ which vanish for $t = T$ (the proof is left to the reader). For a fixed $\Phi \in \mathfrak{M}_p$ in (3.19) we can take the limit of the subsequence $\{u^{N_k}\}$ chosen above, starting with $N_k \geq p$. As a result, we obtain (3.15) for u, with $\eta = \Phi \in \mathfrak{M}_p$. But since $\bigcup_{p=1}^{\infty} \mathfrak{M}_p$ is dense in $\hat{W}^1_{2,0}(Q_T)$, it is not hard to verify that (3.15) holds for all $\eta \in \hat{W}^1_{2,0}(Q_T)$, that is, $u(x, t)$ is actually a generalized solution in $\mathring{W}^{1,0}_2(Q_T)$ of (3.1)–(3.2).

By the same token, we have proved

Theorem 3.1. *If the hypotheses* (3.3)–(3.5) *are fulfilled, then the problem* (3.1)–(3.2) *has at least one generalized solution in* $\mathring{W}^{1,0}_2(Q_T)$.

Let us now investigate this solution $u(x, t)$. To do this, let us consider it as a generalized solution in $L_2(Q_T)$ of (2.1)–(2.3) with f in (3.1) replaced by $f - b_i u_{x_i} - au \equiv \tilde{f}$ and f_i replaced by $f_i + a_{ij}u_{x_j} + a_i u - u_{x_i} \equiv \tilde{f}_i$. This is possible since $\tilde{f} \in L_{2,1}(Q_T)$, $\tilde{f}_i \in L_2(Q_T)$, and (3.15) can be transformed to the form (2.21) for $\eta \in W^{\Delta,1}_{2,0}(Q_T)$ and $\eta(x, T) = 0$. But then it follows from Theorems 2.2 and 2.3 that $u(x, t)$ is a generalized solution of (2.1)–(2.3) in $V^{1,0}_2(Q_T)$, so that it is in $\mathring{V}^{1,0}_2(Q_T)$, and so that (2.22) and (2.23) hold for it, where f must be replaced by $\tilde{f}$ and f_i by $\tilde{f}_i$.

The relation (2.22) can be rewritten in the form (3.6) and the identity (2.23) in the form

$$\int_\Omega u(x, t)\eta(x, t)\, dx - \int_\Omega \varphi\eta(x, 0)\, dx$$

$$+ \int_{Q_t} (-u\eta_t + a_{ij}u_{x_j}\eta_{x_i} + a_i u\eta_{x_i} + b_i u_{x_i}\eta + au\eta)\, dx\, dt$$

$$= \int_{Q_t} (f\eta - f_i\eta_{x_i})\, dx\, dt, \tag{3.20}$$

in which η is an arbitrary element of $W^1_{2,0}(Q_T)$ and t is any number in $[0, T]$. We have proved that every generalized solution in $\mathring{W}^{1,0}_2(Q_T)$ of (3.1)–(3.2) is a *generalized solution of* (3.1)–(3.2) *in* $\mathring{V}^{1,0}_2(Q_T)$. Such solutions of (3.1)–(3.2) (according to §2) are defined as elements of $\mathring{V}^{1,0}_2(Q_T)$ for which the identity (3.20) and the energy relation (3.6) hold. We shall show that (3.1)–(3.2) cannot have two different solutions in $W^{1,0}_2(Q_T)$. Indeed, if the problem were to have two such solutions u' and u'', then their difference $u = u' - u''$ would be a generalized solution of (3.1)–(3.2) in the class $\mathring{W}^{1,0}_2(Q_T)$ corresponding to the zero initial condition and a zero free term. By what has been proved, u is actually a generalized solution of this problem in the class $\mathring{V}^{1,0}_2(Q_T)$, so that (3.6), with the right-hand side zero, holds for u. But from this it follows that (3.14) holds, with the right-hand side zero. Consequently, $u(x, t)$ must be zero, which proves that u' and u'' coincide.

From these arguments concerning any two generalized solutions u' and u'' of (3.1)–(3.2) in $\mathring{V}_2^{1,0}(Q_T)$ with distinct f, f_i, and φ, it follows that the operator B assigning to $\{f; f_i; \varphi\}$ a generalized solution in $\mathring{V}_2^{1,0}(Q_T)$ is linear, and that the energy-balance equation (3.6) is a consequence of (3.20) under the assumptions on the coefficients of $\mathscr{M}$ and the functions f, f_i, and φ that were indicated in Theorem 3.1.

Thus we have proved the following theorem.

Theorem 3.2. *If the assumptions* (3.3) *and* (3.5) *are fulfilled, then any generalized solution of* (3.1), (3.2) *from* $W_2^{1,0}(Q_T)$ *is the generalized solution in* $\mathring{V}_2^{1,0}(Q_T)$ *and it is unique in* $W_2^{1,0}(Q_T)$.

§4. Other Boundary Value Problems. The Method of Fourier and Laplace. The Second Fundamental Inequality

1. Unbounded Domains Ω. The boundedness of the domain Ω in §§2 and 3 was used essentially only where we described the domain of definition $\mathscr{D}(\Delta)$ of the operator Δ (cf. §2, (2.12)). For an unbounded domain Ω it is necessary to take the operator $\Delta - E$ instead of Δ. For $\Delta - E$ the domain of definition $\mathscr{D}(\Delta - E)$ will consist of generalized solutions of the problem

$$\Delta u - u = f, \qquad u|_S = 0, \tag{2.12'}$$

as f ranges over all of $L_2(\Omega)$. These solutions u will be elements of $\mathring{W}_2^1(\Omega)$. If the boundary of Ω is sufficiently "nice," then u will be in $W_{2,0}^2(\Omega)$. This can be proved in the same way as in the case of a bounded domain Ω in §§6 and 7, Chapter II. As the operator $\mathscr{M}_0 u$ it is better to take $\mathscr{M}_0 u = u_t - \Delta u + u$ and rederive for it all the proofs, remembering that, for an unbounded Ω, the norm in $\mathring{W}_2^1(\Omega)$ is not equivalent to $\|u_x\|_{2,\Omega}$. All the bounds, arguments, and results of §3 go over to unbounded domains.

2. The Second and Third Initial-Boundary Value Problems. These problems consist of finding a solution of the equation

$$\hat{\mathscr{M}} u \equiv u_t - \frac{\partial}{\partial x_i}(a_{ij}(x, t)) + b_i(x, t)u_{x_i} + a(x, t)u = f(x, t) \tag{4.1}$$

(we take for brevity $a_i = f_i \equiv 0$) satisfying the conditions

$$u|_{t=0} = \varphi(x), \qquad \frac{\partial u}{\partial N} + \sigma(s, t)u|_{S_T} = 0. \tag{4.2}$$

Let us first define a generalized solution in the space $W_2^{1,0}(Q_T)$. We recall that this space consists of functions $v(x, t)$ with a finite norm

$$\|v\|_{2,Q_T}^{(1,0)} = \left(\int_{Q_T} (v^2 + v_x^2)\,dx\,dt\right)^{1/2},$$

where the scalar product in $W_2^{1,0}(Q_T)$ is defined by

$$(v, w)_{2,Q_T}^{(1,0)} = \int_{Q_T} (vw + v_x w_x)\,dx\,dt.$$

A generalized solution of the problem (4.1)–(4.2) *in* $W_2^{1,0}(Q_T)$ is defined to be a function $u(x, t)$ in $W_2^{1,0}(Q_T)$ satisfying the integral identity

$$\hat{\mathscr{M}}(u, \eta) \equiv \int_{Q_T} (-u\eta_t + a_{ij}u_{x_j}\eta_{x_i} + b_i u_{x_i}\eta + au\eta)\,dx\,dt$$

$$+ \int_{S_T} \sigma u\eta\,ds\,dt = \int_\Omega \varphi\eta(x, 0)\,dx + \int_{Q_T} f\eta\,dx\,dt \qquad (4.3)$$

for all $\eta \in W_2^1(Q_T)$ vanishing at $t = T$.

The existence of at least one such solution can be proved, for example, by means of Galerkin's method in the same way as was done in the second half of §3 for the first boundary condition. It is necessary only that the functions $\varphi_k(x)$ $(k = 1, 2, \ldots)$ be chosen in such a way that they form a fundamental system in $W_2^1(\Omega)$, and that it be assumed that the boundary of Ω not be too bad (such as in §5, Chapter II). The energy bound for the solutions of (4.1)–(4.2) has the same form (3.14) as it does for the problem (3.1)–(3.2). It can be derived from the energy-balance equation, which differs from (3.6) only by the presence of the additional term $\int_{S_t} \sigma u^2\,ds\,dt$, which is not essential for the estimates. This equation can be obtained from the equality

$$\int_{Q_t} \hat{\mathscr{M}}u \cdot u\,dx\,dt = \int_{Q_t} fu\,dx\,dt$$

by integration by parts and by the conditions (4.2), and has the form

$$\tfrac{1}{2}\|u(\cdot, t)\|_{2,\Omega}^2 + \int_{Q_t} (a_{ij}u_{x_j}u_{x_i} + b_i u_{x_i}u + au^2)\,dx\,dt$$

$$+ \int_{S_t} \sigma u^2\,ds\,dt = \tfrac{1}{2}\|u(\cdot, 0)\|_{2,\Omega}^2 + \int_{Q_t} fu\,dx\,dt. \qquad (4.4)$$

The bound (3.14) can be derived from (4.4) in the same way as from (3.6). If $\sigma \geq 0$, the term $j_1(t) \equiv \int_{S_t} \sigma u^2\,ds\,dt$ can simply be discarded from (4.4) to obtain an inequality which implies (3.14). In the opposite case, $|j_1(t)|$ must be bounded as follows:

$$|j_1(t)| \leq c\int_{S_t} u^2\,ds\,dt \leq c_2 \int_{Q_t} \left(\varepsilon u_x^2 + \frac{1}{\varepsilon}u^2\right) dx\,dt, \qquad (4.5)$$

for arbitrary $\varepsilon \in (0, 1]$. Here we have assumed that $|\sigma| \le c$ and have used the inequality (6.24) in Chapter I. The remainder of the derivation of (3.14) from (4.4) coincides with that carried out in §3. Hence we can assume that for any $u(x, t)$ satisfying (4.4) for all $t \in [0, T]$ there is a bound of the form (3.14). On the basis of this existence at least one generalized solution of (4.1)–(4.2) in $W_2^{1,0}(Q_T)$ can be proved in the same way as in §3 for the problem (3.1)–(3.2).

At this stage we can proceed in two ways: Either we can prove "head-on" the uniqueness theorem in the class of such generalized solutions by following the first of our proposed methods of proving uniqueness theorems for non-stationary problems (cf. [4, Chapter III]), or we can carry out a lengthier argument using the head equation, which is analogous to the arguments of §§2 and 3. In the first, and shorter, way we must assume the existence of bounded derivatives of a_{ij}, b_i, and σ with respect to t. In the second way no new restrictions on the data of the problem are required. We shall describe these methods in somewhat more detail.

Let the problem (4.1)–(4.2) have two generalized solutions in $W_2^{1,0}(Q_T)$. Then their difference $u(x, t)$ is in $W_2^{1,0}(Q_T)$ and satisfies

$$\mathscr{M}(u, \eta) = 0 \tag{4.6}$$

for all $\eta \in W_2^1(\Omega)$ and $\eta(x, T) = 0$. As $\eta(x, t)$ let us take the function

$$\eta(x, t) = \begin{cases} 0, & t \in [b, T], \\ \displaystyle\int_b^t u(x, \tau)\, d\tau, & t \in [0, b], \end{cases} \tag{4.7}$$

where b is any fixed number on $[0, T]$.

It is not hard to verify that this function is admissible for (4.6). We substitute (4.7) into (4.6) and write the result in the form

$$\int_0^b \int_\Omega (-\eta_t^2 + a_{ij}\eta_{tx_j}\eta_{x_i} + b_i\eta_{tx_i}\eta + a\eta_t\eta)\, dx\, dt + \int_{S_b} \sigma\eta_t\eta\, ds\, dt = 0, \tag{4.8}$$

which is possible since $u = \eta_t$ for $t \in (0, b)$. Let us write $a_{ij}\eta_{tx_j}\eta_{x_i}$ in the form $\frac{1}{2}(\partial/\partial t)(a_{ij}\eta_{x_j}\eta_{x_i}) - \frac{1}{2}(\partial a_{ij}/\partial t)\eta_{x_j}\eta_{x_i}$ and $\sigma\eta_t\eta$ in the form $\frac{1}{2}(\partial/\partial t)(\sigma\eta^2) - \frac{1}{2}(\partial\sigma/\partial t)\eta^2$, and perform on the second, third, and last terms of (4.8) an integration by parts in the following form:

$$\begin{aligned} &\int_0^b \int_\Omega \left(-\eta_t^2 - \frac{1}{2}\frac{\partial a_{ij}}{\partial t}\eta_{x_j}\eta_{x_i} - b_i\eta_{x_i}\eta_t - \frac{\partial b_i}{\partial t}\eta_{x_i}\eta + a\eta_t\eta\right) dx\, dt \\ &\quad + \int_\Omega (\tfrac{1}{2}a_{ij}\eta_{x_j}\eta_{x_i} + b_i\eta_{x_i}\eta)\, dx\Big|_{t=0}^{t=b} + \int_{S_b} -\frac{1}{2}\frac{\partial\sigma}{\partial t}\eta^2\, ds\, dt \\ &\quad + \frac{1}{2}\int_S \sigma\eta^2\, ds\Big|_{t=0}^{t=b} = 0. \end{aligned} \tag{4.9}$$

Let us now utilize our assumptions about the coefficients of $\hat{\mathscr{M}}$ and about σ, and also take into account the fact that $\int_\Omega \cdots dx$ in (4.9) vanishes for $t = b$ by our choice of the function η in (4.7). Then we can deduce from (4.9), after changing signs, the inequality

$$\begin{aligned}\tfrac{1}{2}\nu \int_\Omega \eta_x^2(x, 0)\,dx + \int_{Q_b} \eta_t^2\,dx\,dt &\le c \int_{Q_b} \left[\varepsilon_1 \eta_t^2 + \left(1 + \frac{1}{\varepsilon_1}\right)\eta_x^2 + \left(1 + \frac{1}{\varepsilon_1}\right)\eta^2\right] dx\,dt \\ &\quad + c \int_\Omega \left(\varepsilon_2 \eta_x^2(x, 0) + \frac{1}{\varepsilon_2}\eta^2(x, 0)\right) dx \\ &\quad + c \int_{S_b} \eta^2\,ds\,dt + c \int_S \eta^2(s, 0)\,ds, \end{aligned} \tag{4.10}$$

where the ε_i are arbitrary positive numbers and the constant c is defined by the majorants of the coefficients of $\hat{\mathscr{M}}$, of σ and of their derivatives with respect to t. We can bound the integrals over S_b and S by using the inequality (6.24), Chapter I, as follows:

$$\int_S \eta^2(s, 0)\,ds \le c_1 \int_\Omega \left[\left(1 + \frac{1}{\varepsilon_3}\right)\eta^2(x, 0) + \varepsilon_3 \eta_x^2(x, 0)\right] dx, \tag{4.11}$$

$$\int_{S_b} \eta^2(s, t)\,ds\,dt \le c_1 \int_{Q_b} \left[\left(1 + \frac{1}{\varepsilon_3}\right)\eta^2 + \varepsilon_3 \eta_x^2\right] dx\,dt. \tag{4.12}$$

Moreover, from (4.7) or, what is the same thing, from the representations

$$\eta(x, t) = \int_b^t \eta_t(x, \tau)\,d\tau, \qquad t \in [0, b],$$

we have the inequalities

$$\left.\begin{aligned} \eta^2(x, t) &\le b \int_0^b \eta_t^2(x, \tau)\,d\tau, \\ \int_\Omega \eta^2(x, 0)\,dx &\le b \int_{Q_b} \eta_t^2\,dx\,dt, \\ \int_{Q_b} \eta^2\,dx\,dt &\le b^2 \int_{Q_b} \eta_t^2\,dx\,dt. \end{aligned}\right\} \tag{4.13}$$

Let us substitute (4.11) and (4.12) into (4.10) and then combine similar terms, putting $\eta_x^2(x, 0)$ and η_t^2 on the left-hand side; then let us choose the ε_i so small that the coefficients of $\eta_x^2(x, 0)$ and η_t^2 are equal to $\nu/4$ and $\frac{1}{2}$, respectively. After this, we use the inequalities (4.13) for getting rid of the terms $\eta^2(x, 0)$ and η^2 from the right-hand side of the resulting inequality. As a result, we have

$$\frac{\nu}{4}\int_\Omega \eta_x^2(x, 0)\,dx + \frac{1}{2}\int_{Q_b} \eta_t^2\,dx\,dt \le c_2 \int_{Q_b} (\eta_x^2 + b\eta_t^2)\,dx\,dt. \tag{4.14}$$

We shall take b so small that

$$c_2 b \le \tfrac{1}{4}. \tag{4.15}$$

For such b it follows from (4.14) that

$$\int_\Omega \eta_x^2(x, 0)\,dx + \frac{1}{\nu}\int_{Q_b} \eta_t^2\,dx\,dt \le c_3 \int_{Q_b} \eta_x^2\,dx\,dt. \tag{4.16}$$

We use now the fact that b was chosen arbitrarily and show explicitly that the function $\eta(x, t)$ which we chose depends on b. For this we introduce the notation

$$\int_0^t u(x, \tau)\,d\tau = y(x, t).$$

Then $\eta(x, t)$ is $y(x, t) - y(x, b)$ for $t \in [0, b]$. Let us substitute this expression for η into (4.16), discarding there the second term of (4.16), and then majorize the right-hand side of the resulting inequality as follows:

$$\begin{aligned}\int_\Omega y_x^2(x, b)\,dx &\le c_3 \int_{Q_b} [y_x(x, t) - y_x(x, b)]^2\,dx\,dt \\ &\le 2c_3 \int_{Q_b} [y_x^2(x, t) + y_x^2(x, b)]\,dx\,dt \\ &= 2c_3 b \int_\Omega y_x^2(x, b)\,dx + 2c_3 \int_{Q_b} y_x^2(x, t)\,dx\,dt. \end{aligned} \tag{4.17}$$

For

$$b \le 1/(4c_3) \tag{4.18}$$

we obtain from (4.17) the inequality

$$\int_\Omega y_x^2(x, b)\,dx \le 4c_3 \int_{Q_b} y_x^2(x, t)\,dx\,dt, \tag{4.19}$$

which holds for all $b \in [0, b_1]$, where $b_1 = \min\{1/(4c_2); 1/(4c_3)\}$. Since $y(x, 0) = 0$, and hence $y_x(x, 0) = 0$, it follows from (4.19) that $y_x(x, b) \equiv 0$ for $b \in [0, b_1]$ (cf. Lemma 1.1). But then $\eta_x(x, t) = y_x(x, t) - y_x(x, b) \equiv 0$ for $t \in [0, b_1]$, so that we may conclude from (4.16) that $\eta_t(x, t) = u(x, t) \equiv 0$ for $t \in [0, b_1]$. Thus we have proved that two solutions u' and u'' coincide in the cylinder $Q_{b_1} = \Omega \times [0, b_1]$. If we repeat this argument for the cylinders

$\Omega \times [b_1, 2b_1]$, $\Omega \times [2b_1, 3b_1]$, and so on, we shall exhaust the entire cylinder Q_T, thus proving the desired uniqueness theorem.

We have proved the following theorem

Theorem 4.1. *Let the coefficients of* (4.1) *satisfy the conditions*

$$\nu\xi^2 \le a_{ij}(x,t)\xi_i\xi_j \le \mu\xi^2, \qquad \nu = \text{const} > 0, \qquad |b_i|, |a| \le \mu_1, \tag{4.20}$$

and let $|\sigma| \le \mu_1$. *Then the problem* (4.1)–(4.2) *has at least one generalized solution in* $W_2^{1,0}(Q_T)$ *if* $\varphi \in L_2(\Omega)$ *and* $f \in L_{2,1}(Q_T)$. *If, in addition,*

$$\left|\frac{\partial a_{ij}}{\partial t}\right|, \quad \left|\frac{\partial b_i}{\partial t}\right|, \quad \left|\frac{\partial \sigma}{\partial t}\right| \le \mu_2,\dagger \tag{4.21}$$

then a uniqueness theorem holds for generalized solutions of (4.1)–(4.2) *in* $W_2^{1,0}(Q_T)$. *The boundary of* Ω *is subject to the requirement given in* §5, *Chapter II.*

An analogous theorem can be proved for the first boundary condition in the same way as we proved Theorem 4.1. However, it is more crude than Theorem 3.2 in which we did not require the existence of the derivatives of the coefficients in the equation. The restrictions (4.21) can be removed also for the boundary conditions (4.2); for this, we must carry out arguments much like those of §§2 and 3. We shall give the sketch of this without going into details, assuming that $\sigma(s, t)$ is independent of t.

At first we must consider the auxiliary problem

$$\left.\begin{aligned} \hat{\mathscr{M}}_0 u \equiv u_t - \Delta u = f(x,t) + \sum_{i=1}^n \frac{\partial f_i(x,t)}{\partial x_i}, \\ u|_{t=0} = \varphi(x), \quad \left.\frac{\partial u}{\partial n} + \sigma u + f_i \cos(\mathbf{n}, \mathbf{x}_i)\right|_{S_T} = 0, \end{aligned}\right\} \tag{4.22}$$

assuming that $\varphi \in W_2^1(\Omega)$, $f \in L_2(Q_T)$, $f_i \in W_2^{1,0}(Q_T)$; for this problem we must prove theorems like Theorems 2.1 and 2.2 (let us denote them as Theorems 4.2 and 4.3) using the same steps as in §2. Then we must introduce generalized solutions of the problem (4.22) in the space $V_2^{1,0}(Q_T)$, which differs from $\mathring{V}_2^{1,0}(Q_T)$ only in that its elements do not vanish on S_T. Such solutions $u(x, t)$ are elements of $V_2^{1,0}(Q_T)$ satisfying both the identity

$$\int_\Omega u(x,t)\eta(x,t)\,dx - \int_\Omega \varphi\eta(x,0) + \int_{Q_t} (-u\eta_t + u_{x_i}\eta_{x_i})\,dx\,dt$$
$$+ \int_{S_t} \sigma u\eta\,ds\,dt = \int_{Q_t} (f\eta - f_i\eta_{x_i})\,dx\,dt \tag{4.23}$$

† The condition $|\partial b_i/\partial t| \le \mu_2$ can be replaced by the condition $|\sum_{i=1}^n \partial b_i/\partial x_i| \le \mu_2$.

for all $t \in [0, T]$ and all $\eta \in W_2^1(Q_T)$, and the energy equation

$$\tfrac{1}{2}\|u(\cdot, t)\|_{2,\Omega}^2 + \|u_x\|_{2,Q_t}^2 + \int_{S_t} \sigma u^2\, ds\, dt$$
$$= \int_{Q_t} (fu - f_i u_{x_i})\, dx\, dt + \tfrac{1}{2}\|u(\cdot, 0)\|_{2,\Omega}^2, \qquad t \in [0, T]. \tag{4.24}$$

By means of Theorem 4.2 (cf. comments following (4.22)) we can prove the existence of such solutions for any $\varphi \in L_2(\Omega)$, $f \in L_{2,1}(Q_T)$, and $f_i \in L_2(Q_T)$. After that we can go over to the general problem (4.1)–(4.2), for which we already know the existence of at least one generalized solution $u(x, t)$ in $W_2^{1,0}(Q_T)$ (cf. Theorem 4.1). This solution satisfies (4.3), which can be written in the form

$$\int_{Q_T} (-u\eta_t + u_{x_i}\eta_{x_i})\, dx\, dt + \int_{S_t} \sigma u \eta\, ds\, dt$$
$$= \int_\Omega \varphi \eta(x, 0)\, dx + \int_{Q_T} (\mathscr{F}\eta - f_i \eta_{x_i})\, dx\, dt, \tag{4.25}$$

where $\mathscr{F} = f - b_i u_{x_i} - au \in L_{2,1}(Q_T)$ and $f_i = a_{ij}u_{x_j} - u_{x_i} \in L_2(Q_T)$, and where $\eta \in W_2^1(Q_T)$ and $\eta(x, T) = 0$. Because of this, we can assert that $u(x, t)$ is a generalized solution in $L_2(Q_T)$ of the problem (4.22) with these the free terms $\mathscr{F}$ and f_i. But, on the other hand, (4.22) has a generalized solution $v(x, t)$ in $V_2^{1,0}(Q_T)$ which by Theorem 4.3 (cf. comment following (4.22)), must coincide with $u(x, t)$. Hence $u(x, t) \in V_2^{1,0}(Q_T)$, and (4.23) and (4.24) hold for $u(x, t)$; these equations, when we substitute the expressions for $\mathscr{F} = f - b_i u_{x_i} - au$ and $f_i = a_{ij}u_{x_j} - u_{x_i}$, assume the form

$$\int_\Omega u(x, t)\eta(x, t)\, dx - \int_\Omega \varphi \eta(x, 0)\, dx$$
$$+ \int_{Q_t} (-u\eta_t + a_{ij}u_{x_j}\eta_{x_i} + b_i u_{x_i}\eta + au\eta)\, dx\, dt$$
$$+ \int_{S_t} \sigma u \eta\, ds\, dt = \int_{Q_t} f\eta\, dx\, dt \tag{4.26}$$

and (4.4), respectively.

Now, if the problem (4.1)–(4.2) had two generalized solutions $u'(x, t)$ and $u''(x, t)$ in $W_2^{1,0}(Q_T)$, then their difference $u(x, t)$ would be a generalized solution in $W_2^{1,0}(Q_T)$ of the same problem, but with $f \equiv 0$ and $\varphi \equiv 0$. It follows from what has been proved that u is in $V_2^{1,0}(Q_T)$ and satisfies (4.4) with $f \equiv \varphi \equiv 0$. But, as we indicated above, (4.4) implies the inequality (3.14), in which the right-hand side is zero, so that $u(x, t)$ is equal to zero, so that u' and u'' coincide. In this way it proves that any generalized solution of (4.1)–(4.2) in $W_2^{1,0}(Q_T)$ is actually a generalized solution in $V_2^{1,0}(Q_T)$, and the uniqueness theorem holds in these classes. Thus we have dispensed with the restrictions (4.21) under the assumption, to be sure, that σ is independent of t.

3. The Fourier Method and the Method of the Laplace Transform. For parabolic equations

$$u_t - \mathscr{L}u = f \tag{4.27}$$

with a symmetric operator $\mathscr{L}$,

$$\mathscr{L}u = \frac{\partial}{\partial x_i}(a_{ij}(x)u_{x_j} + a_i(x)u) - a_i(x)u_{x_j} - a(x)u, \tag{4.28}$$

whose coefficients are independent of t, the initial-boundary value problem under any of the conditions (1.9)–(1.11) (in (1.11) σ should be independent of t) can be solved by the method of Fourier. We shall do this for the problem

$$u_t - \mathscr{L}u = 0, \tag{4.29}$$

$$u|_{t=0} = \varphi(x), \qquad u|_{S_t} = 0, \tag{4.30}$$

under the assumption that Ω is a bounded domain. We shall present, for the time being, the formal side of the method in its classical form (that is, assuming that the coefficients a_{ij}, a_i are differentiable and that the solutions satisfy the equation in its usual form). Let us find all solutions of (4.29) of the form $u = T(t)X(x)$ which satisfy the boundary condition $u|_{S_t} = 0$ for all $t \geq 0$. Obviously, the latter is a condition on X: $X|_S = 0$. Let us set $u = TX$ in (4.29) and divide both sides by TX. This gives us

$$\frac{T'(t)}{T(t)} = \frac{\mathscr{L}X(x)}{X(x)},$$

which implies that both sides are equal to a constant, i.e.,

$$\mathscr{L}X = \lambda X, \qquad X|_S = 0, \tag{4.31}$$

and

$$T' = \lambda T, \tag{4.32}$$

where $\lambda = \text{const}$. The problem (4.31) is a spectral problem which we studied in §4, Chapter II (see also §7, Chapter II). All of its linearly independent solutions can be arranged in a sequence $\{u_k(x)\}$ ($k = 1, 2, \ldots$) which is orthonormal in $L_2(\Omega)$, where the $\{\lambda_k\}$ corresponding to the $\{u_k(x)\}$ are real and arranged in decreasing order with $\lambda_k \to -\infty$ as $k \to \infty$. The eigenfunctions $\{u_k(x)\}$ are also mutually orthogonal in the following ways:

$$[u_k, u_l] \equiv \mathscr{L}(u_k, u_l) + \lambda_0(u_k, u_l)$$

$$\equiv \int_\Omega [a_{ij} u_{kx_j} u_{lx_i} + a_i u_k u_{lx_i} + a_i u_{kx_i} u_l + (a + \lambda_0) u_k u_l]\, dx$$

$$= (\lambda_0 - \lambda_k)\delta_k^l \tag{4.33}$$

and

$$\{u_k, u_l\} \equiv ((\mathscr{L} - \lambda_0)u_k, (\mathscr{L} - \lambda_0)u_l) = (\lambda_0 - \lambda_k)^2 \delta_k^l, \tag{4.34}$$

where λ_0 is chosen in such a way that $\lambda_1 < \lambda_0$ (and hence $\lambda_k \leq \lambda_0$, $k = 2, 3, \ldots$). If the coefficients of $\mathscr{L}$ satisfy (3.3) and (3.5), then the norm $[u, u]^{1/2}$ is equivalent to that of $\mathring{W}_2^1(\Omega)$. If the hypotheses of Theorem 7.2, Chapter II, are satisfied by the coefficients of $\mathscr{L}$ and by the domain Ω, then the norm $\{u, u\}^{1/2}$ is equivalent to that of $W_{2,0}^2(\Omega)$. In the first case (where only (3.3) and (3.5) are fulfilled), the $u_k(x)$ are elements of $\mathring{W}_2^1(\Omega)$ and for these it makes sense to speak only of orthogonality in the form (4.33). In the second case, the $u_k(x)$ are in $W_{2,0}^2(\Omega)$, and the orthogonality also has the form (4.34).

If we take $X = u_k(x)$ as a solution of the problem (4.31), then the corresponding solution $T(t)$ of (4.32) has the form $T_k(x) = a_k e^{\lambda_k t}$ with an arbitrary constant a_k. Let us consider the series

$$u(x, t) = \sum_{k=1}^{\infty} a_k e^{\lambda_k t} u_k(x). \tag{4.35}$$

The sum of any finite number of terms is a solution of (4.29) (in general, a generalized solution) satisfying the boundary condition in (4.30). In order that it satisfy the initial condition (at least formally, without considering the question of convergence), we must take the entire series (4.35), whose coefficients are defined by

$$u|_{t=0} = \varphi(x) = \sum_{k=1}^{\infty} a_k u_k(x)$$

namely,

$$a_k = (\varphi, u_k). \tag{4.36}$$

Our problem consists of studying the character of the convergence of (4.35) under various assumptions about the known functions. This is easy to do, based on the kind of orthogonality and on the fact that $\lambda_k \to -\infty$ as

$k \to \infty$. Let us calculate the following norms of (4.35), assuming for convenience of notation that none of the $\{\lambda_k\}$ is zero:

$$\|u(\cdot, t)\|_{2,\Omega}^2 = \sum_{k=1}^{\infty} a_k^2 e^{2\lambda_k t}, \tag{4.37}$$

$$[u(\cdot, t), u(\cdot, t)] = \sum_{k=1}^{\infty} a_k^2 (\lambda_0 - \lambda_k) e^{2\lambda_k t}, \tag{4.38}$$

$$\int_0^t [u(\cdot, t), u(\cdot, t)]\, dt = \sum_{k=1}^{\infty} a_k^2 \frac{\lambda_0 - \lambda_k}{-2\lambda_k} (1 - e^{2\lambda_k t}), \tag{4.39}$$

$$\{u(\cdot, t), u(\cdot, t)\} = \sum_{k=1}^{\infty} a_k^2 (\lambda_0 - \lambda_k)^2 e^{2\lambda_k t}, \tag{4.40}$$

$$\|u_t(\cdot, t)\|_{2,\Omega}^2 = \sum_{k=1}^{\infty} a_k^2 \lambda_k^2 e^{2\lambda_k t}, \tag{4.41}$$

$$\int_0^t \{u(\cdot, t), u(\cdot, t)\}\, dt = \sum_{k=1}^{\infty} a_k^2 \frac{(\lambda_0 - \lambda_k)^2}{-2\lambda_k} (1 - e^{2\lambda_k t}), \tag{4.42}$$

$$\int_0^t \|u_t(\cdot, t)\|_{2,\Omega}^2\, dt = \frac{1}{2} \sum_{k=1}^{\infty} a_k^2 \lambda_k (e^{2\lambda_k t} - 1). \tag{4.43}$$

If only (3.3) and (3.5) are fulfilled and if $\varphi \in L_2(\Omega)$, then $\sum_{k=1}^{\infty} a_k^2 = \|\varphi\|_{2,\Omega}^2$, and the series on the right-hand sides of (4.37) and (4.39) are convergent, and uniformly convergent for $t \in [0, T]$. Because of this, the sum $u(x, t)$ of the series (4.35) is a generalized solution of the problem (4.29)–(4.30) in $\overset{\circ}{V}{}_2^{1,0}(Q_T)$ for all T. In fact, the convergence mentioned implies that $u \in \overset{\circ}{V}{}_2^{1,0}(Q_T)$. It is further necessary to verify that $u(x, t)$ satisfies the corresponding integral identity (cf. (3.20)). But this is easy to do if we start from the fact that the sums $u^N(x, t)$ of the first N terms of (4.35) satisfy this identity with the initial function $\varphi_N(x) = \sum_{k=1}^{N} a_k u_k(x)$, and then take the limit in the identity as $N \to \infty$. Finally, the validity of the energy-balance equation for $u(x, t)$ follows from its validity for all the $u^N(x, t)$ and the fact that it is still valid after passage to the limit as $N \to \infty$. Thus, for the problem (4.29)–(4.30) we have proved the existence of a generalized solution $u(x, t)$ in $\overset{\circ}{V}{}_2^{1,0}(Q_T)$ and that it can be represented as the sum of the series (4.35), which converges in the norm of $\overset{\circ}{V}{}_2^{1,0}(Q_T)$ for all T. We can say more about this solution, namely, that it is an element of $\overset{\circ}{W}{}_2^1(\Omega)$ for all $t > 0$ and that it is a continuous function of t in the norm of $\overset{\circ}{W}{}_2^1(\Omega)$. This is obvious from (4.38) and from the bounds $0 \le (\lambda_0 - \lambda_k) e^{2\lambda_k t} \le c(t)$ $(k = 1, 2, \ldots)$, where

$$c(t) = \sup_{-\infty < \lambda \le \lambda_0} (\lambda_0 - \lambda) e^{2\lambda t}$$

is a bounded function of t on any interval $[\varepsilon \le t \le T]$, $\varepsilon > 0$. Moreover, analogous calculations and estimates show that the series (4.35) can be differentiated term-by-term any number of times with respect to t, and that the series thus obtained converges for $t > 0$ in the norm of $\mathring{W}_2^1(\Omega)$, uniformly in t on any interval $[\varepsilon \leq t \leq T]$, $\varepsilon > 0$.

If, in addition to (3.3) and (3.5), the hypotheses of Theorem 7.2, Chapter II, are satisfied, then we can use the relations (4.34) and (4.40)–(4.43), which guarantee the convergence of (4.35), as well as the convergence of the series obtained from (4.35) by term-by-term differentiation with respect to t, in the norm of $W_{2,0}^2(\Omega)$ for all $t > 0$, and uniformly for $t \in [\varepsilon, T]$, for all $\varepsilon > 0$. If we require more of $\varphi(x)$, for example, that $\varphi(x) \in \mathring{W}_2^1(\Omega)$, then, as we showed in §4, Chapter II, the numerical series $\sum_{k=1}^{\infty} a_k^2(\lambda_0 - \lambda_k)$ converges, whence (4.38) implies the convergence of (4.35) in the norm $W_2^1(\Omega)$ for all $t \geq 0$, uniformly on $t \in [0, T]$. Moreover, (4.42) and (4.43) imply the convergence of the series in the norms corresponding to the left-hand sides of these relations. The numerical series $\sum_{k=1}^{\infty} a_k^2(\lambda_0 - \lambda_k)^2$ converges for $\varphi(x) \in W_{2,0}^2(\Omega)$, which implies the corresponding convergence of (4.35); each type of convergence guarantees that the sum of the series belongs to the corresponding function space.

The results which we enumerated regarding the convergence of (4.35) reveal a specific property of equations of parabolic type which we have not discussed in the preceding sections. This is the so-called property of "instantaneous smoothing," and it consists of the fact that the solutions $u(x, t)$ of parabolic equations are "better" for $t > 0$ than they are for $t = 0$ (i.e., "better" in the sense of smoothness). "How much better" depends on the smoothness of the coefficients of the equation and of the free term. In the present case the free term is identically zero (i.e., it is an infinitely differentiable function), and the coefficients do not depend on t (i.e., they are infinitely differentiable functions of t); this stipulated the unlimited differentiability of $u(x, t)$ with respect to t for $t > 0$ even for $\varphi(x) = u(x, 0) \in L_2(\Omega)$. The influence of the smoothness of the coefficients of $\mathscr{L}$ in terms of x has been discerned above in the norms of $\mathring{W}_2^1(\Omega)$ and $W_{2,0}^2(\Omega)$.

We can also prove a "localization principle" for parabolic equations similar to that for elliptic equations, but we shall not define it or prove it here.

For non-homogeneous equations (4.27), where $\mathscr{L}u$ has the form (4.28), solutions of the problem (4.27), (4.30) can be represented in the form of a series

$$u(x, t) = \sum_{k=1}^{\infty} a_k e^{\lambda_k t} u_k(x) + \sum_{k=1}^{\infty} \int_0^t f_k(\tau) e^{\lambda_k(t-\tau)} \, d\tau u_k(x), \tag{4.44}$$

where $a_k = (\varphi, u_k)$ and $f_k(t) = (f(\cdot, t), u_k(\cdot))$. It is easily verified that (4.44) satisfies formally all conditions of the problem, and its convergence can be investigated in the same way as the convergence of (4.35). We leave it to the reader to carry out independently the corresponding details.

We go over to a description of the scheme for the method of the Laplace transform, which is applicable to equations of the general type (3.1) if their coefficients are independent of t. We shall assume that this condition is satisfied. Let us write (3.1) in the form

$$\mathscr{M}u = u_t - \mathscr{L}u = \mathscr{F}(x, t), \tag{4.45}$$

where $\mathscr{L}u = (a_{ij}(x)u_{x_j} + a_i(x)u)_{x_i} - b_i(x)u_{x_i} - a(x)u$ and $\mathscr{F}(x, t) = f(x, t) + \partial f_i(x, t)/\partial x_i$. Let us denote the Laplace transforms of the functions $u(x, t)$ and $\mathscr{F}(x, t)$ with respect to t by $v(x, \lambda)$ and $\hat{\mathscr{F}}(x, \lambda)$, i.e.,

$$v(x, \lambda) = \int_0^\infty u(x, t)e^{-\lambda t}\, dt, \qquad \hat{\mathscr{F}}(x, \lambda) = \int_0^\infty \mathscr{F}(x, t)e^{-\lambda t}\, dt.$$

Without regard to the convergence of the integrals introduced and appearing below, we carry out formally the following transformations:

$$\int_0^\infty u_t e^{-\lambda t}\, dt = \lambda \int_0^\infty ue^{-\lambda t}\, dt - u(x, 0) = \lambda v(x, \lambda) - u(x, 0);$$

$$\int_0^\infty \mathscr{L}ue^{-\lambda t}\, dt = \mathscr{L}v(x, \lambda).$$

In view of this, if we multiply (4.45) by $e^{-\lambda t}\, dt$, integrate the result from 0 to ∞, and impose the initial condition (4.30), the result can be written in the form

$$\mathscr{L}v = \lambda v - (\hat{\mathscr{F}}(x, \lambda) + \varphi(x)). \tag{4.46}$$

We must combine this equation with the boundary condition

$$v|_S = 0, \tag{4.47}$$

which is obtained from the condition $u|_{S_\infty} = 0$. If the problem (4.46) – (4.47) is uniquely solvable for all λ in a half-plane of the form $\lambda_1 \equiv \operatorname{Re} \lambda \geq \lambda_0$ and if the solution $v(x, \lambda)$ decreases sufficiently rapidly as $\operatorname{Im} \lambda \to \pm\infty$, then its inverse Laplace transform

$$u(x, t) = \frac{1}{2\pi i} \int_{\lambda_1 - i\infty}^{\lambda_1 + i\infty} v(x, \lambda)e^{\lambda t}\, d\lambda, \qquad \lambda_1 \geq \lambda_0, \tag{4.48}$$

yields the solution of the problem (4.45)–(4.30).

If we use the results of Chapter II and derive the necessary estimates for the norms in $L_2(\Omega)$, $\mathring{W}_2^1(\Omega)$, $W_{2,0}^2(\Omega)$ of the solution v in terms of λ and certain known quantities, it is not hard to prove that this way of solving the problem (4.45)–(4.30) is always possible. We shall not cite here all these arguments, or the final results, which essentially coincide with those obtained above. The reader can do this independently, referring to my book [4], where the more complicated case of hyperbolic equations is treated with general remarks concerning parabolic equations.

4. The Second Fundamental Inequality. For parabolic operators $\mathcal{M}$ of the general type (3.1) we have an inequality similar to the "second fundamental inequality" proved in §6, Chapter II, if the boundary S of Ω and the coefficients of $\mathcal{M}$ have a certain smoothness. In addition to (3.3)–(3.5), the coefficients of $\mathcal{M}$ should satisfy the conditions

$$\left|\frac{\partial a_{ij}}{\partial x_k}\right|, \quad \left|\frac{\partial a_{ij}}{\partial t}\right| \leq \mu_1 \tag{4.49}$$

and $|\partial a_i/\partial x_i| \leq \mu_1$. By this last inequality, $\mathcal{M}u$ can be written in the abbreviated form

$$\mathcal{M}u = u_t - \frac{\partial}{\partial x_i}(a_{ij}(x,t)u_{x_j}) + \hat{a}_i(x,t)u_{x_i} + \hat{a}(x,t) \equiv u_t - \mathcal{L}u,$$

where $|\hat{a}_i, \hat{a}| \leq \mu_2$. The conditions on S are the same as in §6, Chapter II. As we proved in §6, for any function $u \in W^2_{2,0}(\Omega)$ we have the inequality

$$(\|u\|^{(2)}_{2,\Omega})^2 \leq c\|\mathcal{L}u\|^2_{2,\Omega} + c_1\|u\|^2_{2,\Omega}, \tag{4.50}$$

in which the constants are determined by v and μ in (3.5), and by μ_1, μ_2, and S. Let us take the integral $\int_{Q_t}(\mathcal{M}u)^2\,dx\,dt$ for an arbitrary, "sufficiently smooth" function $u(x,t)$, vanishing on S_t, and transform it as follows:

$$\int_{Q_t}(\mathcal{M}u)^2\,dx\,dt = \int_{Q_t}[u_t^2 - 2u_t\mathcal{L}u + (\mathcal{L}u)^2]\,dx\,dt$$

$$= \int_{Q_t}[u_t^2 + 2a_{ij}u_{tx_i}u_{x_j} + 2u_t(\hat{a}_i u_{x_i} + \hat{a}u) + (\mathcal{L}u)^2]\,dx\,dt$$

$$= \int_{\Omega}a_{ij}u_{x_i}u_{x_j}\,dx\Big|_{t=0}^{t=t} + \int_{Q_t}\left[u_t^2 + (\mathcal{L}u)^2 - \frac{\partial a_{ij}}{\partial t}u_{x_i}u_{x_j}\right.$$

$$\left. + 2u_t(\hat{a}_i u_{x_i} + \hat{a}u)\right]dx\,dt \tag{4.51}$$

Hence, by our assumptions about the coefficients of $\mathcal{L}$ and by Cauchy's inequality (1.3), Chapter I, we derive the inequality

$$\int_{\Omega}a_{ij}u_{x_i}u_{x_j}\,dx|_{t=t} + \int_{Q_t}[u_t^2 + (\mathcal{L}u)^2]\,dx\,dt$$

$$\leq \int_{\Omega}a_{ij}u_{x_i}u_{x_j}\,dx|_{t=0} + c_2\int_{Q_t}\left[u_x^2 + \varepsilon u_t^2 + \frac{1}{\varepsilon}(u_x^2 + u^2)\right]dx\,dt$$

$$+ \int_{Q_t}(\mathcal{M}u)^2\,dx\,dt \tag{4.52}$$

for all $\varepsilon > 0$. Let us replace $\int_{Q_t} (\mathscr{L}u)^2\,dx\,dt$ and $\int_\Omega a_{ij}u_{x_i}u_{x_j}\,dx|_{t=t}$ by smaller quantities in accordance with (4.50) and (3.5), and then collect similar terms, taking $\varepsilon = 1/(2c_2)$:

$$\begin{aligned}\nu\int_\Omega u_x^2(x,t)\,dx &+ \int_{Q_t}(\tfrac{1}{2}u_t^2 + c^{-1}u_{xx}^2)\,dx\,dt\\ &\le \mu\int_\Omega u_x^2(x,0)\,dx + c_3\int_{Q_t}(u_x^2+u^2)\,dx\,dt + \int_{Q_t}(\mathscr{M}u)^2\,dx\,dt\\ &\le \mu\int_\Omega u_x^2(x,0)\,dx + c_4\int_{Q_t}u_x^2\,dx\,dt + \int_{Q_t}(\mathscr{M}u)^2\,dx\,dt,\end{aligned} \tag{4.53}$$

where we have used inequality (6.3) of Chapter I. If we now discard the second term on the left-hand side of (4.53), then from the resulting inequality we have the bound

$$\int_\Omega u_x^2(x,t)\,dx \le c_5(t)\left[\mu\int_\Omega u_x^2(x,0)\,dx + \int_{Q_t}(\mathscr{M}u)^2\,dx\,dt\right], \tag{4.54}$$

in which $c_5(t) = \nu^{-1}\exp(c_4\nu^{-1}t)$ (in connection with this, see Lemma 1.1). If we substitute this bound into (4.53), we have the inequality

$$\begin{aligned}\nu\int_\Omega u_x^2(x,t)\,dx &+ \int_{Q_t}(\tfrac{1}{2}u_t^2 + c^{-1}u_{xx}^2)\,dx\,dt\\ &\le c_6(t)\left[\mu\int_\Omega u_x^2(x,0)\,dx + \int_{Q_t}(\mathscr{M}u)^2\,dx\,dt\right],\end{aligned} \tag{4.55}$$

where $c_6(t) = 1 + c_4\int_0^t c_5(\tau)\,d\tau$. The desired inequality now follows from this one on the basis of (4.54) and from (6.3), Chapter I:

$$\begin{aligned}\nu\int_\Omega u_x^2(x,t)\,dx &+ \int_{Q_t}[\tfrac{1}{2}u_t^2 + c^{-1}(u_{xx}^2 + u_x^2 + u^2)]\,dx\,dt\\ &\le c_7(t)\left[\mu\int_\Omega u_x^2(x,0)\,dx + \int_{Q_t}(\mathscr{M}u)^2\,dx\,dt\right],\end{aligned} \tag{4.56}$$

in which $c_7(t)$ has the same growth in t as $c_6(t)$ has. We call this the "second fundamental inequality" (the "first" being the energy inequality (3.14)). This is true for all $u \in W_{2,0}^{2,1}(Q_T)$. With the help of (4.56) it is not difficult to prove the solvability of the first initial-boundary value problem in the space $W_{2,0}^{2,1}(Q_T)$ if $S \in C^2$, $\varphi \in \mathring{W}_2^1(\Omega)$, $\mathscr{F} \equiv f + \partial f_i/\partial x_i \in L_2(Q_T)$, and a_{ij} satisfy the conditions (4.49). For this purpose we can use either the method of continuation by the parameter $\tau \in [0, 1]$ from the operator $\mathscr{M}_0 = \partial/\partial t - \Delta$ to the operator $\mathscr{M}$ (see §7, Chapter II), or the "functional" method which we have utilized in §2 for the heat equation. Analogous facts prove to be true for the boundary conditions $\partial u/\partial l + \sigma u|_{S_T} = 0$, where $\partial u/\partial l$ is a directional derivative which is not tangent to S and which varies smoothly under displacement along S (which, of course, requires S to belong to class C^2).

§5. The Method of Rothe

For the proof of existence theorems and the practical determination of solutions of initial-boundary value problems, one can use the method of Rothe, which essentially reduces such problems to boundary value problems for equations of elliptic type. This method was proposed in 1930 by the German–American mathematician, R. Rothe [1], and was justified by him for a one-dimensional parabolic equation. His proof was given naturally in the framework of classical solvability under the condition of sufficient (but no overly excessive) smoothness on all the data of the problem. For many-dimensional parabolic equations (linear and quasi-linear) this method was investigated in my paper [12] (for this and other investigations, see also [LSU 1]). This was done in some classes of generalized solutions and smooth solutions.

We shall adduce here the proof of Rothe's method in that class of generalized solutions where all the arguments are basically simple, namely, in the class of generalized solutions in $W_2^{1,0}(Q_T)$. We shall consider the problem (3.1)–(3.2) and shall assume, only for economy of space, that $a_i = f_i = 0$. Thus, we shall seek a solution $u(x, t)$ of the problem

$$\mathscr{M}u \equiv \frac{\partial u}{\partial t} - \mathscr{L}u = f, \tag{5.1}$$

$$u|_{t=0} = \varphi(x), \qquad u|_{S_T} = 0, \tag{5.2}$$

where $\mathscr{L}u \equiv (\partial/\partial x_i)(a_{ij}(x, t)(\partial u/\partial x_j)) - b_i(\partial u/\partial x_i) - au$, under the assumption that conditions (3.3)–(3.5) are satisfied. Let us cut up the cylinder $Q_T = \Omega \times (0, T)$ by the planes $t = k\tau$ $(k = 1, \ldots, N)$, $\tau = T/N$, and denote by Ω_k the cross-section of Q_T by the plane $t = k\tau$.

For $k = 1, \ldots, N$ we introduce the functions

$$\left.\begin{aligned}
a_{ij}^\tau(k) &\equiv a_{ij}^\tau(x, k) = \frac{1}{\tau}\int_{(k-1)\tau}^{k\tau} a_{ij}(x, t)\, dt,\\
b_i^\tau(k) &\equiv b_i^\tau(x, k) = \frac{1}{\tau}\int_{(k-1)\tau}^{k\tau} b_i(x, t)\, dt,\\
a^\tau(k) &\equiv a^\tau(x, k) = \frac{1}{\tau}\int_{(k-1)\tau}^{k\tau} a(x, t)\, dt,\\
f^\tau(k) &\equiv f^\tau(x, k) = \frac{1}{\tau}\int_{(k-1)\tau}^{k\tau} f(x, t)\, dt.
\end{aligned}\right\} \tag{5.3}$$

These functions are defined for $k = 1, 2, \ldots, N$ and for almost all x in Ω, where $f^\tau(x, k) \in L_2(\Omega)$, and where the remaining functions are bounded on Ω. Let us replace (5.1) by the differential-difference equations

$$\mathscr{M}^\tau u(k) \equiv u_{\bar t}(k) - \mathscr{L}^\tau u(k) = f^\tau(k), \qquad k = 1, \ldots, N, \tag{5.4}$$

where

$$u(k) \equiv u(x, k\tau),$$

$$\mathscr{L}^{\tau}u(k) \equiv \frac{\partial}{\partial x_i}\left(a_{ij}^{\tau}(k)\frac{\partial u(k)}{\partial x_j}\right) - b_i^{\tau}(k)\frac{\partial u(k)}{\partial x_i} - a^{\tau}(k)u(k), \tag{5.5}$$

$$u_{\bar{t}}(k) \equiv u_{\bar{t}}(x, k\tau) = \frac{1}{\tau}[u(k) - u(k-1)].$$

We shall define the functions $u(k)$ as solutions of (5.4) satisfying the conditions

$$u(0) = \varphi(x), \qquad u(k)|_S = 0, \qquad k = 1, \ldots, N. \tag{5.6}$$

For all sufficiently small τ ($\tau \le \tau_0$) the $u(k)$ are defined uniquely as elements of $\overset{\circ}{W}{}_2^1(\Omega)$ (cf. §3, Chapter II).†

For all $\eta \in \overset{\circ}{W}{}_2^1(\Omega)$ the $u(k)$ satisfy the integral identity

$$\int_{\Omega} u_{\bar{t}}(k)\eta(x)\,dx + \mathscr{L}^{\tau}(u(k), \eta(\cdot)) = (f^{\tau}(k), \eta(\cdot)), \tag{5.7}$$

in which

$$\mathscr{L}^{\tau}(u(k), \eta(\cdot)) \equiv \int_{\Omega}\left[a_{ij}^{\tau}(k)\frac{\partial u(k)}{\partial x_j}\frac{\partial \eta(x)}{\partial x_i} + b_i^{\tau}(k)\frac{\partial u(k)}{\partial x_i}\eta(x) + a^{\tau}(k)u(k)\eta(x)\right]dx. \tag{5.8}$$

We shall obtain an *a priori* bound for $u(k)$ not depending on the step size τ. For this we take $\eta(x) = 2\tau u(k)$ in (5.7) and use the relation

$$2\tau u(k)u_{\bar{t}}(k) = u^2(k) - u^2(k-1) + \tau^2 u_{\bar{t}}^2(k) \tag{5.9}$$

and the inequalities (3.3) and (3.5) which still hold for the averages (5.3) of the coefficients of $\mathscr{L}$. This guarantees the validity of the following inequalities:

$$\begin{aligned}
&\|u\|_{2,\Omega_k}^2 - \|u\|_{2,\Omega_{k-1}}^2 + \tau^2\|u_{\bar{t}}\|_{2,\Omega_k}^2 + 2\nu\tau\|u_x\|_{2,\Omega_k}^2 \\
&\le -2\tau\int_{\Omega_k}(b_i^{\tau}u_{x_i}u + a^{\tau}u^2)\,dx + 2\tau\int_{\Omega_k} f^{\tau}u\,dx \\
&\le 2\mu\tau(\|u_x\|_{2,\Omega_k}\|u\|_{2,\Omega_k} + \|u\|_{2,\Omega_k}^2) + 2\tau\|f^{\tau}\|_{2,\Omega_k}\|u\|_{2,\Omega_k} \\
&\le \nu\tau\|u_x\|_{2,\Omega_k}^2 + \left(\frac{\mu^2}{\nu} + 2\mu\right)\tau\|u\|_{2,\Omega_k}^2 + 2\tau\|f^{\tau}\|_{2,\Omega_k}\|u\|_{2,\Omega_k},
\end{aligned}$$

† As is clear from (5.10), this will happen for $\tau < 1/c$.

whence, for $k = 1, \ldots, N$,

$$\|u\|^2_{2,\Omega_k} - \|u\|^2_{2,\Omega_{k-1}} + \tau^2\|u_{\bar{t}}\|^2_{2,\Omega_k} + \nu\tau\|u_x\|^2_{2,\Omega_k} \leq c\tau\|u\|^2_{2,\Omega_k} + 2\tau\|f^\tau\|_{2,\Omega_k}\|u\|_{2,\Omega_k}, \tag{5.10}$$

where $c = \mu^2/\nu + 2\mu$. Let us discard the third and fourth terms from the left-hand side of (5.10) and divide both sides of the resulting inequality by $\|u\|_{2,\Omega_k} + \|u\|_{2,\Omega_{k-1}}$ if this sum is positive. Replacing the expression $\|u\|_{2,\Omega_k}(\|u\|_{2,\Omega_k} + \|u\|_{2,\Omega_{k-1}})^{-1}$ in the right-hand side by unity, we obtain as a result the inequality

$$\|u\|_{2,\Omega_k} \leq (1 - c\tau)^{-1}\|u\|_{2,\Omega_{k-1}} + 2\tau(1 - c\tau)^{-1}\|f^\tau\|_{2,\Omega_k} \tag{5.11}$$

for $\tau \leq (2c)^{-1}$. This inequality obviously holds for those k for which $\|u\|_{2,\Omega_k} + \|u\|_{2,\Omega_{k-1}} = 0$. It is not hard to verify that (5.11) implies

$$\begin{aligned}\|u\|_{2,\Omega_k} &\leq (1 - c\tau)^{-k}\|u\|_{2,\Omega_0} + 2\tau\sum_{s=1}^{k}(1 - c\tau)^{-k+s-1}\|f^\tau\|_{2,\Omega_s} \\ &\leq (1 - c\tau)^{-k}\left[\|u\|_{2,\Omega_0} + 2\tau\sum_{s=1}^{k}\|f^\tau\|_{2,\Omega_s}\right] \\ &\leq \exp\left\{\frac{c\tau}{1 - c\tau}k\right\}[\cdots] \leq e^{2cT}[\|\varphi\|_{2,\Omega_0} + 2\|f^\tau\|_{2,1,Q_k}],\end{aligned} \tag{5.12}$$

where

$$\|f^\tau\|_{2,1,Q_k} \equiv \tau\sum_{s=1}^{k}\|f^\tau\|_{2,\Omega_s} \tag{5.13}$$

(for this, we used the fact that $c\tau \leq \frac{1}{2}$).

Let us now sum the inequalities (5.10) from $k = 1$ to $k = m \leq N$ and apply the estimate (5.12). This gives us

$$\|u\|^2_{2,\Omega_m} + \nu\tau\sum_{k=1}^{m}\|u_x\|^2_{2,\Omega_k} + \tau^2\sum_{k=1}^{m}\|u_{\bar{t}}\|^2_{2,\Omega_k} \leq c_1[\|\varphi\|^2_{2,\Omega} + \|f^\tau\|^2_{2,1,Q_m}], \tag{5.14}$$

$m = 1, \ldots, N$, where c_1 depends only on ν, μ, and T. The inequalities (5.12) and (5.14) give us a bound for $u(k)$ by means of which we can justify the convergence of Rothe's method as $\tau \to 0$. The quantities $\|f^\tau\|_{2,1,Q_m}$ do not exceed $\|f\|_{2,1,Q_m}$. Let us introduce piecewise constant interpolations in t for $u(k)$ and all remaining functions under consideration. Namely, we shall denote by $\tilde{u}^\tau(x, t)$ the function which is equal to $u(x, k\tau)$ for $t \in ((k-1)\tau, k\tau]$. It is obvious that $\tilde{u}^\tau(x, t) \in W_2^{1,0}(Q_T)$ and that, by (5.14)

$$\|\tilde{u}^\tau\|_{2,Q_T} + \|\tilde{u}^\tau_x\|_{2,Q_T} \leq c_2, \tag{5.15}$$

where c_2 is a constant independent of τ. We now let N tend to infinity. Because of (5.15), we can select from $\{\tilde{u}^\tau\}$ a subsequence which is weakly convergent in $\mathring{W}_2^{1,0}(Q_T)$ to some element $u(x, t)$ in $\mathring{W}_2^{1,0}(Q_T)$. We shall show that $u(x, t)$ satisfies the integral identity (3.15), that is, it is a generalized solution of the problem (5.1)–(5.2) in $\overset{\circ}{W}{}_2^{1,0}(Q_T)$. It is enough to establish (3.15) only for smooth $\eta(x, t)$ vanishing on S_T and for $t = T$. Let $\eta(x, t) \in C^1(Q_{T+\tau})$, and let $\eta(x, t)$ be zero on S_T and for $t \in [T, T + \tau]$. We construct for $\eta(x, t)$ the functions $\eta(k) = \eta(x, k\tau)$, $\tilde{\eta}^\tau(x, t)$, and $(\tilde{\eta}_t)^\tau$, where $\eta_t(k) = \tau^{-1}[\eta(k + 1) - \eta(k)]$. It is easy to verify that $\tilde{\eta}^\tau$, $\partial\tilde{\eta}^\tau/\partial x_i$, and $\tilde{\eta}_t^\tau \equiv (\tilde{\eta}_t)^\tau$ converge uniformly in $\bar{Q}_T$ to η, $\partial\eta/\partial x_i$, and $\partial\eta/\partial t$, respectively, where $\tilde{\eta}^\tau(x, t)$ vanishes for $t \in [T, T + \tau]$. Let us take $\eta = \tau\eta(x, k\tau)$ in (5.7) and sum (5.7) from $k = 1$ to $k = N$. The result can be written in the form

$$-\tau \sum_{k=1}^{N} \int_\Omega u(k)\eta_t(k)\, dx - \int_\Omega \varphi\eta(1)\, dx + \tau \sum_{k=1}^{N} \mathscr{L}^\tau(u(k), \eta(k)) = \tau \sum_{k=1}^{N} (f^\tau(k), \eta(k)), \tag{5.16}$$

if we use the formula for "summation by parts,"

$$\tau \sum_{k=1}^{N} u_{\bar{t}}(k)\eta(k) = -\tau \sum_{k=1}^{N} u(k)\eta_t(k) - u(0)\eta(\ )† \tag{5.17}$$

and take into account the fact that $\eta(N) = \eta(N + 1) = 0$. Let us write (5.16) in the integral form

$$-\int_{Q_T} \tilde{u}^\tau\tilde{\eta}_t^\tau\, dx\, dt + \int_0^T \int_\Omega \left(a_{ij}\frac{\partial\tilde{u}^\tau}{\partial x_j}\frac{\partial\tilde{\eta}^\tau}{\partial x_i} + b_i\frac{\partial\tilde{u}^\tau}{\partial x_i}\tilde{\eta}^\tau + a\tilde{u}^\tau\eta^\tau\right) dx\, dt - \int_\Omega \varphi\eta(1)\, dx = \int_0^T \int_\Omega f\tilde{\eta}^\tau\, dx\, dt. \tag{5.18}$$

We can go to the limit in (5.18) with the subsequence selected above; as a result, it is not hard to verify that we obtain (3.15). Thus, we have proved that the limit function $u(x, t)$ is actually a generalized solution of (5.1)–(5.2) in $\overset{\circ}{W}{}_2^{1,0}(Q_T)$. By the uniqueness theorem (Theorem 3.2) and the estimate (5.15), every subsequence $\{\tilde{u}^\tau\}$ converges weakly in $\overset{\circ}{W}{}_2^{1,0}(Q_T)$ to $u(x, t)$. Thus we have proved

Theorem 5.1. *If the conditions* (3.3)–(3.5) *are fulfilled, the solution of the problem* (5.1)–(5.2) *is the limit of the functions* $\{\tilde{u}^\tau\}$, *which can be calculated from the relations* (5.4)–(5.6).

In establishing Rothe's method, we have proved along the way the existence of a solution $u(x, t)$ of (5.1)–(5.2) in $\overset{\circ}{W}{}_2^{1,0}(Q_T)$. If we have a little more information about $u(x, t)$, it is not difficult to prove the strong convergence of $\tilde{u}^\tau$ to u in $\overset{\circ}{W}{}_2^{1,0}(Q_T)$ and even in $V_2(Q_T)$.

† The relation (5.17) is easy to verify directly.

It is possible to solve other initial-boundary value problems (problems (4.1)–(4.2)) by Rothe's method. In these problems we have to solve the elliptic equations (5.4) under the second or third boundary condition. The unique solvability of such problems in $W_2^1(\Omega)$ was established in §5, Chapter II. On the basis of these results, the solvability of the problems (4.1)–(4.2) in $W_2^{1,0}(Q_T)$ can be proved in the same way as was done for the problem (5.1)–(5.2).

Supplements and Problems

1. In this chapter it is shown that for initial-boundary value problems (and for Cauchy's problem too) uniqueness theorems are true for arbitrary coefficients $a(x, t)$ of u if only the coefficients and solutions are not "very bad." Thus, unlike elliptic equations, homogeneous initial-boundary value problems for parabolic equations have only zero solutions. This is true for equations $\mathcal{M}u + \lambda u = 0$ with any λ. The fact that such an inclusion of λ in the equation $\mathcal{M}u = 0$ does not influence the solvability is evident because of the following reason: Instead of $u(x, t)$ we may introduce the new unknown function $v(x, t) = u(x, t)e^{-\lambda t}$. If u was a solution of the equation $\mathcal{M}u = f$ then v would be a solution of $\mathcal{M}v + \lambda v = fe^{-\lambda t}$ and vice versa. By this substitution of the unknown function we can obtain any sign of the coefficient of v if a is a bounded function. This substitution should be kept in mind when studying parabolic equations.

2. Theorem 2.1 can be proved more easily with the help of the expansion in eigenfunctions $\{u_k(x)\}_{k=1}^{\infty}$ of the Laplace operator Δ under the first boundary condition (see subsection 3, §4 of this chapter). The solution $u^m(x,t)$ of the problem (2.1)–(2.3) for $f_i(x, t) \equiv 0$,

$$f(x, t) = \sum_{k=1}^{m} f_k(t)u_k(x) \quad \text{and} \quad \varphi(x) = \sum_{k=1}^{m} a_k u_k(x),$$

is given by (4.44) with $a_k = 0, f_k(\tau) \equiv 0, k \geq m + 1$. It belongs, obviously, to $W_{2,0}^{\Delta,1}(Q_T)$, if $f_k \in L_2((0, T))$. Arbitrary functions $f \in L_2(Q_T)$ and $\varphi \in \mathring{W}_2^1(\Omega)$ we approximate by

$$f^m(x, t) = \sum_{k=1}^{m} (f(\cdot, t), u_k(\cdot))u_k(x) \quad \text{and} \quad \varphi^m(x) = \sum_{k=1}^{m} (\varphi(\cdot), u_k(\cdot))u_k(x)$$

in the norm of $L_2(Q_T)$ and $\mathring{W}_2^1(\Omega)$, respectively. The solutions $u^m(x, t)$, $m = 1, 2, \ldots$ corresponding to f^m and φ^m converge to a solution u corresponding to f and φ in the norm (2.7), and even in the norm $\|\cdot\|_{W_T}$, so that $u \in W_{2,0}^{\Delta,1}(Q_T)$. This is proved using the equalities (2.15).

Theorem 2.2 can also be proved with the help of the functions $\{u_k(x)\}_{k=1}^{\infty}$. Indeed, let us take, in (2.21), with $f \equiv f_i \equiv 0, \eta(x, t) = c(t)u_k(x)$ where $c(t)$ is the solution of the problem

$$\frac{dc(t)}{dt} + \lambda_k c(t) = b(t), \qquad c(\tau) = 0,$$

with an arbitrary smooth function b. It gives

$$\int_0^T b(t)\left(\int_\Omega u(x, t)u_k(x)\, dx\right) dt = 0$$

and, therefore, $(u(\cdot, t), u_k(\cdot)) = 0, k = 1, 2, \ldots$ and $u(x, t) \equiv 0$.

3. While studying parabolic equations it seemed useful to introduce and make use of several function spaces. Some of them are Hilbert spaces and it is easier to work in such spaces. The main relation of the energy balance (equality (3.6)), however, generates the Banach, but not Hilbert, norm $|\cdot|_Q$ (see (1.14)) and enables us to estimate $|u|_{Q_T}$ by the data of the problem. To use this estimate wholly we introduce the Banach spaces $V_2(Q_T)$ and $V_2^{1,0}(Q_T)$ and their subspaces $\overset{\circ}{V}_2(Q_T)$ and $\overset{\circ}{V}{}_2^{1,0}(Q_T)$, suitable for the study of the first initial-boundary value problem. We could have used the Hilbert spaces $W_2^{1,1/2}(Q_T)$ and $\overset{\circ}{W}{}_2^{1,1/2}(Q_T)$ which are related most closely to the spaces $V_2^{1,0}(Q_T)$ and $\overset{\circ}{V}{}_2^{1,0}(Q_T)$. Their elements $u(\cdot)$ belong to $L_2(Q_T)$ and have the derivatives u_{x_i} and $D_t^{1/2}u$ belonging to $L_2(Q_T)$. The generalized solutions we found are also elements of $W_2^{1,1/2}(Q_T)$ (for details see Chapter III of [LSU 1]).

Essentially, to comprehend the material of this chapter one must master all these spaces, and first of all the following things;

(a) Prove that the set of all smooth functions on $\bar{Q}_T$ equal to zero near S_T is dense in $\overset{\circ}{V}{}_2^{1,0}(Q_T)$ and the set of all smooth functions on $\bar{Q}_T$ is dense in $V_2^{1,0}(Q_T)$ (in the second case the domain Ω must be such that $W_2^1(\Omega) = \tilde{W}_2^1(\Omega)$). Moreover, from these sets it is sufficient to take only functions of the type $u(x, t) = \varphi(t)\psi(x)$. The density is meant in the norm (1.14). (The prefix "ess" may be discarded.)

(b) In §2 we used the fact that the generalized solutions of the problem $\Delta u = f(x)$, $u|_S = 0$, with $f \in L_2(\Omega)$, have the generalized derivatives $u_{xx} \in L_2(\Omega')$, $\forall \Omega' \subset \Omega$. The simple proof of this is done in No. 12 of the Supplement to Chapter II. If $S \subset C^2$ (or even satisfies the hypothesis of §7, Chapter II), then $u \in W_{2,0}^2(\Omega)$. Prove that in this case $W_{2,0}^2(\Omega) = W_2^2(\Omega) \cap \overset{\circ}{W}{}_2^1(\Omega)$ and the set of all functions from $C^2(\bar{Q}_T)$ equal to zero on S_T is dense in $W_{2,0}^{2,1}(Q_T)$. In (2.4) and in (2.5) the terms $u_x v_x$ may be discarded and for bounded domains Ω, the terms uv may be discarded as well.

4. In subsection 4 of §4 we derived the second fundamental inequality. It is true for any element u of $W_2^{2,1}{}_0(Q_T)$. Using this fact prove the following theorem:

Theorem 1. *If the conditions of regularity for the coefficients of* $\mathscr{M}$ *and* S *indicated in subsection* 4 *of* §4 *are fulfilled then the problem* (3.1)–(3.2) *has a unique solution belonging to* $W_{2,0}^{2,1}(Q_T)$ *for every* $\varphi \in \mathring{W}_2^1(\Omega)$ *and*

$$F \equiv f + \sum_{i=1}^{n} \frac{\partial f_i}{\partial x_i} \in L_2(Q_T).$$

It may be carried out by different methods:

(a) By the method of continuation by a parameter, taking the family $\mathscr{M}_\tau u = (1-\tau)\mathscr{M}_0 u + \tau \mathscr{M} u$, $\tau \in [0, 1]$, where $\mathscr{M}_0 u = u_t - \Delta u$, and using the results of §2 about the solvability of the problem in $W_{2,0}^{2,1}(Q_T)$ for $\mathscr{M}_0$.

(b) One can use the method which we used in §2 for $\mathscr{M}_0$. The set $W_{2,0}^{2,1}(Q_T)$ may be taken as the domain of definition of the operator $Au = \{\mathscr{M}u; u(\cdot, 0)\}$. One may take as $\mathscr{D}(A)$ only those elements u of $W_{2,0}^{2,1}(Q_T)$ which have the derivatives u_{tx_i} in $L_2(Q_T)$. This method was proposed in [LA 6]. In [LA 8(a), (b)] it was realized for abstract equations in the Hilbert space and in [LA 8(c)] it was shown that the conditions of theorems from [LA 8(b)] are fulfilled for some initial-boundary value problems, the problem (3.1), (3.2) included.

(c) In the frame of the method indicated in (b) the absence of an orthogonal complement in $\mathscr{W} = L_2(Q_T) \times \mathring{W}_2^1(\Omega)$ to the range $R(A)$ of operator A can be proved in a different way. To do this it is convenient from the beginning to make the substitution $u(x, t) = v(x, t)e^{\lambda_0 t}$, $\lambda_0 \gg 1$, of No. 1 and to deal with the operator $\mathscr{M} + \lambda_0 I$, instead of $\mathscr{M}$, considering the operator $Au = \{\mathscr{M}u + \lambda_0 u; u(\cdot, 0)\}$. The proof is based on the inequality

$$\int_{Q_T} (\mathscr{M}u + \lambda_0 u)(\mathscr{M}'u + \lambda_0 u)\, dx\, dt \geq c\|u\|_{W_T}^2 - c_1\|u_x(\cdot, 0)\|_{2,\Omega}^2, \quad (1)$$

which is true for arbitrary elements u of $\mathscr{D}(A)$ and arbitrary parabolic operators $\mathscr{M} = \partial/\partial t - \mathscr{L}$ and $\mathscr{M}' = \partial/\partial t - \mathscr{L}'$ provided λ_0 is large enough.

The positive constants c, c_1 and λ_0 are effectively calculated and they are determined by the constants ν and ν' of ellipticity of $\mathscr{L}$ and $\mathscr{L}'$ and by the majorants μ and μ' of the coefficients of both $\mathscr{L}$, $\mathscr{L}'$ and their derivatives indicated in (3.3), (3.5) and (4.49). These constants depend also on some characteristics of $S \subset C^2$ as in inequality (8.17) of Chapter II.

Thus first we must prove (1), choosing an appropriate λ_0. The principal part of the proof of Theorem 1 consists of the proof that $w \equiv 0$ if w belongs to $L_2(Q_T)$ and satisfies the identity

$$\int_{Q_T} w(v_t - \mathscr{L}v + \lambda_0 v)\, dx\, dt = 0, \tag{2}$$

where v is an arbitrary element of $\mathscr{D}(A)$ and $v(x, 0) \equiv 0$. Take as v the solution of the problem

$$\mathscr{M}'v + \lambda_0 v \equiv v_t - \Delta v + \lambda_0 v = w, \qquad v|_{S_T} = 0, \qquad v|_{t=0} = 0 \tag{3}$$

(whose solvability in $W_{2,0}^{2,1}(Q_T)$ is proved in §2). Substituting the expression $\mathscr{M}'v + \lambda_0 v$ in (2) instead of w and using (1) we obtain that $v \equiv 0$ and, therefore, $w \equiv 0$.

5. We obtained estimate (4.56) by analyzing the integral $\int_{Q_t} (\mathscr{M}u)^2\, dx\, d\tau$. As it can be seen from this analysis the estimate

$$\begin{aligned} &\nu \int_\Omega u_x^2(x, t)\, dx + \int_{Q_t} u_\tau^2(x, \tau)\, dx\, d\tau \\ &\quad\le c_1(t)\left[\int_\Omega (u_x^2 + u^2)|_{t=0}\, dx + \int_{Q_t} (\mathscr{M}u)^2\, dx\, d\tau\right] \end{aligned} \tag{4}$$

could be derived from considering the integral $\int_{Q_t} u_\tau \mathscr{M}u\, dx\, d\tau$. The terms $\int_\Omega u^2(x, t)\, dx + \int_{Q_t}(u_x^2 + u^2)\, dx\, d\tau$ may be inserted into the left-hand side of (4). The usefulness of (4) lies in the fact that (4) is true for any domain Ω, with no smoothness required for the boundary S and for a_{ij} non-differentiable in x, $\partial a_{ij}/\partial t$ being the only derivatives used. Note that in all our estimations the domain Ω may be unbounded if $\int_\Omega u_x^2|_{t=0}\, dx$ in their right members is replaced by $\int_\Omega (u_x^2 + u^2)|_{t=0}\, dx$, as has been done for inequality (4). Inequality (6.3) in Chapter I could not be used.

Show that all theorems in Chapter III except for the Fourier representations are true for unbounded domains.

Now let us return to (4.56). Inequality (4) is also fulfilled for the operator $\mathscr{M}u \equiv \alpha(x, t)u_t - \mathscr{L}u$ if $0 < \alpha_0 \le \alpha(x, t) \le \alpha_1$.

The derivatives u_{xx} can be estimated by $\mathscr{M}u$ if one considers the integral $-\int_{Q_t} \Delta u \cdot \mathscr{M}u\, dx\, d\tau$ and uses (8.17) of Chapter II and also (3.14). This estimate for bounded domains Ω has the form

$$\begin{aligned} &\int_\Omega u_x^2(x, t)\, dx + \nu \int_{Q_t} u_{xx}^2\, dx\, d\tau \\ &\quad\le c_2(t)\left[\int_\Omega u_x^2(x, 0)\, dx + \int_{Q_t} (\mathscr{M}u)^2\, dx\, dt\right]. \end{aligned} \tag{5}$$

It is valid provided S and the coefficients of $\mathscr{L}$ satisfy the conditions of §7 of Chapter II. The existence of $\partial a_{ij}/\partial t$ is not necessary. Thus we have separated

(4.56) in two parts, (4) and (5), and showed that each of them is true when is only a part of the conditions is fulfilled. This partition turns out to be useful for the study of linear equations with "bad" coefficients and even for some classes of quasilinear parabolic equations.

6. In different branches of mathematics, physics, and mechanics there are different kinds of degenerated parabolic equations similar to those which are described in the Supplement of Chapter II for elliptic equations. A simple enumeration of these cases would have taken a lot of space and therefore we omit it. However, we want to point out that it is the elliptic part $\mathscr{L}$ of the parabolic operator $\mathscr{M}$ which corresponds to the boundary condition on S_T and the term u_t (or αu_t) that responds to the initial condition. The methods and estimates presented in this chapter are often useful for the investigation of those equations. Another set of estimates is based on the fact that for second-order parabolic equations the maximum principle is valid. However, this principle and its generalizations and consequences are not presented and not used here.

7. Show that the solution of problem (3.2) for the equation $\mathscr{M}u + \lambda_0 u = f$, $\lambda_0 \gg 1$, can be obtained as the limit of the solutions u^ε of the Dirichlet problem

$$\mathscr{M}_\varepsilon u \equiv -\varepsilon u_{tt} + \mathscr{M}u + \lambda_0 u = f, \qquad u|_{S_T} = 0, \qquad u|_{t=0} = \varphi, u|_{t=T} = 0, \tag{6}$$

in Q_T as $\varepsilon \to 0+$. (Here $\mathscr{M}$ is of the form (3.1).) For this purpose, obtain estimates considering the equalities

$$\int_{Q_T} u\mathscr{M}_\varepsilon u \, dx\, dt = \int_{Q_T} fu \, dx\, dt$$

and

$$\int_{Q_T} \zeta(t, \tau)u_t(x, t)\mathscr{M}_\varepsilon u(x, t)\, dx\, dt = \int_{Q_T} \zeta(t, \tau)f(x, t)u(x, t)\, dx\, dt,$$

where $\zeta(t, \tau) = 1$ for $t \in [0, \tau - \delta]$, $\zeta(t, \tau) = (\tau - t)\delta^{-1}$ for $t \in [\tau - \delta, \tau]$ and $\zeta(t, \tau) = 0$ for $t \in [\tau, T]$ and $0 < \delta \leq \tau \leq T$.

8. Find the limit of the solutions u_ε of the problems

$$\mathscr{M}^\varepsilon u = u_t - \varepsilon\frac{\partial}{\partial x_i}(a_{ij}u_{x_j}) + b_i u_{x_i} + au = f, \qquad u|_{S_T} = 0, u|_{t=0} = \varphi,$$

as $\varepsilon \to 0+$ (see [LA 10]).

9. Initial-boundary value problems for parabolic equations are uniquely solvable in $Q_T = \{(x, t): t \in (0, T), x \in \Omega(t)\}$ with domains $\Omega(t)$ changing in time. If the external normals to the lateral surface $S^T = \{(x, t): t \in [0, T], x \in \partial\Omega(t)\}$ form non-zero angles with the t-axis then the boundary condition is prescribed on the whole of S^T. Prove the unique solvability of the first boundary value problem in such Q^T by putting $u = 0$ on ∂Q^T except for the set $\{(x, t): t = T, x \in \Omega(T)\}$. It is convenient to use the method of Rothe.

10. Examine in detail the methods of Fourier and Laplace (see subsection 3, §4).

11. We can show using representation (4.35) that $u(x, t)$ converges to zero as $t \to +\infty$ not slower than $e^{\lambda_1 t}$ if all λ_k are negative. This fact can also be proved without this representation. Moreover, it remains valid for the solutions of the equations

$$u_t - \mathscr{L}u = 0, \qquad u|_{S_\infty} = 0, \tag{7}$$

in which the coefficients of $\mathscr{L}$ depend not only on x but also on t and $\mathscr{L} \neq \mathscr{L}^*$. Let us suppose that $\mathscr{L}$ satisfies

$$\mathscr{L}(v, v) \geq \nu_1 \|v\|^2_{2,\Omega}, \qquad \nu_1 = \text{const.} > 0, \tag{8}$$

for all $v \in \overset{\circ}{W}{}^1_2(\Omega)$. Then for the solutions of (7) the following inequality holds:

$$\frac{1}{2}\frac{d}{dt}\|u(\cdot, t)\|^2_{2,\Omega} + \nu_1\|u(\cdot, t)\|^2_{2,\Omega} < 0,$$

and therefore

$$\|u(\cdot, t)\|_{2,\Omega} \leq e^{-\nu_1 t}\|u(\cdot, 0)\|_{2,\Omega} \tag{9}$$

(for $\mathscr{L}$ in (4.29), $\nu_1 = -\lambda_1$). For the solutions of the problem

$$\mathscr{M}u \equiv u_t - \mathscr{L}u = f(x, t), \qquad u|_{S_\infty} = 0, \tag{10}$$

the inequality

$$\frac{1}{2}\frac{d}{dt}\|u(\cdot, t)\|^2_{2,\Omega} + \nu_1\|u(\cdot, t)\|^2_{2,\Omega} \leq \|f(\cdot, t)\|_{2,\Omega}\|u(\cdot, t)\|_{2,\Omega} \tag{11}$$

holds. From (11) we can easily deduce that

$$\|u(\cdot, t)\|_{2,\Omega} \leq e^{-\nu_1 t}\|u(\cdot, 0)\|_{2,\Omega} + \int_0^t e^{-\nu_1(t-s)}\|f(\cdot, s)\|_{l,\Omega}\, ds. \tag{12}$$

If, in particular, f is independent of t, then (12) implies the uniform boundedness of $\|u(\cdot, t)\|_{2,\Omega}$. If, moreover, the coefficients of $\mathscr{L}$ do not depend on t, then all solutions of (10) converge (in the $\mathscr{L}_2(\Omega)$ norm) to the solution $u_0(x)$ of the stationary problem $-\mathscr{L}u_0 = f(x)$, $u_0|_{\partial\Omega} = 0$.

Investigate the behavior of solutions of problems (7) and (10) in stronger norms; for example, in the norm of $W^1_2(\Omega)$.

12. The proof of estimate (3.14) is a bit too long. It is easier to majorize $|u|_{Q_t}$ by $\|\varphi\|_{2,\Omega}$ and $\|f + \partial f_i/\partial x_i\|_{2,Q_t}$ if $F \equiv f + \partial f_i/\partial x_i$ belongs to $L_2(Q_t)$. To do that let us write the last term in (3.6) in the form $\int_Q Fu\,dx\,d\tau$ and majorize it by $\|F\|_{2,Q_t}\|u\|_{2,Q_t}$. The other terms are estimated as in (3.8). Continue this approach and calculate $c(t)$ in the inequality

$$|u|_{Q_t} \leq c(t)[\|u(\cdot, 0)\|_{2,\Omega} + \|F\|_{2,Q_t}]. \tag{13}$$

13. We have proved in §2 and in subsection 2 of §4 the existence of a generalized solution in $W_2^{1,0}(Q_T)$ for all three classical initial-boundary value problems using the Galerkin approximations. Do the same for the boundary conditions described in No. 4 of the Supplement to Chapter II.

In subsection 2 of §4 we have showed how the uniqueness theorems for such a class of generalized solutions can be proved. However, this proof is not as simple as the proof of existence in this class. If all coefficients in (3.1) have bounded derivatives with respect to t and $f_i, f_{it} \in L_2(Q_T)$ then it is possible to obtain an estimate for $|u_t^N|_{Q_T}$. This estimate will be uniform in N if the initial function $\varphi = u|_{t=0}$ is sufficiently "good" (find out the conditions on φ). The estimate for $|u_t^N|_{Q_T}$ is derived from equalities (3.16) which admit differentiation with respect to t. A limit function u of sequence $\{u^N\}_{N=1}^{\infty}$ will have the derivatives $u_{tx_i} \in L_2(Q_T)$. For such generalized solutions, identity (3.20), as well as (3.6) and (3.14), are derived directly from (3.16), (3.17). Uniqueness in this class of solutions follows immediately from (3.14).

In the framework of Galerkin's method it is not difficult to watch step-by-step the improvements of regularity of solutions in connection with the regularity of data and compatibility of the boundary condition with the initial data and the equation. In [LA 4] this was done for hyperbolic equations. The same approach is applicable to parabolic equations. At the beginning one must follow the increasing of smoothness in t (and obtain the estimations for $D_t^l u$ and $D_x D_t^l u$) simultaneously. This is possible since the boundary condition admits differentiation with respect to t. If, for example, the coefficients of $\mathscr{L}$ in (10) do not depend on t then $v \equiv D_t^l u$ satisfies the equations

$$\mathscr{M}v \equiv v_t - \mathscr{L}v = D_t^l f, \qquad v|_{S_\infty} = 0, \tag{14}$$

and therefore the energy relation and the corresponding energy estimate are valid for v. The derivatives $D_t^l u^N$ of the Galerkin approximations u^N satisfy the same equations as does u^N and for them we can derive estimates analogous to those for u^N. The initial data for $D_t^l u$ and $D_t^l u^N$ are derived from the initial data for u and the equations for u and u^N, respectively.

If the coefficients of $\mathscr{L}$ (and of σ in the case of the third boundary condition) depend on t, then the procedure becomes slightly more complicated, but the principal line of reasoning remains the same. For the investigation of the smoothness of u in x it is necessary to use results about the solutions of elliptic problems considering the terms u_t and v_t in (10) and (14) as known functions. It can also be done in the framework of Galerkin's approximations.

It is useful to do this all in detail. In [LA 4] we used finite-difference approximations for the investigation of the smoothness of the generalized solutions.

Sometimes it is desirable to have a better convergence of u^N to u; for example, the convergence to zero of the residue $\mathscr{M}(u - u^N)$ in the norm of $L_2(Q_T)$. To obtain this, one should use one of the "modified" Galerkin methods offered in [LA 11]. We described these modifications for the elliptic equations in §8 of Chapter II. For the parabolic equations they are analogous.

14. Let us note that the Rothe method allows to obtain solutions belonging to various function spaces. For example, in [LA 12] it was used for the demonstration of the classical solvability of the first initial value problem for quasilinear parabolic equations.

15. In subsection 3 of §4 we have presented the Fourier method for equations (4.27) with a symmetric operator $\mathscr{L}$. If we add to $\mathscr{L}u$ some lower terms lu and consider the corresponding parabolic equation $\mathscr{M}_1 u \equiv u_t - \mathscr{L}u - lu = f$ then it may be useful to represent the solutions u of this equation satisfying a boundary condition as the series $u(x, t) = \sum_{k=1}^{\infty} c_k(t)u_k(x)$, where $\{u_k(x)\}_{k=1}^{\infty}$ are the eigenfunctions of the operator $\mathscr{L}$ under the same boundary condition. For the coefficients $c_k(t)$ we can obtain, with the help of $\mathscr{M}_1 u = f$, an infinite system of ordinary differential equations. Analysis of this system may give some information about the solutions $u(x, t)$.

16. In Chapter V and in the Supplement to it we explain how the methods of this chapter can be applied to equations of higher orders and to systems of parabolic type. Moreover, these methods are applicable as well to "abstract parabolic" equations embracing some systems or differential equations which are not among the traditional parabolic systems.

CHAPTER IV

Equations of Hyperbolic Type

§1. General Considerations. Posing the Fundamental Problems

Let us recall the information that we shall need from the general theory of hyperbolic equations. The equation

$$\sum_{i,j=1}^{n+1} a_{ij}(x)u_{x_ix_j} + \sum_{i=1}^{n+1} a_i(x)u_{x_i} + a(x)u = f(x) \tag{1.1}$$

is called hyperbolic at the point x^0 if all the eigenvalues of the quadratic form $\sum_{i,j=1}^{n+1} a_{ij}(x^0)\xi_i\xi_j$ are not zero, and if one of them differs in sign from the others; as always, $a_{ij} = a_{ji}$. At the point x^0 such an equation can be represented in the form

$$\sum_{i=1}^{n+1} \lambda_i(x^0)u_{y_iy_i} + \sum_{i=1}^{n+1} b_i(x^0)u_{y_i} + b(x^0)u = f(x^0), \tag{1.2}$$

where $\lambda_{n+1}(x^0) > 0$ and $\lambda_i(x^0) < 0$ $(i = 1, \ldots, n)$, by means of a change of variables $y = y(x)$ (we may take as y an orthogonal system of coordinates with origin at x^0, so that the $\lambda_i(x^0)$ will be the eigenvalues of the form $\sum_{i,j=1}^{n+1} a_{ij}(x^0)\xi_i\xi_j$). If (1.1) is hyperbolic at all points of some domain $\mathscr{D} \subset R^{n+1}$, then we shall say that it is hyperbolic in the domain $\mathscr{D}$. The reduction of such equations to the form (1.2) in a domain is impossible in the general case if $n + 1 > 2$, even locally (that is, in a small neighborhood of the point $x^0 \in \mathscr{D}$).

Instead of (1.2), we can transfer (1.1) to the form

$$u_{y_{n+1}y_{n+1}} - \sum_{i,j=1}^{n} b_{ij}(y)u_{y_iy_j} + \sum_{i=1}^{n+1} b_i(y)u_{y_i} + b(y)u = \tilde{f}(y), \tag{1.3}$$

in which the form $\sum_{i,j=1}^{n} b_{ij}(y)\xi_i\xi_j$ is positive definite. Such a representation can be achieved by a special choice of the variables $y = y(x)$ and also, in general, in a small neighborhood of the point $x \in \mathscr{D}$. By a simpler change of variables $y = y(x)$, we can transform (1.1) to the form

$$u_{y_{n+1}y_{n+1}} + \sum_{i=1}^{n} b_{i,n+1}u_{y_i y_{n+1}} - \sum_{i,j=1}^{n} b_{ij}(y)u_{y_i y_j} + \sum_{i=1}^{n+1} b_i(y)u_{y_i} + b(y)u = \tilde{f}(y), \tag{1.3'}$$

in which the form $\sum_{i,j}^{n} b_{in}\xi_i\xi_j$ is positive definite. But this transformation cannot also be realized in an arbitrary domain $\mathscr{D}$. We shall study equations which have been reduced to the form (1.3); a typical representative of such equations is the wave equation

$$u_{tt} - c^2 \Delta u = f. \tag{1.4}$$

The methods which will be presented in this chapter are applicable to equations of the form (1.3'); the terms $\sum_{i=1}^{n} b_{i,n+1}u_{y_i y_{n+1}}$ appearing in (1.3') have essentially no influence on the line of argument, nor on the final results. The class of equations (1.3) is sufficiently broad: it comprises all second-order equations of hyperbolic type encountered in the problems of mechanics and physics. The variable y_{n+1} is called the time and is denoted by t, while the variables $y_1, \ldots, y_n$ are called space variables and are most often denoted by $x_1, \ldots, x_n$. Thus, our object will be to study equations of the type

$$u_{tt} - \sum_{i,j=1}^{n} a_{ij}(x,t)u_{x_i x_j} + \sum_{i=1}^{n+1} a_i(x,t)u_{x_i} + a(x,t)u = f(x,t), \tag{1.5}$$

in which the form $a_{ij}(x,t)\xi_i\xi_j$ is positive definite and $u_{x_{n+1}} \equiv u_t$. In the space R^{n+1} of the variables (x, t), three types of smooth surfaces are distinguished with respect to (1.5): *space-like oriented surfaces*, *time-like oriented surfaces*, and *characteristic surfaces*. Let a surface be given by the equation $\mathscr{F}(x_1, \ldots, x_{n+1}) = 0$. It will be called a *space-like oriented surface* if

$$\omega(\mathscr{F}_x) \equiv \mathscr{F}^2_{x_{n+1}} - a_{ij}(x,t)\mathscr{F}_{x_i}\mathscr{F}_{x_j} > 0. \tag{1.6}$$

The surface will be called a *characteristic surface* if it satisfies the equation

$$\omega(\mathscr{F}_x) = 0. \tag{1.7}$$

On surfaces with *time-like orientation*, the condition

$$\omega(\mathscr{F}_x) < 0.$$

must be satisfied. The planes $\mathscr{F} \equiv t - c = 0$ obviously are oriented in a space-like way, and cylindrical surfaces formed by "vertical" lines (we assume that the axis $t = x_{n+1}$ is directed "upwards"), that is, by the lines

$(x, t) = \{x_i = x_i^0, i = 1, \ldots, n, t \in (-\infty, \infty)\}$, are oriented in a time-like way. Both nappes of the cone

$$(t - t^0)^2 = c^{-2} \sum_{i=1}^{n} (x_i - x_i^0)^2$$

are characteristic surfaces for the equation (1.4).

We now formulate the Cauchy problem for the equation (1.5). The variant of this problem which is most frequently encountered in mathematical physics consists of determining a solution of (1.5) for $t \in [t_0, T]$ and $x \in R^n$ satisfying the Cauchy conditions

$$u|_{t=t_0} = \varphi(x), \qquad u_t|_{t=t_0} = \psi(x). \tag{1.8}$$

The functions $\varphi(x)$ and $\psi(x)$ prescribing the initial conditions are independent of each other and of the equation (1.5). If we replace t by $\tau = t - t_0$ we can always be assured that the initial instant of time is $\tau = 0$. Moreover, if we replace t by $\tau = -t$, we shall not contradict the properties listed above, so that the problems of finding solutions for $t > t_0$ and for $t < t_0$ are the same. It is not hard to verify that, by means of (1.8) and (1.5), we can calculate all the derivatives of the desired solution for $t = t_0$ if only they exist and if the coefficients of (1.5), as well as f, φ, and ψ, are infinitely differentiable. This fact shows that the conditions (1.8) are not in contradiction with (1.5) and seem to be sufficient for the determination of the function $u(x, t)$.

The Cauchy problem in its general formulation can be stated as follows: The initial conditions (1.8) are prescribed, not on the plane $t = t_0$, but on some space-like surface $\mathscr{F}$ which separates the whole space R^{n+1} into two parts. In one of these parts it is required to find a solution of (1.5) satisfying, on $\mathscr{F}$, the conditions

$$u|_{\mathscr{F}} = \varphi, \qquad u_t|_{\mathscr{F}} = \psi. \tag{1.9}$$

We shall restrict ourselves to a study of the problem (1.5), (1.8). By a change of variables the general case can be reduced to the problem (1.8) for equations of the form (1.3′) and can be analyzed in a way similar to (1.5), (1.8). Initial-boundary value problems are more difficult, but are more frequently encountered in applications. In such problems a solution $u(x, t)$ of (1.5) is sought in some domain $\Omega \subset R^n$ of the variable x for $t \in [t_0, T]$. For $x \in \Omega$ and $t = t_0$ this solution should satisfy the initial conditions (1.8), and, for $x \in S$, $t \in [t_0, T)$, the boundary condition

$$u|_{S_T} = \chi(s, t), \tag{1.10}$$

or

$$\frac{\partial u}{\partial N} + \sigma(s, t)u|_{S_T} = \chi(s, t), \tag{1.11}$$

where $S_T = S \times [t_0, T]$, $S = \partial\Omega$, $\sigma(x, t)$, and $\chi(s, t)$ are given functions, and $\partial u/\partial N = a_{ij}u_{x_j}\cos(\mathbf{n}, \mathbf{x}_i)$, $\mathbf{n}$ being the outer normal to S. In the general formulation, these problems can be posed as follows: In the space R^{n+1} we are given a domain $\tilde{Q}_T$ which is homeomorphic to the cylinder $Q_T = \{(x, t): x \in \Omega,\ t \in (t_0, T)\}$. Its lower and upper bases corresponding to the lower and upper bases of Q_T lie on space-like surfaces, while the lateral surface corresponding to S_T is oriented in a time-like way. We seek a solution of (1.5) in $\tilde{Q}_T$ satisfying on the lower base the initial condition (1.9) and on the lateral surface either the condition (1.10) or the condition (1.11). If $\tilde{Q}_T$ can be represented in Q_T by a sufficiently smooth change of variables, then this problem reduces to the corresponding problem in the cylinder Q_T for the equation (1.3′), which can be treated in the same way as (1.5).

We shall give special attention to the problem (1.5), (1.8), (1.10) in Q_T, where we shall assume that the boundary condition (1.10) has been reduced to a homogeneous condition. The problem (1.5), (1.8), (1.11) can be treated similarly. Having solved the problem (1.5), (1.8), (1.10) for a bounded domain Ω, we can then prove the solvability of the Cauchy problem and the problem (1.5), (1.8), (1.10) for unbounded domains. This is connected with a very important (characteristic) property of (1.5): The finiteness of the speed of propagation of perturbations allows us to solve the indicated problems "by parts" by dividing the domain in question into bounded subdomains of a special form. We begin the following section with a proof of these facts.

§2. The Energy Inequality. Finiteness of the Speed of Propagation of Perturbations. Uniqueness Theorem for Solutions in W_2^2

Suppose that we are given the equation

$$\mathscr{L}u \equiv u_{tt} - \sum_{i,j=1}^{n} \frac{\partial}{\partial x_i}(a_{ij}(x, t)u_{x_j}) + \sum_{i=1}^{n+1} a_i(x, t)u_{x_i} + a(x, t)u = f(x, t) \tag{2.1}$$

in the strip $\Pi_T^+ = \{(x, t): x \in R^n,\ t \in [0, T]\}$ (here, as above, we denote u_t by $u_{x_{n+1}}$ for the convenience of writing the sum as $\sum_{i=1}^{n+1} a_i u_{x_i}$), where the conditions

$$a_{ij} = a_{ji}, \qquad \nu\xi^2 \le a_{ij}(x, t)\xi_i\xi_j \le \mu\xi^2, \qquad \nu > 0, \tag{2.2}$$

$$\left|\frac{\partial a_{ij}}{\partial t}, a_i, a\right| \le \mu_1, \qquad \left|\frac{\partial a_{ij}}{\partial x_i}\right| \le \mu', \tag{2.3}$$

are fulfilled in the strip $(x, t) \in \Pi_T^+$, $f \in L_{2,1}(Q_T)$, $Q_T = \Omega \times (0,T)$, and where Ω is an arbitrary bounded domain in R^n.†

We shall assume that $u \in W_2^2(Q_T)$ for all Ω. We take in Π_T^+ a bounded domain $\mathscr{D}_\tau$, $\tau \leq T$, which is homeomorphic to the cylinder $K_1 \times (0, \tau) = \{(x, t): |x| \leq 1, t \in (0, \tau)\}$ and possesses the following properties: Its lower base (i.e., the one corresponding to the lower base of $K_1 \times (0, \tau)$) lies in the plane $t = 0$, its upper base lies in the plane $t = \tau$; its lateral surface is oriented at all points in a space-characteristic-like way, and the normals $\mathbf{n} = (\alpha_1, \ldots, \alpha_{n+1})$ to the lateral surface directing outside of $\mathscr{D}_\tau$ form acute angles with the t-axis. Such a domain $\mathscr{D}_\tau$ is naturally called a "cone" narrowing upwards (i.e., in the direction of the t-axis). The conditions on the lateral surface S_τ of the "cone" $\mathscr{D}_\tau$ are of the form:

$$\omega(\mathbf{n}) = \alpha_{n+1}^2 - a_{ij}\alpha_i\alpha_j \geq 0 \tag{2.4}$$

and

$$\alpha_{n+1} = \cos(\mathbf{n}, t) > 0, \tag{2.5}$$

where we assume that S_τ is a smooth surface. In the remainder the domain $\mathscr{D}_\tau$ is arbitrary.

Let us denote by $\mathscr{D}_t$ and S_t the intersections of $\mathscr{D}_\tau$ and S_τ, respectively, with the strip Π_t^+, $t \leq \tau$, and by Ω_{t_1} the cross-section of $\mathscr{D}_\tau$ by the plane $t = t_1$. Let us multiply (2.1) by $2u_t$ and integrate the result over $\mathscr{D}_t$, $t \leq \tau$:

$$2\int_{\mathscr{D}_t} \mathscr{L}u \cdot u_t \, dx\, dt = 2\int_{\mathscr{D}_t} f \cdot u_t \, dx\, dt. \tag{2.6}$$

By integrating by parts, we can write the left-hand side of (2.6) in the following form:

$$\begin{aligned} 2\int_{\mathscr{D}_t} \mathscr{L}u \cdot u_t \, dx\, dt &= \int_{\mathscr{D}_t} \left(\frac{\partial u_t^2}{\partial t} + 2a_{ij}u_{x_j}u_{tx_i} + 2a_i u_{x_i}u_t + 2auu_t\right) dx\, dt \\ &\quad - \int_{S_t} 2a_{ij}u_{x_j}u_t\alpha_i \, ds = \int_{\Omega_t} (u_t^2 + a_{ij}u_{x_i}u_{x_j})\, dx \Big|_{t=0}^{t=t} \\ &\quad + \int_{\mathscr{D}_t} \left(-\frac{\partial a_{ij}}{\partial t} u_{x_i}u_{x_j} + 2a_i u_{x_i}u_t + 2auu_t\right) dx\, dt \\ &\quad + \int_{S_t} [(u_t^2 + a_{ij}u_{x_i}u_{x_j})\alpha_{n+1} - 2a_{ij}u_{x_j}u_t\alpha_i]\, ds. \end{aligned} \tag{2.7}$$

† Instead of $|\partial a_{ij}/\partial t| \leq \mu_1$ and $|\partial a_{ij}/\partial x_i| \leq \mu'$, it is enough to require the local boundedness of $|\partial a_{ij}/\partial t|$ and $|\partial a_{ij}/\partial x_i|$ and also $\sum_{i,j=1}^n (\partial a_{ij}/\partial t)\xi_i\xi_j \leq \mu_1\xi^2$, where $\xi_1, \ldots, \xi_n$ here, and in (2.2), are arbitrary real numbers.

We write (2.6) in the form

$$y(t) + \int_{S_t} j\,ds = y(0)$$
$$+ \int_{\mathscr{D}_t} \left(\frac{\partial a_{ij}}{\partial t} u_{x_i} u_{x_j} - 2a_i u_{x_i} u_t - 2auu_t + 2fu_t \right) dx\,dt, \quad (2.8)$$

where

$$y(t) = \int_{\Omega_t} (u_t^2 + a_{ij} u_{x_i} u_{x_j})\,dx, \quad (2.9)$$

and where j denotes the expression under the integral sign $\int_{S_t} \ldots ds$ in (2.7). By our choice of the domain $\mathscr{D}_\tau$, the quantity $j|_{S_\tau} \geq 0$, for j can be written in the form

$$j = \alpha_{n+1}^{-1}[a_{ij}(u_{x_i}\alpha_{n+1} - u_t\alpha_i)(u_{x_j}\alpha_{n+1} - u_t\alpha_j)$$
$$+ u_t^2(\alpha_{n+1}^2 - a_{ij}\alpha_i\alpha_j)]$$

to which we can apply the assumptions (2.2), (2.4), and (2.5). We discard the integral $\int_{S_t} j\,ds$ from the left-hand side of (2.8) and the integral $\int_{\mathscr{D}_t} \ldots dx\,dt$ on the right-hand side of (2.8) and majorize by a simpler quantity using (2.2) and (2.3). This leads us to the inequality

$$y(t) \leq y(0) + c\int_0^t y(t)\,dt + c_1 \int_{\mathscr{D}_t} u^2\,dx\,dt + 2\int_0^t \| f \|_{2,\Omega_t} y^{1/2}(t)\,dt, \quad (2.10)$$

in which the constants are determined by the constants ν and μ_1 from (2.2) and (2.3). Let us estimate the integral of $u^2(x, t)$ in terms of the integrals of $u^2(x, 0)$ and $y(t)$, starting with the equality

$$u(x, t) = u(x, 0) + \int_0^t u_\xi(x, \xi)\,d\xi. \quad (2.11)$$

Let us square both sides of (2.11) and integrate over Ω_t, then bring in the following bound:

$$\int_{\Omega_t} u^2(x, t)\,dx \leq 2\int_{\Omega_t} u^2(x, 0)\,dx + 2\int_{\Omega_t} \left(\int_0^t u_t(x, t)\,dt \right)^2 dx$$
$$\leq 2\int_{\Omega_0} u^2(x, 0)\,dx + 2t\int_{\mathscr{D}_t} u_t^2\,dx\,dt$$
$$\leq 2\int_{\Omega_0} u^2(x, 0)\,dx + 2t\int_0^t y(t)\,dt. \quad (2.12)$$

If we add this inequality to (2.10) and overestimate somewhat the right-hand side of the inequality obtained, we can write the result in the form

$$z(t) \equiv \int_{\Omega_t} (u^2 + u_t^2 + a_{ij}u_{x_i}u_{x_j})\,dx$$
$$\leq 2z(0) + (c + c_1 + 2t)\int_0^t z(t)\,dt + 2\int_0^t \|f\|_{2,\Omega_t} z^{1/2}(t)\,dt. \quad (2.13)$$

Let us denote $\max_{0\leq\xi\leq t} z(\xi) = \hat{z}(t)$. It is clear that (2.13) implies the inequality

$$\hat{z}(t) \leq 2z(0) + (c + c_1 + 2t)t\hat{z}(t) + 2\|f\|_{2,1,\mathscr{D}_t}\hat{z}^{1/2}(t),\dagger$$

from which we obtain

$$\hat{z}^{1/2}(t) \leq 4z^{1/2}(0) + 4\|f\|_{2,1,\mathscr{D}_t} \quad (2.14)$$

for $t \leq \min(t_1, \tau)$, where $t_1 > 0$ is defined by the equality $(c + c_1 + 2t_1)t_1 = \frac{1}{2}$.

If $t_1 < \tau$, then, by choosing $t = t_1$ as the initial instant, our previous arguments allow us to assert that

$$\hat{z}^{1/2}(t) \leq 4z^{1/2}(t_1) + 4\|f\|_{2,1,\mathscr{D}_{t_1,t}} \quad (2.14')$$

for $t \leq \min(2t_1, \tau)$, where $\mathscr{D}_{t_1,t}$ is the part of $\mathscr{D}_t$ situated between the planes $t = t_1$ and $t = t$ $(t_1 < t)$.

The inequality (2.14') holds for any t_1 and $t > t_1$ in $[0, \tau]$ provided only that $t - t_1$ satisfy the condition $(c + c_1 + 2(t - t_1))(t - t_1) \leq 1/2$. This implies that, for all $t \in [0, \tau]$,

$$\hat{z}^{1/2}(t) \equiv \max_{0\leq\xi\leq t}\left(\int_{\Omega_\xi}(u^2 + u_t^2 + a_{ij}u_{x_i}u_{x_j})\,dx\right)^{1/2}$$
$$\leq c_2(t)z^{1/2}(0) + c_3(t)\|f\|_{2,1,\mathscr{D}_t}, \quad (2.15)$$

where $c_2(t)$ and $c_3(t)$ are defined by the constants ν and μ_1 from (2.2) and (2.3) and by the quantity t. This is the *energy inequality* which allows us to estimate the energy norm of the solution $u(x, t)$ in terms of the initial Cauchy data and the force $f(x, t)$. For the wave equation

$$u_{tt} - c^2\Delta u = 0 \quad (2.16)$$

the inequality (2.10), from which (2.15) was obtained, takes the form

$$\int_{\Omega_t}(u_t^2 + c^2u_x^2)\,dx \leq \int_{\Omega_0}(u_t^2 + c^2u_x^2)\Big|_{t=0}\,dx, \quad (2.17)$$

† Remember that $\|f\|_{2,1,\mathscr{D}_t} = \int_0^t \|f\|_{2,\Omega_\xi}\,d\xi$.

where we can take $\mathscr{D}_\tau$ to be, for example, the lower half of any characteristic cone $\{(x, t) \in R^{n+1}: |x - x_0| \leq c(\tau - t), 0 \leq t \leq \tau\}$.† The bound for

$$\int_{\Omega_t} (u^2 + u_t^2 + c^2 u_x^2)\, dx$$

follows from (2.17) and (2.12).

Let us return to the general case of (2.1) and the inequality (2.15) corresponding to it, which gives an *a priori* bound for arbitrary solution $u(x, t) \in W_2^2(\mathscr{D}_\tau)$. We draw two conclusions from it. The first is a uniqueness theorem for the Cauchy problem:

Theorem 2.1. *If the conditions* (2.2) *and* (2.3) *are satisfied for equation* (2.1), *then the Cauchy problem*

$$\mathscr{L}u = f, \qquad u|_{t=0} = \varphi(x), \qquad u_t|_{t=0} = \psi(x) \tag{2.18}$$

for (2.1) *cannot have more than one solution in* Π_T^+ *in the class of functions which are locally square-summable, together with their derivatives of first and second order, in* $(x, t) \in \Pi_T^+$.

In fact, if u' and u'' are two solutions of the problem (2.18) with the properties indicated in the theorem, then their difference $u = u' - u''$ is a solution of the homogeneous problem, $\mathscr{L}u = 0$, $u|_{t=0} = 0$, $u_t|_{t=0} = 0$. If we apply the bound (2.15) to u, we see that $z(t)$ is equal to zero, that is $u \equiv 0$. This is true for any domain of the form $\mathscr{D}_\tau$, $\tau \leq T$, and consequently for the entire strip Π_T^+. Moreover, we can draw a stronger conclusion from (2.15): u' and u'' coincide in a domain of type $\mathscr{D}_\tau$ if the initial values for u' and u'' and the free terms corresponding to them in equation (2.1) coincide only on the lower base of $\mathscr{D}_\tau$ and in $\mathscr{D}_\tau$ itself, respectively, irrespective of their behavior outside $\mathscr{D}_\tau$.

This phenomenon is related to another property of equations of hyperbolic type and expresses the fact that perturbations which can be described by such equations are propagated with a finite velocity. Let us explain this. We shall assume that the initial perturbations, that is, the functions φ and ψ, are different from zero only in some bounded domain $\hat{\Omega}$ of the space x, and that $f = 0$. Then the solution $u(x, t)$ corresponding to them (which we shall assume to exist in Π_T^+) for all $t \in [0, T]$ will vanish everywhere except for $x \in \hat{\Omega}(t)$—a "$\sqrt{\mu} t$ neighborhood" of the domain $\hat{\Omega}$ (i.e., the totality of points of $\hat{\Omega}$ and of all points x whose distance from $\hat{\Omega}$ is less than or equal to

† If the derivatives of $u(x, t)$ decrease sufficiently rapidly as $|x| \to \infty$, then we can take $\mathscr{D}_\tau$ to be the strip $\{(x, t) \in R^{n+1}, 0 \leq t \leq \tau\}$, for which (2.17) assumes the form of the equality

$$\int_{t=t_1} (u_t^2 + c^2 u_x^2)\, dx = \int_{t=0} (u_t^2 + c^2 u_x)\, dx,$$

which expresses the law of conservation of energy. This is the basis of the name "energy" for the inequalities (2.17) and (2.15).

$\sqrt{\mu}t$). In fact, if the point $\tilde{x}$ lies at a positive distance d from $\hat{\Omega}(\tau)$, then we can construct a domain of type $\mathscr{D}_\tau$ for the point (x, τ) whose lower base does not intersect $\hat{\Omega}$ and whose upper base (for $t = \tau$) contains the point $(\tilde{x}, \tau)$. We can take as $\mathscr{D}_\tau$ a truncated spherical cone with axis lying on the line $\{(x, t): x = \tilde{x}\}$ with upper base $\{(x, t): |x - \tilde{x}| < d, t = \tau\}$ and lower base $\{(x, t): |x - \tilde{x}| < d + \sqrt{\mu}\tau, t = 0\}$. If we now apply (2.15) to our solution in this domain $\mathscr{D}_\tau$, then, because $f \equiv 0$ and $z(0) = 0$, we have that $u \equiv 0$ in $\mathscr{D}_\tau$ and hence for $x = \tilde{x}$ and $t \in [0, \tau]$.

Thus we have shown that an initial perturbation concentrated in the domain $\hat{\Omega}$ will be propagated in time τ not more than a distance $\sqrt{\mu}\tau$ from the domain $\hat{\Omega}$. But this means that the perturbation, which can be described by the solution $u(x, t)$ of (2.1) with $f \equiv 0$, is propagated with finite velocity, where this velocity does not exceed $\sqrt{\mu}$, where μ is given in (2.2). The same is true for perturbations caused by the presence of a force $f(x, t)$. For the wave equation (2.16) this argument shows that the velocity of propagation of a perturbation does not exceed c. On the other hand, the Poisson formula, which expresses the solution of the Cauchy problem for (2.16), shows that the velocity of propagation of a perturbation for (2.16) is equal to c. This shows the precision of our conclusion from the inequality (2.15) for the whole family of hyperbolic equations (2.1). Let us formulate what we have proved in the form of a theorem.

Theorem 2.2. *The hyperbolic equations* (2.1) *describe the propagation of perturbations whose velocity does not exceed* $\sqrt{\mu}$, *where* μ *is given by* (2.2).

Because of the properties established in Theorems 2.1 and 2.2, the solution $u(x, t)$ of the Cauchy problem (2.18) in the strip Π_T^+ can be found "by parts" by subdividing Π_T^+ into finite domains of type $\mathscr{D}_\tau$. If, say, we are interested in $u(x, t)$ for (x, t) in some bounded domain $Q \subset \Pi_T^+$, then it is not necessary (in contrast to the case of elliptic and parabolic equations) to find $u(x, t)$ in all of Π_T^+. It is enough to enclose Q by a bounded domain of type $\mathscr{D}_\tau \subset \Pi_T^+$ with base on the plane $t = 0$ and to determine $u(x, t)$ only in $\mathscr{D}_\tau$. Moreover, $\mathscr{D}_\tau$ can, in turn, be enclosed in a cylinder $Q_\tau = \{\Omega \times [0 \le t \le \tau]\}$ whose lower base contains as an interior subdomain the lower base of $\mathscr{D}_\tau$, and then, instead of the problem (2.18), we can solve the first initial-boundary value problem

$$\mathscr{L}u = f, \quad u|_{t=0} = \tilde{\varphi}(x), \quad u_t|_{t=0} = \tilde{\psi}(x), \quad u|_{S_\tau} = 0 \quad (2.19)$$

with functions $\tilde{\varphi}(x)$ and $\tilde{\psi}(x)$ which coincide with $\varphi(x)$ and $\psi(x)$ on the lower base of $\mathscr{D}_\tau$ and which vanish in the vicinity of $\partial\Omega$. By Theorem 2.1 the solution of (2.19) will coincide in $\mathscr{D}_\tau$ (and indeed in Q) with the desired solution of (2.18).

This method of solving (2.18) is not always the best, especially if the question is that of actually calculating $u(x, t)$. For the simplest equations of hyperbolic type, for example, for (1.4), there are relatively simple formulas for determining the solution $u(x, t)$ of the Cauchy problem in terms of f, φ, and ψ (for the homogeneous equation (1.4) there is the Poisson formula, and for the non-homogeneous equation the Poisson–Kirchkoff formula).

There is no need to use the initial-boundary value problem for solving the Cauchy problem by the method of finite differences, which we shall present in Chapter VI. Moreover, the Cauchy problem admits other "classical" methods of investigation, among them the method of analytic approximation (see the well-known papers of J. Schauder [2], I. G. Petrovski [1], and others) and far-reaching generalizations of potential theory, which reduce the problem to a study of integral equations (see the famous investigations of the Cauchy problem by J. Hadamard [1], the "generalized Kirchkoff method" proposed by S. L. Sobolev [1], and others).

The Cauchy problem has been studied, not only for the case of a single hyperbolic equation, but also for equations and systems of hyperbolic type of any order (see the works of R. Courant, K. O. Friedrichs and H. Lewy in Courant and Hilbert [1], I. G. Petrovski [1], J. Leray [2], L. Gårding [2], Ladyzhenskaya [1], etc.)

However, for a second-order hyperbolic equation of the form (1.3), the way we chose to prove the unique solvability of the Cauchy problem as a special case of the first initial-boundary value problem has definite advantages, if only of a methodological character.

§3. The First Initial-Boundary Value Problem. Solvability in $W_2^1(Q_T)$

In this section we shall study the solvability of the first initial-boundary value problem for (2.1), that is, the problem of finding a function $u(x, t)$ in $Q_T = \Omega \times (0, T)$, where Ω is an arbitrary bounded domain in R^n, and where $u(x, t)$ satisfies the requirements

$$\mathscr{L}u = f \quad \text{for } (x, t) \in Q_T, \qquad u|_{t=0} = \varphi(x) \quad \text{and} \quad u_t|_{t=0} = \psi(x) \quad \text{for } x \in \Omega,$$
$$u|_{S_T} = 0 \quad \text{for } (x, t) \in S_T \equiv \partial\Omega \times [0, T]. \tag{3.1}$$

As everywhere, we assume that the boundary condition has been reduced to the homogeneous case. Let

$$f \in L_{2,1}(Q_T), \qquad \varphi \in \overset{\circ}{W}{}_2^1(\Omega), \qquad \psi \in L_2(\Omega), \tag{3.2}$$

and let (2.2) and (2.3) be satisfied for $\mathscr{L}$. We shall show that for solutions $u(x, t) \in W_2^2(Q_T)$ of problem (3.1), under the hypotheses of Theorem 2.1, we can give an *a priori* bound of the type (2.15), namely,

$$z^{1/2}(t) = \left(\int_{\Omega_t} (u^2 + u_t^2 + a_{ij} u_{x_i} u_{x_j})\, dx \right)^{1/2} \leq c_2(t) z^{1/2}(0) + c_3(t) \| f \|_{2,1,Q_t}, \tag{3.3}$$

$t \in [0, T]$, with the same functions $c_2(t)$ and $c_3(t)$ appearing in (2.15). In this section Ω_{t_1} will be a cross-section of the cylinder Q_T by the plane $t = t_1$. For the proof of (3.3) we consider the equality

$$2\int_{Q_t} \mathscr{L}u \cdot u_t \, dx\, dt = 2\int_{Q_t} fu_t \, dx\, dt.$$

This gives us, as (2.6) did in §2, equation (2.8) with the difference that the integral over S_t vanishes† by virtue of the condition $u|_{S_T} = 0$. All further deductions from (2.8), §2, go through and lead to (3.3).

The uniqueness theorem follows from (3.3) for solutions of (3.1) in the space $W_2^2(Q_T)$. However, we want to establish a stronger uniqueness theorem for generalized solutions from $W_2^1(Q_T)$. This is due to the fact that it is in this class that it is easiest to prove the solvability of (3.1). Corresponding to the approach to the boundary value problems described in Chapters II and III, we shall call a function $u(x, t)$ a *generalized solution from the space* $W_2^1(Q_T)$ *of the problem* (3.1) if it belongs to $W_{2,0}^1(Q_T)$, if it is equal to $\varphi(x)$ for $t = 0$, and if it satisfies the identity

$$\int_{Q_T} (-u_t\eta_t + a_{ij}u_{x_j}\eta_{x_i} + a_i u_{x_i}\eta + au\eta)\, dx\, dt$$
$$- \int_\Omega \psi\eta(x, 0)\, dx = \int_{Q_T} f\eta \, dx\, dt \tag{3.4}$$

for all $\eta \in \hat{W}_{2,0}^1(Q_T)$. By $W_{2,0}^1(Q_T)$ we mean here the closure in the norm of $W_2^1(Q_T)$ of the set of all smooth functions in $\bar{Q}_T$ which vanish in the vicinity of S_T; $\hat{W}_{2,0}^1(Q_T)$ consists of those elements of $W_{2,0}^1(Q_T)$ which vanish for $t = T$.

The fact that this definition is meaningful and is actually an extension of the concept of the classical solution can be verified in the same way as in Chapter II. For this, we should recall the properties of the elements of $W_2^1(Q_T)$ and $W_{2,0}^1(Q_T)$ enumerated in Chapter I, §6, in particular, the fact that these elements are defined on every cross-section Ω_t as elements of $L_2(\Omega)$ and are continuous in t in the norm of $L_2(\Omega)$. We shall prove the following theorem of uniqueness:

Theorem 3.1. *Let the coefficients of* $\mathscr{L}$, *additionally to* (2.2), *satisfy the conditions*

$$\max_{Q_T} \left| \frac{\partial a_{ij}}{\partial t}, \frac{\partial a_i}{\partial x_i}, a_i, a \right| \le \mu_1. \tag{3.5}$$

Then the problem (3.1) *can not have more than one generalized solution from* $W_2^1(Q_T)$. *The condition* $|\partial a_i/\partial x_i| \le \mu_1$ *may be replaced by the condition* $|\partial a_i/\partial t| \le \mu_1$.

† For a "bad" boundary Ω, the vanishing of u and u_t must be understood in the sense that u and u_t belong to the space $W_{2,0}^1(Q_T)$ to be defined below.

Suppose that (3.1) has two generalized solutions u' and u'' from $W_2^1(Q_T)$. Then the difference $u = u' - u'' \in W_{2,0}^1(Q_T)$, satisfies (3.4) with $f = \psi = 0$, and vanishes for $t = 0$. In the resulting identity let us take

$$\eta(x, t) = \begin{cases} 0 & \text{for } \; b \le t \le T, \\ \int_b^t u(x, \tau)\, d\tau & \text{for } \; 0 \le t \le b. \end{cases} \tag{3.6}$$

It is not hard to show that, if we use the definition of generalized derivatives, $\eta(x, t)$ is in the class $\hat{W}_{2,0}^1(Q_T)$ and has the generalized derivatives $\eta_{tx} = u_x$ in $Q_b = \Omega \times (0, b)$, which are square-summable over Q_b. Moreover, η, η_x, and u are the elements of the space $L_2(\Omega)$ continuously depending on $t \in [0, T]$. Let us substitute η from (3.6) into (3.4) with $f = \psi = 0$, and express u, u_t, and u_x in terms of η and its derivatives. After changing signs in (3.4), we obtain

$$\int_{Q_b} [\eta_t \eta_{tt} - a_{ij} \eta_{tx_j} \eta_{x_i} - a_i \eta_{tx_i} \eta - a\eta_t \eta]\, dx\, dt = 0. \tag{3.7}$$

If we integrate the first three terms, observing that $\eta_t|_{t=0} = u|_{t=0} = 0$ and $\eta_x|_{t=b} = 0$, we obtain the relation

$$\frac{1}{2} \int_\Omega [\eta_t^2(x, b) + a_{ij} \eta_{x_i} \eta_{x_j}|_{t=0}]\, dx$$

$$= -\int_{Q_b} \left[\frac{1}{2} \frac{\partial a_{ij}}{\partial t} \eta_{x_i} \eta_{x_j} + a_i \eta_t \eta_{x_i} + \left(\frac{\partial a_i}{\partial x_i} - a \right) \eta_t \eta \right] dx\, dt,$$

which, together with (3.5), implies that

$$\int_\Omega [\eta_t^2(x, b) + a_{ij} \eta_{x_i} \eta_{x_j}|_{t=0}]\, dx \le c \int_{Q_b} (\eta_x^2 + \eta_t^2 + \eta^2)\, dx\, dt, \tag{3.8}$$

where c depends only on μ_1. Let us introduce the functions

$$v(x, t) = \int_t^0 u(x, \tau)\, d\tau$$

and

$$v_{x_i}(x, t) = \int_t^0 u_{x_i}(x, \tau)\, d\tau \equiv v_i(x, t).$$

Using (3.6),

$$\eta_{x_i}(x, t) = \int_b^t u_{x_i}(x, \tau)\, d\tau = v_i(x, b) - v_i(x, t), \qquad t \le b.$$

Moreover, for almost all $x \in \Omega$,

$$\int_0^b \eta^2\, dt = \int_0^b \left(\int_b^t u\, d\tau \right)^2 dt \le \int_0^b (b - t) \int_t^b u^2\, d\tau\, dt \le b^2 \int_0^b u^2\, d\tau.$$

In view of this, it follows from (3.8) that

$$\int_\Omega [u^2(x, b) + a_{ij}(x, 0)v_i(x, b)v_j(x, b)]\, dx$$

$$\leq c \int_{Q_b} \left[\sum_{i=1}^n (v_i(x, b) - v_i(x, t))^2 + (1 + b^2)u^2 \right] dx\, dt$$

$$\leq 2cb \int_\Omega \sum_{i=1}^n v_i^2(x, b)\, dx + c \int_{Q_b} \left[2 \sum_{i=1}^n v_i^2 + (1 + b^2)u^2 \right] dx\, dt, \quad (3.9)$$

from which we obtain, by using (2.2),

$$\int_\Omega u^2(x, b)\, dx + (\nu - 2bc) \int_\Omega \sum_{i=1}^n v_i^2(x, b)\, dx$$

$$\leq c_1 \int_{Q_b} \left(\sum_{i=1}^n v_i^2 + u^2 \right) dx\, dt, \quad (3.10)$$

where $c_1 = c(2 + b^2)$. Let us now use the fact that b can be chosen arbitrarily. For $b \in [0, \nu/4c]$ the coefficient $\nu - 2bc \geq \nu/2$, and $u(x, b)$ and $v_i(x, b)$ vanish for $b = 0$. This guarantees, on the basis of Lemma 1.1 of Chapter III, that (3.10) implies that $u(x, b) = 0$ and $v_i(x, b) = 0$ for $b \in [0, \nu/4c]$.

If we repeat the argument for $t \in [\nu/4c, \nu/2c]$, we see that $u(x, t) = 0$ on this interval. Thus, after a finite number of steps, we shall have proved that $u(x, t)$ vanishes for all $t \in [0, T]$. If the condition $|\partial a_i/\partial x_i| \leq \mu_1$ is replaced by the condition $|\partial a_i/\partial t| \leq \mu_1$, then the term $j \equiv \int_{Q_b} (-a_i \eta_{tx_i} \eta)\, dx\, dt$ in (3.7) must be transformed in the following way:

$$j = \int_{Q_b} (a_i \eta_{x_i} \eta_t + a_{it} \eta_{x_i} \eta)\, dx\, dt + \int_\Omega a_i \eta_{x_i} \eta \bigg|_{t=0} dx.$$

Moreover, let us add

$$\int_\Omega \lambda \eta^2(x, 0)\, dx + \int_{Q_b} 2\lambda \eta_t \eta\, dx\, dt = 0$$

to both sides of (3.7), where the constant λ is chosen in such a way that, for any η and η_{x_i}, we have the inequality

$$a_{ij} \eta_{x_i} \eta_{x_j} + a_i \eta_{x_i} \eta + \lambda \eta^2 \geq \nu_1 (\eta_x^2 + \eta^2), \qquad \nu_1 > 0.$$

This reduces to an inequality

$$\int_\Omega [\eta_t^2(x, b) + \nu_1 \eta_x^2(x, 0) + \nu_1 \eta^2(x, 0)]\, dx \leq c \int_{Q_b} (\eta_x^2 + \eta_t^2 + \eta^2)\, dx\, dt$$

of the form (3.8). From this it follows, as we have shown above, that $u(x, t) \equiv 0$.

Now let us go over to the theorem of existence.

Theorem 3.2. *Let the coefficients of $\mathscr{L}$ satisfy the conditions* (2.2) *and the condition*

$$\max_{Q_T} \left| \frac{\partial a_{ij}}{\partial t}, a_i, a \right| \leq \mu_1. \tag{3.11}$$

Then the problem (3.1) *has a generalized solution from $W_2^1(Q_T)$ for $f \in L_{2,1}(Q_T)$, $\varphi \in \mathring{W}_2^1(\Omega)$, and $\psi \in L_2(\Omega)$.*

We shall use Galerkin's method for the proof. Let $\{\varphi_k(x)\}$ be a fundamental system in $\mathring{W}_2^1(\Omega)$ with $(\varphi_k, \varphi_l) = \delta_k^l$.† We seek an approximate solution $u^N(x, t)$ in the form $u^N = \sum_{k=1}^N c_k^N(t)\varphi_k(x)$ from the relations

$$\left(\frac{\partial^2 u^N}{\partial t^2}, \varphi_l\right) + \int_\Omega (a_{ij} u_{x_j}^N \varphi_{lx_i} + a_i u_{x_i}^N \varphi_l + a u^N \varphi_l)\, dx$$
$$= (f, \varphi_l), \qquad l = 1, \ldots, N, \tag{3.12}$$

and

$$\left.\frac{d}{dt} c_k^N(t)\right|_{t=0} = (\psi, \varphi_k), \qquad c_k^N(0) = \alpha_k^N, \tag{3.13}$$

where the α_k^N are the coefficients in the sum $\varphi^N(x) = \sum_{k=1}^N \alpha_k^N \varphi_k(x)$ which approximates $\varphi(x)$ in the norm of $W_2^1(\Omega)$ as $N \to \infty$. The equations (3.12) form a system of linear ordinary differential equations of second order in t for the unknowns $c_k^N(t)$, $k = 1, \ldots, N$, which have been solved for $d^2c_k^N/dt^2$. The coefficients of the system are bounded functions, and the free terms $f_l \equiv (f(\cdot, \cdot), \varphi_l(\cdot)) \in L_1(0, T)$. This system is uniquely solvable for the initial data (3.13), where $d^2c_k^N/dt^2 \in L_1(0, T)$. For u^N the estimate (3.3) holds. Actually, if we multiply each of the equations in (3.12) by the appropriate $dc_l^N(t)/dt$ and sum them from $l = 0$ to $l = N$, we have the equation

$$\left(\frac{\partial^2 u^N}{\partial t^2}, \frac{\partial u^N}{\partial t}\right) + \int_\Omega (a_{ij} u_{x_j}^N u_{tx_i}^N + a_i u_{x_i}^N u_t^N + a u^N u_t^N)\, dx = (f, u_t^N),$$

from which we can deduce (3.3) (cf. §2). The right-hand side of (3.3) for u^N can be majorized by a constant depending neither on N nor on $t \in [0, T]$, so that

$$\int_{\Omega_t} [(u^N)^2 + (u_x^N)^2 + (u_t^N)^2]\, dx \leq c, \qquad t \in [0, T], \tag{3.14}$$

and, therefore,

$$\|u^N\|_{2, Q_T}^{(1)} \leq c_1. \tag{3.15}$$

† Remember that (,) is the scalar product in $L_2(\Omega)$.

Because of (3.15) we can extract from the sequence $\{u^N\}$ $(N = 1, 2, \ldots)$ a subsequence (for which we shall use the same notation) which converges weakly in $W_2^1(Q_T)$ (i.e., u^N, u_t^N, and $u_{x_i}^N$ converge weakly in $L_2(Q_T)$) and uniformly in $t \in [0, T]$ in the norm of $L_2(\Omega)$ to some element $u \in W_{2,0}^1(Q_T)$ (cf. §1 and §5, Chapter I). We shall show that $u(x, t)$ is a generalized solution of (3.1). The initial condition $u|_{t=0} = \varphi(x)$ will be fulfilled because of the convergence of $u^N(x, t)$ to $u(x, t)$ in $L_2(\Omega)$, as just pointed out, and because $u^N(x, 0) \to \varphi(x)$ in $L_2(\Omega)$. To prove that (3.4) holds for $u(x, t)$, we proceed exactly as we did in §3, Chapter III; namely, we multiply each of the relations (3.12) by a function $d_l(t) \in W_2^1(0, T)$, $d_l(T) = 0$, add up the equalities obtained from $l = 1$ to $l = N$, and then integrate over t from 0 to T. We then integrate the first term by parts, by transferring $\partial/\partial t$ from u^N to $\eta \equiv \sum_{l=1}^N d_l(t)\varphi_l(x)$. This gives the identity

$$\int_{Q_T} (-u_t^N \eta_t + a_{ij} u_{x_j}^N \eta_{x_i} + a_i u_{x_i}^N \eta + a u^N \eta)\, dx\, dt$$

$$- \int_\Omega u_t^N \eta \Big|_{t=0} dx = \int_{Q_T} f \eta \, dx\, dt, \tag{3.16}$$

which holds for all η of the form $\sum_{l=1}^N d_l(t)\varphi_l(x)$. Let us denote by $\mathfrak{M}_N$ the totality of all such η. We can take the limit in (3.16) along the subsequence chosen above, for a fixed η from some $\mathfrak{M}_{N_i}$. This leads to the identity (3.4) for the limit function u for all $\eta \in \mathfrak{M}_{N_i}$. But $\bigcup_{N=1}^\infty \mathfrak{M}_N$ is dense in $\hat{W}_{2,0}^1(Q_T)$ and $u \in W_{2,0}^1(Q_T)$, so that (3.4) holds for $u(x, t)$ for all $\eta \in \hat{W}_{2,0}^1(Q_T)$.

Thus we have shown that the limit-function $u(x, t)$ is a generalized solution of the problem (3.1) in $W_2^1(Q_T)$. If, instead of (3.11), certain stronger hypotheses of Theorem 3.1 are satisfied (so that the uniqueness theorem (Theorem 3.1) holds for the problem (3.1)), then it is not hard to see that all sequences $\{u^N\}$, $N = 1, 2, \ldots$, will converge weakly to u in $W_2^1(Q_T)$.

The inequality (3.15) holds for the solution $u(x, t)$ that we obtained, or, what is the same thing, the inequality

$$\|u\|_{2,Q_T}^{(1)} \le c(T)(\|\varphi\|_{2,\Omega}^{(1)} + \|\psi\|_{2,\Omega} + \|f\|_{2,1,Q_T}). \tag{3.17}$$

§4. On the Smoothness of Generalized Solutions

In §3 we established the existence of generalized solutions of the problem (3.1) from $W_2^1(Q_T)$. We shall show that, under certain stronger assumptions on the data of the problem, these solutions are in $W_2^2(Q_T)$. To be precise, let the coefficients of the equation (2.1) satisfy in $Q_T = \Omega \times (0, T)$, in addition to (2.2) and (3.11), the conditions

$$|a_{ijt}, a_{it}, a_t, a_{ijx}| \le \mu_2, \tag{4.1}$$

let the free term f have a derivative f_t with

$$f_t \in L_{2,1}(Q_T), \tag{4.2}$$

and let the initial functions have the properties

$$\varphi \in W_2^2(\Omega) \cap \overset{\circ}{W}{}_2^1(\Omega), \qquad \psi \in \overset{\circ}{W}{}_2^1(\Omega). \tag{4.3}$$

Then we have the theorem:

Theorem 4.1. *If the conditions* (2.2), (3.11), *and* (4.1)–(4.3) *are fulfilled, then the generalized solution* $u(x, t)$ *from* $W_2^1(Q_T)$ *of the problem* (3.1) *possesses derivatives* u_{tt} *and* u_{tx_i} *in* $L_2(Q_T)$, *and* $\|u_t(\cdot, t) - \psi\|_{2,\Omega} + \|u(\cdot, t) - \varphi\|_{2,\Omega}^{(1)} \to 0$ *as* $t \to 0$.

We shall prove the existence of a generalized solution of the problem (3.1) having the properties specified in Theorem 4.1. By the uniqueness theorem (Theorem 3.1), this solution will coincide with the generalized solution from $W_2^1(Q_T)$ which is guaranteed by Theorem 3.2. To prove the existence of such a solution we again start with Galerkin's method, taking as the basis $\{\varphi_k(x)\}$ a fundamental system in $W_2^2(\Omega) \cap \overset{\circ}{W}{}_2^1(\Omega)$ which has been orthogonalized in $L_2(\Omega)$. Thus, we seek an approximate solution $u^N(x, t)$ in the form $u^N = \sum_{k=1}^N c_k^N(t)\varphi_k(x)$ from the relations (3.12) and the initial conditions

$$c_k^N(0) = \alpha_k^N, \qquad \left.\frac{dc_k^N}{dt}\right|_{t=0} = \beta_k^N, \tag{4.4}$$

where the α_k^N and β_k^N are chosen in such a way that the sums

$$\varphi^N(x) = \sum_{k=1}^N \alpha_k^N \varphi_k(x), \qquad \psi^N(x) = \sum_{k=1}^N \beta_k^N \varphi_k(x),$$

approximate $\varphi(x)$ and $\psi(x)$ in the norms of the spaces $W_2^2(\Omega)$ and $W_2^1(\Omega)$, respectively. For the $u^N(x, t)$ constructed in this way the norms $\|u^N(\cdot, 0)\|_{2,\Omega}^{(2)}$ and $\|u_t^N(\cdot, 0)\|_{2,\Omega}^{(1)}$ are uniformly bounded with respect to N. By this and by the hypotheses of Theorem 4.1, it follows from (3.12) that $\|u_{tt}^N(\cdot, 0)\|_{2,\Omega}$ are also uniformly bounded with respect to N. This is not hard to show if we multiply each equality in (3.12) by $d^2c_l^N/dt^2$ and sum over l from 1 to N. The coefficients $c_l^N(t)$, which can be determined uniquely from (3.12) and (4.4), have derivatives in t up to third order inclusive (this is not hard to verify by using the hypotheses of the theorem). If we differentiate (3.12) with respect to t, then multiply by $d^2c_l^N/dt^2$ and sum over l from 1 to N, we obtain the equality

$$\begin{aligned}(u_{ttt}^N, u_{tt}^N) + \int_\Omega (&a_{ij}u_{tx_j}^N u_{ttx_i}^N + a_{ijt}u_{x_j}^N u_{ttx_i}^N + a_i u_{tx_i}^N u_{tt}^N \\ &+ a_{it}u_{x_i}^N u_{tt}^N + au_t^N u_{tt}^N + a_t u^N u_{tt}^N)\, dx = (f_t, u_{tt}^N).\end{aligned} \tag{4.5}$$

Let us write the first three terms in the form

$$(u_{ttt}^N, u_{tt}^N) = \frac{1}{2}\frac{d}{dt}\|u_{tt}^N\|^2;$$

$$(a_{ij}u_{txj}^N, u_{ttx_i}^N) = \frac{1}{2}\frac{d}{dt}(a_{ij}u_{tx_j}^N, u_{tx_i}^N) - \tfrac{1}{2}(a_{ijt}u_{tx_j}^N, u_{tx_i}^N),$$

$$(a_{ijt}u_{x_j}^N, u_{ttx_i}^N) = \frac{d}{dt}(a_{ijt}u_{x_j}^N, u_{tx_i}^N) - (a_{ijtt}u_{x_j}^N + a_{ijt}u_{tx_j}^N, u_{tx_i}^N).$$

If we substitute these into (4.5) and then integrate (4.5) over t from 0 to t, we have the relation

$$\tfrac{1}{2}[\|u_{tt}^N\|^2 + (a_{ij}u_{tx_j}^N, u_{tx_i}^N)]|_{t=0}^{t=t} + (a_{ijt}u_{x_j}^N, u_{tx_i}^N)|_{t=0}^{t=t}$$
$$= \int_0^t \Big[\frac{1}{2}(a_{ijt}u_{tx_j}^N, u_{tx_i}^N) + (a_{ijtt}u_{x_j}^N + a_{ijt}u_{tx_j}^N, u_{tx_i}^N)$$
$$- (a_i u_{tx_i}^N + a_{it}u_{x_i}^N + au_t^N + a_t u^N - f_t, u_{tt}^N)\Big]\, dt. \tag{4.6}$$

Here the first term has the same form for the function $v \equiv u_t$ as the first term on the left-hand side of (2.8) has for the function u. From (4.6) and the energy inequality (3.14) we deduce the inequality

$$\|u_{tt}^N\|_{2,\Omega_t}^2 + \|u_{tx}^N\|_{2,\Omega_t}^2 \le c, \qquad t \in [0, T], \tag{4.7}$$

and the further inequality

$$\|u_{tt}^N\|_{2,Q_T} + \|u_{tx}^N\|_{2,Q_T} \le c, \tag{4.8}$$

where the constant c is independent of N. This derivation is similar to the derivation of the energy inequality (2.15) from (2.8), indeed somewhat simpler, for the integrals over the lateral surfaces do not come into (4.6). To be sure, in (4.6) there is, in contrast to (2.8), the "superfluous" term $j \equiv (a_{ijt}u_{x_j}^N, u_{tx_i}^N)$. We bound it so $|j| \le \varepsilon\|u_{tx}^N\| + c_\varepsilon\|u_x^N\|$, taking $\varepsilon = \nu/4$ and then use the bound (3.14) already shown. In order to write the majorant c in (4.7), we use the fact that the left-hand side of (4.7) is majorized, for $t = 0$, by a constant which is independent of N. The bound (4.8), along with the bound (3.15), allows us to take the limit as $N \to \infty$ and thus to conclude that a certain subsequence $\{u^{N_k}\}$ converges in $L_2(Q_T)$, together with its first-order derivatives and the derivatives $u_{tt}^{N_k}$ and $u_{tx_i}^{N_k}$, to the desired solution $u(x, t)$ of the problem, and possesses the properties asserted in the theorem. By the uniqueness theorem (Theorem 3.1), every sequence $\{u^N\}$ must converge to this solution. By (4.8) the inequality

$$\|u_{tt}\|_{2,Q_T} + \|u_{tx}\|_{2,Q_T} \le c(T)[\|\varphi\|_{2,\Omega}^{(2)} + \|\psi\|_{2,\Omega}^{(1)} + \|f\|_{2,1,Q_T} + \|f_t\|_{2,1,Q_T}] \tag{4.9}$$

holds for $u(x, t)$, as well as the inequalities (3.17) and (3.3).

Corollary 4.1. *If, to the hypotheses of Theorem 4.1, we add the assumption that $\partial\Omega \in C^2$, then the solution $u(x, t)$ of (3.1) possesses derivatives $u_{x_i x_j}$ in $L_2(Q_T)$, so that $u \in W_2^2(Q_T)$. If the boundary of Ω is not assumed to be smooth, then the derivatives $u_{x_i x_j}$ are square-summable over all $Q'_T = \Omega' \times (0, T)$ such that $\overline{\Omega}' \subset \Omega$. Such a solution satisfies (2.1) for almost all $(x, t) \in Q_T$.*

Actually, the solution $u(x, t)$, which is guaranteed by Theorem 4.1, satisfies the equation $\mathscr{L}u = f$ in the form (3.4), as well as in the form

$$\int_{Q_T} (u_{tt}\eta + a_{ij}u_{x_j}\eta_{x_i} + a_i u_{x_i}\eta + au\eta)\, dx\, dt = \int_{Q_T} f\eta\, dx\, dt, \quad (4.10)$$

where η is an arbitrary element of the space $\overset{\circ}{W}{}_2^{1,0}(Q_T)$, which was defined in §1, Chapter III. The identity (4.10) can be derived from (3.4) or from the fact that (4.10) holds for u^N and for all $\eta(x, t)$ of the form $\sum_{l=1}^N d_l(t)\varphi_l(x)$, $d_l(t) \in L_2(0, T)$.

Let us take $\eta(x, t) = \chi(t)\Phi(x)$ in (4.10), where $\chi(t)$ is an arbitrary element of $L_2(0, T)$ and $\Phi(x)$ an arbitrary element of $\overset{\circ}{W}{}_2^1(\Omega)$, and let us write the integrals $\int_{Q_T} \ldots dx\, dt$ as iterated integrals of the form $\int_0^T \chi(t)(\int_\Omega \ldots dx)\, dt$. By the arbitrariness of the choice of $\chi(t)$ in (4.10) it will follow that, for almost all t in $[0, T]$ ("nice t"), the following identity will hold:

$$\int_{\Omega_t} (a_{ij}u_{x_j}\Phi_{x_i} + a_i u_{x_i}\Phi + au\Phi)\, dx = \int_{\Omega_t} (-u_{tt} + f)\Phi\, dx. \quad (4.11)$$

But this identity means that, for such a "nice t," $u(x, t)$ is a generalized solution in $\overset{\circ}{W}{}_2^1(\Omega)$ for the corresponding elliptic equation with free term $-u_{tt} + f$. By Theorem 7.3 and Remark 7.1, Chapter II, this solution actually has derivatives $u_{x_i x_j}$, where $u_{x_i x_j} \in L_2(\Omega)$ in the case $\partial\Omega \in C^2$, and where, for an arbitrary $\partial\Omega$, $u_{x_i x_j} \in L_2(\Omega')$ for all $\overline{\Omega}' \subset \Omega$. Moreover, the solution satisfies the equation in the ordinary form, i.e., in the form (2.1). It follows from (2.1), (3.3), and (4.9) that $\|u_{xx}\|_{2, Q_T}$ and $\|u_{xx}\|_{2, Q'_T}$ can be majorized by those norms of φ, ψ, and f appearing in (3.3) and (4.9), respectively.

Corollary 4.2. *Let the coefficients of $\mathscr{L}$ satisfy conditions (2.2), (2.3), and (4.1) in the strip $\Pi_T^+ = \{(x, t) : x \in R^n, t \in [0, T]\}$, let the functions $f(x, t)$ and $f_t(x, t)$ be in $L_{2,1}(Q_T^R)$, and let $\varphi(x) \in W_2^2(K_R)$, $\psi \in W_2^1(K_R)$, where $Q_T^R = \{(x, t) : x \in K_R,\ t \in [0, T]\}$, where $K_R = \{x : |x| < R\}$, and where R is an arbitrary positive number. Then the Cauchy problem (2.18) has a unique solution $u(x, t)$ in $W_2^2(Q_T^R)$ for all $R > 0$.*

This corollary follows from Corollary 4.1 and the results of §2. The constants ν^{-1}, μ, and μ_i in the inequalities (2.2), (2.3), and (4.1) can increase in an unbounded way, along with the radius R of the domain Q_T^R in which the arguments (x, t) of the coefficients of $\mathscr{L}$ are chosen in these inequalities. It is only necessary to assume that domains of the type $\mathscr{D}_\tau$ (cf. §2) exist and that the totality of these domains exhausts all of the strip Π_T^+ (this imposes a restriction on the dependence of μ in (2.1) on R).

Thus we have shown how it is possible to prove the existence of second-order derivatives of the generalized solution of the problem (3.1) if the corresponding smoothness of the data of the problem and the necessary order of agreement of initial and boundary conditions of the equation are known. In the case of Theorem 3.2 the condition of zero-order agreement was used, which was expressed by the requirement that $\varphi(x)$ should vanish on S—the condition that $\varphi(x) \in \mathring{W}_2^1(\Omega)$. This is in agreement with the fact that, by the boundary condition, u vanishes on S_T. The next order of agreement assumes the vanishing of u_t on $S \times \{0\} \equiv S_0$, which we expressed by the requirement that $\psi(x) \in \mathring{W}_2^1(\Omega)$. In the second-order agreement the question is the consistency on the set S_0 of the boundary condition, the initial conditions, and the equation (2.1) itself. Indeed, it follows from the boundary conditions that $u_{tt}|_{S_T} = 0$; on the other hand, on the lower base, including S_0, u_{tt} is determined from (2.1) and the initial data, so that we must require the vanishing on S_0 of the expression

$$(a_{ij}\varphi_{x_j})_{x_i} - \sum_{i=1}^{n} a_i\varphi_{x_i} - a_{n+1}\psi - a\varphi + f|_{S_0} = u_{tt}|_{S_0} = 0.$$

All of the following orders of agreement lead to corresponding requirements on the data of the problem caused by the vanishing on S_T of the derivatives $D_t^m u$. Analogously, we can investigate the further increasing of the smoothness of the solution of (3.1) in dependence on the increasing of the smoothness of the data of the problem and the order of agreement. The general scheme of this is the following: We first prove the increased smoothness with respect to t, and then with respect to x_i (for which we must bring in the corresponding results for elliptic equations). We shall not deduce the results which can be obtained by this method (cf. Ladyzhenskaya [4, Chapter 3]). Instead, we shall investigate another class of generalized solutions of (3.1) which goes by the natural name of "energy class." It is somewhat narrower than the class of generalized solutions in $W_2^1(Q_T)$, but wider than the class of generalized solutions in Theorem 4.1. This class is of interest because of two facts. First, we are able to establish in it the specific property of hyperbolic equations that a solution $u(x, t)$ has exactly the same differential properties which are assumed to be fulfilled at the initial moment of time (we recall that for hyperbolic equations both directions of time are equivalent; all problems can be solved equally for increasing or decreasing t). In other words, in terms of this class we obtain "extendable initial conditions." This is not the only class of generalized solutions with this property; "extendable initial conditions" can be expressed in terms of a whole scale of spaces $W_2^l(\Omega)$ (that is, by the requirements $u(\cdot, 0) \in W_2^l(\Omega)$, $u_t(\cdot, 0) \in W_2^{l-1}(\Omega)$, plus the corresponding conditions concerning the vanishing of $u(x, 0)$ and $u_t(x, 0)$ on S). However, the energy class of generalized solutions is, in some sense, the principal class among all others, a fact which is connected to its second property, related directly to the law of "conservation of energy." We use quotation marks here, for in the general case the energy of the system described by (2.1) is not preserved: the external force $f(x, t)$ and the dissipative terms $a_i u_{x_i}$ are in the system. We admit, moreover, that the characteristics of the medium could vary with time, that is, the coefficients in (2.1) could depend on t. But if $f \equiv 0$, $a_i \equiv 0$, and the coefficients are independent of t, then the energy of the system, which can be expressed by the integral

$$j(t) = \frac{1}{2}\int_{\Omega_t} (u_t^2 + a_{ij}u_{x_i}u_{x_j} + au^2)\,dx, \tag{4.12}$$

does not depend on t.

Because of this, it is natural to use the term *generalized solution of the problem* (3.1) *in the energy class* for a function $u(x, t)$ belonging to $\overset{\circ}{W}{}_2^1(\Omega)$ for all $t \in [0, T]$ and varying continuously in t in the norm of $W_2^1(\Omega)$; in addition, its derivative u_t should exist as an element of $L_2(\Omega)$ for all $t \in [0, T]$ and vary continuously in t in the norm of $L_2(\Omega)$. The initial conditions should start fulfilled in the spaces $\overset{\circ}{W}{}_2^1(\Omega)$ and $L_2(\Omega)$, respectively, i.e.,

$$\|u(\cdot, t) - \varphi(\cdot)\|_{2,\Omega}^{(1)} + \|u_t(\cdot, t) - \psi(\cdot)\|_{2,\Omega} \to 0 \quad \text{as} \quad t \to +0,$$

and the equation should be satisfied either in the sense of (3.4), or (what is the same thing) in the sense of the identity

$$\int_{\Omega_t} u_t\eta\,dx + \int_{Q_t} (-u_t\eta_t + a_{ij}u_{x_j}\eta_{x_i} + a_iu_{x_i}\eta + au\eta)\,dx\,dt$$

$$-\int_{\Omega_0} \psi\eta\,dx = \int_{Q_t} f\eta\,dx\,dt, \tag{4.13}$$

for all $t \in [0, T]$, where η is an arbitrary element of $W_{2,0}^1(Q_T)$. Finally, the energy relation must be fulfilled (which, in the general case of (2.1), is equivalent to the law of conservation of energy):

$$\frac{1}{2}\int_{\Omega_t} (u_t^2 + a_{ij}u_{x_i}u_{x_j})\,dx\Big|_{t=0}^{t=t}$$

$$+ \int_{Q_t} (-\tfrac{1}{2}a_{ijt}u_{x_i}u_{x_j} + a_iu_{x_i}u_t + auu_t)\,dx\,dt$$

$$= \int_{Q_t} fu_t\,dx\,dt, \tag{4.14}$$

for all $t \in [0, T]$. From this equality we can derive the "energy estimate"

$$\int_{\Omega_t} (u^2 + u_t^2 + u_x^2)\,dx \le c(t)\left[\int_{\Omega_0} (u^2 + u_t^2 + u_x^2)\,dx + \left(\int_0^t \|f\|_{2,\Omega_t}\,dt\right)\right]. \tag{4.15}$$

This can be done in a fashion similar to that by which we obtained the inequality (2.15) from (2.7). In general, the continuous function $c(t)$ increases monotonically with increasing t and is defined only by the constants in the hypotheses of Theorem 3.2.

It is easy to see that the generalized solutions guaranteed by Theorem 4.1 are generalized solutions of this class; to verify this, it is only necessary to derive (4.14) from (4.10), and for this we must take as $\eta(x, t)$ in (4.10) a function which is equal to zero for $t \in [t_1, T]$ and equal to $u_t(x, t)$ for $t \in [0, t_1]$. This leads to an equation from which (4.14) can be obtained as a result of transforming the principal terms, similar to the transformation carried out in (2.7). The generalized solutions from the energy class are generalized solutions in $W_2^1(Q_T)$. The existence of generalized solutions of (3.1) in the energy class can be proved under hypotheses somewhat stronger than the hypotheses of Theorem 3.2 but weaker than those of Theorem 4.1.

Theorem 4.2. *Let the coefficients of the equation* (3.1) *satisfy* (2.2) *and* (3.11), *and*

$$|a_{ijt}, a_{ijx}| \leq \mu_2, \tag{4.16}$$

and let

$$\varphi \in \overset{\circ}{W}{}_2^1(\Omega), \qquad \psi \in L_2(\Omega), \qquad f \in L_{2,1}(Q_T). \tag{4.17}$$

Then the problem (3.1) *has a generalized solution in the energy class, and in the class of generalized solutions from* $W_2^1(Q_T)$ *the uniqueness theorem holds.*

Theorem 3.2 guarantees the existence of at least one generalized solution $u(x, t)$ of (3.1) in the class $W_2^1(Q_T)$. Let us denote by $\mathscr{F}(x, t)$ the expression $-a_i u_{x_i} - au + f$ (it is obvious that $\mathscr{F} \in L_{2,1}(Q_T)$) and rewrite the equation $\mathscr{L}u = f$ in the form

$$\mathscr{L}_0 u \equiv u_{tt} - (a_{ij} u_{x_j})_{x_i} = \mathscr{F}(x, t). \tag{4.18}$$

Our function $u(x, t)$ is a generalized solution in $W_2^1(Q_T)$ for the equation (4.18) with the initial-boundary conditions

$$u|_{t=0} = \varphi, \qquad u_t|_{t=0} = \psi, \qquad u|_{S_T} = 0, \tag{4.19}$$

where, by Theorem 3.1, it is the unique solution for the problem (4.18)–(4.19) in the class of generalized solutions in $W_2^1(Q_T)$. Let us show that it possesses the properties asserted in Theorem 4.2. Thus, let us approximate $\mathscr{F}$, φ, and ψ by functions $\mathscr{F}^{(m)}$, $\varphi^{(m)}$, and $\psi^{(m)}$ satisfying the hypotheses of Theorem 4.1 in the norms of $L_{2,1}(Q_T)$, $W_2^1(\Omega)$, and $L_2(\Omega)$, respectively. By Theorem 4.1, there correspond the solutions $u^{(m)}(x, t)$ of the problem (4.18)–(4.19) in which $\mathscr{F}$, φ, and ψ have been replaced by $\mathscr{F}^{(m)}$, $\varphi^{(m)}$, and $\psi^{(m)}$. For $u^{(m)}$ we have the energy relations, which, for (4.18), have the form

$$\tfrac{1}{2}\|u_t^{(m)}\|_{2,\Omega_t}^2 + \tfrac{1}{2}(a_{ij}u_{x_j}^{(m)}, u_{x_i}^{(m)})_{\Omega_t} - \frac{1}{2}\int_0^t (a_{ijt}u_{x_j}^{(m)}, u_{x_i}^{(m)})\,dt$$

$$= \tfrac{1}{2}\|u_t^{(m)}\|_{2,\Omega_0}^2 + \tfrac{1}{2}(a_{ij}u_{x_j}^{(m)}, u_{x_i}^{(m)})_{\Omega_0} + \int_0^t (\mathscr{F}^{(m)}, u_t^{(m)})\,dt. \tag{4.20}$$

The differences $u^{m,p} = u^{(m)}(x,t) - u^{(p)}(x,t)$ are solutions of the same problem (4.18)–(4.19), but with $\mathscr{F}$, φ, and ψ replaced by $\mathscr{F}^{m,p} = \mathscr{F}^{(m)} - \mathscr{F}^{(p)}$, $\varphi^{m,p} = \varphi^{(m)} - \varphi^{(p)}$, and $\psi^{m,p} = \psi^{(m)} - \psi^{(p)}$. Relations of the form (4.20) also hold for these functions.

It is easy to obtain from these relations the estimate

$$\max_{0 \le t \le T} (\|u_t^{m,p}\|_{2,\Omega_t}^2 + \|u_x^{m,p}\|_{2,\Omega_t}^2) \le c[\|u_t^{m,p}\|_{2,\Omega_0}^2 + \|u_x^{m,p}\|_{2,\Omega_0}^2 + \|\mathscr{F}^{m,p}\|_{2,1,Q_T}^2] \tag{4.21}$$

with a constant c independent of m and p. It follows from this that $u^m(x,t)$ converges to some function $v(x,t)$ in the norm defined by the left-hand side of (4.21). It is not hard to verify that this function is a generalized solution of (4.18)–(4.19) in the energy class, and, in particular, satisfies a relation of the form (4.20). But, on the other hand, we know that (4.18)–(4.19) has a generalized solution $u(x,t)$ in $W_2^1(Q_T)$, which, by Theorem 3.1, coincides with $v(x,t)$ and hence possesses those smoothness properties required by Theorem 4.2. If we now recall the expression for $\mathscr{F}$, it is easy to see from what we have said that $u(x,t)$ is a generalized solution of (3.1) in the energy class. All that remains to be proved is the last assertion of Theorem 4.2, the uniqueness theorem. This does not follow from Theorem 3.1, since we have not assumed the existence of the derivatives a_{it}. Let (3.1) have two solutions u' and u'' in the class $W_2^1(Q_T)$. Then their difference $v(x,t)$ can be considered as a generalized solution of (4.18)–(4.19) in $W_2^1(Q_T)$ with $\mathscr{F} = -a_i v_{x_i} - av$, $\varphi = 0$, and $\psi = 0$. But every such solution, as we established in the proof of the first part of the theorem, is a generalized solution of the problem in the energy class. In particular, it satisfies an energy relation of the form (4.20), that is,

$$\|v_t\|_{2,\Omega_t}^2 + (a_{ij} v_{x_j}, v_{x_i})_{\Omega_t} - \int_0^t (a_{ijt} v_{x_j}, v_{x_i})\, dt = 2 \int_0^t (-a_i v_{x_i} - av_t, v_t)\, dt. \tag{4.22}$$

Here we have incorporated the expressions for $\mathscr{F}$, φ, and ψ. But (4.22) coincides with (4.14) in which f, $u(x,0)$, and $u_t(x,0)$ have been set equal to zero, and (4.15) follows from (4.14), whence $v \equiv 0$, that is, u' and u'' coincide. This proves Theorem 4.2.

§5. Other Initial-Boundary Value Problems

The initial-boundary value problems

$$\mathscr{L}u = f, \quad u|_{t=0} = \varphi, \quad u_t|_{t=0} = \psi, \quad \frac{\partial u}{\partial N} + \sigma u|_{S_T} = 0, \tag{5.1}$$

where $\partial u/\partial N = a_{ij}u_{x_j}\cos(\mathbf{n}, \mathbf{x}_i)$, can be solved, like the problem (3.1), by Galerkin's method. The only difference in constructing the approximations $u^N = \sum_{k=1}^N c_k^N(t)\varphi_k(x)$ consists of the fact that this time the functions $\{\varphi_k(x)\}$ should form a basis (or a fundamental system) in $W_2^1(\Omega)$ rather than in $\mathring{W}_2^1(\Omega)$. The entire remaining part, relating to the calculation of u^N and the proof that u^N converges to u, is analogous to that in §3. In the same method it proved the uniqueness theorem for the generalized solution of (5.1) in the class $W_2^1(Q_T)$.

A generalized solution $u(x, t)$ of (5.1) *in the space $W_2^1(Q_T)$* is defined as an element of $W_2^1(Q_T)$ which is equal to $\varphi(x)$ when $t = 0$ and which satisfies the identity

$$\int_{\Omega_T} (-u_t\eta_t + a_{ij}u_{x_i}\eta_{x_j} + a_i u_{x_i}\eta + au\eta)\,dx\,dt$$

$$+ \int_{S_T} \sigma u\eta\,dx\,dt - \int_{\Omega} \psi\eta(x, 0)\,dx = \int_{Q_T} f\eta\,dx\,dt \tag{5.2}$$

for any function $\eta(x, t)$ in $W_2^1(Q_T)$ which is zero for $t = T$. If $\sigma \neq 0$, then we must require a certain regularity of the boundary of Ω (the same as in §5, Chapter II).

Let us sum up what has been said in the form of a theorem:

Theorem 5.1. *Let the coefficients of $\mathcal{L}$ fulfill the conditions* (2.2) *and* (3.11) *in Q_T, and let the function $\sigma(s, t)$ in the boundary condition be bounded, together with its derivative σ_t. Then the problem* (5.1) *has a generalized solution $u(x, t)$ in $W_2^1(Q_T)$ for any f in $L_{2,1}(Q_T)$, φ in $W_2^1(\Omega)$, and ψ in $L_2(\Omega)$. If, in addition to these hypotheses, the a_i have bounded derivatives $\partial a_i/\partial t$ or $\partial a_i/\partial x_i$ in Q_T, then the uniqueness theorem holds for* (5.1) *in the class of generalized solutions in $W_2^1(Q_T)$.*

Sketch of the Proof of Theorem 5.1. In the process of deriving the energy inequality and proving the uniqueness theorem, we have the integral $\int_{S_t} \sigma u u_t\,ds\,dt$, which must be transformed into

$$\int_{S_t} \sigma u u_t\,ds\,dt = -\frac{1}{2}\int_{S_t} \sigma_t u^2\,ds\,dt + \frac{1}{2}\int_{\partial\Omega_t} \sigma u^2\,ds\Big|_{t=0}^{t=t},$$

and then to estimate $\int_{\partial\Omega_t} u^2\,ds$, we must use the inequalities

$$\int_{\partial\Omega_t} u^2\,ds \leq \int_{\Omega_t} (\varepsilon u_x^2 + c_\varepsilon u^2)\,dx$$

$$\leq \varepsilon\int_{\Omega_t} u_x^2\,dx + 2c_\varepsilon\int_{\Omega_0} u^2\,dx + 2c_\varepsilon t\int_{Q_t} u_t^2\,dx\,dt, \tag{5.3}$$

in which we can take ε to be any positive number (cf. (6.24), Chapter I and (2.12), Chapter IV).

§6. The Functional Method of Solving Initial-Boundary Value Problems

We can use the method of finite differences for actually finding the solutions of problems (3.1) and (5.1) when $\mathscr{L}$ is of general type; this method will be explained in Chapter VI. Here we shall describe the "functional" method of solving (3.1), which is analogous to the method of §2, Chapter III. It is based on the proof of the uniqueness theorem for generalized solutions in $L_2(Q_T)$, from which the theorem on the existence of generalized solutions in the energy class follows "almost as a gift." This method, which we presented in [LA 6, 8], is applicable not only to a wide class of differential equations of arbitrary order and systems of hyperbolic type, but also to the Cauchy problem for operator equations of the form

$$\frac{d^2(S_1(t)u)}{dt^2} + \frac{d(S_2(t)u)}{dt} + S_3(t)u = f(t), \qquad u|_{t=0} = \varphi, \quad u_t|_{t=0} = \psi, \quad (6.1)$$

where the $S_i(t)$ are unbounded operators in a Hilbert space H which depend on a parameter t; $u(t)$ and $f(t)$ are elements of H depending on t; and φ and ψ are elements of H. Under specific conditions on the $S_i(t)$ which guarantee the hyperbolic character of (6.1) and the smooth dependence of the $S_i(t)$ on t, we have proved the unique solvability of the problem (6.1) (cf. [8(a), (b)]). The conditions on the $S_i(t)$ are such that the Cauchy problem and the first initial-boundary problem for the hyperbolic equations (2.1) and for the so-called strongly hyperbolic systems (to which, for example, the non-stationary system of equations of the theory of elasticity is related) are special cases of (6.1); consequently, their unique solvability follows from that of the abstract Cauchy problem (6.1) which was established in [8(a), (b)] (in this connection see [8(c)], where these conclusions were drawn on the basis of [8(a), (b)]). An analogous method was given by us ([8]) for non-stationary equations of other types: for a parabolic equation, for an equation of Schrödinger type, and for a linearized system of the Navier–Stokes equations. It allowed us to study initial-boundary problems for the Maxwell system equations (Bychovskii [1]) and for a number of other systems of mechanics (Derguzov and Yacubovich [1], Fomin [1], etc.).

We shall illustrate this method with the example of the first initial-boundary problem for the wave equation

$$\mathscr{L}u \equiv u_{tt} - \Delta u = f, \qquad u|_{t=0} = \varphi, \qquad u_t|_{t=0} = \psi, \qquad u|_{S_T} = 0, \quad (6.2)$$

in $Q_T = \Omega \times (0, T)$, where Ω is a bounded domain, and where $f \in L_2(Q_T)$. Let the boundary of Ω possess the properties described in Theorem 7.2, Chapter II, which guarantee the unique solvability in $W_2^2(\Omega)$ of the problem

$$\Delta u = \chi(x), \qquad u|_S = 0, \quad (6.3)$$

for all $\chi \in L_2(\Omega)$.† Let us write (6.2) in the form of an operator equation

$$Au = \{f; \varphi; \psi\}, \tag{6.4}$$

where A is an unbounded linear operator in $L_2(Q_T)$ which assigns to each function $u(x, t)$ in its domain of definition $\mathscr{D}(A)$ an element $\{\mathscr{L}u(x, t); u(x, 0), u_t(x, 0)\}$ of the Hilbert space $\mathscr{W} = L_2(Q_T) \times \overset{\circ}{W}{}_2^1(\Omega) \times L_2(\Omega)$. We define the scalar product in $\mathscr{W}$ to be

$$(\{f_1; \varphi_1; \psi_1\}, \{f_2; \varphi_2; \psi_2\})_{\mathscr{W}} = (f_1, f_2)_{2,Q_T} + (\varphi_1, \varphi_2)_{2,\Omega}^{(1)} + (\psi_1, \psi_2)_{2,\Omega},$$

and we denote the norm in $\mathscr{W}$ by $\|\cdot\|_{\mathscr{W}}$.

We take $\mathscr{D}(A)$ to be $W_{2,0}^2(Q_T)$, the totality of elements of $W_2^2(Q_T)$ which vanish on the lateral surface of Q_T. The object of the subsequent arguments is the proof of the fact that the operator A admits a closure $\bar{A}$, that $\bar{A}$ has an inverse, and that the range of $\bar{A}$ coincides with all of $\mathscr{W}$. The function $u = \bar{A}^{-1}\{f; \varphi, \psi\}$ will be a generalized solution of (6.2). The proof for the closure of A is the same as in §2, Chapter III. For the proof of the remaining facts, as well as a description of the operator $\bar{A}^{-1}$, we recall the energy inequality (3.3). Applied to the wave operator in (6.2), it can, with a little effort, be written in the form

$$\begin{aligned} |u|_{Q_T} &\equiv \max_{0 \le t \le T} \left(\int_{\Omega_t} (u_t^2 + u_x^2 + u^2)\, dx \right)^{1/2} \\ &\le c[(\|u(\cdot, 0)\|_{2,\Omega}^{(1)})^2 + \|u_t(\cdot, 0)\|_{2,\Omega}^2 + \|\mathscr{L}u\|_{2,Q_T}^2]^{1/2} \end{aligned} \tag{6.5}$$

which holds for all u in $\mathscr{D}(A)$. Because of this, the convergence of $Au_m = \{f_m; \varphi_m; \psi_m\}$ $(m = 1, 2, \ldots)$, $u_m \in \mathscr{D}(A)$, in the norm of $\mathscr{W}$ to some element $\{f; \varphi; \psi\} \in \mathscr{W}$ implies the convergence of the functions u_m in the norm $|\cdot|_{Q_T}$ of (6.5) (and indeed in the norm of $L_2(Q_T)$) to some $u \in C([0, T]; \overset{\circ}{W}{}_2^1(\Omega))$. This means that $\bar{A}$ has a bounded inverse and that the range $\mathscr{R}(\bar{A})$ of $\bar{A}$ is a closed subspace of $\mathscr{W}$. The limiting element u is adjoined, by definition, to $\mathscr{D}(A)$, and $\bar{A}u$ is set equal to $\{f; \varphi; \psi\}$; this is how $\mathscr{D}(\bar{A})$ is constructed. It follows from (6.5) that for u in $\mathscr{D}(\bar{A})$, the norm $|u|_{Q_T}$ is finite, $u(x, t)$ is continuous with respect to $t \in [0, T]$ in the norm

$$\left(\int_{\Omega} (u_t^2 + u_x^2 + u^2)\, dx \right)^{1/2},$$

and $u \in \overset{\circ}{W}{}_2^1(\Omega)$ for all $t \in [0, T]$.

However, the elements of $\mathscr{D}(\bar{A})$ which were adjoined to $\mathscr{D}(A)$ need not have second-order derivatives, so that we cannot assert that, for all elements

† In point of fact, for the equation (6.2), as well as for hyperbolic equations whose coefficients a_{ij} do not depend on t, this condition concerning the solvability in $W_2^2(\Omega)$ of the Dirichlet problem $(\partial/\partial x_i)(a_{ij}u_{x_j}) \equiv \chi$, $u|_S = 0$ may be discarded, and we may work in the space $W_2^1(\Omega)$, taking Ω to be an arbitrary domain. If, however, the coefficients a_{ij} depend on t, we have to deal with the space $W_2^2(\Omega)$ in order to be certain that the solutions of the above Dirichlet problem, when χ ranges over all of $L_2(\Omega)$, fill out a certain subset of $L_2(\Omega)$ which is independent of t (that is, a subset which is independent of the coefficients $a_{ij}(x, t)$).

u in $\mathscr{D}(\bar{A})$, the element $\bar{A}u$ is equal to $\{\mathscr{L}u; u(\cdot, 0); u_t(\cdot, 0)\}$. In order to understand how to "calculate" $\bar{A}$ for $u \in \mathscr{D}(\bar{A})$, we replace (6.4) for u in $\mathscr{D}(A)$ by the equivalent requirements:

(1) $u(x, 0) = \varphi(x)$, and (2) $u(x, t)$ satisfies the identity

$$\int_\Omega u_t(x, t)\eta(x, t)\,dx + \int_{Q_t} (-u_t\eta_t + u_x\eta_x)\,dx\,dt$$

$$- \int_\Omega \psi\eta(x, 0)\,dx = \int_{Q_t} f\eta\,dx\,dt \tag{6.6}$$

for any $\eta \in \overset{\circ}{W}{}_{2,0}^1(Q_T)$, that is, $\eta \in W_2^1(Q_T)$ and $\eta|_{S_T} = 0$. The form of requirements (1) and (2) is preserved for $u \in \mathscr{D}(\bar{A})$. In fact, if $u_m \in \mathscr{D}(A)$ and if $Au_m = \{f_m; \varphi_m; \psi_m\}$ converges in the norm of $\mathscr{W}$ to $\{f; \varphi; \psi\}$, then, as we indicated above, u_m converges to u in the norm $|\cdot|_{Q_T}$, and hence if we write $u = u_m$, $f = f_m, \psi = \psi_m$ in (6.6), we can take the limit as $m \to \infty$. As a result, we obtain (6.6) for the limiting functions u, f, and ψ. The equality $u(x, 0) = \varphi(x)$ will also hold for the limiting functions u and φ.

Thus we have shown that, if $u \in \mathscr{D}(\bar{A})$ and

$$\bar{A}u = \{f; \varphi; \psi\}, \tag{6.7}$$

then the relations (1) and (2) hold for u, f, φ, and ψ.

We shall now prove that $\mathscr{R}(\bar{A}) = \mathscr{W}$. Since $\mathscr{R}(\bar{A}) = \overline{\mathscr{R}(\bar{A})}$, that is, $\mathscr{R}(\bar{A})$ is a closed subspace of $\mathscr{W}$, the failure of $\mathscr{R}(\bar{A})$ to coincide with $\mathscr{W}$ would mean that there is an element $\{\tilde{f}; \tilde{\varphi}; \tilde{\psi}\}$ in $\mathscr{W}$ which is orthogonal to every element in $\mathscr{R}(A)$, that is, such that

$$\int_{Q_T} \tilde{f}\mathscr{L}v\,dx\,dt + \int_\Omega \tilde{\varphi}_x v_x(x, 0)\,dx + \int_\Omega \tilde{\psi} v_t(x, 0)\,dx = 0 \tag{6.8}$$

for all $v \in \mathscr{D}(A)$. We shall show that it follows from (6.8) that $\tilde{f}$, $\tilde{\varphi}$, and $\tilde{\psi}$ vanish identically. Let us take as v in (6.8) the function $v(x, t) = 0$ for $0 \le t \le t_1$ and $v(x, t) = \int_{t_1}^t d\xi \int_{t_1}^\xi \Phi(x, \eta)\,d\eta$, $t \in [t_1, T]$, where $\Phi(x, t)$ is the solution of the problem

$$\Delta\Phi(x, t) = \int_T^t \tilde{f}(x, \tau)\,d\tau, \qquad \Phi|_S = 0, \qquad t \in [t_1, T]. \tag{6.9}$$

It is not hard to see that such a v belongs to $\mathscr{D}(A)$. Let us substitute it into (6.8) and express v and $\tilde{f}$ in terms of Φ. This gives us the relation

$$\int_{t_1}^T \int_\Omega \Delta\Phi_t \left(\Phi - \int_{t_1}^t d\xi \int_{t_1}^\xi \Delta\Phi(x, \eta)\,d\eta\right) dx\,dt = 0,$$

which, when integrated by parts and using $\Phi|_S = 0$ and $\Phi|_{t=T} = 0$, yields

$$\frac{1}{2}\int_\Omega \Phi_x^2(x, t_1)\,dx + \int_{t_1}^T \int_\Omega \Delta\Phi(x, t) \int_{t_1}^t \Delta\Phi(x, \eta)\,d\eta\,dx\,dt = 0,$$

whence

$$\frac{1}{2}\int_\Omega \Phi_x^2(x, t_1)\, dx + \frac{1}{2}\int_\Omega \left(\int_{t_1}^T \Delta\Phi(x, \eta)\, d\eta\right)^2 dx = 0. \tag{6.10}$$

Because $t_1 \in [0, T]$ in (6.10) is arbitrary, we conclude that $\Phi \equiv 0$, and hence $\tilde{f} \equiv 0$. Because of this, (6.8) takes the form

$$\int_\Omega \tilde{\varphi}_x v_x(x, 0)\, dx + \int_\Omega \tilde{\psi} v_t(x, 0)\, dx = 0. \tag{6.11}$$

Since (6.11) holds for all $v \in \mathscr{D}(A)$, that is, for all v in $W^2_{2,0}(Q_T)$, it is easy to see that it follows from (6.11) that $\tilde{\varphi} \equiv 0$ and $\tilde{\psi} \equiv 0$. Thus we have proved that $\mathscr{R}(\bar{A})$ is all of $\mathscr{W}$, that is, that (6.7) has a solution $u \in \mathscr{D}(\bar{A})$ for each $\{f; \varphi; \psi\} \in \mathscr{W}$. The solution is unique in the class $\mathscr{D}(\bar{A})$ and satisfies the estimate (6.5), i.e.,

$$|u|_{Q_T} \leq c[(\|\varphi\|^{(1)}_{2,\Omega})^2 + \|\psi\|^2_{2,Q_T} + \|f\|^2_{2,Q_T}]^{1/2}, \tag{6.12}$$

as well as the energy equality

$$\frac{1}{2}\int_{Q_T} (u_t^2 + u_x^2)\, dx \Big|_{t=0}^{t=t} = \int_{Q_T} f u_t\, dx\, dt.$$

In fact, (6.5) and the last equality hold for any $u(x, t)$ in $\mathscr{D}(A)$, and then, by "closure in the norm $|\cdot|_{Q_T}$," they hold for all u in $\mathscr{D}(\bar{A})$.

It follows from everything that has been said about the functions in $\mathscr{D}(\bar{A})$ that $u = \bar{A}^{-1}\{f; \varphi; \psi\}$ is a generalized solution of (6.2) in the energy class. Thus we have proved the following theorem.

Theorem 6.1. *The problem* (6.2) *has a unique generalized solution in the energy class for all* $f \in L_2(Q_T)$, $\varphi \in \overset{\circ}{W}{}^1_2(\Omega)$, *and* $\psi \in L_2(\Omega)$.

It follows from the arguments of this section that the uniqueness of the solution of (6.2) holds in $\mathscr{D}(\bar{A})$. However, by Theorem 3.1, the uniqueness also holds in the class $W^1_2(Q_T)$. Moreover, in the course of the proof of Theorem 6.1 we actually proved the uniqueness theorem for (6.2) in the class of generalized solutions in $L_2(Q_T)$. By a *generalized solution of the problem* (6.2) *in the space* $L_2(Q_T)$ we shall mean a function $u \in L_2(Q_T)$ which satisfies the identity

$$\int_{Q_T} u(\eta_{tt} - \Delta\eta)\, dx\, dt - \int_\Omega \psi\eta(x, 0)\, dx$$

$$+ \int_\Omega \varphi\eta_t(x, 0)\, dx = \int_{Q_T} f\eta\, dx\, dt \tag{6.13}$$

for all $\eta \in W^2_{2,0}(Q_T)$ such that $\eta|_{t=T} = 0$ and $\eta_t|_{t=T} = 0$.

Theorem 6.2. *For the problem* (6.2) *the uniqueness theorem holds in the class of generalized solutions in* $L_2(Q_T)$.

In fact, let u be in $L_2(Q_T)$ and satisfy (6.13) with $\varphi = \psi = f = 0$. Let us replace the variable t by $T - \tau$ in (6.13), so that (6.13) assumes the form

$$\int_{Q_T} u(\eta_{\tau\tau} - \Delta\eta)\,dx\,d\tau = 0, \qquad Q_T = \{(x, \tau): x \in \Omega, \tau \in (0, T)\}, \quad (6.14)$$

where $\eta \in W^2_{2,0}(Q_T)$, $\eta|_{\tau=0} = 0$, and $\eta_\tau|_{\tau=0} = 0$. But (6.14) carries the same information as (6.8) with $\tilde{f} = u$ and $\tilde{\varphi} = \tilde{\psi} = 0$, and that is why from (6.14) it follows that $u \equiv 0$. This proves Theorem 6.2. If we use (3.3) for the equation in (6.2), it is easy to show that the assertion of Theorem 6.1 is true for all $f \in L_{2,1}(Q_T)$.

As we said at the beginning of this section, Theorems 6.1 and 6.2 are valid for equations (2.1) of general type.

The remarkable feature of the argument of this section consists of the fact that we were able to derive the existence theorem for the problem (6.2) from the uniqueness theorem for the adjoint problem (the argument, which was connected with the proof that there was no orthogonal complement of $\mathscr{R}(A)$ in $\mathscr{W}$, is a uniqueness theorem in $L_2(Q_T)$ for the adjoint problem to (6.2)). This fact by itself is almost trivial and can be formulated equally well for linear algebraic systems in a finite-dimensional Euclidean space as in infinite-dimensional spaces for the equations $Au = f$ with a bounded or unbounded linear operator A. The non-trivial part of the argument consists of the fact that we succeeded in proving directly the uniqueness theorem for the adjoint problem in the class of generalized solutions in $L_2(Q_T)$ (we remark that the problem (6.2) and the problem that is formally adjoint to it,

$$\mathscr{L}^*u = f, \qquad u|_{t=T} = \varphi; \qquad u_t|_{t=T} = \psi, \qquad u|_{S_T} = 0, \qquad (6.2')$$

simply coincide in the case of the wave equation if we replace t by $T - \tau$ in (6.2′), and are of the same type for the general equations (2.1) since they differ in the lower-order, inessential terms).

The existence theorems are derived from such uniqueness theorems without any analytic representations for the desired solutions and even without constructing any approximations to them. Such methods of proving existence theorems we call "functional"; their structure is typical for arguments carried out in functional analysis.

§7. The Methods of Fourier and Laplace

Initial-boundary problems (as well as the Cauchy problem) for the equations (2.1) with coefficients independent of t can be reduced by means of the Laplace transform to the solution of the corresponding boundary value problems for elliptic equations with a parameter. This is done in the following way. Let $u(x, t)$ be the solution of (3.1), where the coefficients of $\mathscr{L}$ ($\mathscr{L}$ is written out in (2.1)) do not depend of t. We multiply the equation $\mathscr{L}u = f$ by $e^{-\lambda t}$ and integrate over t from 0 to ∞. If $u(x, t)$ and its derivatives which appear in $\mathscr{L}$ do not increase as fast as $e^{\mathrm{Re}\,\lambda t}$ as $t \to \infty$, then all the integrals converge. We introduce the notation:

$$\int_0^\infty u(x, t)e^{-\lambda t}\,dt = v(x, \lambda), \qquad \int_0^\infty f(x, t)e^{-\lambda t}\,dt = \mathscr{F}(x, \lambda).$$

Then

$$\int_0^\infty u_{x_i} e^{-\lambda t}\,dt = v_{x_i}, \qquad \int_0^\infty u_{x_i x_j} e^{-\lambda t}\,dt = v_{x_i x_j},$$

and

$$\int_0^\infty u_{tt} e^{-\lambda t}\,dt = \lambda^2 \int_0^\infty u e^{-\lambda t}\,dt - (u_t + \lambda u)|_{t=0} = \lambda^2 v - \psi - \lambda\varphi,$$

whence $\int_0^\infty \mathscr{L}u\, e^{-\lambda t}\,dt = \int_0^\infty f e^{-\lambda t}\,dt$ yields the equation

$$\mathscr{L}_0 v = \lambda^2 v - (\psi(x) + \lambda\varphi(x) + \mathscr{F}(x, \lambda)), \tag{7.1}$$

where $\mathscr{L}_0 v = (\partial/\partial x_i)(a_{ij}(x)v_{x_j}) - a_i(x)v_{x_i} - a(x)v$ is the elliptic part of the operator $\mathscr{L}$ in (2.1). The boundary condition for $u(x, t)$ in (3.1) gives the boundary condition for v:

$$v|_S = 0. \tag{7.2}$$

Thus, the problem (3.1) has been reduced to the problem (7.1)–(7.2), which is uniquely solvable in the half-plane $\lambda_1 = \mathrm{Re}\,\lambda \geq \lambda_1^0$ for sufficiently large λ_1^0 (cf. Chapter II); its solution depends analytically on λ and tends to zero as $|\lambda| \to \infty$ in the norm of one space or another. In order to attain the optimal decrease of v with respect to λ, we must first reduce (3.1) to a problem with homogeneous initial conditions (or even to the case where $\varphi = \psi = 0$ and $\partial^l f/\partial t^l|_{t=0} = 0$, $l = 0, 1, \ldots, N$). The solution of (3.1) can be recovered from $v(x, t)$ by the inverse Laplace transform:

$$u(x, t) = \frac{1}{2\pi i}\int_{\lambda_1 - i\infty}^{\lambda_1 + i\infty} v(x, \lambda)e^{\lambda t}\,d\lambda, \qquad \lambda_1 \geq \lambda_1^0. \tag{7.3}$$

The justification of this device in the class of classical solutions was given in Chapter IV of our first book [4].† On the basis of the solvability of the

† It should be kept in mind that when this book was being written (1950–51), we had not yet obtained the results which were present, in part, in §7 of Chapter II. At that time, the only method of studying the smoothness of generalized solutions was the method of finite differences, which is given a central role in this book. Questions of relaxing the assumptions of smoothness of the functions appearing in the problem and of the smoothness of the boundary of the domain Ω were not given special attention: the basic results of the book were new even for sufficiently smooth data.

problem (7.1)–(7.2), which was proved there, in the class $\overset{\circ}{W}{}_2^1(\Omega)$, and estimates for the norms $\|v\|_{2,\Omega}$ and $\|v\|_{2,\Omega}^{(1)}$ of its solutions, we justified the use of Laplace's method in the class of generalized solutions in the energy class (cf. [4, p. 273]). If we apply to the problem (7.1)–(7.2) the results of §§6–7, Chapter II, on its solvability in $W_2^2(\Omega)$, it is not difficult to obtain the corresponding results on the solvability of (3.1) in the space $W_2^2(Q_T)$. More complete results for elliptic equations, presented in Ladyzhenskaya and Ural'čeva [1], allow us to obtain, in an analogous way, the corresponding results on the solvability of (3.1) in other functional spaces.

We go over now to the method of Fourier. A complete study in all the spaces $W_2^l(\Omega)$, $l = 1, 2, \ldots$, for all three classical boundary conditions was given in [LA 4]. We shall give here a presentation of the method for the case of the first boundary condition for $l = 1$ and $l = 2$. Let us apply the Fourier method to equations of the form

$$\mathscr{L}u \equiv u_{tt} - \mathscr{L}_0 u = f(x, t), \tag{7.4}$$

where $\mathscr{L}_0 u = (\partial/\partial x_i)(a_{ij}(x)u_{x_j}) + a(x)u$. For the sake of convenience in what follows, take $a(x) \leq 0$. (If $a(x)$ does not satisfy this condition, then some of the first eigenvalues corresponding to the operator $\mathscr{L}_0$ turn out to be positive or zero. The particular solutions of (7.4) corresponding to these eigenvalues will have a different form from the solutions that correspond to the negative eigenvalues, but this fact will not introduce any changes in the arguments below concerning the convergence of the infinite series which form the solution of (7.4): The presence of a few first terms of a different form in these series will not affect the character of the convergence of these series.)

We begin with a study of the homogeneous problem

$$\mathscr{L}u \equiv u_{tt} - \mathscr{L}_0 u = 0, \quad u|_{S_T} = 0, \quad u|_{t=0} = \varphi, \quad u_t|_{t=0} = \psi. \tag{7.5}$$

The equation in (7.5) has particular solutions of the form $u = T(t)X(x)$. If we substitute this u in (7.5) and divide both sides by $u = TX$, we obtain

$$\frac{T''}{T} - \frac{\mathscr{L}_0 X}{X} = 0,$$

from which it is clear that $\mathscr{L}_0 X/X$ and T''/T must be constant, that is,

$$\mathscr{L}_0 X = \lambda X \tag{7.6}$$

and

$$T'' = \lambda T, \tag{7.7}$$

where λ is a constant. Let us take only those solutions $u = TX$ which satisfy the boundary condition for all $t \geq 0$ which was posed in the investigation of the problem, that is, in this case

$$X|_S = 0. \tag{7.8}$$

The problem (7.6), (7.8) is a spectral problem which was studied earlier in §§4 and 7, Chapter II. It has non-trivial solutions $\varphi_k(x)$ $(k = 1, 2, \ldots)$ only for $\lambda = \lambda_k$, which form the spectrum of the problem. Let the system of all linearly independent eigenfunctions $\{\varphi_k(x)\}$, $k = 1, 2, \ldots$, be orthonormalized in $L_2(\Omega)$, so that

$$(\varphi_k, \varphi_l) = \delta_k^l. \tag{7.9}$$

Then

$$[\varphi_k, \varphi_l] \equiv \int_\Omega (a_{ij}\varphi_{kx_j}\varphi_{lx_i} - a\varphi_k\varphi_l)\,dx = -\lambda_k\delta_k^l. \tag{7.10}$$

It follows from the assumption $a(x) \le 0$ that all eigenvalues are negative. We find it convenient to introduce the notation $\lambda_k = -\mu_k^2, 0 < \mu_1 \le \mu_2 \le \ldots$. The numbers $\mu_k \to \infty$ as $k \to \infty$.

Let us return to (7.7). Its general solution T_k for $\lambda = -\mu_k^2$ is equal to $a_k \cos \mu_k t + b_k \sin \mu_k t$, where the constants a_k and b_k are arbitrary. Hence the series

$$u = \sum_{k=1}^{\infty} (a_k \cos \mu_k t + b_k \sin \mu_k t)\varphi_k(x) \tag{7.11}$$

formally satisfies the first two equations of (7.5) for any numbers a_k and b_k. These numbers may be calculated uniquely from the initial conditions (3.1), namely,

$$u|_{t=0} = \sum_{k=1}^{\infty} a_k\varphi_k(x) = \varphi(x), \qquad u_t|_{t=0} = \sum_{k=1}^{\infty} \mu_k b_k \varphi_k(x) = \psi(x),$$

whence, by (7.9),

$$a_k = (\varphi, \varphi_k), \qquad b_k = \mu_k^{-1}(\psi, \varphi_k). \tag{7.12}$$

The basis of the Fourier method consists of investigating when the series (7.11), with a_k and b_k given by (7.12), converges in the norm in one or another functional space, and in what sense the sum of the series is the solution of the problem (7.5). We have the following theorem.

Theorem 7.1. *Let Ω be an arbitrary bounded domain, let $\mathscr{L}$ satisfy the conditions (2.2), let $0 \ge a(x) \ge c$, and let $\varphi(x)$ and $\psi(x)$ be elements of the spaces $\mathring{W}_2^1(\Omega)$ and $L_2(\Omega)$, respectively. Then the series (7.11), with a_k and b_k given by (7.12), and the series which are obtained by differentiating (7.11) once with respect to any of the x_i or t, converge in the norm of $L_2(\Omega)$, uniformly with respect to $t \in [0, \infty)$. The sum of the series is a generalized solution of (7.5) in the energy class.*

If the coefficients of $\mathscr{L}$ in (7.5) and the domain Ω satisfy the hypotheses of Theorem 7.2, Chapter II, if $a(x) \le 0$, and if $\varphi \in W_{2,0}^2(\Omega)$ and $\psi \in \mathring{W}_2^1(\Omega)$, then the series (7.11) with a_k and b_k as in (7.12) may be differentiated twice with respect to x_i and t, and all the series converge in $L_2(\Omega)$, uniformly with respect *to $t \in [0, \infty)$. The sum of the series is a generalized solution of (7.5) in the class $W_2^2(Q_T)$ and satisfies the equation for all t and for almost all x in Ω.*

Theorem 7.1 can be deduced easily from Theorem 4.1 and Theorem 7.4, Chapter II. In fact, by the assumption that $\varphi \in \mathring{W}_2^1(\Omega)$, the series $\varphi(x) = \sum_{k=1}^{\infty} a_k \varphi_k(x)$ converges in the norm of $W_2^1(\Omega)$ where $\|\varphi\|_{2,\Omega}^2 = \sum_{k=1}^{\infty} a_k^2$ and $[\varphi, \varphi] = \sum_{k=1}^{\infty} \mu_k^2 a_k^2$. The series $\psi(x) = \sum_{k=1}^{\infty} \mu_k b_k \varphi_k(x)$ converges in $L_2(\Omega)$ and $\|\psi\|_{2,\Omega}^2 = \sum_{k=1}^{\infty} \mu_k^2 b_k^2$, for $\psi \in L_2(\Omega)$. Hence the series (7.11) converges, as was formulated in the first part of the theorem, where for its sum we have the following relations:

$$\|u\|_{2,\Omega}^2 = \sum_{k=1}^{\infty} (a_k \cos \mu_k t + b_k \sin \mu_k t)^2 \leq 2 \sum_{k=1}^{\infty} (a_k^2 + b_k^2) < \infty,$$

$$[u, u] = \sum_{k=1}^{\infty} \mu_k^2 (a_k \cos \mu_k t + b_k \sin \mu_k t)^2 \leq 2 \sum_{k=1}^{\infty} \mu_k^2 (a_k^2 + b_k^2) < \infty,$$

$$\|u_t\|_{2,\Omega}^2 = \sum_{k=1}^{\infty} \mu_k^2 (-a_k \sin \mu_k t + b_k \cos \mu_k t)^2 \leq 2 \sum_{k=1}^{\infty} \mu_k^2 (a_k^2 + b_k^2) < \infty.$$

When the conditions of the second part of Theorem 7.1 are fulfilled, the eigenfunctions $\varphi_k(x)$ are in $W_{2,0}^2(\Omega)$ and form a basis in $W_{2,0}^2(\Omega)$, where

$$\{\varphi_k, \varphi_l\} \equiv (\mathscr{L}_0 \varphi_k, \mathscr{L}_0 \varphi_l) = \lambda_k^2 \delta_k^l = \mu_k^4 \delta_k^l. \tag{7.13}$$

We recall that the usual norm in $W_{2,0}^2(\Omega)$ is equivalent to the norm $|u|_2 = \{u, u\}^{1/2}$. The norm $|\cdot|_2$ for the series (7.11) is easy to estimate, starting from (7.13), namely,

$$|u|_2^2 = \sum_{k=1}^{\infty} \mu_k^4 (a_k \cos \mu_k t + b_k \sin \mu_k t)^2 \leq 2 \sum_{k=1}^{\infty} \mu_k^4 (a_k^2 + b_k^2). \tag{7.14}$$

The quantities $[u_t, u_t]$ and $\|u_{tt}\|_{2,\Omega}^2$ can be majorized by the same $j \equiv 2 \sum_{k=1}^{\infty} \mu_k^4 (a_k^2 + b_k^2)$. Now the numerical series j converges, for it follows from the condition $\varphi \in W_{2,0}^2(\Omega)$ that

$$a_k = (\varphi, \varphi_k) = \frac{1}{\lambda_k} (\varphi, \mathscr{L}_0 \varphi_k) = \frac{1}{\lambda_k} (\mathscr{L}_0 \varphi, \varphi_k)$$

and $\sum_{k=1}^{\infty} \lambda_k^2 a_k^2 = \sum_{k=1}^{\infty} \mu_k^4 a_k^2 = \|\mathscr{L}_0 \varphi\|_{2,\Omega}^2 < \infty$; the convergence of the series $\sum_{k=1}^{\infty} \mu_k b_k \varphi_k(x)$ to $\psi(x)$ in the norm $[\ ,\ ]^{1/2}$ follows from the condition $\psi \in \mathring{W}_2^1(\Omega)$, where, by (7.10),

$$[\psi, \psi] = \sum_{k=1}^{\infty} \mu_k^4 b_k^2 < \infty.$$

Thus, we have proved the assertions of Theorem 7.1 concerning the convergence of the series (7.11). The fact that its sum is a generalized solution of (7.5) in the corresponding function space is easy to verify if we start from the definition of these solutions. If the conditions of the second part of the theorem are fulfilled, the sum of the series (7.11) satisfies the equation $\mathscr{L}u = 0$ for all $t \geq 0$ and for almost all x in Ω, so that it is automatically a generalized solution of (7.5) in the class $W_2^2(Q_T)$. The initial conditions are fulfilled in the first case in the form

$$\|u(\cdot, t) - \varphi(\cdot)\|_{2,\Omega}^{(1)} \to 0, \qquad \|u_t(\cdot, t) - \psi(\cdot)\|_{2,\Omega} \to 0,$$

and in the second case in the form

$$\|u(\cdot, t) - \varphi(\cdot)\|_{2,\Omega}^{(2)} \to 0, \qquad \|u_t(\cdot, t) - \psi(\cdot)\|_{2,\Omega}^{(1)} \to 0$$

as $t \to 0$.

Let us now consider the case of the non-homogeneous equation. The initial conditions can be taken to be homogeneous:

$$\mathscr{L}u \equiv u_{tt} - \mathscr{L}_0 u = f(x, t), \qquad u|_{S_T} = 0, \qquad u|_{t=0} = 0, \qquad u_t|_{t=0} = 0. \tag{7.15}$$

The sum of the solutions of problems (7.5) and (7.15) clearly yields the solution of the problem

$$\mathscr{L}u = f, \qquad u|_{S_T} = 0, \qquad u|_{t=0} = \varphi, \qquad u_t|_{t=0} = \psi. \tag{7.16}$$

Let $f \in L_2(Q_T)$, and let us expand f into a series of the eigenfunctions $\varphi_k(x)$:

$$f(x, t) = \sum_{k=1}^{\infty} f_k(t)\varphi_k(x), \qquad f_k(t) = (f(\cdot, t), \varphi_k(\cdot)). \tag{7.17}$$

This series obviously converges in $L_2(\Omega)$ for almost all $t \in [0, T]$. Let us seek solutions of (7.15) in the form of a series

$$u(x, t) = \sum_{k=1}^{\infty} T_k(t)\varphi_k(x). \tag{7.18}$$

If we substitute this into the equation $\mathscr{L}u = f$, we obtain the equations

$$T_k'' + \mu_k^2 T_k = f_k(t).$$

for determining the functions $T_k(t)$. Their solutions satisfying the homogeneous initial conditions for $t = 0$ are the functions

$$T_k(t) = \frac{1}{\mu_k}\int_0^t f_k(\tau) \sin \mu_k(t - \tau)\, d\tau. \tag{7.19}$$

Thus the series (7.18) with $T_k(t)$ given by (7.19) formally satisfies all the conditions of the problem (7.15). It converges in $W_2^1(\Omega)$, uniformly in $t \in [0, T]$, for

$$\left[\sum_{k=m}^{m+N} T_k\varphi_k, \sum_{k=m}^{m+N} T_k\varphi_k\right] = \sum_{k=m}^{m+N} \mu_k^2 T_k^2(t)$$

$$\leq \sum_{k=m}^{m+N}\left(\int_0^t |f_k(\tau)|\, d\tau\right)^2 \leq T\sum_{k=m}^{m+N}\int_0^T f_k^2(\tau)\, d\tau$$

and

$$\sum_{k=1}^{\infty} \int_0^T f_k^2(\tau)\, d\tau = \|f\|_{2,Q_T}^2 < \infty.$$

Moreover, it is easy to verify that the sum of the series (7.18) is a generalized solution of (7.15) in the energy class. To show that the series (7.18) is twice differentiable, we must guarantee, as we saw above, the convergence of the series $j = \sum_{k=1}^{\infty} T_k^2(t)\mu_k^4$. If we assume that $f(x, t)$ has a derivative f_t in $L_2(Q_T)$, then the series j can be majorized for all $t \in [0, T]$ by a convergent numerical series, that is, j converges uniformly with respect to $t \in [0, T]$. In fact, $T_k(t)$ can be represented in the form

$$\begin{aligned} T_k(t) &= \mu_k^{-2} \int_0^t f_k(\tau) d \cos \mu_k(t - \tau) \\ &= \mu_k^{-2} \left[- \int_0^t f'_k(\tau) \cos \mu_k(t - \tau)\, d\tau + f_k(t) - f_k(0) \cos \mu_k t \right] \\ &= \mu_k^{-2} \left[\int_0^t f'_k(\tau)(1 - \cos \mu_k(t - \tau))\, d\tau + f_k(0)(1 - \cos \mu_k t) \right], \end{aligned}$$

whence it follows that

$$T_k^2(t) \le 8\mu_k^{-4} \left[T \int_0^T (f'_k(\tau))^2\, d\tau + f_k^2(0) \right]$$

and

$$\sum_{k=m}^{m+N} T_k^2(t)\mu_k^4 \le 8T \sum_{k=m}^{m+N} \int_0^T (f'_k(\tau))^2\, d\tau + 8 \sum_{k=m}^{m+N} f_k^2(0) \to 0$$

as $m, N \to \infty$. Thus we have the following result.

Theorem 7.2. *Let Ω be an arbitrary bounded domain, let $\mathscr{L}$ satisfy the conditions (2.2), let $0 \ge a(x) \ge c$, and let $f \in L_2(Q_T)$. Then the series (7.18) with T_k given by (7.19), as well as the series obtained by term-by-term differentiation of (7.18) in x_i and t, converge in $L_2(\Omega)$, uniformly in $t \in [0, T]$, and its sum is a generalized solution of (7.15) in the energy class. If the coefficients of $\mathscr{L}$ and the domain Ω satisfy the hypotheses of Theorem 7.2, Chapter II, if $0 \ge a(x)$, and if f and f_t are in $L_2(Q_T)$, then the series (7.18), as well as the series obtained by one and two term-by-term differentiations of (7.18) in x_i and t, converge in $L_2(\Omega)$, uniformly in $t \in [0, T]$. The sum of the series is a generalized solution of (7.15) in $W_2^2(Q_T)$ and satisfies (7.15) for all $t \in [0, T]$ for almost all x in Ω.*

Remark 7.1. As we said earlier, the assumption that $0 \ge a(x)$ in Theorems 7.1 and 7.2 can be discarded, for this can affect the form of only the first few terms in the series (7.11) and (7.18). Thus the conclusions of Theorems 7.1 and 7.2 concerning the convergence of the series need not be changed.

In the second part of Theorem 7.2 the condition of existence of the derivative f_t of f can be replaced, for example, by the condition of the existence of $f_{x_i} \in L_2(Q_T)$ and by $f|_{S_T} = 0$.

Analogous theorems are true for the boundary conditions from (5.1).

Supplements and Problems

1. Verify that for hyperbolic equations of the form (1.3′) the theorems, analogous to those of the form (1.3), are valid.

2. Obtain the "energy" estimate

$$z(t) \equiv \int_{\Omega_t} (u^2 + u_t^2 + a_{ij}u_{x_i}u_{x_j})\, dx \leq c_2(t)z(0) + c_3(t) \int_{\mathscr{D}_t} f^2\, dx\, dt$$

for solutions of equations (1.3) and (1.3′) using Lemma 1.1 of Chapter III. Find out what determines the growth of majorants $c_2(t)$ and $c_3(t)$ as t increases. Obtain the best possible expressions for $c_k(t)$, $k = 2, 3$. Do the same for majorants $c_k(t)$ in inequality (2.15). Single out the cases when $c_k(t)$ can be majorized by constants on the semiaxis $\{t: t \geq 0\}$. Pay attention to the role of the terms $a_{n+1}u_{x_{n+1}} \equiv a_{n+1}u_t$ and au. For what homogeneous equations of the form (1.3) is the law of conservation of energy valid (see footnote on page 154).

3. Investigate in more detail the solvability of the second and third initial-boundary value problems (see §5).

4. Prove the unique solvability of the first initial-boundary value problem (3.1) by the functional method described in §6 and carried out in §6 for the wave equation. Try, with the help of this method, to prove the unique solvability of the Cauchy problem (6.1) for abstract hyperbolic equations (see §6 of this chapter and §§2, 3 of Chapter V).

5. Justify the Laplace method for equations (1.3) with coefficients independent of t.

6. Investigate the unique solvability of the basic initial-boundary value problems for the system

$$u_t - \mathscr{L}_1 u + l_1 v = f_1, \qquad v_{tt} - \mathscr{L}_2 v + l_2 u = f_2, \tag{1}$$

where

$$\mathscr{L}_k u = \sum_{i,j=1}^{n} \frac{\partial}{\partial x_i}(a_{ij}^{(k)}(x,t)u_{x_j}) + \sum_{i=1}^{n} a_i^{(k)}(x,t)u_{x_i} + a^{(k)}(x,t)u, \qquad k = 1, 2, \quad \text{are elliptic operators,}$$

$$l_1 v = \sum_{i=1}^{n} b_i^{(1)}(x,t)v_{x_i} + b_0(x,t)v_t + b_{(x,t)}^{(1)} v \quad \text{and}$$

$$l_2 u = \sum_{i=1}^{n} b_i^{(2)}(x,t)u_{x_i} + b^{(2)}(x,t)u.$$

The initial conditions have the form

$$u|_{t=0} = \varphi_1, \quad v|_{t=0} = \varphi_2, \quad v_t|_{t=0} = \varphi_3.$$

7. Let the domain Ω be divided by a $(n-1)$-dimensional surface Γ into two subdomains Ω_1 and Ω_2, so that $\Omega = \Omega_1 \cup \Omega_2 \cup \Gamma$. Investigate the solvability in $Q_T = \Omega \times (0, T)$ of the following diffraction problem:

$$u_t - \mathscr{L}_1 u = f_1 \quad \text{for} \quad (x,t) \in \Omega_1 \times (0,T),$$

$$u_{tt} - \mathscr{L}_2 u = f_2 \quad \text{for} \quad (x,t) \in \Omega_2 \times (0,T)$$

$$u(x,0) = \varphi_1(x) \quad \text{for} \quad x \in \Omega_1, \; u(x,0) = \varphi_2(x) \quad \text{for} \quad x \in \Omega_2,$$

$$u_t(x,0) = \varphi_3(x) \quad \text{for} \quad x \in \Omega_2,$$

$$u|_{\partial\Omega \times [0,T]} = 0, \qquad u|_{\Gamma^{(1)} \times [0,T]} = u|_{\Gamma^{(2)} \times [0,T]},$$

$$\left.\frac{\partial u}{\partial N_1}\right|_{\Gamma^{(1)} \times [0,T]} = \left.\frac{\partial u}{\partial N_2}\right|_{\Gamma^{(2)} \times [0,T]}.$$

Here $\mathscr{L}_k$ are the same elliptic operators as in **6**, $\partial u/\partial N_k = a_{ij}^{(k)} u_{x_j} \cos(\mathbf{n}, \mathbf{x}_i)$, $k = 1, 2$, $\mathbf{n}$ is the normal to Γ directed to Ω_2, and the symbol $w|_{\Gamma^{(k)} \times [0,T]}$ means the limiting value (trace) of the function $w(x,t)$, $(x,t) \in \Omega_k \times [0,T]$ on $\Gamma \times [0,T]$. This and some other problems of such type (with several domains Ω_k with elliptic, parabolic, or hyperbolic equations in each of them) were considered by Stupjalis and myself (see [LSS 1; 2]), by Stupjalis [SS 1–3] and by others.

CHAPTER V

Some Generalizations

In this chapter we shall be concerned with a number of more complicated problems which can be treated by the methods discussed in Chapters II–IV.

§1. Elliptic Equations of Arbitrary Order. Strongly Elliptic Systems

Let us consider in a bounded domain $\Omega \subset R^n$ the system of equations

$$\mathscr{L}\mathbf{u} \equiv \sum_{|i|,\,|j| \leq m} (-1)^{|i|}\mathscr{D}^i(A^{(i,j)}(x)\mathscr{D}^j\mathbf{u}) = \mathbf{f}, \tag{1.1}$$

where $\mathscr{D}^i = \partial^{|i|}/\partial x_1^{i_1} \ldots \partial x_n^{i_n}$, $i = (i_1, \ldots, i_n)$, $|i| = i_1 + \ldots + i_n$, $\mathbf{u}$ and $\mathbf{f}$ are the vector-functions $(u_1, \ldots, u_N)$ and $(f_1 \ldots, f_N)$, and $A^{(i,j)}(x)$ are square matrices of order N with $A^{(i,j)}(x) = A^{(j,i)}(x)$ when $|i| = |j| = m$. The system (1.1) is called *strongly elliptic at the point* x if, for any real vector $\boldsymbol{\zeta} = (\zeta_1, \ldots, \zeta_N) \neq 0$ and for arbitrary real numbers $\xi_1, \ldots, \xi_n$ with $\sum_{i=1}^n \xi_i^2 \neq 0$, the quadratic form

$$A(x;\xi)\boldsymbol{\zeta} \cdot \boldsymbol{\zeta} \equiv \sum_{|i|,\,|j| = m} A^{(i,j)}(x)\xi_1^{i_1} \ldots \xi_n^{i_n}\xi_1^{j_1} \ldots \xi_n^{j_n}\boldsymbol{\zeta} \cdot \boldsymbol{\zeta} > 0. \tag{1.2}$$

Here the symbol $\boldsymbol{\eta} \cdot \boldsymbol{\zeta}$ denotes the scalar product of N-dimensional vectors, that is, $\sum_{i=1}^N \eta_i\zeta_i$, and $A(x;\xi)\boldsymbol{\zeta}$ is a vector. We shall assume, as in most of the preceding sections, that all the quantities are real. The inequality (1.2) imposes a restriction only on the symmetric part of the matrix $A^{(i,j)}$, for if we represent $A^{(i,j)}$ in the form

$$A^{(i,j)} = \hat{A}^{(i,j)} + \tilde{A}^{(i,j)},$$

where $\hat{A}^{(i,j)} = \frac{1}{2}A^{(i,j)} + \frac{1}{2}(A^{(i,j)})^*$ is the symmetric part of $A^{(i,j)}$ and $\tilde{A}^{(i,j)} = \frac{1}{2}A^{(i,j)} - \frac{1}{2}(A^{(i,j)})^*$ is the skew-symmetric part (an asterisk on a matrix denotes its transpose), then

$$A(x, \xi)\zeta \cdot \zeta = \sum_{|i|, |j| = m} \hat{A}^{(i,j)}(x)\xi_1^{i_1} \ldots \xi_n^{i_n} \cdot \xi_1^{j_1} \ldots \xi_n^{j_n}\zeta \cdot \zeta \equiv \hat{A}(x, \xi)\zeta \cdot \zeta.$$

The equations considered in Chapter II are obviously special cases of strongly elliptic systems of the form (1.1). An elliptic equation of arbitrary order satisfies the condition (1.2), and consequently is a strongly elliptic system. Conversely, the class of general elliptic systems singled out by I. G. Petrovskii [2] and also the systems which are elliptic in the sense of Douglis and Nirenberg [1], are broader than the class of strongly elliptic systems defined by M. I. Višik [3]. It is not hard to show that if $A^{(i,j)}$ with $|i| = |j| = m$ is independent of x, then (1.2) implies that

$$\int_\Omega \sum_{|i|, |j| = m} A^{(i,j)}\mathscr{D}^i\mathbf{u} \cdot \mathscr{D}^j\mathbf{u}\, dx \geq \nu \int_\Omega \sum_{|i| = m} (\mathscr{D}^i\mathbf{u})^2\, dx, \qquad \nu = \text{const.} > 0, \quad (1.3)$$

for any vector function $\mathbf{u}$ in $\mathring{W}_2^m(\Omega)$ (the space $\mathring{W}_2^m(\Omega)$, consisting of scalar or vector functions, was defined in §5, Chapter I). For this it is necessary to extend $\mathbf{u}(x)$ by zero on some cube and then to take the Fourier transform with respect to x. If the $A^{(i,j)}$ depend on x and are continuous functions of $x \in \bar{\Omega}$, then, for any function $\mathbf{u}(x) \in \mathring{W}_2^m(\Omega)$, we have the Gårding inequality

$$\int_\Omega \sum_{|i|, |j| = m} A^{(i,j)}(x)\mathscr{D}^i\mathbf{u} \cdot \mathscr{D}^j\mathbf{u}\, dx \geq \nu \int_\Omega \sum_{|i| = m} (\mathscr{D}^i\mathbf{u})^2\, dx - \mu_1 \int_\Omega \mathbf{u}^2\, dx \quad (1.4)$$

for certain $\nu = \text{const} > 0$ and $\mu_1 = \text{const} \geq 0$ (cf. Gårding [1] and F. F. Browder [1], [2]). We shall not go through the proof of this fact here, but we shall assume it as a basis of the definition of a strongly elliptic system (1.1). Let us give a short sketch of the proof of the Fredholm solvability of the Dirichlet problem for the system (1.1). For this the boundary conditions are of the form

$$\mathbf{u}\Big|_S = \frac{\partial \mathbf{u}}{\partial n}\Big|_S = \cdots = \frac{\partial^{m-1}\mathbf{u}}{\partial n^{m-1}}\Big|_S = 0. \quad (1.5)$$

The case of non-homogeneous Dirichlet boundary conditions can be reduced easily to that of the homogeneous, as always. As the free term $\mathbf{f}$ we can take any sum of the form

$$\mathbf{f}(x) = \mathbf{f}^0(x) + \sum_{1 \leq |i| \leq m} \mathscr{D}^i\mathbf{f}^i,$$

where $\mathbf{f}^0$ and $\mathbf{f}^i$ are square-summable over Ω (it is sufficient to require the summability of the $\mathbf{f}^i$ with the powers $q_i \geq 2n/[n + 2(m - |i|)]$ if $n > 2(m - |i|)$, with arbitrary $q_i > 1$ if $n = 2(m - |i|)$, and with $q_i = 1$ if $n < 2(m - |i|)$).

Let us take $\mathbf{f}^i \equiv 0$ only for the sake of economy of space. We shall assume that all elements of the matrices $A^{(i,j)}(x)$ are bounded functions; for $A^{(i,j)}(x)$ with $|i| + |j| < 2m$ this condition may be weakened. We restrict our attention to the problem (1.1), (1.5) without the inclusion of a complex parameter λ. The general case, i.e., the problem

$$\mathscr{L}\mathbf{u} = \lambda \mathbf{u} + \mathbf{f}, \qquad \mathbf{u}\bigg|_S = \cdots = \frac{\partial^{m-1}\mathbf{u}}{\partial n^{m-1}}\bigg|_S = 0, \tag{1.6}$$

where λ is a complex number and $\mathbf{u}(x)$ is the complex-valued function that we seek, can be treated in a similar way.

A generalized solution $\mathbf{u}(x)$ *of the problem* (1.1), (1.5) *in the class* $W_2^m(\Omega)$ is an element of $\mathring{W}_2^m(\Omega)$ satisfying the integral identity

$$\mathscr{L}(\mathbf{u}, \boldsymbol{\eta}) \equiv \int_\Omega \sum_{|i|,\,|j| \leq m} A^{(i,j)} \mathscr{D}^j\mathbf{u} \cdot \mathscr{D}^i\boldsymbol{\eta}\, dx = \int_\Omega \mathbf{f} \cdot \boldsymbol{\eta}\, dx \tag{1.7}$$

for all $\boldsymbol{\eta} \in \mathring{W}_2^m(\Omega)$.

We introduce into $\mathring{W}_2^m(\Omega)$ a new scalar product by means of the equality

$$[\mathbf{u}, \boldsymbol{\eta}] = \int_\Omega \hat{A}^{(i,j)} \mathscr{D}^i\mathbf{u} \cdot \mathscr{D}^j\boldsymbol{\eta}\,\bigg|_{\substack{|i|=m\\|j|=m}} dx + \mu_1 \int_\Omega \mathbf{u}\boldsymbol{\eta}\, dx. \tag{1.8}$$

By (1.4) and the inequalities

$$\|\mathscr{D}^j\mathbf{u}\|_{2,\Omega}^2 \leq c_j \sum_{|i|=m} \|\mathscr{D}^i\mathbf{u}\|_{2,\Omega}^2, \qquad 0 \leq |j| < m, \tag{1.9}$$

which are easily proved for all $\mathbf{u} \in \mathring{W}_2^m(\Omega)$, the norm $\|\mathbf{u}\|_m = \sqrt{[\mathbf{u}, \mathbf{u}]}$ is equivalent to the original norm $\|\mathbf{u}\|_{2,\Omega}^{(m)}$ of $\mathring{W}_2^m(\Omega)$. Let us write the matrix $A^{(i,j)}$, $|i| = |j| = m$, as the sum $\hat{A}^{(i,j)} + \tilde{A}^{(i,j)}$ and rewrite (1.7) in the form

$$[\mathbf{u}, \boldsymbol{\eta}] + \int_\Omega \sum_{|i|,\,|j|=m} \tilde{A}^{(i,j)} \mathscr{D}^j\mathbf{u} \cdot \mathscr{D}^i\boldsymbol{\eta}\, dx$$

$$+ \int_\Omega \sum_{|i|+|j|<2m} A^{(i,j)} \mathscr{D}^j\mathbf{u} \cdot \mathscr{D}^i\boldsymbol{\eta}\, dx = \int_\Omega \mathbf{f} \cdot \boldsymbol{\eta}\, dx. \tag{1.10}$$

Each term of the left-hand side of (1.10) with an arbitrary fixed element $\mathbf{u} \in \mathring{W}_2^m(\Omega)$ is easily seen to be a linear (that is, additive and bounded) functional on $\boldsymbol{\eta}$ in $\mathring{W}_2^m(\Omega)$, and hence, by the Riesz theorem, can be represented as the scalar product [,] of $\boldsymbol{\eta}$ with some element of $\mathring{W}_2^m(\Omega)$ depending on $\mathbf{u}$. The right-hand side of (1.10) can be represented in the form $[\mathscr{F}, \boldsymbol{\eta}]$, where $\mathscr{F}$ is an element of $\mathring{W}_2^m(\Omega)$ defined by $\mathbf{f}$. Thus (1.10) can be rewritten in the form of the identity

$$[\mathbf{u}, \boldsymbol{\eta}] + [\mathscr{K}\mathbf{u}, \boldsymbol{\eta}] + [B\mathbf{u}, \boldsymbol{\eta}] = [\mathscr{F}, \boldsymbol{\eta}]$$

or, what is the same thing, in the form of the operator equation in $\mathring{W}_2^m(\Omega)$:

$$\mathbf{u} + \mathscr{K}\mathbf{u} + B\mathbf{u} = \mathscr{F}. \tag{1.11}$$

In the same way as in §3, Chapter II, it can be proved that $\mathscr{K}$ is a bounded operator in $\mathring{W}_2^m(\Omega)$ and that B is completely continuous. Moreover, because $(\tilde{A}^{(i,j)})^* = -\tilde{A}^{(i,j)}$, the operator $\mathscr{K}$ is skew-symmetric:

$$[\mathscr{K}\mathbf{u}, \boldsymbol{\eta}] = [\mathbf{u}, \mathscr{K}^*\boldsymbol{\eta}] = -[\mathbf{u}, \mathscr{K}\boldsymbol{\eta}] = -[\mathscr{K}\boldsymbol{\eta}, \mathbf{u}].$$

The spectrum of any skew-symmetric operator lies on the imaginary axis, for in the complex space $\mathring{W}_2^m(\Omega)$ the operator $i\mathscr{K}$ is self-adjoint. Hence the operator $E + \mathscr{K}$ is invertible, and $(E + \mathscr{K})^{-1}$ is a bounded operator defined on all of $\mathring{W}_2^m(\Omega)$. Thus (1.11) is equivalent to the equation

$$\mathbf{u} + (E + \mathscr{K})^{-1}B\mathbf{u} = (E + \mathscr{K})^{-1}\mathscr{F}. \tag{1.12}$$

The operator $(E + \mathscr{K})^{-1}B$ is completely continuous since it is the product of a bounded operator and a completely continuous operator. Consequently, Fredholm's theorem holds for (1.12). In particular, the solvability of (1.12) for any right-hand side follows from the uniqueness. Since (1.12) is equivalent to (1.7), this assertion means that if the uniqueness theorem holds for the problem (1.1), (1.5) in the class of generalized solutions in $\mathring{W}_2^m(\Omega)$, then the problem has a unique solution in $W_2^m(\Omega)$ for all $\mathbf{f} \in L_2(\Omega)$. The uniqueness theorem for (1.1), (1.5) holds automatically if the matrix $A^{(0,0)}$ multiplying $\mathbf{u}$ in (1.1) satisfies the inequality $(A^{(0,0)}\mathbf{u}, \mathbf{u}) \geq \nu_1(\mathbf{u}, \mathbf{u})$ for sufficiently large $\nu_1 = \text{const}$.

Let us formulate what we have proved in the form of a theorem.

Theorem 1.1. *The problem* (1.1), (1.5) *is Fredholm-solvable in* $W_2^m(\Omega)$ *if* $\mathbf{f} \in L_2(\Omega)$,† *if the elements of the matrices* $A^{(i,j)}(x)$ *are bounded functions in* Ω, *and if the condition* (1.4) *is fulfilled.*

This theorem was proved by M. I. Višik [3] with the help of a more complicated analysis and special constructions. We can prove in a similar way the Fredholm solvability of the problem (1.6) in the complex space $W_2^m(\Omega)$ for any complex λ. We can study just as easily the solvability in $W_2^m(\Omega)$ of those boundary value problems for systems (1.1) for which a generalized solution in $W_2^m(\Omega)$ is defined by an identity of the form

$$\mathscr{L}_\Omega(\mathbf{u}, \boldsymbol{\eta}) + \mathscr{L}_S(\mathbf{u}, \boldsymbol{\eta}) = \int_\Omega \mathbf{f} \cdot \boldsymbol{\eta}\, dx.$$

Here $\mathscr{L}_\Omega(\mathbf{u}, \boldsymbol{\eta})$ is an integral over Ω of a sum of products of derivatives of $\mathbf{u}$ and $\boldsymbol{\eta}$ of order not higher than m, $\mathscr{L}_S(\mathbf{u}, \boldsymbol{\eta})$ is an integral over the boundary

† It is enough to require that $\mathbf{f} \in L_q(\Omega)$ with $q \geq 2n/(n + 2m)$ for $n > 2m$, with any $q > 1$ for $n = 2m$, and with $q = 1$ for $n < 2m$ and $A^{(i,j)} \in L_{q_{ij}}(\Omega)$ with some q_{ij}.

S of a sum of products of derivatives of $\mathbf{u}$ and $\boldsymbol{\eta}$ of order not higher than $m-1$, and $\boldsymbol{\eta}(x)$ is an arbitrary element of some subspace of $W_2^m(\Omega)$ containing the desired solution $\mathbf{u}(x)$.

For these problems the other analogies of the theorems of Chapter II are also true. In particular, for the operator $\mathscr{L}$ in (1.1) with the boundary condition (1.5), *coercitivity* estimates in the norm of $W_2^{2m}(\Omega)$ take place, and each generalized solution of the problem (1.1), (1.5) in the space $W_2^m(\Omega)$ belongs to $W_2^{2m}(\Omega)$ if only $S \in C^{2m}$ and the coefficients of $\mathscr{L}$ have the bounded derivatives forming $\mathscr{L}$. But the proof of such estimates for (1.1), (1.5) is more complicated than for the scalar operator of second order. At first, it was done by O. V. Guseva [1] with the help of some special constructions. After that the proof was simplified at the expense of one remarkable Korn–Schauder idea which reduces the problem to the proof of some estimates for equations with constant coefficients. This idea also turned out to be very important for the derivation of coercitivity estimates in the spaces W_p^l, $l \geq 2m$, $p > 1$, and $H^{l+\alpha}(\bar{\Omega})$, $l \geq 2m$, for all classes of elliptic systems under Lopatinskii's boundary conditions (cf. A. Douglis and L. Nirenberg [1], [2], V. A. Solonnikov [4–7], etc.).

As was pointed out above, an elliptic equation of order $2m$ belongs to the class of systems considered here. In particular, the biharmonic equation

$$\Delta^2 u = f$$

satisfies (1.2), for its left-hand side (1.2) has the form $(\sum_{i=1}^n \xi_i^2)^2 \zeta^2$ (in this case ζ is a scalar). Equations of the form $\mathscr{L}u \equiv \mathscr{M}^{(m)}(u) = f$ also satisfy (1.2), where $\mathscr{M}$ is an arbitrary elliptic operator of second order with principal part $\sum_{i,j=1}^n a_{ij}u_{x_i x_j}$ and $\mathscr{M}^{(m)}$ is its m-th iterate, for the left-hand side of (1.2) is of the form $(\sum_{i,j=1}^n a_{ij}\xi_i\xi_j)^m \zeta^2$.

Another example of the class of systems considered here is the system of equations of the theory of elasticity. In the isotropic case it has the form

$$\sum_{k=1}^{3} \frac{\partial \tau_{ik}(\mathbf{u})}{\partial x_k} = -f_i(x), \qquad i = 1, 2, 3, \tag{1.13}$$

where

$$\tau_{ik}(\mathbf{u}) = 2\mu\varepsilon_{ik}(\mathbf{u}) + \delta_i^k \lambda \sum_{l=1}^{3} \varepsilon_{ll}(\mathbf{u}), \qquad \varepsilon_{ik}(\mathbf{u}) = \frac{1}{2}\left(\frac{\partial u_i}{\partial x_k} + \frac{\partial u_k}{\partial x_i}\right),$$

$\mathbf{u} = (u_1, u_2, u_3)$, and λ and μ are positive constants. Here the left-hand side of (1.2) is equal to

$$\frac{\mu}{2}\sum_{i,k=1}^{3} (\xi_k\zeta_i + \xi_i\zeta_k)^2 + \lambda\left(\sum_{i=1}^{3} \xi_i\zeta_i\right)^2 = \mu\xi^2\zeta^2 + (\mu + \lambda)\left(\sum_{i=1}^{3} \xi_i\zeta_i\right)^2.$$

so that the condition (1.2) is fulfilled.

The first boundary value problem may be solved more simply whenever the displacements $\mathbf{u}$ are given on S. Under the hypotheses of rigid attachment we have $\mathbf{u}|_S = 0$. The integral identity (1.7) for this case may be conveniently written in the form

$$\int_\Omega \tau_{ik}(\mathbf{u}) \cdot \varepsilon_{ik}(\boldsymbol{\eta})\, dx = \int_\Omega \mathbf{f}\boldsymbol{\eta}\, dx. \tag{1.14}$$

which should be fulfilled for all $\boldsymbol{\eta} \in \mathring{W}_2^1(\Omega)$, and the generalized solution $\mathbf{u}(x)$ should lie in $\mathring{W}_2^1(\Omega)$. It is easy to see that the uniqueness theorem holds in this class (because for $\mathbf{f} \equiv 0$ and $\boldsymbol{\eta} = \mathbf{u}$ it follows from (1.14) and $\mathbf{u}|_S = 0$ than $\mathbf{u} = 0$) and consequently the existence theorem holds for all $\mathbf{f} \in L_2(\Omega)$.

Let us consider one more example, which, although not coming under the case of a strongly elliptic system, can be studied by the same method used for an elliptic equation. Let R^3 denote three-dimensional Euclidean space, and let Ω be a bounded domain in R^3, which, for the sake of simplicity, we shall assume to be simply connected (in the sense that any closed path can be shrunk to a point in Ω). Let us denote by $\mathbf{u}(x) = (u_1(x), u_2(x), u_3(x))$ a vector function and by $p(x)$ a scalar function. It is required to find $\mathbf{u}$ and p satisfying the system

$$\nu\Delta\mathbf{u} + a_i(x)\mathbf{u}_{x_i} = \operatorname{grad} p + \mathbf{f}(x), \tag{1.15$_1$}$$

the condition of incompressibility

$$\operatorname{div} \mathbf{u} = 0, \tag{1.15$_2$}$$

and the boundary condition

$$\mathbf{u}|_S = 0. \tag{1.16}$$

In (1.15$_1$) $\mathbf{a} = (a_1, a_2, a_3)$ and $\mathbf{f} = (f_1, f_2, f_3)$ are given vector functions, where $\mathbf{a}(x)$ is in $W_4^1(\Omega)$, $\operatorname{div} \mathbf{a} = 0$, and $\mathbf{f} \in L_2(\Omega)$. The system (1.15$_i$) is called an Oseen system, and, when $\mathbf{a} = 0$, a Stokes system. Both of these are variants of linearizations of a system of the Navier–Stokes equations (in the case that $\mathbf{u}$ and p do not depend on t) describing the motion of a viscous fluid. The function p is the pressure, $\mathbf{u}$ is the vector velocity, $\nu > 0$ is the coefficient of kinematic viscosity, $\mathbf{f}$ is the vector of mass forces, and (1.16) is the condition of adherence when the container Ω does not move. The pressure p, according to its physical meaning, can be determined up to an arbitrary constant. We shall seek a solution for which $\mathbf{u}$ is in $W_2^1(\Omega)$ ($W_2^1(\Omega)$ is a vector space). By (1.16) $\mathbf{u}$ will be in $\mathring{W}_2^1(\Omega)$. We consider (1.15$_2$) as the condition that chooses a subspace of $\mathring{W}_2^1(\Omega)$ which we denote by $H(\Omega)$. Thus, $H(\Omega)$ *is the subspace of* $\mathring{W}_2^1(\Omega)$ *consisting of all vector functions* $\mathbf{u}(x)$ *satisfying* (1.15$_2$) (vector functions which satisfy the equation (1.15$_2$) are usually called *solenoidal*). It is clear that $H(\Omega)$ is complete. We define the scalar product in $H(\Omega)$ by

$$[\mathbf{u}, \mathbf{v}] = \int_\Omega \mathbf{u}_x \cdot \mathbf{v}_x\, dx \equiv \int_\Omega \sum_{i,k=1}^{3} u_{ix_k} v_{ix_k}\, dx. \tag{1.17}$$

The corresponding norm $|\cdot|_{H(\Omega)}$ is equivalent to the norm of $\mathring{W}_2^1(\Omega)$. In order to define a generalized solution of the problem in the energy class, we take the scalar product of (1.15_1) with an arbitrary smooth vector function $\boldsymbol{\eta}(x)$ and integrate over Ω:

$$\nu \int_\Omega \Delta \mathbf{u} \cdot \boldsymbol{\eta}\, dx + \int_\Omega a_i \mathbf{u}_{x_i} \cdot \boldsymbol{\eta}\, dx = \int_\Omega \operatorname{grad} p \cdot \boldsymbol{\eta}\, dx + \int_\Omega \mathbf{f} \cdot \boldsymbol{\eta}\, dx. \tag{1.18}$$

Integrating both sides by parts, we can reduce (1.18) to the form

$$\nu \int_\Omega \mathbf{u}_x \cdot \boldsymbol{\eta}_x\, dx - \nu \int_S \frac{\partial \mathbf{u}}{\partial n} \cdot \boldsymbol{\eta}\, ds - \int_\Omega a_i \mathbf{u}_{x_i} \cdot \boldsymbol{\eta}\, dx$$
$$= \int_\Omega p \cdot \operatorname{div} \boldsymbol{\eta}\, dx - \int_S p\mathbf{n} \cdot \boldsymbol{\eta}\, ds - \int_\Omega \mathbf{f} \cdot \boldsymbol{\eta}\, dx. \tag{1.19}$$

In order to get rid of the term $\int_S (\partial \mathbf{u}/\partial n) \cdot \boldsymbol{\eta}\, ds$, which has no meaning for arbitrary elements $\mathbf{u} \in W_2^1(\Omega)$, we impose on $\boldsymbol{\eta}$ the condition $\boldsymbol{\eta}|_S = 0$. Moreover, if $\boldsymbol{\eta}$ is subjected to the second condition, $\operatorname{div} \boldsymbol{\eta} = 0$, which the desired solution $\mathbf{u}$ should satisfy, then (1.19) takes the form

$$\nu \int_\Omega \mathbf{u}_x \cdot \boldsymbol{\eta}_x\, dx - \int_\Omega a_i \mathbf{u}_{x_i} \cdot \boldsymbol{\eta}\, dx = - \int_\Omega \mathbf{f} \cdot \boldsymbol{\eta}\, dx. \tag{1.20}$$

The function $p(x)$ does not appear in this identity. We call a vector function $\mathbf{u}(x)$ *a generalized solution of the problem* (1.15_i)–(1.16) *in the energy class* if it is in $H(\Omega)$ and satisfies (1.20) for any $\boldsymbol{\eta} \in H(\Omega)$. Our arguments show that a sufficiently smooth solution $\mathbf{u}, p$ of the problem (1.15_i)–(1.16) satisfies (1.20). We shall show below that (1.20) determines uniquely a velocity field $\mathbf{u}(x)$. Thus the transition to such a generalized treatment allows us to separate the finding of $\mathbf{u}$ from the finding of the pressure p. If $\mathbf{u}(x)$, as defined by (1.20), is a sufficiently smooth function, for example, an element of $W_2^2(\Omega)$ (and this will be the case if $\mathbf{f} \in L_2(\Omega)$ and $S \in C^2$), then, by integration by parts, (1.20) can be put in the form

$$\int_\Omega (\nu \Delta \mathbf{u} + a_i \mathbf{u}_{x_i} - \mathbf{f}) \cdot \boldsymbol{\eta}\, dx = 0. \tag{1.21}$$

We shall show that it follows from (1.21), which holds for all $\boldsymbol{\eta} \in H(\Omega)$, that $\nu \Delta \mathbf{u} + a_i \mathbf{u}_{x_i} - \mathbf{f}$ is the gradient of some single-valued function, which is the pressure to within some arbitrary term c ($c = \text{const}$). Thus, let the vector function $\mathbf{v}(x) \in L_2(\Omega)$ satisfy the identity

$$\int_\Omega \mathbf{v} \cdot \boldsymbol{\eta}\, dx = 0 \tag{1.22}$$

for all $\boldsymbol{\eta} \in H(\Omega)$. Let us show that $\mathbf{v}$ is a gradient. For this we take as $\boldsymbol{\eta}$ the function $\boldsymbol{\eta} = \operatorname{curl} \boldsymbol{\Phi}_\rho$, where $\boldsymbol{\Phi}(x)$ is an arbitrary smooth function vanishing near S, and $\boldsymbol{\Phi}_\rho$ is its average of the form (4.9) of Chapter I. For ρ less than the distance from the support of $\mathbf{\Phi}$ to S, $\boldsymbol{\eta}$ will be an element of $H(\Omega)$. We substitute this into (1.22) and write the left-hand side as

$$0 = \int_\Omega \mathbf{v} \operatorname{curl} \mathbf{\Phi}_\rho \, dx = \int_\Omega \mathbf{v}_\rho \operatorname{curl} \mathbf{\Phi} \, dx = \int_\Omega \operatorname{curl} \mathbf{v}_\rho \mathbf{\Phi} \, dx. \tag{1.23}$$

Here we have used well-known properties of the averaging operation and of the operation curl. Because the choice of $\mathbf{\Phi}$ in (1.23) was arbitrary, it follows that $\operatorname{curl} \mathbf{v}_\rho(x) = 0$ for $x \in \overline{\Omega}' \subset \Omega$ and for ρ less than the distance of Ω' to S. By a well-known theorem of analysis, $\mathbf{v}_\rho(x)$ is the gradient of some smooth function $\varphi(x, \rho)$, which is single-valued because of our assumption that Ω is simply connected, and which is defined (up to a constant term) by the curve integral: $\varphi(x, \rho) = \int_{x_0}^{x} \sum_{k=1}^{3} (v_k)_\rho \, dx_k$. We fix $\overline{\Omega}' \subset \Omega$ and take the limit in this equality as $\rho \to 0$, assuming that x_0 is a fixed point of Ω' and x an arbitrary point of $\overline{\Omega}'$. Since $\mathbf{v}_\rho$ converges to $\mathbf{v}(x)$ in $L_2(\Omega')$, it is easy to see that $\varphi(x, \rho)$ converges in $L_2(\Omega')$ to the function $\varphi(x) = \int_{x_0}^{x} \sum_{k=1}^{3} v_k \, dx_k$ and $\partial\varphi(x, \rho)/\partial x_k = (v_k)_\rho$ to the function $\partial\varphi(x)/\partial x_k = v_k$. Since Ω' is an arbitrary subdomain inside Ω, the equality $\mathbf{v} = \operatorname{grad} \varphi$ holds in all of Ω. This proves our assertion. Returning to (1.21), we see that $\nu\Delta\mathbf{u} + a_i \mathbf{u}_{x_i} - \mathbf{f}$ is the gradient of some single-valued function $p(x)$, which, together with $\mathbf{u}(x)$, gives the solution of the problem (1.15$_i$)–(1.16).

Thus (1.15$_i$)–(1.16) has been reduced to finding a $\mathbf{u}(x)$ in $H(\Omega)$ which satisfies (1.20) for all $\boldsymbol{\eta} \in H(\Omega)$. This problem can be solved completely in the same way (cf. §3, Chapter II) as the problem of finding the generalized solution of a scalar elliptic equation. The identity (1.20) reduces to an operator equation of the form

$$\nu\mathbf{u} + A\mathbf{u} = \mathscr{F} \tag{1.24}$$

in the space $H(\Omega)$. Here A is a completely continuous linear operator defined by

$$[A\mathbf{u}, \boldsymbol{\eta}] = -\int_\Omega a_i \mathbf{u}_{x_i} \cdot \boldsymbol{\eta} \, dx, \tag{1.25}$$

for all $\boldsymbol{\eta} \in H(\Omega)$, and $\mathscr{F}$ is an element of $H(\Omega)$ defined by

$$[\mathscr{F}, \boldsymbol{\eta}] = -(\mathbf{f}, \boldsymbol{\eta}), \tag{1.26}$$

for all $\boldsymbol{\eta} \in H(\Omega)$. For the solvability of (1.24) for arbitrary $\mathscr{F}$ in $H(\Omega)$, it is necessary and sufficient (according to the first theorem of Fredholm) that the homogeneous equation (1.24) have only the null solution. But such an equation is equivalent to the identity

$$\nu \int_\Omega \mathbf{u}_x \cdot \boldsymbol{\eta}_x \, dx - \int_\Omega a_i \mathbf{u}_{x_i} \cdot \boldsymbol{\eta} \, dx = 0, \tag{1.27}$$

for all $\boldsymbol{\eta} \in H(\Omega)$. Let us take $\boldsymbol{\eta} = \mathbf{u}$ in (1.27), let us write the product $-a_i \mathbf{u}_{x_i} \mathbf{u}$ as $-\frac{1}{2} a_i (\partial \mathbf{u}^2/\partial x_i)$ ($\mathbf{u}^2 \equiv \sum_{i=1}^{3} u_i^2$), and transform the second integral of (1.27) to the form $\frac{1}{2} \int_\Omega \mathbf{u}^2 \operatorname{div} \mathbf{a} \, dx$, where we have taken (1.16) into account.

This is equal to zero under the hypothesis that div $\mathbf{a} = 0$, so that (1.27) with $\boldsymbol{\eta} = \mathbf{u}$ yields

$$v \int_\Omega \mathbf{u}_x \cdot \mathbf{u}_x \, dx = 0,$$

which, together with (1.16), guarantees that $\mathbf{u}(x)$ is identically zero. Thus we have proved the uniqueness theorem for (1.24), and along with it the following theorem:

Theorem 1.2. *If* $\mathbf{a} \in W_2^1(\Omega)$ *and* div $\mathbf{a} = 0$, *then the problem* (1.15_i)–(1.16) *has a unique generalized solution* $\mathbf{u}$ *in* $H(\Omega)$ *for any* $\mathbf{f} \in L_2(\Omega)$.

Our investigation of the problem (1.15_i)–(1.16), which is taken from [LA 15], is an example of a non-standard use of the method developed in §3, Chapter II, in connection with an elliptic second-order equation. The non-standard part consisted of the fact that part of the equations of the system (namely, the equation (1.15_2) and (1.16)) was included in the definition of that subspace of the fundamental space $W_2^1(\Omega)$ in which we considered the integral identity (1.20) replacing the remaining equations of the system (here, equation (1.15_1)), and which contain the solution of the problem (more precisely, part of the solution: the function $\mathbf{u}(x)$). The same method of approach has turned out to be useful in studying a number of other boundary value problems of mathematical physics, for example, various problems for systems in magnetohydrodynamics (Ladyzshenkaya and Solonnikov [1], Solonnikov [3]) and for a system of Maxwell's equations (Byhovskii [1]).

§2. Strongly Parabolic and Strongly Hyperbolic Systems

We call systems of the form

$$\mathbf{u}_t + \mathscr{L}\mathbf{u} = \mathbf{f}, \tag{2.1}$$

strongly parabolic if $\mathscr{L}$ is the strongly elliptic operator described in §1 (its coefficients can depend on t as well as on x). Such systems are well-posed for the Cauchy problem, where the vector function $\mathbf{u}(x, t)$ is given at the initial instant $t = t_0$ and is to be found for all $x \in R^n$ and for $t \geq t_0$, and also for various initial-boundary value problems where we find $\mathbf{u}(x, t)$ in the domain $Q = \{(x, t): x \in \Omega \subset R^n, t > t_0\}$ as the solution of (2.1), satisfying the initial condition $\mathbf{u}|_{t=t_0} = \boldsymbol{\varphi}$ along with certain boundary conditions (for example, the Dirichlet conditions (1.5)) on the surface $\Gamma = \{(x, t): x \in \partial\Omega, t \geq t_0\}$. The Cauchy problem and the initial-boundary value problems under the boundary conditions indicated in §1 can be investigated in the same way as in

Chapter III for the case of one parabolic equation of second order. Initial-boundary value problems for the systems (2.1) under other boundary conditions require more complicated considerations.

At the present time the classes of systems which are called parabolic (for example, in the sense of I. G. Petrovskii, or T. Sirota and V. A. Solonnikov) are significantly wider than those of (2.1). Their study requires the develop ment of other techniques (see [SL 6, 8, 9], [EI 1], etc.).

Let us consider systems of the form

$$\mathbf{u}_{tt} + \mathscr{L}\mathbf{u} = \mathbf{f}, \tag{2.2}$$

where $\mathscr{L}$ is the strongly elliptic operator of §1 whose principal part lacks the skew-symmetric part (that is, $\tilde{A}^{(i,j)} = 0$ for $|i| = |j| = m$); the coefficients of $\mathscr{L}$ can depend on both x and t. For the case $m = 1$ we call them strongly hyperbolic systems. For such systems the Cauchy problem and certain initial-boundary value problems (those that were discussed in §1) can be investigated by the same methods used for a hyperbolic equation in Chapter IV. We shall not carry out these investigations here since they are similar to those in Chapter IV. The dynamical equations of elasticity theory form a particular case of the strongly hyperbolic systems.

The non-classical approach to initial-boundary value problems and the methods of investigating them, which were studied in Chapters III and IV, allowed us to investigate more general things: equations of the form

$$\frac{du}{dt} + S_1(t)u = f(t), \tag{2.3}$$

$$\frac{d^2u}{dt^2} + S_1(t)\frac{du}{dt} + S_2(t)u = f(t) \tag{2.4}$$

together with some generalizations of them. Here $u(\cdot)$ and $f(\cdot)$ are functions of t with values in a Hilbert space H and the $S_i(t)$ are linear and, in general, unbounded operators in H which depend on the parameter t. We add to (2.3) and (2.4) the initial Cauchy conditions:

$$u|_{t=0} = u_0 \tag{2.5}$$

for (2.3), and

$$u|_{t=0} = u_0, \qquad \left.\frac{du}{dt}\right|_{t=0} = u_1 \tag{2.6}$$

for (2.4). We shall show that the initial-boundary value problems (and the Cauchy problem) for equations and systems (2.1) and (2.2) can be considered as special cases of the Cauchy problem for the operator equations (2.3) and (2.4), respectively. Let us take, to be definite, the first boundary condition, and, for the sake of clarity, assume that the coefficients of the differential operator $\mathscr{L}$ are sufficiently smooth functions of x and t. As the basic Hilbert space H we choose the space $L_2(\Omega)$, so that we can treat the solution $\mathbf{u}(x, t)$ and the free term $\mathbf{f}(x, t)$ as functions of $t \geq 0$ with values in $L_2(\Omega)$. The role of the operator $S_1(t)$ in (2.3) will be played by the differential operator $\mathscr{L}$ in (2.1) (which has the form (1.1), but the elements $A^{(i,j)}$ can depend on t as well as on x) defined on all those elements of $W_2^{2m}(\Omega)$ which satisfy the boundary conditions (1.5). This operator is unbounded in $L_2(\Omega)$ but is defined on a set which is dense in $L_2(\Omega)$. If the coefficients of $\mathscr{L}$ depend on t, then so does S_1. The domain of definition of $S_1(t)$ for the boundary conditions (1.5) does not depend on t. Thus, the problem

$$\mathbf{u}_t + \mathscr{L}\mathbf{u} = \mathbf{f}, \qquad \mathbf{u}|_{t=0} = \boldsymbol{\varphi}, \qquad \mathbf{u}|_{S_T} = \cdots = \left.\frac{\partial^{m-1}\mathbf{u}}{\partial n^{m-1}}\right|_{S_T} = 0 \tag{2.7}$$

is actually a special case of the Cauchy problem (2.3), (2.5). The boundary conditions for it were concealed when we indicated the domain of definition of the operator $S_1(t)$. For (2.7) the operator $S_1(t)$ defined above has an important property: it is semibounded. It turns out that if, in the abstract problem (2.3), (2.5), we impose the condition of semiboundedness on the operator $S_1(t)$, then this, along with certain other "natural" restrictions (smoothness in t, together with independence of t either of the domain of definition of $S_1(t)$ itself or of a degree of self-adjointness in part of $S_1(t) + \lambda_0 I$, $\lambda_0 \geq 0$), is enough to prove the unique solvability of the problem (2.3), (2.5) in one or another class of "the generalized" solutions. The first results of this type were established in papers by the author [6], [8] and M. I. Višik [5], [6] (see also the survey [1] of Višik and myself). From these results the unique solvability of (2.7) follows as a special case. Another cycle of papers on the investigation of the problem (2.3), (2.5) stem from the work of T. Kato [1], who considered (2.3) in a Banach space. The monographs of J. L. Lions [1] and S. G. Krein [1] are also devoted to the equations (2.3), (2.4).

As one of the applications of the results on the problem (2.3), (2.5), we mention the problem of hydrodynamics,

$$\left.\begin{aligned} &\mathbf{u}_t - \nu\Delta\mathbf{u} + a_i(x, t)\mathbf{u}_{x_i} = -\operatorname{grad} p + \mathbf{f}(x, t), \\ &\operatorname{div}\mathbf{u} = 0, \qquad \mathbf{u}|_{t=0} = \varphi(x), \qquad \mathbf{u}|_{S_T} = 0 \end{aligned}\right\} \tag{2.8}$$

for viscous incompressible fluids. Its unique solvability can be derived from these results and from the results on the stationary problem (1.15_i)–(1.16) (see [LA 8, 15]).

Initial-boundary value problems for strongly hyperbolic systems are included in the Cauchy problem (2.4), (2.6) just as in the parabolic case. Their analysis suggested those restrictions on the operators $S_1(t)$ and $S_2(t)$ for which the problem (2.4), (2.6) is uniquely solvable and which are so broad as to include as special cases a series of initial-boundary value problems (and the Cauchy problem) for the systems (2.2). We shall not cite here the final results for the problems (2.3, (2.5) and (2.4), (2.6), and their generalizations of the form

$$\left.\begin{aligned} &\frac{d^k S_0(t)u}{dt^k} + \frac{d^{k-1}S_1(t)u}{dt^{k-1}} + \cdots + S_k(t)u = f, \\ &u|_{t=0} = \varphi_0, \quad \left.\frac{du}{dt}\right|_{t=0} = \varphi_1, \ldots, \left.\frac{d^{k-1}u}{dt^{k-1}}\right|_{t=0} = \varphi_{k-1}, \end{aligned}\right\} \tag{2.9}$$

but fall back on the special literature; nor shall we list the requirements to be imposed on the $S_k(t)$ as functions of t. We point out only that, for these problems, various lines of investigation have been proposed, where both the hypotheses on $S_i(t)$ and the final results are somewhat different. These differences are sometimes connected with the choice of the class of generalized solution of the problems, sometimes they are dictated by the desire to handle some uninvestigated initial-boundary value problem of interest to the author, and sometimes they are chosen simply by the method selected to solve the problem or by the desire to weaken some hypotheses of the work of earlier investigators.

One of the methods by which we investigated the solvability of the problems (2.3), (2.5) and (2.4), (2.6), as well as the Cauchy problem for the Schrödinger equations and some of their generalizations, and which was one of the very first methods to permit us to investigate all these problems in a unified way, is laid out in §2, Chapter III, and in §6, Chapter IV, in connection with the heat equation and the wave equation. It is taken from our note [6].

§3. Schrödinger-Type Equations and Related Equations

Equations of the form

$$\frac{du}{dt} + iS_1(t)u = f(t), \tag{3.1}$$

are called equations of Schrödinger type if $S_1(t)$ is a self-adjoint operator acting in a complex Hilbert space H. It can depend on the parameter t; here and in the sequel we shall not cite restrictions to be imposed on the nature of the dependence of $S_1(t)$ on t. For such equations we usually consider the Cauchy problem, that is, the problem of determining a solution $u(t)$ of (3.1) satisfying the initial condition

$$u|_{t=0} = \varphi. \tag{3.2}$$

The solution $u(\cdot)$ and the free term $f(\cdot)$ are functions of t with values in H, and φ is a prescribed element of H. Usually $f(t) \equiv 0$ in quantum mechanics.

The unique solvability of the problem (3.1), (3.2) in some classes under appropriate hypotheses on φ and $f(t)$, along with the dependence of $S_1(t)$ on t, was proved by various methods in my paper [8]. The first way of doing this is similar to the method given in §2, Chapter III and in §6, Chapter IV, in connection with equations of parabolic and hyperbolic types (for the other method see *n*.5, §11, Chapter VI). It appears to be applicable to certain more general equations, namely to equations of the form

$$S_0(t)\frac{du}{dt} + S_1(t)u + S_2(t)u = f(t), \tag{3.3}$$

where $S_0(t)$ is a bounded, self-adjoint, positive-definite operator in the Hilbert space H, where $iS_1(t)$ is a self-adjoint operator whose domain of definition $\mathscr{D}(S_1)$ is dense in H and does not depend on t, and where $S_2(t)$ is a bounded operator in H. All the operators $S_i(t)$ depend smoothly on t. The Cauchy problem for this case was studied in the paper [1] by Byhovskii, where it was proved that various initial-boundary value problems for a system of Maxwell's equations fall under this case. We remark that the abstract problem (3.3), (3.2) is not much different from the problem (3.1), (3.2), for it can be reduced to a problem of the form

$$\frac{dv}{dt} + \hat{S}_1(t)v + \hat{S}_2(t)v = \hat{f}, \qquad v|_{t=0} = S_0^{1/2}(0)\varphi, \tag{3.4}$$

where

$$v = S_0^{1/2}(t)u, \qquad \hat{S}_1(t) = S_0^{-1/2}(t)S_1(t)S_0^{-1/2}(t),$$

$$\hat{S}_2(t) = S_0^{-1/2}(t)S_2(t)S_0^{-1/2}(t) - \frac{dS_0^{1/2}(t)}{dt}S_0^{-1/2}(t), \qquad \hat{f}(t) = S_0^{-1/2}(t)f(t).$$

By the conditions that were imposed on $S_i(t)$, the operators $S_0^{1/2}$ and $S_0^{-1/2}$ exist and are bounded, $i\hat{S}_1(t)$ is self-adjoint, and $\hat{S}_2(t)$ is bounded. The unique solvability of (3.4) can be proved in essentially the same way as the unique solvability of (3.1), (3.2). However, the proof of the fact that initial-boundary value problems for the system of Maxwell's equations satisfy conditions under which the solvability of the problem (3.3), (3.2) was established was not trivial and required an analysis of those problems having a common thread with the short description in §1 for the problem (1.15_i)–(1.16).

For other generalizations of the results for the problem (3.1), (3.2) we refer to the papers [1] of V. I. Derguzov and V. A. Yacubovich and [1] of V. N. Fomin. Thus, in the second paper which is devoted to the parametric resonance of elastic systems with distributed parameters, the author encountered the Cauchy problem for equations of the form

$$i\frac{d}{dt}(S_0u) = (E + S_1(t))u, \tag{3.5}$$

where S_0 is a self-adjoint, completely continuous operator in H with an unbounded inverse, E is the identity operator, and $S_1(t)$ is a symmetric operator; moreover, the operator $S_1(t)S_0^{-1}$ is bounded in H and depends smoothly on t. Such a problem can be reduced to the form (3.4) if we introduce a new unknown $v = S_0 u$ in place of u. Actually, v will be a solution of the problem

$$\frac{idv}{dt} = S_0^{-1}v + S_1(t)S_0^{-1}v, \qquad v|_{t=0} = S_0 u|_{t=0}. \tag{3.6}$$

To investigate this problem, we must bear in mind that the initial value for v is a "very nice element of H" and has the form $S_0 u$, $u \in H$, and we must prove that this property (i.e., that $v(t)$ lies in the range of S_0) is preserved for all $t > 0$. We remark that similar reductions of the original equation to an equation whose principal terms are the same as (3.1) are not always convenient, and that it is sometimes more advantageous to investigate the problem in its original form and to adapt to the problem the arguments underlying the investigation of the problem (3.1), (3.2).

§4. Diffraction Problems

From the mathematical point of view, diffraction problems consist of solving boundary value problems—stationary or non-stationary—for equations or systems whose coefficients have discontinuities of the first kind. On those surfaces Γ where the coefficients are discontinuous we impose the so-called conjugacy conditions which express the continuity of the medium and the equilibrium of the forces acting on it. The discontinuity of the coefficients of the equation corresponds to the fact that the medium consists of two or more physically different materials. One of the first problems of this kind is the following:

$$a\Delta u = f, \tag{4.1}$$

$$u|_S = 0, \tag{4.2}$$

$$[u]|_\Gamma = \left[a\frac{\partial u}{\partial n}\right]\Bigg|_\Gamma = 0. \tag{4.3}$$

A solution $u(x)$ is sought in a domain Ω with boundary S. The surface Γ (of dimension $n - 1$) divides Ω into two domains Ω_1 and Ω_2, in one of which the coefficient a is equal to a constant $a_1 > 0$ and in the other to a constant $a_2 > 0$. The symbol $[u]$ in (4.3) denotes the difference between the limiting values of $u(x)$ on Γ, calculated for approach to Γ from Ω_1 and Ω_2. The symbol $[a(\partial u/\partial n)]$ is to be interpreted similarly. The first of the conditions (4.3) requires the continuity of $u(x)$, while the second prescribes the jump of the derivative $\partial u/\partial n$, where the normal $\mathbf{n}$ to Γ is directed towards the interior of Ω_2. The problem (4.1)–(4.3) is encountered, for example, in the diffraction of electromagnetic waves.

In this formulation, problems of this type differ essentially from the usual boundary value problems (when the medium has smoothly varying characteristics), and the investigation of them by classical methods (i.e., by the methods of potential theory), even in the simplest problems with piecewise-constant coefficients, required the development of the theory of integral equations; in these problems there arise the so-called boaded integral equations. Diffraction problems for equations of hyperbolic type are even harder for the classical methods.

However, the analysis of these problems from the point of view of the theory of generalized solutions, which permeates the entirety of this book, shows that they can be considered as special cases of the usual boundary value problems. Thus, the problem (4.1)–(4.3) is equivalent to the problem of finding a function $u(x)$ in $\overset{\circ}{W}{}_2^1(\Omega)$ satisfying the integral identity

$$\int_\Omega a u_{x_i} \eta_{x_i}\, dx = -\int_\Omega f\eta\, dx \tag{4.4}$$

for all $\eta \in \overset{\circ}{W}{}_2^1(\Omega)$. Such a function $u(x)$ is naturally called a *generalized solution of the problem* (4.1)–(4.3). If $u(x)$ is the classical solution of (4.1)–(4.3), then, if we multiply (4.1) by a smooth function $\eta(x)$ vanishing on S and integrate over Ω, we obtain

$$\int_\Omega a\Delta u\eta\, dx = \int_\Omega f\eta\, dx, \tag{4.5}$$

which, after integration by parts (which should be carried out separately for each of the Ω_i and the results combined), can be transformed to

$$-\int_\Omega a u_{x_i}\eta_{x_i}\, dx + \int_\Gamma \left[a\frac{\partial u}{\partial n}\right]\eta\, ds = \int_\Omega f\eta\, dx. \tag{4.6}$$

But this identity coincides with (4.4) if we take into account the second of the conditions (4.3). Thus, the classical solution of (4.1)–(4.3) is a generalized solution in the sense of the definition given above. The converse is true: If $u(x)$ is a generalized solution of the problem, and if it is sufficiently smooth in each of the domains Ω_i (for example, if $u \in C^2(\bar{\Omega}_i)$, $i = 1, 2$), then it satisfies all the conditions of (4.1)–(4.3) in their classical form. Actually, for such a $u(x)$, the identity (4.4) can be transformed to (4.6), from which (4.1) follows, as well as the second of the conditions (4.3), by the arbitrariness in the choice of $\eta(x)$. The two remaining conditions are fulfilled because $u(x) \in \overset{\circ}{W}{}_2^1(\Omega)$.

Thus we have shown that (4.4) implies (4.1) and the second of the conditions (4.3). The two remaining conditions of the problem are expressed by the requirement that $u(x)$ belongs to $\overset{\circ}{W}{}_2^1(\Omega)$. We shall show that such an extension is permissible from the point of view of the uniqueness theorem (it is easy to show that the uniqueness theorem holds for classical solutions).

In fact, the difference of two such possible solutions will be an element of $\overset{\circ}{W}{}_2^1(\Omega)$ satisfying the identity

$$\int_\Omega a u_{x_i} \eta_{x_i}\, dx = 0$$

for all $\eta \in \overset{\circ}{W}{}_2^1(\Omega)$. If we set $\eta = u$, we see that $u \equiv 0$, that is, the uniqueness theorem goes over to our class of generalized solutions. The existence of such solutions $u(x)$ was proved in Chapter II, for this problem is a very special case of the Dirichlet problem studied in Chapter II. In Chapter II we have also given some approximation methods of actually calculating $u(x)$. In Chapter VI we shall study another approximation method for finding the solutions, the method of finite differences. There remains one question concerning problems of type (4.1)–(4.3): When do the generalized solutions possess some sort of smoothness; in particular, when are they classical? For equations and systems of elliptic and parabolic type the smoothness of their solutions is a local property. In that part of Ω where all the data of the problem are smooth, the generalized solutions have the corresponding smoothness, in particular, if the coefficients and the free term are infinitely differentiable, then so are the solutions.

The proof of these facts is based on local *a priori* estimates of solutions. In §7 of Chapter II we spoke about such estimates for solutions of elliptic equations in the norms of $W_2^2(\Omega')$. With their help it is easy to get local estimates in the norms of all $W_2^l(\Omega')$, $l \geq 3$. For solutions of parabolic equations the simplest estimates are derived in the norms of $W_2^{2l,l}(\Omega' \times (\varepsilon, T))$, $l \geq 1$, $\varepsilon > 0$. Local estimates in the norms of other spaces (in particular of Hölder spaces) are proved in a more complicated way.

In diffraction problems there is the additional question of the behavior of generalized solutions near the surface Γ. This question can be studied in much the same way as the question of the behavior of generalized solutions of boundary value problems near the boundary S. In general terms the result is this: Near the smooth pieces Γ_i of the surface Γ the generalized solutions are smooth functions up to Γ_i, from each side of Γ_i, and consequently satisfy the conjugacy conditions in the classical sense. For the problem (4.1)–(4.3) it was proved, for example, that if $f(\cdot) \in H^\alpha(\overline{\Omega}_i)$, $i = 1, 2$, $S \in H^{2+\alpha}$, $\Gamma \in H^{2+\alpha}$, and if Γ has no points in common with S, then $u(x) \in C(\overline{\Omega}) \cap H^{2+\alpha}(\overline{\Omega}_i)$, $i = 1, 2$. The same result has been established for the case of variable coefficients, provided that they are elements of $H^\alpha(\overline{\Omega}_i)$.

We shall show how to apply the plan we described for studying diffraction problems to the following cases. Let it be required to find a function $u(x)$ satisfying the equation

$$\mathscr{L}u \equiv \frac{\partial}{\partial x_i}\left(a_{ij}(x)\frac{\partial u}{\partial x_j}\right) + b_i(x)u_{x_i} + a(x)u = f(x), \tag{4.7}$$

the boundary condition

$$u|_S = 0 \tag{4.8}$$

and the conjugateness conditions

$$[u]_\Gamma = \left[\frac{\partial u}{\partial N}\right]_\Gamma = 0. \tag{4.9}$$

We seek a *generalized solution* of this problem as an element of $\mathring{W}_2^1(\Omega)$ satisfying the integral identity

$$\mathscr{L}(u, \eta) \equiv \int_\Omega (a_{ij} u_{x_j} \eta_{x_i} - b_i u_{x_i} \eta - au\eta)\, dx = -\int_\Omega f\eta\, dx \tag{4.10}$$

for any $\eta \in \mathring{W}_2^1(\Omega)$. The question of the existence and uniqueness of such a solution was studied in §3, Chapter II. If the $a_{ij}(x)$ have bounded generalized derivatives $a_{ijx_k}(x)$ in each of the domains Ω_l ($l = 1, 2$) into which the surface Γ separates Ω, and if $f \in L_2(\Omega)$, then the generalized solution $u(x)$ belongs to $W_2^2(\Omega_l')$ ($l = 1, 2$) in every $\bar{\Omega}_l' \subset \Omega_l$. Furthermore, if $S \in C^2$, if $\Gamma \in C^2$, and if S and Γ have no points in common, then $u(x) \in W_2^2(\Omega_l)$, $l = 1, 2$. Hence it follows from the embedding theorems that not only $u(x)$, but also its derivatives $\partial u/\partial x_i$, have traces on Γ (these traces are elements of $L_2(\Gamma)$), which are to be understood as the limits from the sides of Ω_1 and Ω_2. The traces of the function $u(x)$ on Γ, calculated as the limiting values of $u(x)$ from the sides of Ω_1 and Ω_2, coincide (as elements of $L_2(\Gamma)$) because $u \in \mathring{W}_2^1(\Omega)$. However, the traces of $\partial u/\partial x_i$ admit jumps as x passes through Γ, but $[\partial u/\partial N]_\Gamma = 0$. The latter property follows from (4.10) and from the fact that $u(x) \in W_2^2(\Omega_l)$, $l = 1, 2$. From those properties of $u(x)$ it also follows that, in each of the Ω_l, $u(x)$ satisfies (4.7), which can be rewritten in Ω_l as

$$a_{ij} u_{x_i x_j} + \left(\frac{\partial a_{ij}}{\partial x_i} + b_i\right) u_{x_j} + au = f, \qquad x \in \Omega_l, \quad l = 1, 2.$$

If S and Γ have a non-empty intersection, then a special analysis of $u(x)$ is required in the vicinity of this intersection.

Similarly, we can study the problem (4.7), (4.9) under the condition $\partial u/\partial N + \sigma u|_S = 0$ instead of (4.8), as well as diffraction problems for (4.7) under a somewhat more general conjugacy conditions; such problems are encountered in determining an electromagnetic field. In the latter the conditions (4.9) are replaced by the requirements

$$[u]_\Gamma = 0, \qquad \left[c\frac{\partial u}{\partial N} + \sigma u\right]_\Gamma = 0, \qquad c(x) \geq \nu_1 > 0, \tag{4.11}$$

where c and σ are prescribed smooth functions having discontinuities of the first kind on Γ. We define a *generalized solution* $u(x)$ *of the problem* (4.7), (4.8), (4.11) to be an element of $\overset{\circ}{W}{}_2^1(\Omega)$ satisfying the identity

$$\begin{aligned}\int_\Omega (a_{ij}cu_{x_j}\eta_{x_i} + a_{ij}c_{x_i}u_{x_j}\eta - b_i cu_{x_i}\eta - acu\eta)\,dx \\ + \int_\Gamma [\sigma]u\eta\,ds = -\int_\Omega cf\eta\,dx\end{aligned} \tag{4.12}$$

for all $\eta \in \overset{\circ}{W}{}_2^1(\Omega)$. This identity comes from the relation $-\int_\Omega \mathscr{L}u \cdot c\eta\,dx = -\int_\Omega fc\eta\,dx$ by a formal integration by parts of the first term on the left-hand side and by observing (4.11) (and also from the fact that $[\eta]_\Gamma = 0$).

The proof of solvability of (4.12) in the class $\overset{\circ}{W}{}_2^1(\Omega)$ does not differ essentially from the proof of the solvability of the third boundary value problem, which was considered in §5, Chapter II. For its validity it is necessary to impose upon the surface Γ the same smoothness conditions as upon S in §5, Chapter II. The smoothness of σ is not so important; it is enough to know, for example, that σ is bounded.

Sometimes it happens that the conjugateness conditions are not homogeneous, i.e., certain known functions replace the zero on the right-hand sides of (4.9) or (4.11). Such cases can be reduced to those already considered by replacing $u(x)$ by a new unknown function $v(x) = u(x) - \Phi(x)$, where $\Phi(x)$ is chosen in such a way that the adjointness condition for v becomes homogeneous.

The method described for studying diffraction problems for elliptic equations can be applied to diffraction problems for equations of parabolic and hyperbolic types. The results on generalized solutions in the energy classes, which were studied in Chapters III and IV, guarantee the unique solvability of these problems (as well as the unique solvability of the problems for the systems described in this chapter). In Chapter VI we shall construct difference schemes for various boundary value and initial-boundary value problems which will serve for diffraction problems as well. The inclusion of diffraction problems in the class of ordinary boundary value problems is possible because of the fact that the existence of generalized solutions in the energy classes, uniqueness theorems for them, and various approximation methods permitting us to calculate such solutions were proved by us without any assumptions concerning the smoothness in x of the coefficients of the equations. The results in Chapters II–IV allow us to study the question of when the generalized solutions of these diffraction problems have, in each of the domains Ω_l, generalized derivatives appearing in the equation, and when they satisfy all conditions "almost everywhere" (more precisely, they satisfy the equations for almost all $x \in \Omega$ and $t \in [0, T]$ and the boundary conditions and the conjugateness conditions for almost all x on S and Γ for almost all t in $[0, T]$). In this book we shall not go into questions concerning a high degree of smoothness of solutions (in this connection, see my books with Ural'čeva and Solonnikov). The material in this section in no way covers the vast literature on diffraction problems dealing with the detailed investigation of these or other properties (most often with the asymptotic properties with respect to t or some other parameter). Here we cite only my note [5] in which it was shown that these problems are included in a well-developed theory of generalized solutions for ordinary boundary value problems, and what this theory yields for diffraction problems.

Supplements and Problems

1. In Section 1 of the Supplements to Chapter II there was described a class of second-order equations with complex coefficients for which the three fundamental boundary value problems can be studied by methods used in Chapter II. These methods are applicable to systems cited in §1 of Chapter V, as well as to systems of the same form with complex matrix coefficients if only the inequality

$$\mathrm{Re} \int_\Omega \sum_{|i|,\,|j|=m} A^{(i,j)}(x)\mathscr{D}^i\mathbf{u}\cdot\overline{\mathscr{D}^j\mathbf{u}}\,dx$$

$$\geq \nu \int_\Omega \sum_{|i|=m} (\mathscr{D}^i\mathbf{u})^2\,dx - \mu_1\int_\Omega \mathbf{u}^2\,dx, \qquad \nu = \text{const.} > 0, \quad \mu_1 = \text{const.} \geq 0, \tag{1.4†}$$

is fulfilled for all complex-valued vector functions $\mathbf{u} \in \overset{\circ}{W}{}_2^m(\Omega)$ in the case of the first boundary condition (1.5), and for all $\mathbf{u} \in W_2^m(\Omega)$ in the case of the "second" and "third" boundary conditions. According to Chapter II, it is natural to define the "second" boundary value problem for (1.1) as the problem corresponding to the identity (1.7) fulfilled for all $\boldsymbol{\eta} \in W_2^m(\Omega)$. Find out the explicit form of this and the "third" boundary condition. Prove that the inequality (1.4†) guarantees Fredholm solvability of these problems for systems (1.1) with complex-valued bounded matrices $A^{(i,j)}$. How should inequalities (1.2) be modified for complex-valued $A^{(i,j)}$ in order that the inequality (1.4†) follows from them?

One has to distinguish between the cases $A^{(i,j)} = A^{(j,i)}$ and $A^{(i,j)} \neq A^{(j,i)}$, $i \neq j$ for $|i| = |j| = m$. For m-times differentiable $A^{(i,j)}$ this is not necessary, as for such $A^{(i,j)}$ we can rearrange the terms of (1.1) in such a way that the matrices $\tilde{A}^{(i,j)}$, $|i| = |j| = m$, satisfy the condition $\tilde{A}^{(i,j)} = \tilde{A}^{(j,i)}$. The invariance of $A^{(i,j)}$ with respect to the permutation of indices $(i_1, \ldots, i_n)$ and $(j_1, \ldots, j_n)$ in i and j can be assumed in all cases.

2. Prove the inequality (1.3) for $A^{(i,j)}$ independent of x. For this it is reasonable to extend the elements of $\overset{\circ}{W}{}_2^m(\Omega)$ by zero outside Ω and consider them as elements of $\overset{\circ}{W}{}_2^m(\Pi)$ where Π is a cube containing Ω. Inequality (1.3) for $\mathbf{u} \in \overset{\circ}{W}{}_2^m(\Pi)$ can easily be deduced from condition (1.2) with the help of Fourier transform.

† The dot "·" in (1.4†) denotes the scalar product: $\mathbf{v}\cdot\boldsymbol{\eta} = \sum_{i=1}^N v_i\eta_i$.

3. The scalar equation

$$\sum_{|i|\le k} a_i(x)\mathscr{D}^i u \equiv \sum_{i_1+\cdots+i_n\le k} a_{(i_1\ldots i_n)}(x)\partial_{x_1}^{i_1}\ldots\partial_{x_n}^{i_n}u = f$$

with real coefficients a_i is called elliptic at x^0 if

$$\sum_{|i|=k} a_i(x^0)\xi^i \equiv \sum_{i_1+\cdots+i_n=k} a_{(i_1\ldots i_n)}(x^0)\xi_1^{i_1}\ldots\xi_n^{i_n} \ne 0$$

for any non-zero $\xi = (\xi_1,\ldots,\xi_n)\in R^n$. Show that this is possible only for even k. Prove also that an elliptic equation of order $k = 2m$ with smooth a_i belongs to the class of strongly elliptic systems and the methods described in Chapters II and V are applicable to it.

4. In §1 of the present chapter we employed the fact that there is a bounded operator $(I + K)^{-1}$ for any bounded skew-symmetric K. In Section 1 of the Supplements to Chapter II we described a way by which this might be proved.

5. Our analysis of principal boundary value problems for second-order elliptic equations was based on Green's formula

$$\int_\Omega v\mathscr{L}_0 u\,dx = \int_\Omega u\mathscr{L}_0 v\,dx + \int_S\left(\frac{\partial u}{\partial N}v - u\frac{\partial v}{\partial N}\right)ds, \tag{1}$$

where $\mathscr{L}_0 u = (a_{ij}u_{x_j})_{x_i}$ and $\partial u/\partial N = a_{ij}\cos(\mathbf{n}\cdot\mathbf{x}_j)u_{x_i}$.

This is true for arbitrary elements u and v of $W_2^2(\Omega)$ and for S not "too bad" if $a_{ij}\in W_\infty^1(\Omega)$. Moreover, the very formulation of these problems, i.e., the choice of boundary conditions, was dictated by this formula. This definiteness is lost for systems and even for equations of order higher than 2. Let us explain this by two examples. For the biharmonic operator Δ^2 we have the identity

$$\int_\Omega v\Delta^2 u\,dx = \int_\Omega u\Delta^2 v\,dx + \int_S\left(\frac{\partial\Delta u}{\partial n}v - \Delta u\frac{\partial v}{\partial n} + \frac{\partial u}{\partial n}\Delta v - u\frac{\partial\Delta v}{\partial n}\right)ds, \tag{2}$$

which can be visualized as Green's formula for Δ^2. According to this, it is "natural" to define, as the "first" boundary problem for the equation $\Delta^2 u = f$, the problem with the boundary conditions

$$u\Big|_S = 0, \qquad \frac{\partial u}{\partial n}\Big|_S = 0 \tag{3}$$

(as everywhere in our text, we consider only homogeneous conditions), and, as the "second" problem, that with the boundary conditions

$$\Delta u\Big|_S = 0, \qquad \frac{\partial\Delta u}{\partial n}\Big|_S = 0. \tag{4}$$

Both these problems are Fredholm solvable. On the other hand, the equation $\Delta^2 u = f$ can be rewritten as a system of second-order equations, namely

$$\mathscr{M}\mathbf{u} \equiv \begin{pmatrix} \Delta & -1 \\ 0 & \Delta \end{pmatrix} \begin{pmatrix} u \\ u_1 \end{pmatrix} = \begin{pmatrix} 0 \\ f \end{pmatrix} \equiv \mathbf{f}. \tag{5}$$

Its principal part is the matrix differential operator $\mathscr{M}_0 = \begin{pmatrix} \Delta & 0 \\ 0 & \Delta \end{pmatrix}$. For this operator the identity of type (1) has the form:

$$\int_\Omega \mathbf{v} \cdot \mathscr{M}_0 \mathbf{u}\, dx = \int_\Omega \mathbf{u} \cdot \mathscr{M}_0 \mathbf{v}\, dx + \int_S \left(\frac{\partial \mathbf{u}}{\partial n} \cdot \mathbf{v} - \mathbf{u} \cdot \frac{\partial \mathbf{v}}{\partial n} \right) ds. \tag{6}$$

According to this, it is "natural" to define as the "first" boundary value problem for system (5), the problem with the boundary conditions

$$\mathbf{u}|_S = 0,$$

which corresponds to the conditions $u|_S = 0$, $\Delta u|_S = 0$ for the equation $\Delta^2 u = f$, and, as the "second" problem, that with the boundary conditions $\partial u/\partial n|_S = 0$, $\partial \Delta u/\partial n|_S = 0$. Solvability of these two problems follows directly from the solvability of Dirichlet and Neumann problems for the Laplacian; we shall find first u_1 from $\Delta u_1 = f$ and $u_1|_S = 0$ (or $\partial u_1/\partial n|_S = 0$) and, after that, u from $\Delta u = u_1$ and $u|_S = 0$ (or $\partial u/\partial n|_S = 0$). All four problems arise very often in applications and one is at liberty to enumerate them voluntarily.

6. Much more diversity is provided by systems and sometimes it is impossible to understand, without special analysis, what identity of types (2) and (6) would be "useful" for their investigation and would be worthy to be called "Green's formula".

This difficulty arose, for example, in the study of boundary value problems for Stokes' system

$$\nu \Delta \mathbf{u} - \nabla p = \mathbf{f}, \qquad \operatorname{div} \mathbf{u} = 0, \qquad x \in \Omega \subset R^3. \tag{7_1}$$

The most important boundary value problem for (7_1) is associated with the boundary condition

$$\mathbf{u}|_S = 0 \tag{7_2}$$

(a short account about (7_1), (7_2) is given in §1 of Chapter V). Let us try to find a Green's formula for (7_1) by following the standard procedure used for scalar second-order equations. The differential matrix operator corresponding to (7_1) has the form

$$\mathscr{L}U \equiv \begin{pmatrix} \Delta & 0 & 0 & -\partial x_1 \\ 0 & \Delta & 0 & -\partial x_2 \\ 0 & 0 & \Delta & -\partial x_3 \\ \partial x_1 & \partial x_2 & \partial x_3 & 0 \end{pmatrix} \begin{pmatrix} u_1 \\ u_2 \\ u_3 \\ p \end{pmatrix}.$$

We take two arbitrary four-vectors $\mathbf{U} = \binom{\mathbf{u}}{p}$ and $\mathbf{V} = \binom{\mathbf{v}}{q}$ and transform the integral $\int_\Omega \mathscr{L}\mathbf{U} \cdot \mathbf{V}\, dx$, using standard integration by parts, in the following way:

$$\int_\Omega \mathscr{L}\mathbf{U} \cdot \mathbf{V}\, dx = \int_\Omega \mathbf{U} \cdot \mathscr{L}\mathbf{V}\, dx + \int_S \left[\left(\nu \frac{\partial \mathbf{u}}{\partial n} - p\mathbf{n}\right) \cdot \mathbf{v} - \left(\nu \frac{\partial \mathbf{v}}{\partial n} - q\mathbf{n}\right) \cdot \mathbf{u}\right] ds. \tag{8}$$

According to this identity, it is "natural" to define the first boundary value problem for (7_1) as the problem with the boundary conditions (7_2), and the second boundary value problem for (7_1) as the problem with the boundary conditions

$$\nu \frac{\partial \mathbf{u}}{\partial n} - p\mathbf{n} = 0. \tag{9}$$

It was that pair of problems (more precisely, two pairs of problems, (7_1), (7_2) and (7_1), (9), for Ω and for $R^3 \backslash \Omega$) which was first studied following the classical scheme of investigation of Dirichlet and Neumann problems for the Laplace operator in potential theory. The solutions were represented through hydrodynamical potentials with unknown densities and integral equations were obtained for these densities. But to the great surprise of their authors (Lichtenstein and Ossen) these integral equations turned out to be singular (non-Fredholm) and their study was an open problem. Some years later Odquist considered another pair of problems for (7_1), having boundary conditions (7_2) and

$$\sum_{j=1}^{3} T_{ij}(\mathbf{u}, p) n_j = 0, \qquad i = 1, 2, 3, \tag{10}$$

where $T_{ij}(\mathbf{u}, p) = \nu(\partial_{x_j} u_i + \partial_{x_i} u_j) - \delta_i^j p$ are components of the so-called "stress tensor" and $\mathbf{n} = (n_1, n_2, n_3)$ is a unit vector of an exterior normal to S. He used for a "Green's formula" for (7_1) the identity

$$\int_\Omega [(\nu \Delta u_i - p_{x_i}) v_i - u_i(\nu \Delta v_i - q_{x_i})]\, dx$$
$$= \int_S [T_{ij}(\mathbf{u}, p) n_j v_i - u_i T_{ij}(\mathbf{v}, q) n_j]\, ds, \tag{11}$$

which is valid for arbitrary solenoidal fields $\mathbf{u}$ and $\mathbf{v}$ (i.e., $\operatorname{div} \mathbf{u} = \operatorname{div} \mathbf{v} = 0$) and any p and q. This fact was known in hydrodynamics. The left side of (11) is equal to $\int_\Omega (\mathscr{L}\mathbf{U} \cdot \mathbf{V} - \mathbf{U} \cdot \mathscr{L}\mathbf{V})\, dx$ if u and v are solenoidal. It is easy to see from (11) that the problems, (7_1), (7_2) and (7_1), (10), are self-adjoint (at last formally). It turned out that they are a "good pair" for the theory of hydrodynamical potentials. The corresponding integral equations for densities are of Fredholm type. With the help of this theory, the solvability of the interior and exterior problems, (7_1), (7_2) and (7_1), (10), was proved simultaneously as they occur in the classical potential theory. (Now the problem (7_1), (10) is usually referred to as the second boundary value problem for Stokes and Navier–Stokes systems.) It was the first method which allowed investigation of the problem (7_1), (7_2) [OD 1] (the extended exposition of this material is presented in [LA 15]). I suggested an alternative "functional" method, described briefly in §1 of Chapter V and in detail in [LA 15]. This method is much simpler and is applicable to arbitrary domains (both bounded and unbounded) with non-smooth boundaries. This method also made it possible to investigate the full non-linear Navier–Stokes system: to prove existence of at least one solution of the first and second problems without any restrictions on the magnitude of the data (for arbitrary Reynolds numbers, as hydrodynamicists say). The arguments are close to those of Chapter II for elliptic equations. In the case of a bounded domain Ω, the first boundary value problem

$$\nu\Delta\mathbf{u} - u_k\mathbf{u}_{x_k} - \nabla p = \mathbf{f}, \qquad \operatorname{div}\mathbf{u} = 0, \qquad \mathbf{u}|_S = 0, \tag{12}$$

for the Navier–Stokes system is reduced to an equation of the form

$$u + A(\mathbf{u}) = \mathbf{F} \tag{13}$$

in the Hilbert space $H(\Omega)$ with a completely continuous non-linear operator A (for the definition of $H(\Omega)$ see §1, Chapter V). The solvability of (13) is established with the help of Schauder or Leray–Schauder fixed-point theorems for compact mappings. For unbounded domains Ω a solution is found by a passage to a limit over a sequence of bounded domains Ω_m exhausting Ω. The principle estimate at the heart of the "functional" method is an *a priori* estimate of the norm $\mathbf{u}$ in $H(\Omega)$. Similar results on the solvability of boundary value problems for Stokes and Navier–Stokes systems can also be proved by Galerkin's method. (All these results are presented in [LA 15].)

But let us return to the linear problem (7_1), (7_2) and to some useful exercises connected with it. The method of investigation of (7_1), (7_2) described in §1 of Chapter V is based on the decomposition of the vector space $\mathbf{L}_2(\Omega)$ into two orthogonal subspaces $\mathring{J}(\Omega)$ and $G(\Omega)$. For bounded domains Ω the subspace $G(\Omega)$ consists of all potential fields, i.e., all fields of the form $\nabla\varphi(\cdot)$, where $\varphi \in W_2^1(\Omega)$ and $\mathring{J}(\Omega)$ is the closure (in the norm of $\mathbf{L}_2(\Omega)$ of the set of all smooth solenoidal (divergence-free) vector fields $\mathbf{v}$ with $\mathbf{v}\cdot\mathbf{n}|_S = 0$. The decomposition (which is often cited in literature as the Weil decomposition) was promoted by the identity

$$\int_\Omega \nabla\varphi\cdot\mathbf{v}\,dx = -\int_\Omega \varphi\operatorname{div}\mathbf{v}\,dx + \int_S \varphi\mathbf{v}\cdot\mathbf{n}\,ds \tag{14}$$

and was known for a long time in mechanics for smooth vector fields $\mathbf{u}(\cdot)$ and for domains with smooth boundaries. Prove this decomposition, using classical results on the Neumann problem for the Laplace equation, when $\partial\Omega$

is smooth, and using methods of Chapter II when $\partial\Omega$ is not smooth (see in this connection Section 4 of the Supplements to Chapter II).

But for the investigation of Stokes and Navier–Stokes systems there is required additional information about $\mathring{J}(\Omega)$, namely: in $\mathring{J}(\Omega)$ the set $\mathring{J}^\infty(\Omega)$, of all infinitely smooth solenoidal vector fields with compact supports belonging to Ω, is dense. Try to prove this fact for a possibly larger class of Ω (note, that for unbounded Ω, the functions φ for elements $\nabla\varphi$ of $G(\Omega)$ need not be square-integrable over all of Ω, but they have to be single-valued, locally square-integrable functions). The fact that $\mathring{J}^\infty(\Omega)$ is dense in $\mathring{J}(\Omega)$ allows us to remove $p(\cdot)$ from the problem (7_1), (7_2) without any hypothesis about the behavior of $p(\cdot)$ near $\partial\Omega$, and reduces the problem (7_1), (7_2) to finding $\mathbf{u}$ as an element of $H(\Omega)$ satisfying the identity

$$\nu \int_\Omega \mathbf{u}_x \cdot \boldsymbol{\eta}_x \, dx = -\int_\Omega \mathbf{f} \cdot \boldsymbol{\eta} \, dx$$

for all $\boldsymbol{\eta} \in H(\Omega)$. It is important that, according to the definition of $H(\Omega)$, $\mathring{J}^\infty(\Omega)$ is dense in $H(\Omega)$. Show that the closure (completion) of $\mathring{J}^\infty(\Omega)$ in the norm of $H(\Omega)$ does not give rise to a joining together of different vector fields (even in the case $\Omega = R^3$), and that elements of $H(\Omega)$ are square-integrable over any bounded $\Omega' \subset \Omega$.

7. The methods described in this book are also helpful for a study of boundary value problems associated with equations of the theory of elasticity. Classical methods of potential theory are applicable only to the case of homogeneous media bounded by smooth surfaces (i.e., to the equations with constant coefficients and, to a lesser extent, to diffraction problems governed by equations with piecewise constant coefficients). In §1 of Chapter V we pointed out that the basic system of the theory of elasticity is symmetric and strongly elliptic, so the methods of this book are applicable to it. In Chapter V this system was written only in the case of a homogeneous medium and only the first boundary value problem was mentioned. For this system, just as for a biharmonic equation and Stokes system, there can be specified different "Green's formulas" and self-adjoint boundary problems associated with them. In mechanics it is preferred that we employ the following identity:

$$-\int_\Omega A\mathbf{u} \cdot \mathbf{v} \, dx = 2 \int_\Omega W(\mathbf{u}, \mathbf{v}) \, dx - \int_S \mathbf{t}(\mathbf{u}) \cdot \mathbf{v} \, ds. \tag{15}$$

Here, $A\mathbf{u} = \sum_{i,k=1}^3 (\partial/\partial x_k)\tau_{ik}(\mathbf{u})\mathbf{e}_i$ is the left-hand side of the fundamental system of the theory of elasticity written in vector form

$$A\mathbf{u} = \mathbf{f}, \tag{16}$$

$\mathbf{e}_i$ is a unit vector directed along the x_i-axis,

$$W(\mathbf{u}, \mathbf{v}) = \tfrac{1}{2} \sum_{i,k=1}^3 \varepsilon_{ik}(\mathbf{u})\tau_{ik}(\mathbf{v})$$

is the density of the potential energy of the elastic deformation, and

$$\mathbf{t}(\mathbf{u}) = \sum_{i,k=1}^{3} \tau_{ik}(\mathbf{u}) n_i \mathbf{e}_k$$

is the vector of stresses at the point $x \in S$. Hooke's law, connecting the components of a stress tensor $\tau_{ik}(\mathbf{u})$ with those of a deformation tensor $\varepsilon_{ik}(\mathbf{u}) = \frac{1}{2}(\partial_{x_k} u_i + \partial_{x_i} u_k)$ for a general case of inhomogeneous and anisotropic medium, has the form

$$\tau_{ik}(\mathbf{u}) = \sum_{j,l=1}^{3} c_{ikjl} \varepsilon_{jl}(\mathbf{u}),$$

where c_{ikjl} are functions of $x \in \Omega$ satisfying the symmetry relations

$$c_{ikjl} = c_{jlik} = c_{kijl},$$

so that $\tau_{ik}(\mathbf{u}) = \tau_{ki}(\mathbf{u})$. From (15) it follows that

$$\int_\Omega (A\mathbf{u} \cdot \mathbf{v} - \mathbf{u} \cdot A\mathbf{v})\, dx = \int_S [\mathbf{t}(\mathbf{u}) \cdot \mathbf{v} - \mathbf{u} \cdot \mathbf{t}(\mathbf{v})]\, ds \tag{17}$$

for any **u** and **v**. This is the basic "Green's formula" for (16) which is used for formulation and investigation of the principal boundary value problems of elasticity. This identity, as well as (15), was derived by Betti and is referred to as third Betti's formula. The principal boundary value problems for (16) are: the first, with the conditions

$$\mathbf{u}|_S = 0, \tag{18}$$

and the second, with the conditions

$$\mathbf{t}(\mathbf{u})|_S = 0. \tag{19}$$

(There are some other interesting boundary value problems for (16) but we shall not describe them here.) Both problems are self-adjoint, they can be formulated in a variational form and their solvability can be proved by a "direct method" of the calculus of variations (see Ritz's method in subsection 2, §8, Chapter II), or by the "functional" method used in sections 2 and 3 of Chapter II for the second-order elliptic equations. This possibility is stipulated by properties of the operator A in (16) and by the following hypothesis about c_{ikjl} in Hooke's law: for arbitrary $\mathbf{u} \in W_2^1(\Omega)$, the inequality

$$W(\mathbf{u}, \mathbf{u}) \geq \nu_0 \sum_{i,k=1}^{3} \varepsilon_{ik}^2(\mathbf{u}), \qquad \nu_0 = \text{const.} > 0, \tag{20}$$

is valid. Moreover, $c_{ikjl}(\cdot)$ are supposed to be smooth (to obtain generalized solutions it is sufficient to assume them bounded). The basic analytic fact that makes it possible to prove the solvability of the problems mentioned above is the so-called Korn's inequality:

$$\int_\Omega \mathbf{u}_x^2\,dx \equiv \int_\Omega \sum_{i,k=1}^{3} u_{ix_k}^2 dx \le \mu \int_\Omega \sum_{i,k=1}^{3} \varepsilon_{ik}^2(\mathbf{u})\,dx. \tag{21}$$

For treatment of the first boundary value problem we need to have (21) only for $\mathbf{u} \in \overset{\circ}{W}{}_2^1(\Omega)$. In this case the proof of (21) is very simple; it is sufficient to write $\varepsilon_{ik}^2(\mathbf{u})$ as functions of u_{jx_l} and transform the integrals $\int_\Omega u_{jx_l} u_{lx_j}\,dx, j \neq l$, to $\int_\Omega u_{jx_j} u_{lx_l}\,dx$. Do that and then prove the unique solvability of the problem (16), (18) in $\overset{\circ}{W}{}_2^1(\Omega)$. Investigation of the problem (16), (19) is more complicated; it is connected with the fact that the problem has infinitely many solutions, since the homogeneous equations (16), (19) are satisfied by vector fields $\mathbf{u}(x)$ corresponding to the displacements of the domain Ω as a rigid body. These solutions $\mathbf{u}(\cdot)$ form a finite-dimensional subspace $\mathbf{M}$ of $\mathbf{L}_2(\Omega)$, and in order to determine the problem it is necessary to introduce additional conditions excluding elements of $\mathbf{M}$. (We faced a similar phenomenon in the Neumann problem for the Laplace equation, see Section 4 in the Supplements to Chapter II.) These conditions can be taken in the form

$$\int_\Omega \mathbf{u}(x)\,dx = 0, \qquad \int_\Omega [\mathbf{x}, \mathbf{u}(x)]\,dx = 0, \tag{22}$$

where $\mathbf{x}$ is a vector directed from the origin to the point x and $[\mathbf{x}, \mathbf{u}(x)]$ is the vector product of $\mathbf{x}$ and $\mathbf{u}(x)$, or in the equivalent form

$$\int_\Omega \mathbf{u}(x)\,dx = 0, \qquad \int_\Omega \operatorname{curl} \mathbf{u}(x)\,dx = 0. \tag{23}$$

The first condition excludes translations of Ω and the second excludes rotations. The inequality (21) turns out to be true for $\mathbf{u} \in W_2^1(\Omega)$ satisfying either (22) or (23), and the constant μ in (21) is determined only by Ω. This statement is the main analytical basis for a proof of the unique solvability of (16), (19), (22) (or (23)) in $W_2^1(\Omega)$. Its proof, given by K. O. Friedrichs in [FR 5] (see also [MI 4]), is rather complicated. Try to simplify it and to find out how μ depends on Ω. Having (21) at our disposal, it is not difficult to establish solvability of the problem by following the approach of Chapter II. In this connection it is necessary to bear in mind that the problem is solvable not for all $\mathbf{f} \in \mathbf{L}_2(\Omega)$ but for those which satisfy certain restrictions. Find conditions which are necessary and sufficient for the solvability of (16), (19), (21), (or (23)). This problem was investigated by Friedrichs ([FR 5]). In the note [ED 1] of D. M. Eidus, the mixed problem for (16) was considered (the extended exposition of Eidus' investigations in [MI 4]). The paper [FI 3] by G. Fichera is devoted to various boundary value problems for (16).

8. As was pointed out in the beginning of Chapter V, strongly elliptic systems do not embrace the whole class of elliptic systems, defined as follows: a system of differential equations

$$\mathscr{L}\mathbf{u} \equiv \sum_{|s|=0}^{m} A_s(x)\mathscr{D}^s\mathbf{u} = \mathbf{f}, \tag{24}$$

where $s = (s_1, \ldots, s_n)$, $|s| = s_1 + \cdots + s_n$, $A_s(x)$ are complex-valued matrices, $N \times N$, $\mathbf{u}(x) = (u_1(x), \ldots, u_N(x))$, $x \in R^n$, is called elliptic in Ω if

$$\det \sum_{|s|=m} A_s(x)\boldsymbol{\xi}^s \neq 0, \qquad \forall x \in \overline{\Omega}, \quad \forall \boldsymbol{\xi} \in R^N\backslash\{0\},$$

where $\boldsymbol{\xi}^S \equiv \xi_1^{s_1} \ldots \xi_n^{s_n}$. For each system of that kind, a large class of boundary value problems was found which are Nöether solvable in spaces $W_p^l(\Omega)$ and $H^{l+\alpha}(\overline{\Omega})$. It should be noted that the class depends on the system and, for instance, for certain systems this class does not contain the first boundary problem (see [SL 7]).

9. In §§2 and 3 we gave a short description of some classes of systems of differential and abstract-operator equations for which various initial-boundary value problems and the Cauchy problem can be studied by the methods presented in Chapters III and IV. Try to give exact formulations and proofs of existence and uniqueness theorems for operator equations pointed out in these sections.

10. In all three chapters devoted to evolution equations we did not study separately the Cauchy problem and the extracted existence theorems as only a consequence of existence theorems for the first initial-boundary value problem. For hyperbolic equations these results are quite valuable due to boundedness of the domain of dependence $u(x, t)$ on the data. Let us give some instructions as to how to prove existence theorems for the systems

$$\mathscr{L}\mathbf{u} \equiv \mathbf{u}_t + \sum_{k=1}^{n} A_k(x, t)\mathbf{u}_{x_k} + B(x, t)\mathbf{u} = \mathbf{f}(x, t) \tag{25}$$

for real symmetric matrices $A_k(x, t)$ (i.e. $A_k^*(x, t) = A_k(x, t)$). For such systems the speed of propagation of perturbations is finite, so it is natural to refer them to the class of hyperbolic systems despite the fact that the roots of the characteristic equation $\det(\lambda I + \sum_{k=1}^n A_k\xi_k) = 0$ can be multiple. The roots are real because the matrices $A_k(x, t)$ are symmetric. Conditions on the smoothness both of $A_k(\cdot)$, $B(\cdot)$ and initial data

$$\mathbf{u}|_{t=0} = \boldsymbol{\varphi} \tag{26}$$

depend on the smoothness of solutions which we wish to obtain. The basic "energy" inequality for (25) has the form

$$\int_{\Omega_t} \mathbf{u}^2(x, t)\,dx \leq c(t)\left[\int_{\Omega_0} \mathbf{u}^2(x, 0)\,dx + \int_{\mathscr{D}_t} \mathbf{f}^2(x, \tau)\,dx\,d\tau\right], \tag{27}$$

where $\mathscr{D}_t$ is a domain in $(R^{n+1})^+$ diffeomorphic to a frustrum of a cone and contracting upwards "rapidly enough." Inequality (27) is derived from the relation

$$\int_{\mathscr{D}_t} \mathscr{L}\mathbf{u} \cdot \mathbf{u}\, dx\, d\tau = \int_{\mathscr{D}_t} \mathbf{f} \cdot \mathbf{u}\, dx\, d\tau, \tag{28}$$

if elements of A_k, B and first derivatives of A_k are bounded. The symmetry of A_k implies the relation

$$A_k \mathbf{u}_{x_k} \cdot \mathbf{u} = \frac{1}{2}\frac{\partial}{\partial x_k}(A_k \mathbf{u} \cdot \mathbf{u}) - \frac{1}{2}\frac{\partial A_k}{\partial x_k} \mathbf{u} \cdot \mathbf{u},$$

which makes it possible to write the principal terms of (28) in the form of a divergence. The solvability of (25), (26) in $\mathbf{L}_2(\mathscr{D}_t)$ can be proved on the basis of (27) only. But the proof of uniqueness in this class of generalized solutions requires, as in the case of scalar equations (see §6, Chapter V), some "dual" inequality together with some additional hypothesis about smoothness of the coefficients. It is simpler to deal with generalized solutions possessing square-integrable first derivatives. The uniqueness theorem for them follows directly from (27). But to prove an existence theorem in this class of solutions we have to estimate the integrals $j_1(t) \equiv \int_{\Omega_t} [\mathbf{u}_t^2(x, t) + \mathbf{u}_x^2(x, t)]\, dx$. To this end we consider the relations

$$\int_{\mathscr{D}_t} [(\mathscr{L}\mathbf{u})_\tau \cdot \mathbf{u}_\tau + \sum_{l=1}^{n} (\mathscr{L}\mathbf{u})_{x_l} \cdot \mathbf{u}_{x_l}]\, dx\, d\tau = \int_{\mathscr{D}_t} [\mathbf{f}_\tau \cdot \mathbf{u}_\tau + \sum_{l=1}^{n} \mathbf{f}_{x_l} \cdot \mathbf{u}_{x_l}]\, dx\, d\tau. \tag{29}$$

If $A_k(\cdot)$ and $B(\cdot)$ are independent of x and t then $\mathbf{u}_t$ and $\mathbf{u}_{x_t}$ satisfy system (25) with the right-hand sides $\mathbf{f}_t$ or $\mathbf{f}_{x_l}$, respectively, and the desired estimate for $j_1(t)$ follows from (27). The dependence of $A_k(\cdot)$ and $B(\cdot)$ on x and t leads to some difficulties, but they are easily overcome, similarly for the case of scalar equations. Thus the *a priori* estimate

$$\int_{\Omega_t} (\mathbf{u}_t^2 + \mathbf{u}_x^2)\, dx \le c_1(t) \left[\int_{\Omega_0} (\mathbf{u}^2 + \mathbf{u}_t^2 + \mathbf{u}_x^2)\Big|_{t=0} dx + \int_{\mathscr{D}_t} (\mathbf{f}^2 + \mathbf{f}_t^2 + \mathbf{f}_x^2)\, dx\, d\tau \right. \tag{30}$$

can be obtained from (29) and (27). To prove an existence theorem we can use Galerkin's method with an appropriate basis (see §8, Chapter II). For this purpose let us take a "large" parallelepiped

$$\Pi_{R,T} = \{(x, t) : |x_k| < R, 0 < t < T\}$$

containing the domain $\bar{\mathscr{D}}_T$, where we want to find a solution $\mathbf{u}(\cdot)$, and change $A_k(\cdot)$, $B(\cdot)$, $f(\cdot)$ outside $\bar{\mathscr{D}}_T$, and $\boldsymbol{\varphi}(\cdot)$ outside Ω_0, in such a way as to make these functions $2R$-periodic with respect to all x_k, preserving original smoothness. The basic elements for Galerkin's approximations will be composed of functions $\sin(\pi k_i x_i/R)$ and $\cos(\pi k_i x_i/R)$ with integers k_i, $i = 1, \ldots, n$. This choice of the basis makes it possible to prove for Galerkin's approximations not only (27) but also the estimate (30) and, if as desired, the estimates for $\mathbf{L}_2(\Omega_t)$-norms of higher-order derivatives.

A solution of (25), (26) can also be obtained as a limit of solutions $\mathbf{u}^\varepsilon$ of the equations

$$\mathscr{L}_\varepsilon \mathbf{u} \equiv \mathscr{L}\mathbf{u} - \varepsilon\Delta\mathbf{u} = \mathbf{f}, \qquad \mathbf{u}|_{t=0} = \boldsymbol{\varphi}, \tag{31}$$

as $\varepsilon \to 0+$. For (31), solvability of the Cauchy problem (in R^n), or of an initial-boundary value problem in $\Pi_{R,T}$ (for instance, with the boundary condition $u = 0$, or with the condition of 2R-periodicity with respect to all x_k) are proved by the same methods as are scalar parabolic equations. In the latter case we need not worry about the behavior of $\mathbf{u}^\varepsilon$ near the lateral surface of $\Pi_{R,T}$ since the solution $\mathbf{u}$ of (25), (26) in the general case will not coincide with a limit of $\mathbf{u}^\varepsilon$ outside of $\mathscr{D}_T$.

In the Supplements to Chapter VI we shall describe other methods of solutions for the problem (25), (26) using finite-difference approximations. To motivate one such approximation, we shall take into account that $\mathbf{u}^\varepsilon$ tend in $\mathscr{D}_T$ to the solution $\mathbf{u}$ of (25), (26).

An original method of solution of symmetric-hyperbolic systems was proposed by K. O. Friedrichs in [FR 7]. He has also formulated and studied some correct initial-boundary value problems for them.

It is much more difficult to investigate hyperbolic equations of higher order and systems with variable coefficients possessing no symmetry. One of the most difficult parts is the proof of *a priori* estimates of type (27). The works by I. G. Petrovski [PT 1], O. A. Ladyzenskaja [LA 1], J. Leray [LR 2], L. Gårding [GD 2], and others are devoted to this subject.

11. Consider problems analogous to those described in Sections 6 and 7 of the Supplements to Chapter IV.

12. In §4 of the present chapter we have shown that solvability of some stationary diffractional problems follows from the results of Chapter II. The solvability of some non-stationary (evolution) problems of diffraction for parabolic and hyperbolic equations can be deduced in an analogous way from the results of Chapters III and IV. All this can easily be extended to strongly elliptic, strongly parabolic, and strongly hyperbolic equations. It is also possible to take, instead of the conjugation conditions considered in §4, some more general conditions

$$[u]\Big|_\Gamma = 0, \qquad \left[c\frac{\partial u}{\partial l} + \sigma u\right]\Big|_\Gamma = 0,$$

where $\mathbf{l} = (l_1, \ldots, l_n)$ is a smooth vector field on Γ which is non-tangential to Γ. Consider this problem using the approach described in Section 4 of the Supplements to Chapter II.

CHAPTER VI
The Method of Finite Differences

§1. General Description of the Method. Some Principles of Constructing Convergent Difference Schemes

This method of investigating various problems for differential equations consists of reducing them to systems of algebraic equations in which the unknowns are the values of grid functions u_Δ at the vertices of the grids Ω_Δ, and then examining the limit process when the lengths of the sides of the cells in the grid tend to zero. This brings us to the aim if, in the limit, the functions u_Δ give us the solution of the original problem. Such a reduction of the problem to an infinite sequence of auxiliary finite-dimensional problems defining approximate solutions u_Δ is neither unique nor uniform for problems of various types. In other words, for every problem we can construct different difference schemes converging to it, and for problems of various types such schemes differ essentially from each other. In this chapter we shall consider the same boundary value and initial-boundary value problems for the equations of the basic types that we studied in the earlier chapters, and for each of them construct a few elementary difference schemes, which give us in the limit the solutions of these problems. This will be done in such a way that the existence of these solutions will not be stipulated in advance, but, to the contrary, will be established by the method of finite differences. We shall restrict ourselves to "rectangular" lattices in which the cells will be parallelopipeds with faces parallel to the coordinate planes.

Our investigation of boundary value problems, carried out in Chapters II–IV, was based essentially on certain general facts connected with several function spaces, principally the spaces $\overset{\circ}{W}{}_2^1(\Omega)$ and $W_2^1(\Omega)$. In this chapter we shall lay out from the beginning a number of general properties concerning families of functions defined on contracting networks, including the embedding difference theorems which are needed for the entire chapter. This will be accomplished by means of some of the elementary "interpolations" of grid functions u_Δ which associate to each u_Δ functions which are defined on certain domains connected with the grid Ω_Δ. Such interpolations, which were introduced in my book ([4]), allow us to reduce various questions arising in connection with grid functions to the corresponding questions for ordinary functions, defined in domains that have already been examined.

Thus the statements (theorems) which are laid down in the first two sections will be used in all subsequent sections (in §§3–4 we give their generalizations, which will not be used in this book, but which will be rather useful in a more detailed analysis of problems to be discussed here, as well as for an analysis of other problems and questions connected with difference methods). They have a more general significance not connected with the special character of this or any other boundary value problem, for, with their help, the limiting processes can be carried out in a uniform way in all problems whenever we can establish for the approximate solutions u_Δ of these problems the uniform boundedness (in some norm) with respect to the lengths of the sides of the cells. If there is such a uniform boundedness for a given difference scheme, we say that this scheme is *stable* (in the sense of that norm in which the boundedness of u_Δ is established).

Thus, the theorems, which will be proved in §§1 and 2, allow us to assert that *the stability of a scheme implies its convergence.* In order that the limit function $u(x)$ for $\{u_\Delta\}$ be the solution of the problem, it is necessary, of course, to know that the difference equations of the scheme approximate the corresponding equations of the problem. This property of a scheme can be verified in a standard way and, as difference embedding theorems, in a way that is independent of the type of problem. Ordinarily such a verification can be done separately for each term appearing in the equation, where it is sufficient to do this only on smooth functions. Let us clarify this. Suppose that the derivative $\partial u(x)/\partial x_i$ appears in the equation. It can be obtained (for smooth functions $u(x)$), for example, as the limit of the difference quotients

$$\frac{u(x+\mathbf{e}_i\Delta x_i)-u(x)}{\Delta x_i}\equiv\frac{\Delta u(x)}{\Delta x_i}\quad\text{as } \Delta x_i\to 0,$$

where $\mathbf{e}_i$ is the unit vector along the x_i-axis. In this connection we say that $\Delta u(x)/\Delta x_i$ approximates $\partial u(x)/\partial x_i$. If the differential equation has the form

$$\mathscr{L}u=\sum_{i=1}^{n}a_i\frac{\partial u}{\partial x_i}+au=f,$$

then we shall say that the difference equation

$$\mathscr{L}_\Delta u=\sum_{i=1}^{n}a_i\frac{\Delta u}{\Delta x_i}+au=f$$

is an approximation to it.

Thus the verification whether the difference equations of a scheme approximate the corresponding equations of a problem can be done uniformly for all problems on the basis of well-known facts about derivatives. The situation is the same in the part of the investigation concerning the procedure of the limit process of going from $\{u_\Delta\}$ to $u(x)$ if it is known that the given scheme is stable.

However, this "if" is the main stumbling block in studying a chosen difference scheme. The verification of stability or, what is the same thing (here we have in mind linear problems), finding bounds for u_Δ which are independent of the mesh size, forms the central part of the study of a scheme. If a scheme approximates a problem and is stable, then it leads to the solution $u(x)$ (at least in principle: i.e., for precise or sufficiently precise calculation of u_Δ, which is guaranteed theoretically) of the problem under consideration. On the other hand, for an unstable scheme we can choose the data of the problem (i.e., the free term or the functions appearing in the initial conditions or boundary conditions) in such a way that the $\{u_\Delta\}$ will form an unbounded sequence and thus cannot give in the limit the solution $u(x)$ of the problem.

The above statements of general character concerning the method of finite differences by far require refinements in connection with the approximations and with the stability and convergence, i.e., the designation of those spaces in whose topology they are chosen; here there are a large number of possibilities, and we have chosen the simplest among them.

In §§3 and 4 we prove "embedding theorems for grid functions," which are the analogues of the embedding theorems for the spaces $W_m^l(\Omega)$ presented in Chapter I, §§6–8. From these it follows that whichever stability guarantees whichever convergence, where the question here is not the convergence to the solution of some problem, but rather the interior convergence in the set of grid functions. We shall choose the approximation of the differential equations and boundary conditions in "a weak sense" and shall be concerned essentially with "generalized solutions with finite energy norm." This enables us to include equations with discontinuous coefficients and thus to avoid using certain theorems on the solvability of the problems under discussion. For them we shall establish weak convergence in "energy" spaces. As the example of the Dirichlet problem shows in §11, we can prove strong convergence in the energy norm under the same assumptions on the data of the problem (but for this we need to use some information about the exact solution of the problem).

Let us dwell now on the most important question of all: How do we construct convergent difference schemes? The first thing that comes to mind is to replace all the equations of the problem by difference equations approximating them. But first of all, there are infinitely many such approximations, and which of these gets preference? It seems natural to take that one which is "nicer" (i.e., simpler) than the others (in other words, that one for which the calculation of the solutions u_Δ requires the least time of all), or that one which gives the best stepwise approximation involving, in the replacement of derivatives, a fixed (not large) number of neighboring grid points. But will these approximations yield convergent schemes? It appears that for a large number of problems they will not (to prove this was not at all simple, just as it is not easy to prove many negative assertions). In §6 we shall give examples showing that what would seem to be the most natural schemes do not achieve the desired aim. But we recall what was said above, that convergence is a consequence of the approximation and stability. Thus the moral is this: It is necessary to choose those approximating schemes that are stable. However, in the norms of which spaces might we hope to prove stability? Naturally, in those spaces where the original problem is stable (i.e., stable relative to small perturbations of the coefficients and other data of the problem, including the domain). Such spaces are known for many problems. Thus, for example, for all well-posed problems of mathematical physics these are the "energy spaces" where the squares of the norms express the total energy of the system. Because of this, it was a perfectly natural intention to construct difference schemes for which there is a difference analogue of the "energy estimates" which guarantees stability, and we followed this desire to find convergent difference schemes. The subsequent observations were these: We must analyze the derivations of the energy estimations and find the schemes for which we can imitate this derivation in terms of differences.

As is clear from the corresponding sections of Chapters II–IV, two formulas were used for the proof of the energy bounds in addition to Cauchy's inequality (which is the same for discrete sum as for integrals), namely, the formula for differentiating a product, and, above all, the formula for "integration by parts." Unfortunately, it turns out that both of these formulas are somewhat "tarnished" in their difference analogues (cf. (2.2), (2.3), and (2.11) in this chapter)—there is a "displacement" of the arguments of the factors, and, as will be evident in the sequel, this sort of "damage" is the analytic reason for the divergence of the many "nice" schemes. It is possible to avoid these operations in the case of elliptic equations if we start, not from the differential equation itself, but from the integral identity corresponding to it, as given in Chapter II as the definition of "generalized solution in $W_2^1(\Omega)$" (or what is the same thing, the solution in the energy space). If this identity is approximated by a summation identity in order that the derivatives $\partial u/\partial x_i$, $i = 1, \ldots, n$, be replaced in a uniform way by difference relations, and indeed that the positive definite form $a_{ij}(\partial u/\partial x_i)(\partial u/\partial x_j)$ be replaced by a positive definite form, then it is not hard to see that this identity allows us to obtain an estimate for the difference analogue of the energy norm for the approximate solution u_Δ in the same way as the estimate for the energy norm in the original problem (cf. Chapter II, §2). Thus we draw this conclusion: It is more useful (convenient) to start, not from the differential equation, but from the corresponding integral identity and seek approximations of the latter, not the former. In this way it is easier to understand also how to approximate the boundary conditions, keeping in mind the basic goal of obtaining the analogue of the energy estimate. In §§7 and 8 we shall describe this procedure in detail for all three of the classical boundary value problems for elliptic equations. It will be clear that we shall have a wide choice of constructing difference schemes by this method for which stability, and thus convergence, will hold. Moreover, this approach immediately covers the case of continuous coefficients, including the diffraction problems, by indicating how to construct various accurate approximations in the neighborhood of the singularities of the coefficients. Finally, this approach helps make a natural bridge between the method of finite differences and Galerkin's method on which we can set various variants of the "finite-element method," thus entering the spirit of recent years (for this, see §12).

To realize the idea described here for constructing convergent difference schemes for equations of parabolic and hyperbolic type is more complicated (especially the latter), for the principal terms in the integral identities corresponding to them do not yield positive definite bilinear forms and the derivations of the energy bounds use integration by parts, and, in the hyperbolic case, the differentiation of products. But the relation from which we obtained the energy bound for parabolic equations is the sum of the term $u\,\partial u/\partial t$ and terms of the same type as in the elliptic equations. Hence for all the terms of a parabolic equation except for the first term we can use any of the difference approximations indicated above for equations of elliptic type. Thus it remains only to take care of replacing the one term $\partial u/\partial t$. This question can be investigated beforehand for the elementary equation $\partial u/\partial t = \partial^2 u/\partial x^2$ and the result obtained applied to the general parabolic equation.

The construction of convergent difference schemes for hyperbolic equations is also facilitated essentially by the fact that replacements for the elliptic part of it have already been understood. And with the latter the possible approximations of the boundary conditions are determined (for it is well known that the posing of the boundary conditions is dictated by the elliptic part of the equation). What remains is to understand how to replace the term $\partial^2 u/\partial t^2$. This can be understood through examining the simplest equation of hyperbolic type—the wave equation. The same difference approximations of the term $\partial^2 u/\partial t^2$ will be good and for the general hyperbolic equations of second order, provided that they do not contain terms of the form $a_{i0}\,\partial^2 u/\partial t\,\partial x_i$. It is true that here the dependence of the coefficients on x and t introduces some difficulties, as in this case by the use of (2.2) (which has to be brought in) for there occurs an unravelling of the arguments of the factors. But we can cope with this unpleasantness if we use the fact that the shift of argument in (2.2) can be given up to an arbitrary one of the two factors, and when one of them is a coefficient of the equation (i.e., a known function), then we have to give it up to that one.

From what we have said about equations of parabolic and hyperbolic type it is clear that, in order to find convergent difference schemes, it is reasonable to begin with the simplest versions of these equations, that is, the heat equation and the wave equation. Since questions touching on the replacement of the boundary conditions have been clarified by the case of elliptic equations, and since we have set up a precise plan of action for these, we can take, in a preliminary analysis of the simplest equations of parabolic and hyperbolic type, as boundary conditions those which can be studied in the simplest possible fashion. Such conditions are those of periodicity. Thus we begin our study of non-stationary problems with the heat equation and the wave equation under the condition of periodicity of the space variable. In 1947–48 we presented a "difference analogue of the Fourier method" for constructing convergent difference schemes for such problems ([LA 1]). It allowed us not only to find convergent schemes, but also to study, generally speaking, an arbitrary difference scheme from the point of view of its stability (and, consequently, of its convergence). This method was also successful in proving the divergence of some "reasonable" difference schemes whose convergence, however, could not be proved. We shall present this method in §5, and apply it in §6 to the simplest one-dimensional equations of elliptic, parabolic, and hyperbolic types. It is also useful in analyzing the method of variable directions and the fractional-step method for equations with constant coefficients in rectangular domains. We shall explain the latter in §8 for general parabolic equations in arbitrary domains.

The difference analogue of the Fourier method is suitable for analyzing difference schemes for those problems whose solutions can be represented by ordinary Fourier series (we have in mind the classical expansions in terms of $e^{i(k,x)}$), that is, for equations and systems with constant coefficients (or, more generally, with coefficients independent of the space variables) in domains of parallelopiped type. The method can be used to study difference schemes for equations with variable coefficients, which we did in the paper [LA 1], in connection with a very general topic, the hyperbolic systems of Petrovskii (linear and even quasi-linear), but it is better not to follow this way. As is clear from [LA 1], this approach is too unwieldy and requires unnecessary smoothness of the coefficients. We feel that, to a large extent, it was the unwieldly character of [LA 1] (and also its limited distribution and large delay in publication) which proposed a simple and "neat" method of analyzing difference schemes for equations with constant coefficients, that failed to attract proper attention until the appearance in 1951 of a popular paper [1] of G. G. O'Brien, M. A. Hyman, and S. Kaplan—three American authors who dealt with one-dimensional equations with constant coefficients with a reference to von Neumann as the author of the method. The simplicity of the investigated equations and the description (there are no proofs in the paper) attracted to this method not only the theoreticians but also the specialists in computing. Many papers have appeared since, and, after that, books where the method is applied to a series of equations with constant coefficients see (Richtmayer [1], Godunov and Rajben'kii [1], Marchuk [1], Samarskii [1], Yanenko [1], etc.). It has been extended to the case where x lies in arbitrary domain and where the coefficients of equations depend on x (cf. Rajben'kii and Filippov [1]); this generalization is the difference analogue of the Fourier method presented in Chapter III, §4, and in Chapter IV, §7. Here a solution of the difference equation is sought in the form of a sum of eigenfunctions of the difference operator corresponding to the elliptic part of the equation. However, from a practical point of view, this generalization is not very useful, for the calculation of even one of these eigenfunctions for a non-rectangular domain is usually more laborious than solving the initial-boundary value problem by means of well-known difference schemes.

Thus we have described the basic principles by which we can handle efficiently the construction of stable difference schemes. As a basis we have placed the *a priori* bounds of the solutions of the original problem in the "energy norms" (and their consequences). We could take, instead of these, the bounds obtained from the "maximum principle" (in its extensive interpretations), but we shall not do this for several reasons. One of them is that "one should not embrace the unembraceable," particularly in a course of lectures which, in any case, include a great deal of material. Second, bounds of this kind hold for a significantly smaller circle of problems than do the energy bounds. Third, even for those problems for which such bounds hold, it is not always possible to have an analogue of them for the difference approximations (as an example of this we can exhibit the second-order elliptic equations containing mixed derivatives $\partial^2 u/\partial x_i\, \partial x_j, i \neq j$). Fourth, a complete analysis of finite-difference schemes based on bounds obtained from the maximum principle is, as a rule, more unwieldy and requires greater smoothness than that which we shall present in this chapter. But one should not draw the conclusion from all that has been said that such an analysis is "worse" then what we have chosen. On the contrary, in those cases where it is possible, such an analysis frequently gives more complete information about the approximate solutions and their rate of convergence to the exact solution of the problem. In particular, such a detailed analytic study has been carried out in the widely known fundamental papers of L. A. Lusternik [1], R. Courant, K. O. Friedrichs, and H. Lewy [1], I. G. Petrovskii [3], and Rothe [1] which deal with simplest equations of various types.

This chapter makes no pretense to a complete elucidation of all questions related to the method of finite differences. Thus, for example, we have left aside such an important question as an estimate of the rate of convergence of the approximate solutions to the exact solution. In the papers between the 1930's and the 1950's such estimates were derived for the simplest elliptic and parabolic equations in the metric of the space $C(\overline{\Omega})$. In the 1960's estimates for the rate of convergence were obtained in the energy metrics for various problems under significantly weaker assumptions about the data of the problems and their exact solutions (cf. Rivkind, Oganesjan, Ruhovetz, Dem'jnovich, etc.). Nor do we consider compound difference schemes which sometimes have certain advantages over the simpler schemes. We shall not touch on the extensive domain of the numerical realization of a difference method, nor shall we describe the interesting papers of F. John, P. Lax, and L. Nirenberg on the study of the "principle of freezing coefficients" (cf. [LA 9] for this principle). But it is better to stop here, for an enumeration of what is not here would consume many pages. However, we hope that this chapter will give the attentive reader an idea of the methods of the theoretical investigation of difference schemes, as well as orient him in that vast field which has arisen in connection with the appearance of high-speed computers and with the use of the method of finite differences as one of the fundamental tools for the numerical solution of various applied problems.

§2. The Fundamental Difference Operators and Their Properties

We subdivide the Euclidean space R^n of the variable $x = (x_1, \ldots, x_n)$ by the planes $x_i = k_i h_i$, $h_i > 0$, $i = 1, \ldots, n$, where the k_i are integers, into elementary parallelopipeds (cells) $\omega_{(kh)}$ whose coordinates are defined by the inequalities $k_i h_i < x_i < (k_i + 1)h_i$, $i = 1, \ldots, n$. The vertices of these cells will be called grid points (points of the lattice). Functions which are defined on the grid points (more precisely, on some totality of such points—the domain of definition of these functions) will be denoted by u_h and will be called grid functions. Sometimes, when there is no confusion, we shall omit the index h. We introduce the difference operations

$$u_{x_i}(x) = \frac{1}{h_i}[u(x + h_i \mathbf{e}_i) - u(x)],$$

$$u_{\bar{x}_i}(x) = \frac{1}{h_i}[u(x) - u(x - h_i \mathbf{e}_i)], \tag{2.1}$$

where $\mathbf{e}_i$ is the unit vector along the x_i-axis. (For grid functions u_h we shall use the same notations u_{x_i} and $u_{\bar{x}_i}$ instead of u_{hx_i} and $u_{h\bar{x}_i}$.) Of the first of these we say that it is the right-hand difference quotient (or the right-hand divided difference), and of the second the left-hand difference quotient. In this chapter the partial derivatives of functions $u(x)$ defined in domains Ω (which will always be assumed to be bounded) will be denoted by $\partial u/\partial x_i$. If $u(x)$ is continuously differentiable in Ω, then $u_{x_i}(x)$ and $u_{\bar{x}_i}(x)$ converge to $\partial u(x)/\partial x_i$ as $h_i \to 0$ (uniformly in every $\overline{\Omega}' \subset \Omega$); in other words, the difference quotients u_{x_i} and $u_{\bar{x}_i}$ approximate the derivative $\partial u/\partial x_i$. It is clear that the sums $\alpha u_{x_i} + (1 - \alpha)u_{\bar{x}_i}$ also approximate $\partial u/\partial x_i$ for every α. It is possible to write down more complicated divided differences that converge to $\partial u/\partial x_i$ as $h_i \to 0$, but we shall make no use of them. It is easy to verify that $u_{x_i x_j} \equiv (u_{x_i})_{x_j} = u_{x_j x_i}$; $u_{x_i \bar{x}_j} \equiv (u_{x_i})_{\bar{x}_j} = u_{\bar{x}_j x_i} \equiv (u_{\bar{x}_j})_{x_i}$ and $u_{\bar{x}_i \bar{x}_j} \equiv (u_{\bar{x}_i})_{\bar{x}_j} = u_{\bar{x}_j \bar{x}_i}$ approximate $\partial^2 u/\partial x_i\, \partial x_j$, provided that this last expression exists. In a similar way, we can define the difference quotients $u_{x_{i_1}\dots x_{i_k}\bar{x}_{j_l}\dots\bar{x}_{j_l}}$ of any order. If the derivatives in a differential equation are replaced by their corresponding difference quotients according to the described rule, then we say that the resulting difference equation approximates the given equation. Here the coefficients of the equation can also be replaced by certain functions converging to them as $h_i \to 0$, $i = 1, \dots, n$. Ordinarily, the difference equation is considered only on the grid points. It is customary also to define the concept of the order of approximation (for any derivative and for the entire differential equation). Thus, for example, it is 1 for u_{x_i} for, according to Taylor's formula,

$$u_{x_i}(x) = \frac{\partial u(x)}{\partial x_i} + \tfrac{1}{2}h_i \frac{\partial^2 u(\tilde{x})}{\partial x_i^2}.$$

We shall not dwell on this idea since we shall not use it in the sequel. The difference quotients for the product uv of two grid functions or ordinary functions (the latter defined in a domain) can be represented as

$$\begin{aligned}(uv)_{x_i}(x) &= u_{x_i}(x)v(x) + u(x + h_i\mathbf{e}_i)v_{x_i}(x)\\ &= u_{x_i}(x)v(x) + u(x + h_i\mathbf{e}_i)v_{\bar{x}_i}(x + h_i\mathbf{e}_i),\end{aligned} \tag{2.2}$$

$$\begin{aligned}(uv)_{\bar{x}_i}(x) &= u_{\bar{x}_i}(x)v(x) + u(x - h_i\mathbf{e}_i)v_{\bar{x}_i}(x)\\ &= u_{\bar{x}_i}(x)v(x) + u(x - h_i\mathbf{e}_i)v_{x_i}(x - h_i\mathbf{e}_i).\end{aligned} \tag{2.3}$$

As we see, these formulas are different from the formula for the derivative of a product, for they contain the same unpleasant "displacement" of arguments of which we spoke in the introduction. Because of it, we lose the relationship $u\,\partial u/\partial x_i = \frac{1}{2}\,\partial u^2/\partial x_i$, which was extremely important in deriving all the *a priori* estimates. For none of the approximations $\Delta/\Delta x_i$ to the derivative $\partial/\partial x_i$ do we have the equality $u\,\Delta u/\Delta x_i = \frac{1}{2}\,\Delta(u^2)/\Delta x_i$.

The difference analogue of the formula for integration by parts is given by

$$h\sum_{k=0}^{m-1} u_x(kh)v_h(kh) = -h\sum_{k=1}^{m} u_h(kh)v_{\bar{x}}(kh) + u_h(mh)v_h(mh) - u_h(0)v_h(0). \tag{2.4}$$

Here u_h and v_h are arbitrary grid functions prescribed at the points $x = kh$, $k = 0, \dots, m$, and $(\ )_x$ and $(\ )_{\bar{x}}$ are calculated in accordance with (2.1) with step-size $h_i = h$. (The first term on the right-hand side of (2.4) can be written as $-h\sum_{k=0}^{m-1} u_h((k+1)h)v_x(kh)$, but we prefer the first form where the arguments of both terms are taken at the same point.) We can verify (2.4) by a regrouping of terms (Abel's transformation; (2.4) is also an obvious consequence of the identity $h\sum_{k=0}^{m-1} w_x(kh) = w_h(mh) - w_h(0)$ if we take $w_h = u_h v_h$ and use (2.2)). Here there is also a deviation from the usual formula for integration by parts: On the left-hand and right-hand sides we have difference quotients of different types, so that we lose an important identity whenever we go over to differences, namely the identity $2\int_0^l u(\partial u/\partial x)\,dx$

$= u^2(l) - u^2(0)$, which was used in an essential way in deriving estimates. In place of this we have to use the following:

$$2h \sum_{k=1}^{m} u_{\bar{x}}(kh)u_h(kh) = u_h^2(mh) - u_h^2(0) + h \sum_{k=1}^{m} h(u_{\bar{x}}(kh))^2 \tag{2.5}$$

and

$$2h \sum_{k=0}^{m-1} u_x(kh)u_h(kh) = u_h^2(mh) - u_h^2(0) - h \sum_{k=0}^{m-1} h(u_x(kh))^2. \tag{2.6}$$

These may be obtained by summing the elementary relations

$$2hu_{\bar{x}}(kh)u_h(kh) = u_h^2(kh) - u_h^2((k-1)h) + h^2(u_{\bar{x}}(kh))^2 \tag{2.7}$$

and

$$2hu_x(kh)u_h(kh) = u_h^2((k+1)h) - u_h^2(kh) - h^2(u_x(kh))^2. \tag{2.8}$$

Adding (2.5) and (2.6), we obtain a representation for the difference of the squares of the values $u_h(kh)$ at different points:

$$u_h^2(mh) - u_h^2(0) = h \sum_{k=1}^{m} u_{\bar{x}}(kh)u_h(kh) + h \sum_{k=0}^{m-1} u_x(kh)u_h(kh). \tag{2.9}$$

In the multi-dimensional case the formula for summation by parts can be obtained by simple addition of the equalities (2.4) over all intervals parallel to the x_i-axis for which the difference quotients $(\ )_{x_i}$ and $(\ )_{\bar{x}_i}$ in (2.4) are formed and which lie in the domain of definition of the functions u_h and v_h. We shall use this formula for an important case, when the "outintegral terms" are absent, i.e., when $u_h v_h$ vanishes on the "boundary of the grid domain." Let us clarify this.

Let us use the symbol $\overline{\Omega}_h$ to denote the closed domain consisting of the cells $\bar{\omega}_h$ (or, what is the same thing, $\bar{\omega}_{(kh)}$) belonging to $\overline{\Omega}$. Let us denote its boundary by S_h, and let $\overline{\Omega}_h - S_h = \Omega_h$. We shall use the same symbols Ω_h, S_h, and $\overline{\Omega}_h$ to indicate the sets of vertices (points) of the grid belonging to Ω_h, S_h, and $\overline{\Omega}_h$, respectively. The grid points of Ω_h are called "interior points": together with them, all their "nearest neighbors" belong to $\overline{\Omega}_h$; more precisely, the vertices of all the cells $\omega_{(kh)}$ one of whose vertices is the given point of Ω_h. On the other hand, points of S_h will be called boundary points.

Let us take a grid function w_h on $\overline{\Omega}_h$ which is zero on S_h and define it to be zero outside $\overline{\Omega}_h$. It is easy to verify that

$$\Delta_h \sum_{\Omega_h} w_{x_i} = 0, \qquad i = 1, \ldots, n, \quad \Delta_h \equiv h_1 \cdots h_n. \tag{2.10}$$

If we take w_h in (2.10) to be the product $u_h v_h$ of two grid functions and apply (2.2), we obtain a very important special case of the multi-dimensional formula for "*summation by parts*":

$$\Delta_h \sum_{\Omega_h} u_{x_i} v_h = -\Delta_h \sum_{\Omega_h} u_h v_{\bar{x}_i}. \tag{2.11}$$

Here we recall that $u_h v_h = 0$ outside Ω_h (we can assume, for example, that $u_h = 0$ outside Ω_h and v_h is arbitrary). Let us turn to a convenient notation in (2.11), the sums from left and right extended over the same set $\overline{\Omega}_h$ (in contrast to the general case where $u_h v_h|_{S_h} \neq 0$; cf. (2.4)). This is possible because of our extension of the functions u_h and v_h outside $\overline{\Omega}_h$. But an extension was necessary, for in both sums in (2.11) there are terms containing values of u_h and v_h at points not in $\overline{\Omega}_h$.

Formulas resembling (2.4), (2.10), and (2.11) can be written down for functions u and v given in the domain. Thus, if $w(x)$ is defined and summable on the interval $[0, l]$, then, for $h < l$,

$$\int_0^{l-h} w_x(x)\,dx = \frac{1}{h}\int_{l-h}^{l} w(x)\,dx - \frac{1}{h}\int_0^h w(x)\,dx. \tag{2.12}$$

From (2.12) and from (2.2) for $w(x) = u(x)v(x)$ it follows that

$$\int_0^{l-h} u_x v\,dx = -\int_h^l uv_{\bar{x}}\,dx + \frac{1}{h}\int_{l-h}^{l} uv\,dx - \frac{1}{h}\int_0^h uv\,dx. \tag{2.13}$$

If $u(x)$ and $v(x)$ vanish outside $(0, l)$, then

$$\int_0^l u_x v\,dx = -\int_0^l uv_{\bar{x}}\,dx. \tag{2.14}$$

This holds if only $v(x)$ vanishes for $x < 0$ and for $x > l - h$. In the multi-dimensional case for $u(x)$ and $v(x)$ vanishing outside the domain Ω,

$$\int_\Omega u_{x_i} v\,dx = -\int_\Omega uv_{\bar{x}_i}\,dx. \tag{2.15}$$

Now (2.15) is also valid if only the function v is zero outside Ω and outside the intersection of Ω with the translation of Ω by the vector $(-h_i\mathbf{e}_i)$. In (2.12)–(2.15), $w(x)$, $u(x)$. and $v(x)$ are arbitrary summable functions on their domains of definition.

The following fact is easy to prove by means of (2.15).

Lemma 2.1. *Let $u(x) \in L_2(\Omega)$ and let its difference quotients $u_{x_i}(x)$ converge weakly in $L_2(\Omega')$ to a function $u_i(x)$ as $h_i \to 0$ for all $\overline{\Omega}' \subset \Omega$ (so that $u_i \in L_2(\Omega')$ for all $\overline{\Omega}' \subset \Omega$). Then $u_i(x)$ is the generalized derivative $\partial u/\partial x_i$ of $u(x)$ in Ω.*

We remark that the functions $u_{x_i}(x)$ are defined in $\overline{\Omega}' \subset \Omega$ for all h_i less than the distance from $\overline{\Omega}'$ to $\partial\Omega$. Let us take an arbitrary $\Phi(x) \in \dot{C}^\infty(\Omega)$ and apply (2.15) to $u(x)$ and $\Phi(x)$; the formula is true for all sufficiently small h_i. Now we let h_i tend to zero. The functions $\Phi_{\bar{x}_i}(x)$ tend to $\partial\Phi/\partial x_i$ uniformly in $\overline{\Omega}$, so that we are led to

$$\int_\Omega u_i \Phi\,dx = -\int_\Omega u\frac{\partial\Phi}{\partial x_i}\,dx,$$

which guarantees what we want, namely, $u_i = \partial u/\partial x_i$.

It is obvious from this derivation that instead of the difference quotients $(\ )_{x_i}$ we can take any others which approximate the derivatives $\partial/\partial x_i$.

§3. Interpolations of Grid Functions. The Elementary Embedding Theorems

Let R^n be decomposed into cells $\omega_{(kh)}$ as described at the beginning of §2. Let us take any bounded domain Ω and its corresponding sets Ω_h, $\overline{\Omega}_h$, and S_h as defined in §2. For an arbitrary function u_h prescribed on the grid $\overline{\Omega}_h$ we shall construct several elementary interpolations and determine the connection between them and the u_h. The first of these, the piecewise constant, has the form

$$\tilde{u}_h(x) = u_h(kh), \qquad x \in \omega_{(kh)} \tag{3.1}$$

where we recall that $kh = (k_1 h_1, \ldots, k_n h_n)$ and $\omega_{(kh)} = \{x: k_i h_i < x_i < (k_i + 1)h_i\}$. The second is the multi-linear:

$$\begin{aligned} u_h'(x) = u_h &+ \sum_{r=1}^{n} u_{x_r}(x_r - k_r h_r) + \cdots \\ &+ \sum_{r=1}^{n} u_{x_1 \ldots x_{r-1} x_{r+1} \ldots x_n} \prod_{\substack{i=1 \\ i \neq r}}^{n} (x_i - k_i h_i) \\ &+ u_{x_1 \ldots x_n} \prod_{i=1}^{n} (x_i - k_i h_i) \end{aligned} \tag{3.2}$$

for $x \in \overline{\omega}_{(kh)}$. In (3.2) u_h and all the difference quotients are calculated at the vertex kh. The function $u_h'(x)$ is continuous in $\overline{\Omega}_h$ and within any cell it is linear in each variable x_i. At all vertices of $\overline{\Omega}_h$, $u_h'(x)$ coincides with u_h. It is clear that $u_h'(x) \in W_m^1(\Omega_h)$ for all $m \geq 1$. The third interpolation, more precisely, n interpolations of the third type are defined by

$$\begin{aligned} u_{(m)}(x) = u_h &+ \sum_{\substack{r=1 \\ r \neq m}} u_{x_r}(x_r - k_r h_r) + \cdots \\ &+ u_{x_1 \ldots x_{m-1} x_{m+1} \ldots x_n} \prod_{\substack{r=1 \\ r \neq m}}^{n} (x_r - k_r h_r) \end{aligned} \tag{3.3}$$

for $x \in \omega_{(kh)}$; u_h and all difference quotients are calculated at the vertex kh. The function $u_{(m)}(x)$ is constant in x_m within $\omega_{(kh)}$, is linear on the remaining x_j, and coincides with u_h on those vertices of $\omega_{(kh)}$ lying in the plane $x_m = k_m h_m$. It is not hard to verify that

$$\frac{\partial u_h'(x)}{\partial x_m} = (u_{x_m})_{(m)}(x), \qquad x \in \omega_{(kh)}. \tag{3.4}$$

Here and below interpolations of the form (3.1) for divided differences u_{x_i} of the grid function u_h will be denoted by $(\tilde{u}_{x_i})(x)$, or, briefly, by $\tilde{u}_{x_i}$. Would it be possible, for the same purposes for which we shall use the interpolations described, to introduce others, for example, to take, instead of $u_h'(x)$, another continuous completion in $\overline{\Omega}_h$ which coincides with u_h at all the vertices of $\overline{\Omega}_h$, and which is linear within the elementary simplexes into which we must subdivide each parallelopiped $\omega_{(kh)}$? It would be possible to construct several such interpolations according to the various possible subdivisions of $\omega_{(kh)}$ into simplexes, but we shall not use them (except in §12, where we discuss them in the case $n = 2$) because their analytic expression is not as simple as (3.2).

Let the positive numbers $h_1, \ldots, h_n$ range over some numerical sequences which have limit zero. We consider the lattices corresponding to them and the sets $\overline{\Omega}_h$. Let us assume that there is defined on each of the $\overline{\Omega}_h$ a function u_h. Let us define this function to be zero on all lattice-points not belonging to $\overline{\Omega}_h$ and construct the interpolation functions $\tilde{u}_h(x)$, $u_h'(x)$, and $u_{(m)}(x)$. We prove the following assertion.

Lemma 3.1. *Let*

$$\Delta_h \sum_{\overline{\Omega}_h} u_h^2 \leq c, \qquad \Delta_h = h_1 \cdots h_n. \tag{3.5}$$

Then, if one of the sequences $\{\tilde{u}_h\}$, $\{u_h'\}$, $\{u_{(m)}\}$ converges weakly in $L_2(\Omega)$ to a function $u(x)$ as $h_1, \ldots, h_n \to 0$, then the others also converge to $u(x)$ in the same manner.

Remark. The constant c here, as well as the constants c and c_i to be encountered below, are independent of the step-sizes $h = (h_1, \ldots, h_n)$.

First, we remark that (3.5) implies the uniform boundedness of the norms $\|\tilde{u}_h\|_{2,\Omega}$, $\|u_h'\|_{2,\Omega}$, and $\|u_{(m)}\|_{2,\Omega}$, for the values of the functions $\tilde{u}_h(x)$, $u_h'(x)$, and $u_{(m)}(x)$ in each of the cells lies between the largest and smallest values of u_h at the vertices of $\omega_{(kh)}$. Hence, each of these sequences is weakly compact in $L_2(\Omega)$. Moreover, it follows from this that, for the convergence proofs of interest to us, it is enough to prove the convergence of the projections (in $L_2(\Omega)$) of the corresponding functions only for the functions $\Phi(x)$ in $\dot{C}^\infty(\Omega)$. If $\int_\Omega u_h'\Phi\, dx \to \int_\Omega u\Phi\, dx$, let us prove that $\int_\Omega \tilde{u}_h\Phi\, dx \to \int_\Omega u\Phi\, dx$. The other cases can be handled in a similar way. We construct for $\Phi(x)$ a piecewise-constant function $\tilde{\Phi}_h(x)$ which coincides with $\Phi(x)$ at the lattice-points. It is easy to see that $\int_\Omega u_h'\tilde{\Phi}_h\, dx \to \int_\Omega u\Phi\, dx$. We take h_i, $i = 1, \ldots, n$, so small that $\tilde{\Phi}_h$ is zero on S_h and outside $\overline{\Omega}_h$. Then

$$R \equiv \int_\Omega (u_h' - \tilde{u}_h)\tilde{\Phi}_h\, dx = \sum_{(kh)} \Phi(kh) \int_{\omega(kh)} (u_h' - \tilde{u}_h)\, dx,$$

where $\sum_{(kh)}$ denotes the summation over all $\omega_{(kh)} \subset \bar{\Omega}_h$. It is not hard to calculate that

$$R = \Delta_h \sum_{\bar{\Omega}_h} \Phi_h \left\{ u_{x_1 \ldots x_n} \frac{1}{2^n} \Delta_h + \sum_{r=1}^{n} u_{x_1 \ldots x_{r-1} x_{r+1} \ldots x_n} \frac{1}{2^{n-1}} \prod_{i \neq r} h_i + \cdots + \sum_{r=1}^{n} u_{x_r} \tfrac{1}{2} h_r \right\},$$

where the Φ_h denote the values of Φ at the vertices of the grid.

Let us transfer to each term of the braces one of the difference quotients from u_h to Φ_h and use (2.11). Then

$$R = -\Delta_h \sum_{\bar{\Omega}_h} \left\{ \Phi_{\bar{x}_1} u_{x_2 \ldots x_n} h_2 \ldots h_n \frac{1}{2^n} h_1 + \cdots + \sum_{r=1}^{n} \Phi_{\bar{x}_r} u_h \tfrac{1}{2} h_r \right\}.$$

But

$$(u_{x_{i_1} \ldots x_{i_s}} h_{i_1} \cdots h_{i_s})^2 |_{(kh)} \leq c_1 \sum_{\bar{\omega}_{(kh)}} u_h^2,$$

and

$$\Delta_h \sum_{\bar{\Omega}_h} \sum_{r=1} (\Phi_{\bar{x}_r})^2 \leq c_2,$$

so that $R \to 0$ as $h_1, \ldots, h_n \to 0$. Thus we have shown that

$$\int_\Omega \tilde{u}_h \tilde{\Phi}_h \, dx \to \int_\Omega u\Phi \, dx.$$

Hence

$$\int_\Omega \tilde{u}_h \Phi \, dx = \int_\Omega \tilde{u}_h \tilde{\Phi}_h \, dx + \int_\Omega \tilde{u}_h (\Phi - \tilde{\Phi}_h) \, dx \to \int_\Omega u\Phi \, dx.$$

This proves the lemma.

We introduce the following abbreviated notation which will be used in the sequel

$$u_{x_i}^2 \equiv (u_{x_i})^2, \qquad u_{\bar{x}_i}^2 = (u_{\bar{x}_i})^2,$$

$$u_x^2 \equiv \sum_{i=1}^n u_{x_i}^2, \qquad u_{\bar{x}}^2 \equiv \sum_{i=1}^n u_{\bar{x}_i}^2, \qquad |u_x| \equiv (u_x^2)^{1/2},$$

$$\|u_h\|_{m, \bar{\Omega}_h} \equiv \left(\Delta_h \sum_{\bar{\Omega}_h} |u_h|^m \right)^{1/m}, \tag{3.6}$$

$$\|u_{x_i}\|_{m, \bar{\omega}_{(kh)}} \equiv \left(\Delta_h \sum_{\omega_{(kh)}^{(i)}} |u_{x_i}|^m \right)^{1/m}, \tag{3.7}$$

$$\|u_x\|_{m, \bar{\omega}_{(kh)}} = \left(\sum_{i=1}^n \|u_{x_i}\|_{m, \omega_{(kh)}}^m \right)^{1/m}. \tag{3.8}$$

In (3.7) the symbol $\sum_{\omega^{(i)}_{(kh)}}$ denotes summation over the vertices of the cell $\omega_{(kh)}$ belonging to the face $x_i = k_i h_i$. Furthermore,

$$\|u_{x_i}\|_{m,\bar{\Omega}_h} = \left(\sum_{\bar{\omega}_{(kh)} \in \bar{\Omega}_h} \|u_{x_i}\|^m_{m,\bar{\omega}_{(kh)}} \right)^{1/m}, \tag{3.9}$$

$$\|u_x\|_{m,\bar{\Omega}_h} = \left(\sum_{i=1}^{n} \|u_{x_i}\|^m_{m,\bar{\Omega}_h} \right)^{1/m} \tag{3.10}$$

and

$$\|u_h\|^{(1)}_{m,\bar{\Omega}_h} = [\|u_h\|^m_{m,\bar{\Omega}_h} + \|u_x\|^m_{m,\bar{\Omega}_h}]^{1/m}. \tag{3.11}$$

The quantities $\|u_h\|_{m,\bar{\Omega}_h}$ and $\|u_h\|^{(1)}_{m,\bar{\Omega}_h}$ can naturally be called the norms of the spaces $L_m(\bar{\Omega}_h)$ and $W^1_m(\bar{\Omega}_h)$ of the grid functions u_h defined on $\bar{\Omega}_h$.

There are common terms in the sums (3.9) and (3.10), namely the values $|u_{x_i}|$, calculated at the same point of the grid, but they are repeated no more than 2^{n-1} times. Because of this, we can introduce other, equivalent norms instead of those defined in (3.9)–(3.11), in which such repetitions do not appear (we recall that the number n in our considerations was assumed to be fixed). We shall do this in the following sections. Thus, for example, in the case that the functions u_h vanish on S_h, we shall assume that they have been defined to be zero outside $\bar{\Omega}_h$, and, by the quantity $\|u_x\|^2_{2,\bar{\Omega}_h}$ we shall mean the sum $\Delta_h \sum_{\bar{\Omega}_h} u_x^2$ (and by $\|u_h\|^{(1)}_{2,\bar{\Omega}_h}$ we shall mean $[\Delta_h \sum_{\bar{\Omega}_h} (u_h^2 + u_x^2)]^{1/2}$). But for the sake of convenience in this section, we shall use the norms in the form (3.9)–(3.11).

Now we shall assume that the functions u_h are defined on the grid sets $\bar{\Omega}_h^*$ consisting of the vertices of the cells $\bar{\omega}_{(kh)}$ which meet Ω.

Lemma 3.2. *Let $\|u_h\|^{(1)}_{2,\bar{\Omega}_h^*} \le c$. Then, if one of the sequences $\{\tilde{u}_h\}$, $\{u'_h\}$, $\{u_{(m)}\}$ converges in $L_2(\Omega)$ to some function $u(\cdot)$ as $h_i \to 0$, the other two sequences converge in $L_2(\Omega)$ to $u(\cdot)$.*

First of all, we shall show that the quantity $|u'_h(x) - u_{(m)}(x)|$ assumes its largest value on $\bar{\omega}_{(kh)}$ at one of the vertices of $\omega_{(kh)}$ and that this value is equal to the product $|u_{x_m}| h_m$ taken at one of the vertices of $\omega_{(kh)}$ lying on the plane $x_m = k_m h_m$. Actually, the functions $u'_h(x)$ and $u_{(m)}(x)$ coincide on the face $x_m = k_m h_m$ of the cell $\omega_{(kh)}$ that is taken. Along the interval parallel to the x_m-axis emanating from some point of this boundary the function u'_h is linear in x_m and $u_{(m)}$ is constant. Hence

$$\begin{aligned} u'_h - u_{(m)} &= \left.\frac{\partial u'_h}{\partial x_m}\right|_{x_m = k_m h_m} (x_m - k_m h_m) \\ &= (u_{x_m})_{(m)}|_{x_m = k_m h_m}(x_m - k_m h_m), \end{aligned}$$

and

$$\max_{\bar{\omega}_{(kh)}} |u'_h - u_{(m)}| = h_m \max_{\substack{x_m = k_m h_m \\ k_i h_i \le x_i \le (k_i + 1)h_i \\ i \ne m}} |(u_{x_m})_{(m)}|.$$

But the function $(u_{x_m})_{(m)}$ is linear in each variable x_i, $i \neq m$, on the face $x_m = k_m h_m$ and hence assumes its maximum at one of the vertices of this face, where it is simply equal to u_{x_m}. Hence we have

$$\max_{\bar{\omega}_{(kh)}} |u'_h - u_{(m)}| = h_m \max_{x_m = k_m h_m} |u_{x_m}|, \tag{3.12}$$

where the maximum is taken over the vertices of $\omega_{(kh)}$ lying on the face $x_m = k_m h_m$.

Let us examine one of the cases mentioned in the lemma, for the other cases can be proved analogously.

Let $\int_\Omega (u'_h - u)^2 \, dx \to 0$ as $h_i \to 0$, $i = 1, \ldots, n$. Then

$$R \equiv \int_\Omega (u'_h - u_{(m)})^2 \, dx \leq \sum_{(kh)} \int_{\omega_{(kh)}} (u'_h - u_{(m)})^2 \, dx, \tag{3.13}$$

where the summation $\sum_{(kh)}$ is taken over all $\omega_{(kh)} \in \bar{\Omega}_h^*$, and if we use (3.12), we can majorize the right-hand side of (3.13) by

$$R \leq h_m^2 \|u_{x_m}\|^2_{2,\Omega_h^*} \leq h_m^2 c^2 \to 0, \qquad h_m \to 0.$$

Thus it follows that

$$\int_\Omega (u_{(m)} - u)^2 \, dx \leq 2 \int_\Omega [(u'_h - u_{(m)})^2 + (u'_h - u)^2] \, dx \to 0$$

as $h_1, \ldots, h_n \to 0$.

Lemma 3.3. *Let $\|u_h\|^{(1)}_{2,\bar{\Omega}_h^*} \leq c$, where the $\bar{\Omega}_h^*$ are the same grid sets as in Lemma 3.2. Moreover, let $h_1 = \cdots = h_n = h \to 0$, and let the boundary S of Ω be smooth. Then if $\{u'_h(\cdot)\}$ converges in the norm of $L_2(S)$ to $u(\cdot)$, $\{\tilde{u}_h(\cdot)\}$ also converges in the norm of $L_2(S)$ to $u(\cdot)$.*

This lemma can be proved in the same way as Lemma 3.2, where it is necessary only to remark that $h \int_{S \cap \bar{\omega}_{(kh)}} ds \leq ch^n$, the constant c being independent of h.

Let us go over to the proof of the theorems on the strong compactness of the family of grid functions (more precisely, of their completions) in $L_2(\Omega)$. They are the analogues of the Rellich theorems (Theorems 6.1, 6.2, and 6.6) proved in Chapter I.

Theorem 3.1. *Let the sequences of functions $\{u_h\}$, each defined on its grid $\bar{\Omega}_h$, satisfy*

$$j_h \equiv \Delta_h \sum_{\bar{\Omega}_h} [u_h^2 + u_x^2] \leq c, \tag{3.14}$$

and let the u_h vanish on the boundary points of S_h and outside $\bar{\Omega}_h$. Then the completions $\{u'_h(\cdot)\}$ form a uniformly bounded set in $\mathring{W}_2^1(\Omega)$, which is therefore strongly precompact in $L_2(\Omega)$ and weakly precompact in $\mathring{W}_2^1(\Omega)$.

Important Remark 3.1. We may not assume in this theorem that the edges of the cells of the grids Ω_h tend to zero, although we shall apply it for cases when $h_i \to 0$. If $h_i \to 0$, $i = 1, \ldots, n$, then, from Theorem 3.1 and Lemmas 3.1 and 3.2, we can draw these conclusions: Let $\{u'_{h^\alpha}\}$ be a subsequence of $\{u'_h\}$ converging to some function $u \in \mathring{W}_2^1(\Omega)$, strongly in the norm of $L_2(\Omega)$ and weakly in the norm of $\mathring{W}_2^1(\Omega)$; then all other completions of u_{h^α} will also converge in $L_2(\Omega)$ to $u(x)$ and the functions $\tilde{u}_{x_i}(x)$, $(u_{x_i})_{(m)}(x)$, and $(u_{x_i})'_{h^\alpha}(x)$ formed from u_{h^α} will converge weakly in $L_2(\Omega)$ to $\partial u/\partial x_i$ (to verify this last assertion it is necessary to recall (3.4)).

To prove Theorem 3.1, it is enough to show that (3.14) implies the uniform boundedness of $\|u'_h(x)\|_{2,\Omega}^{(1)}$. For this we remark that the function $u'_h(x)$ can be represented in the parallelopiped $\omega_{(kh)}$ in the form $\sum_{\omega(kh)} u_h(sh) \cdot X_S(x_1/h_1, \ldots, x_n/h_n)$, where the X_S are polynomials of degree n, and where the symbol $\sum_{\omega(kh)}$ denotes summation over all the vertices of $\omega_{(h_h)}$. It is clear from this representation that

$$\int_{\omega(kh)} u'_h(x))^2\,dx = \Delta_h \int_{\omega(k)} (\hat{u}'_h(y))^2\,dy \le c\Delta_h \sum_{\omega(kh)} u_h^2,$$

where $y_i = x_i/h_i$, $i = 1, \ldots, n$, and $\hat{u}'_h(y) = u'_h(x)$. But then

$$\int_{\Omega_h} (u'_h(x))^2\,dx \le 2^n c\Delta_h \sum_{\bar{\Omega}_h} u_h^2. \tag{3.15}$$

Now the derivative $\partial u'_h(x)/\partial x_i$ in $\omega_{(kh)}$ is a sum of the form

$$\sum_{\omega_{(kh)}^{(i)}} u_{x_i}(sh)\tilde{X}_s(x_1/h_i, \ldots, x_n/h_n),$$

where the $\tilde{X}_s$ are polynomials of degree $n - 1$, and the symbol $\sum_{\omega_{(kh)}^{(i)}}$ denotes summation over those vertices of $\omega_{(kh)}$ lying on the face $x_i = k_i h_i$. Hence

$$\int_{\omega(kh)} \left(\frac{\partial u'_h}{\partial x_i}\right)^2 dx \le c_1\Delta_h \sum_{\omega_{(kh)}^{(i)}} (u_{x_i})^2, \tag{3.16}$$

so that

$$\int_{\Omega_h} \left(\frac{\partial u'_h}{\partial x_i}\right)^2 dx \le c_1 \|u_{x_i}\|_{2,\bar{\Omega}_h}^2 \quad\text{and}\quad \int_{\Omega_h} \left(\frac{\partial u'_h}{\partial x}\right)^2 dx \le c_1 \|u_x\|_{2,\bar{\Omega}_h}^2. \tag{3.17}$$

Each of the functions u'_h is in $\mathring{W}_2^1(\Omega)$ and vanishes in $\Omega - \Omega_h$. The inequalities (3.15) and (3.17), along with (3.14), guarantee the uniform boundedness of the norms $\|u'_h\|_{2,\Omega}^{(1)}$.

Thus (3.14) implies the uniform boundedness of the family $\{u'_h\}$ in $W_2^1(\Omega)$, and, since each of u'_h is in $\mathring{W}_2^1(\Omega)$, it now follows from Theorems 6.1 and 1.2 of Chapter I that Theorem 3.1 holds.

Making similar reference to Corollary 6.1 of Chapter I, we can establish the following theorem:

Theorem 3.2. *Corresponding to the cylinder* $\bar{Q}_T = \{(x, x_0): x \in \bar{\Omega}, x_0 \in [0, T]\}$ *let there be a sequence of lattices* $\bar{Q}_{Th} = \bar{\Omega}_h \times [x_0 = k_0 h_0, h_0 = T/N, k_0 = 0, 1, \dots, N]$, $\bar{\Omega}_h \subset \bar{\Omega}$, *with* $h_i \to 0$ $(i = 0, 1, \dots, n)$ *such that on each of them there is a function* u_h *vanishing on* S_h *and outside* Ω_h *for all* $x_0 = k_0 h_0$ $(k_0 = 0, 1, \dots, N)$. *If*

$$\|u_h\|^{(1)}_{2, \bar{Q}_{T_h}} \leq c,$$

then we can choose a subsequence $\{h^\alpha\}$ *from* $\{h\}$ *for which the functions* $\{u'_{h^\alpha}(x, x_0)\}$ *converge to some function* $u(x, x_0)$ *in the norm of* $L_2(Q_T)$, *uniformly with respect to* $x_0 \in [0, T]$ *in the norm of* $L_2(\Omega)$, *and weakly in the space* $W_2^1(Q_T)$. *The limit function* $u(x, x_0)$ *is an element of* $\overset{\circ}{W}{}^1_{2,0}(Q_T)$.†

Theorem 3.1 can be generalized to the case of functions u_h not vanishing on S_h, in the following way. Suppose that there is a sequence of domains $\bar{\Omega}_h^*$ (the same as in Lemma 3.2) containing Ω and satisfying the conditions of Theorem 6.2, Chapter I. Suppose further that the functions u_h are defined on the grid sets $\bar{\Omega}_h^*$ (i.e., on the vertices of the cells $\omega_{(kh)}$ belonging to $\bar{\Omega}_h^*$). For such functions we have the following theorem.

Theorem 3.3. *Let the functions* $\{u_h\}$ *be defined on a sequence of grid sets* $\bar{\Omega}_h^*$, *satisfying those conditions described above, and let*

$$\|u_h\|^{(1)}_{2, \bar{\Omega}_h^*} \leq c. \tag{3.18}$$

Then the family of completions of $\{u'_h(x)\}$ *is uniformly bounded in the norm of* $W_2^1(\Omega)$ *and is hence precompact in* $L_2(\Omega)$. *The family is precompact in* $L_2(\partial\Omega)$ *if* $\partial\Omega$ *satisfies the hypotheses of Theorem 6.6, Chapter I.*

The assertion of the theorem follows from a comparison of the terms on the left-hand side of (3.18) with the norms in $L_2(\Omega)$ and $W_2^1(\Omega)$ for the $u'_h(x)$ carried out above in the proof of Theorem 3.1 (cf. (3.15)–(3.17)).

Remark 3.2. The assertions of Remark 3.1 are valid for those cases considered in Theorems 3.2 and 3.3.

For the grid functions u_h prescribed on $\bar{\Omega}_h$ and equal to zero outside Ω_h we have the inequalities

$$\Delta_h \sum_{\Omega_h} u_h^2 \leq \frac{l_i^2}{2} \Delta_h \sum_{\Omega_h} (u_{x_i})^2, \qquad i = 1, \dots, n, \tag{3.19}$$

† We recall that $\overset{\circ}{W}{}^1_{2,0}(Q_T)$ was obtained as the closure in the norm of $W_2^1(Q_T)$ of the smooth functions vanishing near the lateral surface Γ_T of the cylinder Q_T. It makes sense to say that the elements of $\overset{\circ}{W}{}^1_{2,0}(Q_T)$ vanish on Γ_T.

where l_i is the length of the i-th side of the parallelopiped $\Pi_l \equiv \{x : 0 \leq x_i \leq l_i,$ $i = 1, \ldots, n\}$ containing Ω. They are the difference analogues of the inequality (6.3), Chapter I, and are derived in the same way as (6.3) in Chapter I. It is necessary only to replace the Newton–Leibniz formula (6.5) by its difference equivalent:

$$u_h(mh_i, x_i') = h_i \sum_{k=0}^{m-1} u_{x_i}(kh_i, x_i'), \qquad x_i' = (x_1, \ldots, x_{i-1}, x_{i+1}, \ldots, x_n). \quad (3.20)$$

§4. General Embedding Theorems

For the proof of Theorem 3.1 in §3 we showed that the ratio

$$\|u_h'\|_{2, \omega_{(kh)}} \|u_h\|_{2, \overline{\omega}_{(kh)}}^{-1}$$

is bounded above by a constant which is independent of u_h and h_i. It appears that the inverse ratio $\|u_h\|_{2, \overline{\omega}_{(kh)}} \|u_h'\|_{2, \omega_{(kh)}}^{-1}$ is also bounded above by a constant which is independent of u_h and h_i. Both these facts are true, not just for the L_2 norms, but for any L_m norm with $m \geq 1$. We shall stipulate that such quantities as $\|u_h'\|_{m, \omega_{(kh)}}$ and $\|u_h\|_{m, \overline{\omega}_{(kh)}}$ are to be called equivalent. To show that they are equivalent, we remark first that their ratios are independent of the quantities $h_1, \ldots, h_n$, for, under the change $y_i = x_i h_i^{-1} - k_i$, they are transformed into $\Delta_h^{1/m} \|u_1'\|_{m, \omega_{(0)}}$ and $\Delta_h^{1/m} \|u_1\|_{m, \overline{\omega}_{(0)}}$, respectively, where $\omega_{(0)}$ is the cube $\{y : 0 < y_i < 1\}$, and where the subscript 1 indicates that u_1 is the grid function defined on the vertices of ω_0, and $u_1'(y)$ is its interpolation on $\omega_{(0)}$ calculated according to (3.2) with $h_i = 1$, $k_i = 0$. Let us write $u_1'(y)$ in the form

$$\begin{aligned} u_1'(y_1, y_2) = &\ u(0, 0)(1 - y_1)(1 - y_2) + u(1, 0)(1 - y_2)y_1 \\ &+ u(0, 1)(1 - y_1)y_2 + u(1, 1)y_1 y_2 \dagger \end{aligned} \quad (4.1)$$

for the case $n = 2$ and in a similar way for arbitrary n. It is clear from (4.1) that the ratio $\|u_1'\|_{m, \omega_{(0)}} \|u_1\|_{m, \overline{\omega}_{(0)}}^{-1}$ does not exceed a certain constant c_m which depends only on m. To prove the boundedness of

$$j(u_1) = \|u_1\|_{m, \overline{\omega}_{(0)}} \|u_1'\|_{m, \omega_{(0)}}^{-1}$$

on the set of all grid functions u_1 prescribed on the vertices of $\omega_{(0)}$, we remark that $\sup j(u_1)$ does not vary if this set is further restricted by the requirement that $\|u_1\|_{m, \overline{\omega}_{(0)}} = 1$. We assume that $\sup j(u_1)$ over all such functions is unbounded, or, what is the same thing, that $\inf(j(u_1))^{-1} = 0$. Because of the compactness of the set $\{u_1\}$, this infimum is achieved by some function $\hat{u}_1'(y)$, so that $\|\hat{u}_1'\|_{m, \omega_{(0)}} = 0$. The vanishing of the $L_m(\omega_{(0)})$ norm of $\hat{u}_1'(y)$

---

† For the validity of (4.1) it is enough to verify that the polylinear function defined in this way coincides with the prescribed values at the vertices of the cube $\omega_{(0)}$.

implies that $\hat{u}'_1(y)$ itself vanishes on $\omega_{(0)}$, whence the linear independence of the functions in (4.1) as factors of the values of $\hat{u}'_1(y)$ at the vertices implies that these values are zero. But this contradicts the assumption that $\|\hat{u}_1\|_{m,\bar{\omega}(0)} = 1$, so that $\sup j(u_1)$ must be equal to some finite constant c'_m.†

Thus we have shown the equivalence of the quantities $\|u'_h\|_{m,\omega(kh)}$ and $\|u_h\|_{m,\bar{\omega}(kh)}$. From this follows the equivalence of the quantities $\|u'_h\|_{m,\Omega_h}$ and $\|u_h\|_{m,\bar{\Omega}_h}$. In a similar way, we can prove the equivalence of $\|\partial u'_h/\partial x_i\|_{m,\Omega_h}$ and $\|u_{x_i}\|_{m,\bar{\Omega}_h}$, and therefore of

$$\left\|\frac{\partial u'_h}{\partial x}\right\|_{m,\Omega_h} \equiv \left(\sum_{i=1}^{n}\left\|\frac{\partial u'_h}{\partial x_i}\right\|_{m,\Omega_h}^m\right)^{1/m}$$

and $\|u_x\|_{m,\bar{\Omega}_h}$. These facts allow us to carry over the multiplicative inequalities of §7, Chapter I, to grid functions. Thus, for example, let u_h be given on the grid $\bar{\Omega}_h$ and be equal to zero on S_h. Its completion $u'_h(x)$ will be in $\overset{\circ}{W}{}_2^1(\Omega_h)$ and will therefore satisfy the inequality (7.1) of Chapter I, i.e.,

$$\|u'_h\|_{p,\Omega_h} \leq \beta\left\|\frac{\partial u'_h}{\partial x}\right\|_{m,\Omega_h}^{\alpha}\|u'_h\|_{r,\Omega_h}^{1-\alpha} \tag{4.2}$$

(for the meaning of the parameters, see Theorem 7.1, Chapter I). On the basis of the equivalence of the various norms of the grid function u_h and its completion u'_h, which we have just proved, it follows from (4.2) that

$$\|u_h\|_{p,\bar{\Omega}_h} \leq \hat{\beta}\|u_x\|_{m,\bar{\Omega}_h}^{\alpha}\|u_h\|_{r,\bar{\Omega}_h}^{1-\alpha} \tag{4.3}$$

with the same parameters as in (4.2). We emphasize the fact that the constant $\hat{\beta}$ is defined in terms of these parameters and the dimension n, and that it does not depend on the function u_h, nor on the grid $\bar{\Omega}_h$ (nor, in particular, on the quantities $h_1, \ldots, h_n$).

For the functions u'_h not vanishing at the boundary points of the Ω_h, we have the inequalities (7.7) of Ch. I, as domains of the type Ω_h satisfy the requirements under which we deduced (7.7). However, the constant β appearing in these inequalities depends, in general, on Ω_h. In applications the cases of interest are those where β can be chosen to be the same for the totality of domains Ω_h approximating a given domain Ω. This happens, for example, in the following situation: Let Ω satisfy the hypotheses of Theorem 6.2, Chapter I, and on the sets $\bar{\Omega}_h^*$ consisting of vertices of all cells $\bar{\omega}_{(kh)}$ which are non-empty intersections with Ω, take the completions $u'_h(x)$. We consider the latter to be elements of $W_m^1(\Omega) \cap L_r(\Omega)$ and apply to them the inequalities (7.7) of Chapter I, that is,

$$\|u'_h\|_{p,\Omega} \leq \beta c_{p,m,r}(\Omega)(\|u'_h\|_{m,\Omega}^{(1)})^{\alpha}\|u'_h\|_{r,\Omega}^{1-\alpha}. \tag{4.4}$$

The constant $c_{p,m,r}(\Omega)$ here is defined by Ω, and, as everywhere, by the parameters p, m, r, and n.

† This type of argument appears in the paper [4] by S. L. Sobolev, where a theorem of the type Theorem 4.1 below was proved.

But, as we showed above,

$$\|u_h\|_{p,\bar{\Omega}_h} \le c\|u'_h\|_{p,\Omega_h} \le c\|u'_h\|_{p,\Omega},$$

$$\|u'_h\|_{k,\Omega} \le \|u'_h\|_{k,\Omega^*_h} \le c\|u_h\|_{k,\bar{\Omega}^*_h}, \qquad k = r, m,$$

and

$$\left\|\frac{\partial u'_h}{\partial x}\right\|_{m,\Omega} \le \left\|\frac{\partial u'_h}{\partial x}\right\|_{m,\Omega^*_h} \le c\|u_x\|_{m,\bar{\Omega}^*_h},$$

where the constant c depends only on p, m, r, and n. Thus it follows from (4.4) that

$$\|u_h\|_{p,\bar{\Omega}_h} \le \beta_1(\|u_k\|^{(1)}_{m,\bar{\Omega}^*_h})^{\alpha}\|u_h\|^{1-\alpha}_{r,\bar{\Omega}^*_h}, \tag{4.5}$$

where the constant β_1 depends only on Ω and p, m, r, n. If we use these equivalence properties of the norms of the grid functions u_h and their completions u'_h, along with Theorem 7.3, Chapter I, on the compactness of the embedding of $W^1_m(\Omega)$ into $L_{\tilde{p}}(\Omega)$ for all $\tilde{p} < p$, we obtain the theorem on the precompactness in $L_{\tilde{p}}(\Omega)$ of the set of functions $\{u'_h\}$ for which the grid norms $\|u_h\|^{(1)}_{m,\bar{\Omega}^*_h}$ are uniformly bounded.

It is not hard to show that the norms of other completions of the grid functions u_h considered in §3 of this chapter are majorized by the same norms of u_h and u'_h. Generalizations of the theorems of §3 of this chapter are also valid for any $m > 1$. Our discussion permits us to work simultaneously with all interpolations u_h and to be confident that when the grids contract (i.e., when all $h_1, \ldots, h_n \to 0$), the interpolations lead to the same limit function in the sense of convergence, not only in L_2, but also in any L_p (with p defined by the embedding theorems formulated above).

The interpolations introduced are enough to obtain the grid analogues of the embedding theorems of W^l_m into L_p (cf. §8, Chapter I). We formulate two of them.

Theorem 4.1. *Suppose that, on a sequence of contracting grids* $\bar{\Omega}_h$ ($\bar{\Omega}_h \subset \bar{\Omega}$; $h_1, \ldots, h_n \to 0$), *there are prescribed functions* u_h *(i.e., one function* u_h *on each grid* $\bar{\Omega}_h$) *such that*

$$\sum_{k=0}^{l} \sum_{i_1,\ldots,i_k=1}^{n} \Delta_h \sum_{\bar{\Omega}_h^{(i_1,\ldots,i_k)}} (u_{x_{i_1}\ldots x_{i_k}})^2 \le c, \tag{4.6}$$

where the symbol $\sum_{\bar{\Omega}_h^{(i_1,\ldots,i_k)}}$ *denotes summation only over those points* x *of the lattice* $\bar{\Omega}_h$ *when* $u_{x_{i_1}\ldots x_{i_k}}$ *at* x *is formed by the values of* u_h *at such subsets* $\Gamma(x)$ *of the vertices of* $\bar{\Omega}_h$ *for which the segments connecting the points of this subset* $\Gamma(x)$ *also belong to* $\bar{\Omega}_h$.

Then there exists a subsequence $\{h^\alpha\}$ *for which the functions* u'_{h^α} *and* $(u_{x_{i_1}\ldots x_{i_k}})'_{h^\alpha}$ ($k = 1, 2, \ldots, l-1$; $i_1, \ldots, i_k = 1, \ldots, n$) *converge strongly in* $L_2(\Omega')$ *and the functions* $(u_{x_{i_1}\ldots x_{i_l}})'_{h^\alpha}$ *converge weakly in* $L_2(\Omega')$ *on any sub-*

domain Ω' strictly inside Ω to certain functions u, $u_{i_1\ldots i_k}$, and $u_{i_1\ldots i_l}$, respectively. The limit functions have the following properties:

$$u \in W_2^l(\Omega), \qquad \|u\|_{2,\Omega}^{(l)} \leq c_1,$$

and

$$u_{i_1\ldots i_k} = \frac{\partial^k u}{\partial x_{i_1}\ldots\partial x_{i_k}}, \qquad 0 \leq k \leq l, \quad i_1, \ldots, i_k = 1, \ldots, n.$$

For the proof we take an increasing sequence of sub-domains $\Omega^{(s)}$, $s = 1, 2, \ldots$, strictly inside Ω with smooth $\partial\Omega^{(s)}$, which tends to Ω as $s \to \infty$. For $h_1, \ldots, h_n$ smaller than some $\delta_s > 0$, the functions $(u_{x_{i_1}\ldots x_{i_k}})'_h$, $0 \leq k \leq l$, are defined on all the domains $\Omega^{(s)}$. It follows from (4.6) that

$$\int_{\Omega^{(s)}} \sum_{k=0}^{l} \sum_{i_1, \ldots, i_k=1}^{n} [(u_{x_{i_1}\ldots x_{i_k}})'_h]^2 \, dx \leq c_1, \tag{4.7}$$

where the constant c_1 is determined only by c and n.

By Theorem 1.2, Chapter I, there is a subsequence $h^\alpha(s)$ for which the functions $u'_{h^\alpha(s)}$ and $(u_{x_{i_1}\ldots x_{i_k}})'_{h^\alpha(s)}$ ($k = 1, \ldots, l; i_1, \ldots, i_k = 1, \ldots, n$) converge weakly in $L_2(\Omega^{(s)})$ to functions $u^{(s)}$ and $u^{(s)}_{i_1\ldots i_k}$ which are square-summable in $\Omega^{(s)}$. It then follows from Lemma 3.1 that the completions $(u_{x_{i_1}\ldots x_{i_k}})_{(q)}$ ($q = 1, \ldots, n; k = 0, \ldots, l$) correspond to $(u_{x_{i_1}\ldots x_{i_k}})'_{h^\alpha(s)}$ also converge weakly in $L_2(\Omega^{(s)})$ to $u^{(s)}_{i_1\ldots i_k}$. Hence we can assert (cf. (3.4) in this chapter and Theorem 4.1, Chapter I) that the function $u^{(s)}_{i_1\ldots i_m i_{m+1}}$ is the generalized derivative $\partial u^{(s)}_{i_1\ldots i_m}/\partial x_{i_{m+1}}$ in $\Omega^{(s)}$ (we recall that, because of the permutability of the difference quotients in $u_{x_{i_1}\ldots x_{i_k}}$, the function $u_{i_1\ldots i_k}$ is the same under a permutation of the indices), from which it follows that

$$u^{(s)}_{i_1\ldots i_k} = \frac{\partial^k u^{(s)}}{\partial x_{i_1}\cdots\partial x_{i_k}}$$

in $\Omega^{(s)}$. Moreover, $\|u^{(s)}\|^{(l)}_{2,\Omega^{(s)}} \leq c_1$. We arrange the different terms of the sequence $\{u_{h^\alpha(s)}\}$ so that $\{u_{h^\alpha(s+1)}\}$ lies among $\{u_{h^\alpha(s)}\}$. Then, from the sequences

$$\{u_{h^\alpha(1)}\}, \{u_{h^\alpha(2)}\}, \ldots$$

we choose a diagonal sequence, which we denote by $\{u_{h^\alpha}\}$. The usual argument then leads us to a proof of the fact that the sequences

$$\{u'_{h^\alpha}\}, \quad \{(u_{x_{i_1}\ldots x_{i_k}})'_{h^\alpha}\}, \quad 1 \leq k \leq l, \quad i_i, \ldots, i_k = 1, \ldots, n,$$

converge weakly in $L_2(\Omega')$ for all $\overline{\Omega}' \subset \Omega$ to a certain function $u(x)$ and its generalized derivatives, where $u = u^{(s)}$ in $\Omega^{(s)}$ and $\|u\|^{(l)}_{2,\Omega} \leq c_1$. The assertions of the theorem concerning the strong convergences in $L_2(\Omega')$ follow from what we have proved if we take into account the established connections between the convergence of various completions of grid functions and

Theorem 6.2, Chapter I, which guarantee the strong convergence in $L_2(\Omega')$ of a sequence which is weakly convergent in $W_2^1(\Omega')$. This proves the theorem.†

Similarly, if we use the inequalities (4.3) we can prove the following theorem.

Theorem 4.2. *Let the sequence $\{u_h\}$ of grid functions be defined on the same grids as in Theorem 4.1, and let, in the inequality (4.6), the square term $(u_{x_{i_1}\dots x_{i_k}})^2$ be replaced by $|u_{x_{i_1}\dots x_{i_k}}|^m$, $m \geq 1$. Then there exists a subsequence $\{h^\alpha\}$ such that the functions $\{u'_{h^\alpha}\}$ converge to some function $u \in W_m^l(\Omega)$ in the norms $L_{\tilde{p}}(\Omega')$ for every $\overline{\Omega}' \subset \Omega$, where $\tilde{p} < p \equiv nm/(n-ml)$ for $ml < n$, $\tilde{p}$ is arbitrary for $ml = n$, and $\tilde{p} = \infty$ for $ml > n$. The functions $\{(u_{x_{i_1}\dots x_{i_k}})'_{h^\alpha}\}$, $k = 1, \dots, l-1$, also converge to $\partial^k u/\partial x_{i_1} \dots \partial x_{i_k}$ in the norms $L_{\tilde{p}_k}(\Omega')$ for all $\overline{\Omega}' \subset \Omega$, where $\tilde{p}_k < p_k \equiv nm/[n - m(l-k)]$ for $m(l-k) < n$, $\tilde{p}_k$ is arbitrary for $m(l-k) = n$, and $\tilde{p}_k = \infty$ for $m(l-k) > n$. Finally, the functions $\{(u_{x_{i_1}\dots x_{i_l}})'_{h_\alpha}\}$ converge weakly to $\partial^l u/\partial x_{i_1} \dots \partial x_{i_l}$ in $L_m(\Omega')$ for all $\overline{\Omega}' \subset \Omega$.*

For the proof of this theorem we take a sequence of subdomains $\Omega^{(s)}$, $s = 1, 2, \dots$, similar to that used in the proof of Theorem 4.1, and for each $\Omega^{(s)}$ a function $\zeta^{(s)} \in \dot{C}^\infty(\Omega)$ which is equal to 1 in $\Omega^{(s)}$. Let us introduce the grid functions $v_h^{(s)} = u_h \zeta_h^{(s)}$ ($\zeta_h^{(s)}$ is equal to $\zeta^{(s)}$ on the grid points). These functions coincide with u_h on the nodes $\Omega_h^{(s)}$ and are equal to zero outside Ω_h for all sufficiently small h_k: $h_k \leq h^{(s)}$ ($k = 1, 2, \dots, n$). For these functions the conditions of the theorem guarantee the estimates

$$\sum_{k=0}^{l} \sum_{i_1, \dots, i_k = 1}^{n} \| v^{(s)}_{hx_{i_1}\dots x_{i_k}} \|_{m, R_h^n} \leq c^{(s)}$$

with constant $c^{(s)}$ independent of $v_h^{(s)}$ if only $h_k \leq h^{(s)}$ ($k = 1, 2, \dots, n$) (generally, $c^{(s)} \to \infty$ when $s \to \infty$). It follows from these inequalities and the inequalities (4.3) that there are uniform in h (but not in s!) estimates for $\|v_h^{(s)}\|_{p, R_h^n}$ and $\|v^{(s)}_{hx_{i_1}\dots x_{i_k}}\|_{p_k, R_h^n}$, $k = 1, \dots, l-1$, with the exponents p and p_k indicated in the theorem. They, as well as the relations between different interpolations of grid functions described in §3 (the generalization of Lemma 3.1 for the case of the spaces L_m, $m > 1$, in particular), enable us to prove desirable convergences in each of the domains $\Omega^{(s)}$. Further considerations are similar to those of Theorem 4.1.

To investigate the convergence of difference schemes and especially to find estimates for their speed of convergence, the following result is useful.

† Actually, it follows from $u'_{h^\alpha} \xrightarrow[L_2(\Omega')]{} u$ and $(u_{x_i})'_{h^\alpha} \xrightarrow[L_2(\Omega')]{} \partial u/\partial x_i$ ($i = 1, \dots, n$) that $\partial u'_{h^\alpha}/\partial x_i = (u_{x_i})_{(i)} \xrightarrow[L_2(\Omega')]{} \partial u/\partial x_i$ (see (3.4) and Lemma 3.1). The weak convergence of u'_{h^α} in $W_2^1(\Omega')$ implies strong convergence in $L_2(\Omega')$. Analogous facts hold for $(u_{x_{i_1}\dots x_{i_k}})'_{h^\alpha}$ for $k \leq l-1$.

Lemma 4.1. *Let $u \in L_p(\Omega)$, $p \geq 1$, and let $u(x) = 0$ outside Ω. Let R^n be partitioned into elementary cells $\omega_{(kh)}$, and let u_h be the grid function for $u(x)$ which is equal to the number*

$$u_h = \Delta_h^{-1} \int_{\omega(kh)} u(x)\, dx, \tag{4.8}$$

at the vertex (kh) of the cell $\omega_{(kh)}$. If $\tilde{u}_h(x)$ is the piecewise-constant interpolation for u_h then

$$\|u - \tilde{u}_h\|_{p,R^n} \to 0 \quad \text{for} \quad h_1, \ldots, h_n \to 0.$$

Remark. The result is the same if we set $u_h = 0$ outside the points Ω_h. Then $\tilde{u}_h(x)$ will be zero on S and outside Ω.

For the proof we write $\|\tilde{u}_h - u\|_{p,R^n}^p$ in the form

$$\begin{aligned} j_h &\equiv \|\tilde{u}_h - u\|_{p,R^n}^p \\ &= \sum_{(k)} \int_{\omega(kh)} \left| u(x) - \Delta_h^{-1} \int_{\omega(kh)} u(\xi)\, d\xi \right|^p dx \\ &= \sum_{(k)} \int_{\omega(kh)} \left| \Delta_h^{-1} \int_{\omega(kh)} (u(\xi) - u(x))\, d\xi \right|^p dx, \end{aligned}$$

So that, by Hölder's inequality,

$$\begin{aligned} j_h &\leq \sum_{(k)} \Delta_h^{-1} \int_{\omega(kh)} \int_{\omega(kh)} |u(\xi) - u(x)|^p\, d\xi\, dx \\ &\leq \sum_{(k)} \Delta_h^{-1} \int_{\substack{|\eta_i| \leq h_i \\ i=1,\ldots,n}} d\eta \int_{\omega(kh)} |u(x+\eta) - u(x)|^p\, dx \\ &\leq 2^n \sum_{(k)} \max_{\substack{|\eta_i| \leq h_i \\ i=1,\ldots,n}} \int_{\omega(kh)} |u(x+\eta) - u(x)|^p\, dx \\ &= 2^n \max_{\substack{|\eta_i| \leq h_i \\ i=1,\ldots,n}} \int_{R^n} |u(x+\eta) - u(x)|^p\, dx. \end{aligned}$$

Since $u \in L_p(R^n)$ (note that $u(x) \equiv 0$ outside the bounded domain Ω), it follows from a well-known theorem on the integral continuity of elements of $L_p(R^n)$ (see, for example, V. I. Smiznov, Vol. V) that the right-hand side of the last inequality tends to zero as $(h) \to 0$. This proves the lemma.

If $u \in \overset{\circ}{W}{}_2^1(\Omega)$, then, for the grid functions u_h constructed for u in the same way as in Lemma 4.1 (here we assume that $u(x) \equiv 0$ outside Ω), we can prove that not only does $\|\tilde{u}_h - u\|_{2,\Omega} \to 0$ but also $\|\tilde{u}_{x_i} - \partial u/\partial x_i\|_{2,\Omega} \to 0$. Moreover, if the boundary of Ω is not very bad, then these properties hold for grid functions $\check{u}_h$ which are equal to u_h at points of Ω_h and which vanish on S_h and outside $\overline{\Omega}_h$.

§5. The Finite-Difference Method of Fourier

Let us consider first the one-dimensional case. Let the interval $[0, l] = \hat{l}$ on the x-axis be subdivided into m equal parts, where, for convenience, we take m to be odd, and let us write $l/m \equiv h$. On the grid l_h consisting of the m points $x = kh, k = 0, 1, \dots, m-1$, we consider the totality $\mathscr{E}_h$ of all complex-valued set functions u_h. Each of these functions can be extended to all points on the axis with coordinates $x = kh$, $k = 0, \pm 1, \pm 2, \dots$, so as to obtain an l-periodic grid function; in particular, $u_h(mh) = u_h(0)$. The totality of all such functions will be denoted by $\hat{\mathscr{E}}_h$. We introduce into $\mathscr{E}_h$ the scalar product

$$(u_h, v_h)_{l_h} = h \sum_{k=0}^{m-1} u_h(kh)\bar{v}_h(kh). \tag{5.1}$$

which we can also take as the scalar product in $\hat{\mathscr{E}}_h$. We consider in $\hat{\mathscr{E}}_h$ the difference operator $\Delta/\Delta x \equiv \frac{1}{2}(\ )_x + \frac{1}{2}(\ )_{\bar{x}}$, where $(\ )_x$ and $(\ )_{\bar{x}}$ denote, as before, the right and left difference quotients with ratio h.

The operator $i\,\Delta/\Delta x$ is symmetric on $\hat{\mathscr{E}}_h$, for it follows from the formula (2.4) on summation by parts and from the l-periodicity of the elements u_h and v_h of $\hat{\mathscr{E}}_h$ that

$$\left(i\frac{\Delta u_h}{\Delta x}, v_h\right)_{l_h} \equiv h \sum_{k=0}^{m-1} i\frac{\Delta u_h}{\Delta x}\bar{v}_h = h \sum_{k=0}^{m-1} u_h \overline{\left(i\frac{\Delta v_h}{\Delta x}\right)} \equiv \left(u_h, i\frac{\Delta v_h}{\Delta x}\right)_{l_h}. \tag{5.2}$$

Thus the eigenvalues of the operator $i\,\Delta/\Delta x$ on $\hat{\mathscr{E}}_h$ are real and the corresponding eigenfunctions are mutually orthogonal. There should be m of them, for an element of $\hat{\mathscr{E}}_h$ is defined by its values (complex) at m points of the lattice l_h. The functions $\mu(x)^{(k)} \equiv \exp(2\pi ikx/l)$ $(k = 0, \pm 1, \dots, \pm(m-1)/2)$† are l-periodic, and the grid functions $\mu_h^{(k)}$ corresponding to them (i.e., $\mu_h^{(k)}$ is equal to $\mu^{(k)}(x)$ at the points of $\hat{l}_h = l_h \cup mh$) satisfy the equations

$$\frac{\Delta}{\Delta x}\mu_h^{(k)} = i\alpha_k(h)\mu_h^{(k)}, \tag{5.3}$$

where $\alpha_k(h) = m/l \sin 2\pi k/m$, and are obviously linearly independent. Consequently, the functions $\mu_h^{(k)}$ $(k = 0, \pm 1, \dots, \pm(m-1)/2)$ are the desired eigenfunctions of the operator $i\,\Delta/\Delta x$ (or, what is the same thing, of the operator $\Delta/\Delta x$), where $\bar{\mu}_h^{(k)} = \mu_h^{(-k)}$ and $\alpha_{-k}(h) = -\alpha_k(h)$. By well-known properties of the m-th roots of unity,

$$(\mu_h^{(k)}, \mu_h^{(p)})_{l_h} = h \sum_{s=0}^{m-1} e^{i2\pi(k-p)(s/m)} = mh\delta_k^p = l\delta_k^p, \tag{5.4}$$

† In the role of k we can also take the numbers $0, 1, 2, \dots, m-1$. We shall take them in this form at the end of this section, where the number of points of the subdivision will be even.

where δ_k^p is the Kronecker symbol. Any function u_h in $\hat{\mathscr{E}}_h$ can be expanded into a "Fourier sum"

$$u_h = \sum_{k=-(m-1)/2}^{(m-1)/2} a_{(k)} \mu_h^{(k)}, \tag{5.5}$$

whose coefficients can be calculated via (5.4) as the projections of u_h onto $\mu_h^{(k)}$, that is,

$$a_{(k)} = \frac{1}{l} (u_h, \mu_h^{(k)})_{l_h}. \tag{5.6}$$

The functions $\mu_h^{(k)}$ $(k = 0, \pm 1, \ldots, \pm(m-1)/2)$ are also the eigenfunctions of the operators $(\ )_x$ and $(\ )_{\bar{x}}$ on $\hat{\mathscr{E}}_h$, where

$$(\mu_h^{(k)})_x = \beta_k(h)\mu_h^{(k)}, \qquad \beta_k(h) = \frac{e^{i(2\pi/l)kh} - 1}{h}, \tag{5.7}$$

and

$$(\mu_h^{(k)})_{\bar{x}} = -\bar{\beta}_k(h)\mu_h^{(k)}. \tag{5.8}$$

Hence $\mu_h^{(k)}$ $(k = 0, \pm 1, \ldots, \pm(m-1)/2)$ form a complete system of eigenfunctions of any difference operator $\mathscr{L}_h$ which is the sum of operators of the form $a_{sr} D_h^s \bar{D}_h^r u_h$ with constant coefficients a_{sr} if $\mathscr{L}_h$ is considered as an operator on $\hat{\mathscr{E}}_h$ (here and below we shall denote by $D_h^s u_h$ and $\bar{D}_h^s u_h$ the difference quotients of u_h of order s of the form $u_{xx\ldots x}$ and $u_{\bar{x}\bar{x}\ldots\bar{x}}$, respectively). This permits us to reduce various questions (such as solvability, stability, convergence) for the equations $\mathscr{L}_h u_h = f_h$ (and for systems of such equations) on $\hat{\mathscr{E}}_h$ to purely algebraic form.

Let us go into the question of passage to the limit for grid functions when $h \to 0$. We have from (5.5) and (5.4) that

$$\|u_h\|_{2,l_h}^2 \equiv (u_h, u_h)_{l_h} = l \sum_{k=-(m-1)/2}^{(m-1)/2} |a_{(k)}|^2, \tag{5.9}$$

which is the difference analogue of Parseval's theorem. Similarly, we may calculate

$$\|D_h^p u_h\|_{2,l_h}^2 = l \sum_{k=-(m-1)/2}^{(m-1)/2} |\beta_k(h)|^{2p} |a_{(k)}|^2, \tag{5.10}$$

where

$$|\beta_k(h)| = \frac{2m}{l} \sin\frac{k\pi}{m} = \frac{2}{h} \sin\frac{\pi kh}{l}. \tag{5.11}$$

It is also useful to keep in mind the inequalities

$$\frac{2\pi}{l} |k| \chi \leq |\beta_k(h)| \leq \frac{2\pi}{l} |k| \tag{5.12}$$

for $k = 0, \pm 1, \ldots, \pm(m-1)/2$, where $\chi = \inf_{\alpha\in(0,\pi/2)} \sin\alpha/\alpha > 0$.

Theorem 5.1. *Suppose that, for a sequence of grids l_h with $h = l/m \to 0$ and the corresponding functions u_h in $\hat{\mathscr{E}}_h$,*

$$\sum_{p=0}^{r} \|D_h^p u_h\|_{2,l_h}^2 \leq c. \tag{5.13}$$

Then the trigonometric polynomials

$$\hat{u}_h(x) \equiv \sum_{k=-(m-1)/2}^{(m-1)/2} a_{(k)} \mu^{(k)}(x), \tag{5.14}$$

whose coefficients are given by (5.6) *and where $\mu^{(k)}(x) = \exp(2\pi ikx/l)$, form a bounded set in $W_2^r(\hat{l})$. By* (7.3) *and* (7.4), *Chapter I, the family $\{\hat{u}_h(\cdot)\}$ is precompact in $C^{r-1}(\hat{l})$ if $r \geq 1$, and is weakly precompact in $W_2^r(\hat{l})$.*

The functions $\hat{u}_h(x)$ are the interpolations of grid functions u_h, for it is easy to see that they coincide with u_h at the vertices of $\hat{l}_h$. Theorem 5.1 may be thought of as an embedding theorem for grid functions. For the proof, one has to calculate the norms $\|D^p \hat{u}_h\|_{2,l}$, where D^p is the derivative of order p of $\hat{u}_h$.

It is easy to see that

$$\|D^p \hat{u}_h\|_{2,\hat{l}}^2 = l\left(\frac{2\pi}{l}\right)^{2p} \sum_{k=-(m-1)/2}^{(m-1)/2} |a_{(k)}|^2 |k|^{2p}. \tag{5.15}$$

If we compare this with (5.10) and note (5.12), we obtain

$$\|D^p \hat{u}_h\|_{2,\hat{l}}^2 \leq \chi^{-2p} \|D_h^p u_h\|_{2,l_h}^2. \tag{5.16}$$

and the theorem is proved.

In order to investigate equations whose coefficients depend on x, it is useful to have estimates for the rate of decrease of the numbers $a_{(k)}$ in (5.6) when $k \to \infty$ if it is known u_h is a grid function which coincides with the smooth l-periodic function $u(x)$ at the vertices of the grid $\hat{l}_h$. These bounds for $|a_{(k)}|$ are analogous to the bounds of the usual Fourier coefficients of $u(x)$ with respect to the system of functions $\{e^{2\pi ikx/l}\}$, $k = 0, \pm 1, \pm 2, \ldots$, and depend on the degree of smoothness of $u(x)$. They were obtained in my paper [1] and were used there to investigate difference schemes for equations with variable coefficients. However, as we said at the beginning of this chapter, we shall study the difference analogue of the Fourier method only for equations with constant coefficients and then put aside the bounds mentioned.

The functions $\mu_h^{(k)}$ satisfy (5.7) and (5.8) for any (not necessarily integer) values of k. Hence they can be used in problems where the periodicity conditions are replaced by any other boundary value conditions. In these cases we have to form from these functions another family whose elements are linear combinations of the $\mu_h^{(k)}$ satisfying the boundary conditions that are posed. Let us consider one of these cases which will be used below. Let the interval $[0, l] \equiv \hat{l}$ be subdivided as before into m equal parts of length h, and let us solve the spectral problem

$$-u_{x\bar{x}} = \lambda u_h, \tag{5.17}$$

$$u_h(0) = 0, \qquad u_h(l) = 0. \tag{5.18}$$

The difference equation (5.17) should be fulfilled at all interior points of the grid $\hat{l}_h$: $x = kh$, $k = 1, \dots, m - 1$.

In analogy with the usual Fourier method, we take all grid functions v_h, defined on $\hat{l}_h$ and satisfying (5.18), and continue them first in an odd way to the grid $[l, 2l]_h$, and then in a $2l$-periodic way on the grid (with step-size h) so that we fill out the entire x-axis. The totality of all such functions forms a subset $\hat{\hat{\mathscr{E}}}_h$ of the set $\hat{\mathscr{E}}'_h$ of all $2l$-periodic grid functions. As scalar product in $\hat{\hat{\mathscr{E}}}_h$ we take

$$(u_h, v_h)_{l_h} = h \sum_{k=0}^{m-1} u_h(kh)\bar{v}_h(kh) = h \sum_{k=1}^{m-1} u_h(kh)\bar{v}_h(kh), \tag{5.19}$$

which differs only by a factor $\frac{1}{2}$ from the scalar product in $\hat{\mathscr{E}}'_h$, which is equal to

$$(u_h, v_h)_{(2l)_h} = h \sum_{k=0}^{2m-1} u_h(kh)\bar{v}_h(kh). \tag{5.20}$$

It follows from (5.7) and (5.8) that the functions $\mu_h^{(k)}$, $k = 0, 1, \dots, 2m - 1$, which are equal to $\mu_h^{(k)} = e^{2\pi ikx/2l} = e^{\pi ikx/l}$ at the vertices of the grid, form a complete set of solutions of (5.17) in the space $\hat{\mathscr{E}}'_h$, where the λ_k corresponding to $\mu_h^{(k)}$ is equal to $\lambda_k = |\beta_k(h)|^2 = (2m/l \sin k\pi/2m)^2$. To this value λ_k corresponds a second $2l$-periodic solution $\bar{\mu}_h^{(k)}$ of (5.17). As another pair of linearly independent $2l$-periodic solutions of (5.17) corresponding to λ_k we can take $\operatorname{Re} \mu_h^{(k)}(x) = \cos(\pi/l)kx$ and $\operatorname{Im} \mu_h^{(k)}(x) = \sin(\pi/l)kx$. Of these two solutions only the latter is in $\hat{\hat{\mathscr{E}}}_h$, so that the complete set of solutions of the spectral problem (5.17), (5.18) consists of the functions $\chi_h^{(k)}(x) \equiv \sin(\pi/l)kx$, $k = 1, \dots, 2m - 1$, corresponding to $\lambda = \lambda_k$. But on the grid $\hat{l}_h$ with step-size h, the functions $\chi_h^{(2m-k)}(x) = \sin(\pi/l)(2m - k)x$ and $\chi_h^{(k)} = \sin(\pi/l)kx$ differ only by the factor -1, so that, as a complete set of linearly independent solutions of (5.17), (5.18), we can take the functions

$$\chi_h^{(k)}(x) = \sin \frac{\pi}{l} kx, \qquad k = 1, \dots, m - 1,$$

considered at the points $x = kh$, $k = 0, 1, \dots, m$. There are as many of these functions as there are "interior points" of the lattice $\hat{l}_h$, consisting of the points $x = sh$, $s = 0, 1, \dots, m$. The eigenvalues can be written in the form

$$\lambda_k = \left(\frac{2m}{l} \sin \frac{k\pi}{2m}\right)^2 = \left(\frac{2}{h} \sin \frac{k\pi h}{2l}\right)^2.$$

By (5.4) and by the connection between the scalar products (5.19) and (5.20), the functions $\chi_h^{(k)}$ corresponding to different k are pairwise orthogonal in the sense of (5.19). Moreover, it is not hard to calculate that $(\chi_h^{(k)}, \chi_h^{(k)})_{l_h} = l/2$, so that

$$(\chi_h^{(k)}, \chi_h^{(m)})_{l_h} = \frac{l}{2}\,\delta_k^m. \tag{5.21}$$

Besides this orthogonality, we have

$$h\sum_{s=0}^{m-1}\chi_x^{(k)}(sh)\chi_x^{(m)}(sh) \equiv (\chi_x^{(k)}, \chi_x^{(m)})_{l_h} = \frac{l}{2}\lambda_k\delta_k^m. \tag{5.22}$$

For the proof of (5.22) we must write down (5.17) for $u_h = \chi_h^{(k)}$, multiply it by $h\chi_h^{(m)}$, and sum it over the points $x = sh$, $s = 1, \ldots, m-1$. Then we transform the left-hand side of what we have by the formula for summation by parts into the left-hand side of (5.22), keeping in mind that $\chi_h^{(k)}$ and $\chi_h^{(m)}$ satisfy (5.18). By (5.21) the right-hand side becomes the right-hand side of (5.22).

An arbitrary real function u_h, prescribed on the grid $\hat{l}_h$ and vanishing at the points $x = 0$ and $x = mh$, can be represented in the form of a "Fourier sum":

$$u_h = \sum_{k=1}^{m-1} a_{(k)}\chi_h^{(k)}, \tag{5.23}$$

where, because of (5.21),

$$a_{(k)} = \frac{2}{l}(u_h, \chi_h^{(k)})_{l_h}. \tag{5.24}$$

Moreover,

$$(u_h, u_h)_{l_h} = \frac{l}{2}\sum_{k=1}^{m-1} a_{(k)}^2 \tag{5.25}$$

and

$$(u_x, u_x)_{l_h} = \frac{l}{2}\sum_{k=1}^{m-1} a_{(k)}^2\lambda_k. \tag{5.26}$$

Everything that has been said for the one-dimensional case carries over automatically to the multi-dimensional case, where the periods with respect to different x_i can be taken to be different. This can be done in the same way as for ordinary Fourier series.

§6. The Elementary Equations

1. Equations of Elliptic Type. We begin by considering an elementary boundary value problem of elliptic type, namely,

$$\mathscr{L}u = -\frac{d^2u}{dx^2} + au = f \tag{6.1}$$

on the interval $x \in [0, l] \equiv \hat{l}$ with the periodic conditions

$$u(0) = u(l), \tag{6.2}$$

$$\left.\frac{du}{dx}\right|_{x=0} = \left.\frac{du}{dx}\right|_{x=l}. \tag{6.3}$$

We shall assume that the number a is positive and that the unknown function $u(x)$, as almost everywhere, is real. That a is positive guarantees the unique solvability of the problem (6.1)–(6.3). We did not consider such problems in Chapter II, but they can be investigated in the same way as the Dirichlet problem. Furthermore, the solution of (6.1)–(6.3) can be found easily as an ordinary Fourier series.

We take a grid on the x-axis with vertices at the points $x_k = kh$, $k = 0, \pm 1, \pm 2, \ldots, h = l/m$; here we assume that m is an odd number. To these grids correspond the spaces $\mathscr{E}_h$ and $\hat{\mathscr{E}}_h$ of grid functions which are defined by their values at the points x_k, $k = 0, 1, \ldots, m - 1$ (cf. §5 above).

To compose convergent difference schemes for (6.1)–(6.3) we shall use the ideas described in the introduction, namely, we shall give another form to the equations of the problem (6.1)–(6.3) corresponding to the definition of its "generalized solution with finite energy integral," i.e., a generalized solution in the space W_2^1. We denote by $\hat{W}_2^1(\hat{l})$ the subspace of $W_2^1(\hat{l})$ consisting of the l-periodic elements of $W_2^1(\hat{l})$; $\hat{W}_2^1(\hat{l})$ is the closure in the norm of $W_2^1(\hat{l})$ of the set of sufficiently smooth l-periodic functions. By a *generalized solution of the problem* (6.1)–(6.3) *in* $W_2^1(\hat{l})$ we shall mean a function $u(x)$ in $\hat{W}_2^1(\hat{l})$ satisfying the integral identity

$$\int_0^l \left(\frac{du}{dx}\frac{d\eta}{dx} + au\eta\right) dx = \int_0^l f\eta \, dx \tag{6.4}$$

for all η in $\hat{W}_2^1(\hat{l})$.

The identity (6.4) contains (6.1) and (6.3); (6.2) is included in the definition of $\hat{W}_2^1$ from which we must find a function $u(x)$ satisfying (6.4).

The uniqueness theorem carries over to this class of generalized solutions; in fact, if $f \equiv 0$, then, by setting $\eta = u$ in (6.4), we have that $u \equiv 0$.

In accordance with the chosen principle of forming difference schemes we replace the integrals in (6.4) by Riemann sums corresponding to the subdivision $\hat{l}_h$ and the derivatives du/dx and $d\eta/dx$ by certain difference quotients of the same type which approximate them. The known functions (in this case f) can be "bundled" on the grid in different ways: If they are continuous, we can simply take their values on $\hat{l}_h$, but if they are not, we must replace them by grid functions approximating them in the required metric (more precisely, certain interpolations of the grid functions do the approximating).

Let $f \in L_2(\hat{l})$, and let us construct a grid function f_h so that its piecewise-constant interpolations $\tilde{f}_h(\cdot)$ converge to $f(\cdot)$ in the norm of $L_2(l)$ as $h \to 0$. (For the sequel it is enough to know that $\tilde{f}_h(x)$ converge weakly to $f(x)$ in $L_2(\hat{l})$ as $h \to 0$.) As the function f_h we can take, for example,

$$f_h(x_k) = \frac{1}{h}\int_{x_k}^{x_k+h} f(x)\,dx, \quad k = 0, 1, \ldots, m-1 \tag{6.5}$$

(see the end of §4 in this chapter).

Thus we replace (6.4), for example, by the following "summation identity";

$$h\sum_{k=0}^{m-1} u_x \eta_x + h\sum_{k=0}^{m-1} a u_h \eta_h = h\sum_{k=0}^{m-1} f_h \eta_h, \tag{6.6}$$

which must be fulfilled for all $\eta_h \in \hat{\mathscr{E}}_h$. In this identity appears the values of the unknown grid function u_h at the vertices of the grid $\hat{l}_h$ (i.e., at the points $x_k = kh$, $k = 0, 1, \ldots, m$). We must add to (6.6) the requirement that u_h belongs to $\hat{\mathscr{E}}_h$, that is,

$$u_h(x_0) = u_h(x_m). \tag{6.7}$$

From (6.6) and (6.7) we hope to determine u_h uniquely on $\hat{l}_h$; these relations (6.6) and (6.7) give the difference scheme underlying the investigation. Its stability (more precisely, the *a priori* estimate u_h) in the norm corresponding to the left-hand side of (6.6) is obvious at once: Let us set $\eta_h = u_h$ in (6.6) and estimate the right-hand side by means of Cauchy's inequality. This yields

$$h\sum_{k=0}^{m-1} u_x^2 + ah\sum_{k=0}^{m-1} u_h^2 = h\sum_{k=0}^{m-1} f_h u_h \leq \left(h\sum_{k=0}^{m-1} f_h^2\right)^{1/2}\left(h\sum_{k=0}^{m-1} u_h^2\right)^{1/2}. \tag{6.8}$$

Because a is positive, it follows that

$$\|u_h\|_{2,l_h} \equiv \left(h\sum_{k=0}^{m-1} u_h^2\right)^{1/2} \leq \frac{1}{a}\|f_h\|_{2,l_h} \tag{6.9}$$

and

$$\|u_x\|_{2,l_h}^2 \equiv h\sum_{k=0}^{m-1} u_x^2 \leq \frac{1}{a}\|f_h\|_{2,l_h}^2. \tag{6.10}$$

But

$$\|f\|_{2,l_h}^2 = h\sum_{k=0}^{m-1} f_h^2 = h\sum_{k=0}^{m-1}\left(\frac{1}{h}\int_{x_k}^{x_k+h} f(x)\,dx\right)^2 \leq \int_0^l f^2(x)\,dx \equiv \|f\|_{2,l}^2. \tag{6.11}$$

Therefore (6.9) and (6.10) actually give the uniform bounds (in the sense that they are independent of h and u_h):

$$\|u_h\|_{2,l_h} \leq \frac{1}{a}\|f\|_{2,l}, \tag{6.12}$$

$$\|u_x\|_{2,l_h} \leq \frac{1}{\sqrt{a}}\|f\|_{2,l}. \tag{6.13}$$

Thus the basic part of the proof of stability is settled. Now we must consider more carefully the difference scheme contained in (6.6) and (6.7) and show that it is a system of m linear algebraic equations in m unknowns—the values of u_h at points of the grid l_h (i.e., at the points x_k, $k = 0, 1, \ldots, m - 1$). For this we set u_h at the point $x_{m+1} = (m + 1)h$ equal to u_h at the point $x_1 = h$ and $f_h(m) = f_h(0)$, and then change the first term of (6.6) by the formula (2.4) of "summation by parts" into the form

$$-h\sum_{k=1}^{m} u_{x\bar{x}}\eta_h + u_x\eta_h\big|_0^m, \tag{6.14}$$

and then we group all the terms on the left-hand and right-hand sides of (6.6) in such a way that they assume the form $\sum_{k=1}^{m}\{\cdot\}\eta_h$, keeping in mind that u_h and η_h are subject to (6.7), $u_x(0) = u_x(m)$ and $f_h(m) = f_h(0)$. Finally, if we use the fact that the choice of η_h is arbitrary at the points x_k, $k = 1, \ldots, m$, we equate the quantities $\{\cdot\}$ on both sides in these points. It is easy to see that this leads to the following relations:

$$-u_{x\bar{x}} + au_h = f_h, \tag{6.15}$$

$$u_x\big|_{x_0} = u_x\big|_{x_m}. \tag{6.16}$$

The equation (6.15) should be satisfied at the points x_k, $k = 1, \ldots, m$. Here and in (6.16) appears the value $u_h(x_{m+1}) = u_h(x_1)$. The equations (6.15), (6.16), and (6.7) form a system of $m + 2$ linear algebraic equations in $m + 2$ unknowns, namely, $u_h(x_k)$, $k = 0, 1, \ldots, m + 1$. They give the usual form of writing the difference scheme, equivalent to the description (6.6), (6.7).

If we eliminate the unknowns $u_h(x_m)$ and $u_h(x_{m+1})$ by means of (6.7) and (6.16), the remaining system of m equations is irreducible and connects all m of the unknowns $u_h(x_k)$, $k = 0, 1, \ldots, m - 1$. This corresponds to the known property of elliptic problems: The values of their solutions $u(x)$ at different points x of their domain of existence are interconnected.

For the proof of the unique solvability of the system (6.15), (6.16), (6.7) it is not necessary to calculate the principal determinant, for the solvability follows from the stability, which has already been proved. In fact, if we set $f_h \equiv 0$ in this system (i.e., if we consider the corresponding homogeneous system), then it is obvious from the *a priori* bound (6.9) that the system has only the trivial solution $u_h \equiv 0$. As is well known, this guarantees the unique solvability of the system for any free term, that is, for all f_h. What remains now is to perform the limit process as $h \to 0$ and to show that, in the limit, we obtain the desired solution. This step, as described in the introduction to this chapter, is essentially the same for all types of equations, and is based on propositions proved earlier concerning grid functions which are uniformly bounded in certain norms. We have available to us the bounds (6.12) and (6.13), from which follows the uniform boundedness of the norm $\|\cdot\|_{2,\hat{l}}^{(1)}$ for the linear interpolations $\{u_h'(x)\}$. It is clear that all of the $u_h'(x)$ are in $\hat{W}_2^1(\hat{l})$ and that $\|u_h'(x)\|_{2,\hat{l}}^{(1)} \le c$, that is, the set $\{u_h'(x)\}$ is weakly precompact in $\hat{W}_2^1(\hat{l})$.

Because of the theorems of §§1 and 6 of Chapter I, we can extract convergent subsequences such that $\{u_{h^\alpha}'\}$ converges strongly in $L_2(\hat{l})$ to some element $u \in \hat{W}_2^1(\hat{l})$ and $\{du_{h^\alpha}'/dx\}$ has as its weak limit in $L_2(\hat{l})$ the derivative du/dx. Later we shall show that any such limit element $u(x)$ for $\{u_h'\}$ is a generalized solution in W_2^1 for the problem (6.1)–(6.3). But the problem (6.1)–(6.3) cannot have more than one such solution. Consequently, the set $\{u_h'\}$ has only one limit point $u(x)$, a fact which, together with its uniform boundedness, guarantees the convergence of all sequences $\{u_h'\}$ to $u(x)$.

Thus, let the subsequence $\{u_{h^\alpha}'\}$ converge to $u(x)$ in the way indicated. By what we proved in §3 of this chapter, the piecewise-constant interpolations $\tilde{u}_{h^\alpha}(x)$ converge in $L_2(\hat{l})$ to the same function $u(x)$, and $(\tilde{u}_x)_{h^\alpha}$, equal everywhere except at the grid points to du_{h^α}'/dx, converge weakly in $L_2(\hat{l})$ to du/dx.

We can write (6.6) in the form of the integral identity

$$\int_0^l [(\widetilde{u_x})_h(\widetilde{\eta_x})_h + a\tilde{u}_h\tilde{\eta}_h]\, dx = \int_0^l \tilde{f}_h\tilde{\eta}_h\, dx. \tag{6.17}$$

Let us take an arbitrary smooth l-periodic function $\eta(x)$. Its grid function η_h, which is equal to $\eta(x)$ at the vertices of the grid, is in $\hat{\mathscr{E}}_h$ and generates the functions $\tilde{\eta}_h(x)$ and $(\widetilde{\eta_x})_h$ which converge uniformly to $\eta(x)$ and $d\eta/dx$, respectively, as $h \to 0$. If we let this function play the role of η_h in (6.17), we proceed to the limit in (6.17) with the subsequence $\{h^\alpha\}$ selected above. We then obtain (6.4) for our limit function $u(x)$ and for the smooth l-periodic function $\eta(x)$ that was taken. Since these functions $\eta(x)$ are dense in $\hat{W}_2^1(\hat{l})$ and since $u \in \hat{W}_2^1(\hat{l})$, the identity will hold for all η in $\hat{W}_2^1(\hat{l})$. By the same token, it has been proved that $u(x)$ is the desired generalized solution of the problem. At the same time it has been proved that all sequences $\{u_h'\}$ converge to $u(x)$, just as $\{u_{h^\alpha}'\}$ did.

It is easy to prove a stronger assertion for the problem under investigation, namely, the convergence in $L_2(\hat{l})$ of the functions $\{(u_x)_h'\}$ and $\{(u_{xx})_h'\}$ (as well as $(\widetilde{u_x})_h$ and $(\widetilde{u_{xx}})_h$) to du/dx and d^2u/dx^2, respectively. However, in the general higher-dimensional problem for an arbitrary domain Ω, even with a smooth boundary, it is impossible to prove the latter result, that is, the strong convergence in $L_2(\Omega)$ of some sort of interpolation of difference quotients of $u_{x_i\bar{x}_j}$ of second order to $\partial^2u/\partial x_i\,\partial x_j$. The strong convergence of $(\widetilde{u_x})_h = du_h'/dx$ to du/dx in $L_2(\hat{l})$ can be verified comparatively simply for the higher-dimensional case as well as for the problem (6.1)–(6.3). We shall illustrate this in the first subsection of §11.

The difference scheme (6.15), (6.16), (6.7) is far from the only stable difference scheme which can be constructed, starting with the principle described above, and which can be studied in the same way as the scheme (6.15), (6.16), (6.7). We suggest that the reader show this independently.

We shall now show how to investigate the scheme (6.15), (6.16), (6.7) by means of "finite Fourier sums," as described in §5. For this it will be convenient to construct the grid function f_h for $f(x)$ in a different way, by using the expansion of $f(x)$ into a Fourier series of the functions $\mu^{(k)}(x) = e^{2\pi ikx/l}$, $k = 0, \pm 1, \ldots, \pm(m-1)/2$. Here, let f_n be the value on the grid of the function

$$\check{f}_h(x) = \sum_{k=-(m-1)/2}^{(m-1)/2} f^{(k)}\mu^{(k)}(x), \tag{6.18}$$

where $f^{(k)} = (1/l)\int_0^l f(x)e^{-2\pi ikx/l}\,dx$. It is known that $\check{f}_h(x)$ converges to $f(x)$ in the norm of $L_2(l)$ as $h \to 0$ (i.e., as $m \to \infty$).

Let us represent the solution u_h of the problem (6.15), (6.16), (6.7) in the form

$$u_h = \sum_{k=-(m-1)/2}^{(m-1)/2} a_{(k)}\mu_h^{(k)}, \tag{6.19}$$

where

$$a_{(k)} = \frac{1}{l}(u_h, \mu_h^{(k)})_{l_h}. \tag{6.20}$$

All the grid functions $\mu_h^{(k)}$ satisfy the requirements of (6.7) and (6.16). Moreover, by (5.7) and (5.8),

$$(\mu_h^{(k)})_{x\bar{x}} = -|\beta_k(h)|^2\mu_h^{(k)}, \tag{6.21}$$

where $|\beta_k(h)| = (2/h)\sin(\pi kh/l)$. Let us substitute (6.19) into (6.15) and take into account the fact that the right-hand side of (6.15) is of the form

$$f_h = \check{f}_h = \sum_{k=-(m-1)/2}^{(m-1)/2} f^{(k)}\mu_h^{(k)} \tag{6.22}$$

and that the grid functions $\mu_h^{(k)}$ are linearly independent. This determines all the $a_{(k)}$ uniquely:

$$a_{(k)} = (a + |\beta_k(h)|^2)^{-1}f^{(k)}, \qquad k = 0, \pm 1, \ldots, \pm\frac{m-1}{2}. \tag{6.23}$$

Since we took $f(x)$ to be real-valued, we have that $f^{(-k)} = \bar{f}^{(k)}$, and hence $a_{(-k)} = \bar{a}_{(k)}$, that is, u_h (as well as the interpolations $\hat{u}_h(x)$ to be introduced below) will also be real.

Furthermore, by (5.9)

$$\|u_h\|^2_{2,l_h} = l \sum_{k=-(m-1)/2}^{(m-1)/2} (a + |\beta_k(h)|^2)^{-2} |f^{(k)}|^2$$
$$\leq \frac{l}{a^2} \sum_{k=-(m-1)/2}^{(m-1)/2} |f^{(k)}|^2 = \frac{l}{a^2} \|\check{f}_h\|^2_{2,l_h} \leq \frac{l}{a^2} \|f\|^2_{2,\hat{l}}, \quad (6.24)$$

$$\|u_x\|^2_{2,l_h} = l \sum_{k=-(m-1)/2}^{(m-1)/2} |\beta_k(h)|^2 (a + |\beta_k(h)|^2)^{-2} |f^{(k)}|^2$$
$$\leq \frac{1}{a} \|\check{f}_h\|^2_{2,l_h} \leq \frac{1}{a} \|f\|^2_{2,\hat{l}}, \quad (6.25)$$

and

$$\|u_{x\bar{x}}\|^2_{2,l_h} = \sum_{k=-(m-1)/2}^{(m-1)/2} |\beta_k(h)|^4 (a + |\beta_k(h)|^2)^{-2} |f^{(k)}|^2$$
$$\leq \|\check{f}_h\|^2_{2,l_h} \leq \|f\|^2_{2,\hat{l}}. \quad (6.26)$$

These inequalities prove the stability of the difference scheme under investigation, from which it is already easy to infer its convergence. In this method it is convenient to do this by using the interpolations

$$\hat{u}_h(x) = \sum_{k=-(m-1)/2}^{(m-1)/2} a_{(k)} \mu^{(k)}(x). \quad (6.27)$$

As was shown in §5, (6.24)–(6.26) imply for the $\hat{u}_h(x)$ the uniform boundedness of the norms $\|\hat{u}_h\|^{(2)}_{2,\hat{l}} \leq c$, which in turn guarantees the uniform convergence of $\{\hat{u}_h\}$ and $\{d\hat{u}_h/dx\}$ on $[0, l]$ to some function $u(x)$ and its derivative du/dx, together with the weak convergence of $\{d^2\hat{u}_h/dx^2\}$ to d^2u/dx^2 in $L_2(\hat{l})$ (just as above, this is first established for a subsequence $\{h^\alpha\}$ and then for all sequences $\{h = l/m\}, m = 2p + 1, p = 1, 2, \ldots$). Because of such convergence, u turns out to be continuously differentiable on $[0, l]$ and to satisfy the periodicity conditions (6.2) and (6.3).

The function $u(x)$ satisfies (6.1) almost everywhere on $[0, l]$. This can be shown in various ways. For example, we can use the fact that $\hat{u}_h(x)$ satisfies the difference equation

$$-(\hat{u}_h(x))_{x\bar{x}} + a\hat{u}(x) = \check{f}_h(x) \quad (6.28)$$

for all x. The second and third terms of (6.28) converge in $L_2(\hat{l})$ to $au(x)$ and $f(x)$, respectively. Consequently, $(\hat{u}_h(x))_{x\bar{x}}$ also converges in $L_2(\hat{l})$ to some function, which will be nothing other than d^2u/dx^2.

The finite-difference analogue of the Fourier method enables us to make a similar study of various other difference schemes for the problem (6.1)–(6.3).

In just the same way we can use it to study other boundary value problems for (6.1) (and for equations and systems of arbitrary order with constant coefficients). For example, if, instead of the conditions (6.2), (6.3), we have the first boundary condition

$$u|_{x=0} = 0, \qquad u|_{x=l} = 0, \tag{6.29}$$

then we can take the difference scheme for the problem (6.1), (6.29), defined by (6.15) and the boundary conditions

$$u_h|_{x_0=0} = 0, \qquad u_h|_{x_m=l} = 0. \tag{6.30}$$

The equations (6.15) should be fulfilled at the "interior" grid points $x_k = kh$, $h = l/m$, $k = 1, 2, \ldots, m-1$, while (6.30) should be fulfilled at the boundary points.

If we follow the finite-difference Fourier method, we must seek approximate solutions u_h of the form

$$u_h = \sum_{k=1}^{m-1} a_{(k)} \chi_h^{(k)}, \tag{6.31}$$

where $\chi_h^{(k)}$ are defined at the end of §5, subjecting each term of (6.31) to the same boundary conditions (6.30). The grid functions

$$\chi_h^{(k)}(x_s) \equiv \sin\left(\frac{\pi}{l} k x_s\right), \qquad x_s = \frac{ls}{m},$$

$$s = 0, 1, \ldots, m, \qquad k = 1, 2, \ldots, m-1,$$

form a complete system of eigenfunctions of the difference problem

$$-\chi_{x\bar{x}} = \lambda \chi_h, \qquad \chi_h|_{x_0=0} = \chi_h|_{x_m=l} = 0. \tag{6.32}$$

The difference scheme (6.15), (6.30) possesses the same properties of stability and convergence as the scheme (6.15), (6.7), (6.16) which was investigated above.

Such an analysis of difference schemes can be carried out for higher-dimensional problems if the differential equations have constant coefficients and if the domain of definition of x is a parallelopiped. In more general situations the first method of approach, explained at the beginning of this section, is preferable.

2. The Heat Equation. We shall investigate a number of difference schemes for the heat equation

$$\mathscr{L}u \equiv \frac{\partial u}{\partial t} - \frac{\partial^2 u}{\partial x^2} = 0 \tag{6.33}$$

in the domain $\bar{Q}_T = \{(x, t): x \in [0, l] \equiv \hat{l},\ t \in [0, T]\}$ under the periodicity conditions (6.2) and (6.3), which must be fulfilled for all $t \geq 0$, and under the initial condition

$$u|_{t=0} = \varphi \in L_2(\hat{l}). \tag{6.34}$$

The case when (6.33) is not homogeneous, as well as other classical boundary conditions, may be considered analogously. For the analysis of the difference schemes, we apply the Fourier finite-difference method. The investigations carried out in the first sub-section indicate which difference schemes can be taken for the elliptic part of (6.33) and the boundary conditions. We take that one which was considered in detail, namely, (6.15), (6.16), (6.7). We shall analyze various replacements for the derivative $\partial u/\partial t$ and verify that the conclusions that can be derived by "common sense" from general considerations can be fallacious. To obtain a correct answer requires precise calculations. Let us subdivide the x-axis by the points $x_s = sh$, where $h = l/m, s = 0, \pm 1, \ldots, m$ an odd integer, and the t-axis by the points $t_s = s\tau$, where $s = 0, 1, \ldots$ and τ is some positive number; h is called the step-size in x and τ the step-size in t. We begin with an elementary difference scheme called an *explicit* scheme; it has the form

$$u_t - u_{x\bar{x}} = 0, \qquad (6.35_1)$$

$$u_h|_{x_0} = u_h|_{x_m}, \qquad (6.35_2)$$

$$u_x|_{x_0} = u_x|_{x_m}, \qquad (6.35_3)$$

$$u_h|_{t=0} = \check{\varphi}_h. \qquad (6.35_4)$$

The equations (6.35_2) and (6.35_3) should be satisfied for all $t = t_s$, $s = 0, 1, \ldots, N = [T/\tau]$, (6.35_1) for $x = x_s$, $s = 1, \ldots, m$ and $t = t_s$, $s = 0, 1, \ldots, N$, and (6.35_4) for $x = x_s$, $s = 0, 1, \ldots, m + 1$. It is clear that the equations (6.35_i) approximate the corresponding equations of the problem under consideration.

The function $\check{\varphi}_h$ is formed from $\varphi(x)$ in the same way that $\check{f}_h$ was formed from $f(x)$ in the first section, namely, $\check{\varphi}_h$ coincides at the nodes of the grid with the trigonometrical polynomial

$$\check{\varphi}_h(x) = \sum_{k=-(m-1)/2}^{(m-1)/2} \varphi^{(k)}\mu^{(k)}(x), \qquad (6.36)$$

where

$$\varphi^{(k)} = \frac{1}{l}\int_0^l \varphi(x)e^{-2\pi i kx/l}\,dx.$$

We can depict (6.35_1) symbolically as $\overset{*}{***}$; the four asterisks indicate that (6.35_1) at (x_s, τ_r) connects the values of u_h at the four points, (x_s, τ_r), $(x_s - h, \tau_r)$, $(x_s + h, \tau_r)$, and $(x_s, \tau_r + \tau)$. It is easy to see that the equations (6.35_i), $i = 1, 2, 3, 4$, determine u_h uniquely on the grid

$$Q_\Delta^+ \equiv \{(x_s, \tau_r) : s = 0, 1, \ldots, m + 1; r = 0, 1, \ldots, N + 1\}.$$

We allow h and τ to tend to zero (more precisely, m and N to tend to infinity). "General considerations" suggest that τ should tend to zero faster than h, i.e., in such a way that $\tau/h \to 0$. For otherwise the value of u_h at

(x_s, τ_r) is determined by the values of $\check{\varphi}_h(x_p)$ at the points x_p with $p = s - r$, $s - r + 1, \ldots, s + r$, that is, $u_h(x_s, \tau_r)$ depends on only part of the values of the initial function $\check{\varphi}_h$, if $0 < s - r < s + r < m$. But this contradicts a fundamental property of parabolic equations: they describe processes propagating with infinite velocity, i.e., their solutions $u(x, t)$ at any instant $t = t_0$ and at any point $x = x_0$ depend on the values of $u(x, t)$ with $t < t_0$ taken at all points x of the domain of definition of $u(x, t)$, in particular, they depend on all values of the initial function $u(x, 0)$. If τ and h tend to zero in such a way that $\tau/h \not\to 0$, then it is clear that $u_h(x_s, \tau_r)$ be calculated in a domain of (x_s, τ_r) using only a part of the values of $\varphi(x)$ and could not, in the limit, give the true value of the solution $u(x, t)$, which depends on all of the function $\varphi(x)$. Thus, "general considerations" suggest the necessity of the requirement $\tau/h \to 0$ as h and τ tend to zero. "Common sense" supports the scheme (6.35_i) even without this condition on τ and h. However, as we shall see below, the scheme (6.35_i) is unstable if we do not impose the requirement

$$\frac{\tau}{h^2} \leq \tfrac{1}{2}. \tag{6.37}$$

In order to prove this, we shall look for u_h in the form

$$u_h^r = \sum_{k=-(m-1)/2}^{(m-1)/2} a_{(k)}^r \mu_h^{(k)}, \qquad r = 0, 1, \ldots, N + 1. \tag{6.38}$$

The superscript r indicates the number of the layer in t, i.e., u_h^r is the grid function u_h at the layer $t = t_r = r\tau$. Each term in the sum (6.38) satisfies (6.35_2) and (6.35_3). Substituting (6.38) into (6.35_1) and taking into account the linear independence of the $\mu_h^{(k)}$, we obtain the difference equation for each of the coefficients $a_{(k)}^r$:

$$a_{(k)t}^r \equiv \frac{1}{\tau}[a_{(k)}^{r+1} - a_{(k)}^r] = -|\beta_k(h)|^2 a_{(k)}^r, \tag{6.39}$$

where $r = 0, 1, \ldots$. Let us couple with (6.39) the initial condition for $a_{(k)}^r$,

$$a_{(k)}^0 = \varphi^{(k)}, \tag{6.40}$$

which follows from (6.35_4) and (6.36). From (6.39) we obtain

$$a_{(k)}^{r+1} = (1 - \tau|\beta_k(h)|^2)a_{(k)}^r = (1 - \tau|\beta_k(h)|^2)^{r+1} a_k^0,$$

that is, for arbitrary $r \leq N + 1 = [T/\tau] + 1$,

$$|a_{(k)}^r|^2 = |1 - \tau|\beta_k(h)|^2|^{2r}|\varphi^{(k)}|^2. \tag{6.41}$$

The modulus $|\beta_k(h)| = 2/h \sin(\pi kh/l)$, $h = l/m$, $k = 0, \pm 1, \ldots, \pm(m - 1)/2$; consequently, for $\tau/h^2 \leq \frac{1}{2}$, i.e., under the condition (6.37),

$$|1 - \tau|\beta_k(h)|^2| = \left|1 - \frac{4\tau}{h^2}\sin^2\left(\frac{\pi kh}{l}\right)\right| \leq 1,$$

$$|a_{(k)}^r|^2 \leq |a_{(k)}^0|^2 = |\varphi^{(k)}|^2. \tag{6.42}$$

This, along with (5.9), gives a uniform bound for the norms

$$\|u_h^r\|_{2,l_h}^2 = l \sum_{k=-(m-1)/2}^{(m-1)/2} |a_{(k)}^r|^2 \le l \sum_{k=-(m-1)/2}^{(m-1)/2} |\varphi^{(k)}|^2 = \|\check{\varphi}_h\|_{2,l_h}^2 \le \|\varphi\|_{2,\hat{l}}^2 \tag{6.43}$$

for all r. The inequality (6.43) guarantees the stability of the scheme (6.35$_i$) in the norm of $L_2(\hat{l})$, and it was derived under the condition (6.37). If we use (6.43), we can justify the limit process as $h \to 0$, $\tau \to 0$, under the assumption that $\tau/h^2 \le \frac{1}{2}$ and obtain the generalized solution $u(x, t)$ in the class $L_2(Q_T)$ for the problem under study. In order to guarantee a nicer convergence of the approximate solutions to the exact solution, it is necessry to give a uniform bound for the stronger norms of u_h. This can be done in various ways: either by strengthening the requirement (6.37) on τ/h^2, i.e., requiring, for example, that

$$\frac{\tau}{h^2} \le \frac{1-\varepsilon}{2} \tag{6.44}$$

for some fixed $\varepsilon \in (0, 1)$, or by imposing a more stringent hypothesis on the initial function $\varphi(x)$ by requiring it to belong to some class $\hat{W}_2^l$, $l \ge 1$.

We shall now show that if (6.37) is replaced by

$$\frac{\tau}{h^2} = \frac{1+\varepsilon}{2} \tag{6.45}$$

for some small $\varepsilon > 0$, then the scheme (6.35$_i$) becomes unstable and that the u_h calculated by this scheme will not converge to the solution of our problem. Let, for example, $\varphi^{(0)} = 0$ and $|\varphi^{(k)}| = 1/|k|^\alpha$ for $|k| \ge 1$ with some $\alpha > 1$. It is clear that, with such Fourier coefficients, the $\varphi(x)$ belong to $L_2(\hat{l})$ and, with increasing α, become smoother and smoother l-periodic functions. Let us consider the behavior of the corresponding $a_{(p)}^r$ with $p = (m-1)/2$. It follows from (6.41) and (6.45) that, with some $\delta > 0$,

$$|a_{(p)}^r|^2 = \left|1 - 2(1+\varepsilon)\sin^2\left(\frac{\pi p}{2p+1}\right)\right|^{2r} \cdot \frac{1}{p^{2\alpha}} \ge \frac{(1+\delta)^{2r}}{p^{2\alpha}}$$

for all sufficiently large p. Let us take some $t \in (0, T)$ and

$$r = \left[\frac{t}{\tau}\right] = \left[t\left(\frac{1+\varepsilon}{2}h^2\right)^{-1}\right] = \left[t\left(\frac{1+\varepsilon}{2}\right)^{-1}\left(\frac{2p+1}{l}\right)^2\right].$$

Then

$$\begin{aligned}\|u_h^r\|_{2,l_h}^2 &= l \sum_{k=-p}^{p} |a_{(k)}^r|^2 \ge l|a_{(p)}^r|^2 \\ &\ge \frac{l}{p^{2\alpha}} \exp\left\{2t\left(\frac{1+\varepsilon}{2}\right)^{-1}\left(\frac{2p+1}{l}\right)\ln(1+\delta)\right\}.\end{aligned} \tag{6.46}$$

As $p \to \infty$, the right-hand side of (6.46) tends to ∞ for arbitrary $\alpha > 1$, and, as a result, $\|u_h^r\|_{2,l_h}$ increases without bound, that is, the grid functions u_h cannot actually give in the limit the desired bounded (in whatever L_2 norm) solution $u(x, t)$ to our problem.

Let us go over to a study of other difference schemes for the problem (6.33), (6.34), (6.2), (6.3) and consider the simplest *implicit* scheme. Here the approximations of the boundary and initial conditions are taken in the same way as in the explicit scheme, i.e., in the form (6.35_i), $i = 2, 3, 4$, but (6.35_1) is replaced by the equation

$$u_{\bar{t}} - u_{x\bar{x}} = 0, \tag{6.47}$$

which must be satisfied at the grid points $x = x_s$, $s = 1, 2, \ldots, m$, on the layers $t = t_r$, $r = 1, 2, \ldots, [T/\tau]$. It is clear that (6.47) approximates (6.33). This scheme may be denoted symbolically by $\overset{***}{*}$. It is called implicit because, to determine its solution u_h^r on the layer $t = t_r$ ($r \geq 1$) whenever u_h^{r-1} has already been calculated, we must solve an algebraic system containing the m equations (6.47) with m unknowns, the values of $u_h^r(x_s)$ at the points x_s with $s = 1, 2, \ldots, m$ (the values $u_h^r(x_0)$ and $u_h^r(x_{m+1})$ can be replaced by (6.35_2) and (6.35_3), by the quantities $u_h^r(x_m)$ and $u_h^r(x_1)$). We shall show that these systems are uniquely solvable, where the calculation of u_h^r must be taken layer by layer, beginning with $r = 1$. Moreover, the system is stable no matter how h and τ tend to zero, so that the approximate solutions u_h^r give, in the limit as $h \to 0$ and $\tau \to 0$, the desired solution $u(x, t)$. For the proof of this, we shall again seek u_h^r in the form (6.38). The boundary conditions (6.35_2) and (6.35_3) are satisfied for all $\mu_h^{(k)}$. The coefficients $a_{(k)}^r$ are determined from (6.47) and from the initial condition, more precisely, from (6.35_4). In fact, if we substitute (6.38) into (6.47) and use the linear independence of the $\mu_h^{(k)}$, we obtain

$$a_{(k)\bar{t}}^r \equiv \frac{1}{\tau}\,[a_{(k)}^r - a_{(k)}^{r-1}] = -|\beta_k(h)|^2 a_{(k)}^r \tag{6.48}$$

for $r = 1, 2, \ldots, [T/\tau]$. Now we couple (6.40) to (6.48). From (6.48) we have

$$a_{(k)}^r = (1 + \tau|\beta_k(h)|^2)^{-1} a_{(k)}^{r-1}, \tag{6.49}$$

from which it is already clear that we have the stability of our scheme in the norm of $L_2(\hat{l})$ for arbitrary h and τ. In fact, by (5.9) and (6.49) for arbitrary r,

$$\begin{aligned} \|u_h^r\|_{2,l_h}^2 &= l \sum_{k=-(m-1)/2}^{(m-1)/2} |a_{(k)}^r|^2 \leq l \sum_{k=-(m-1)/2}^{(m-1)/2} |a_{(k)}^{r-1}|^2 \\ &\leq l \sum_{k=-(m-1)/2}^{(m-1)/2} |a_{(k)}^0|^2 = \|\check{\varphi}_h\|_{2,l_h}^2 \leq \|\varphi\|_{2,\hat{l}}^2. \end{aligned} \tag{6.50}$$

We can bound another norm of u_h^r, for example,

$$\|u_x^r\|_{2,l_h}^2 = l \sum_{k=-(m-1)/2}^{(m-1)/2} |\beta_k(h)|^2 |a_{(k)}^r|^2$$
$$= l \sum_{k\neq 0} |\beta_k(h)|^2 \cdot (1+\tau|\beta_k(h)|^2)^{-2r} |a_{(k)}^0|^2, \tag{6.51}$$

whence

$$\tau \sum_{r=1}^{[T/r]} \|u_x^r\|_{2,l_h}^2 = l \sum_{k\neq 0} |a_{(k)}^0|^2 |\beta_k(h)|^2 \sum_{r=1}^{[T/t]} \frac{\tau}{(1+\tau|\beta_k(h)|^2)^{2r}}$$
$$\leq l \sum_{k\neq 0} |a_k^0|^2 \frac{\tau|\beta_k(h)|^2 \dfrac{1}{(1+\tau|\beta_k(h)|^2)^2}}{1-\dfrac{1}{(1+\tau|\beta_k(h)|^2)^2}}$$
$$= l \sum_{k\neq 0} |a_{(k)}^0|^2 \cdot \frac{\tau|\beta_k(h)|^2}{2\tau|\beta_k(h)|^2 + \tau^2|\beta_k(h)|^4}$$
$$\leq \frac{l}{2} \sum_{k\neq 0} |\alpha_{(k)}^0|^2 \leq \frac{1}{2} \|\check{\varphi}_h\|_{2,l_h}^2. \tag{6.52}$$

The inequality (6.52) again gives a uniform bound for u_h^r. In order to obtain uniform bounds for the stronger norms of u_h^r, we must impose more stringent restrictions on the initial function $\varphi(x)$, which we assign to the reader to do independently. In addition, we suggest that he investigate as exercises difference schemes for the problem (6.33), (6.34), (6.2), (6.3) in which (6.33) is replaced by one of the following:

$$u_{\bar{t}}^r - \alpha u_{x\bar{x}}^r - (1-\alpha)u_{x\bar{x}}^{r-1} = 0, \tag{6.53}$$

$$u_t^r - \alpha u_{x\bar{x}}^{r+1} - (1-\alpha)u_{x\bar{x}}^r = 0, \tag{6.54}$$

$$\alpha u_{\bar{t}} + (1-\alpha)u_t - u_{x\bar{x}} = 0, \qquad 0 < \alpha < 1. \tag{6.55}$$

The last of these involves three layers, so that, to use it to calculate u_h^r, it is necessary to bring in some method of determining u_h^r on the layer $t = t_1 = \tau$ (for example, (6.53) or (6.54), $0 \leq \alpha \leq 1$). In the study of this scheme we shall have to encounter the solutions of difference equations of the form

$$a^{(r+1)} + \alpha a^{(r)} + \beta a^{(r-1)} = 0, \qquad r = 1, 2, \ldots, \tag{6.56}$$

where $a^{(0)}$ and $a^{(1)}$ are known, and α and β are given numbers with $\beta \neq 0$. Such equations are called second-order difference equations. In order to study the dependence of $a^{(r)}$ on r, it is helpful to use the known representation of the general solution of (6.56), namely,

$$a^{(r)} = c_1 y_1^r + c_2 y_2^r, \tag{6.57}$$

where the c_k are constants defined by the initial conditions (i.e., $a^{(0)}$ and $a^{(1)}$), where the y_k are the roots of the quadratic

$$y^2 + \alpha y + \beta = 0, \tag{6.58}$$

and where y_k^r is the r-th power of y_k. This can be verified by substituting each of the y_k^r directly into (6.56). The formula (6.57) gives the general solution of (6.56) if $y_1 \neq y_2$. If $y_1 = y_2 = y$, the general solution has the form

$$a^{(r)} = c_1 y^r + c_2(r+1)y^r. \tag{6.59}$$

We have considered the boundary conditions (6.2), (6.3). In a similar way, we can study other boundary conditions for example, the Dirichlet condition (condition (6.29)), where the solution must be sought in the form (6.31). The derivations which were obtained in this subsection under the periodicity conditions remain valid for these boundary conditions.

3. Elementary Equations of Hyperbolic Type. We begin with the first-order equation

$$\frac{\partial u}{\partial t} + a\frac{\partial u}{\partial x} = 0, \qquad a = \text{const} > 0. \tag{6.60}$$

As is the case everywhere in this book, the non-homogeneity of the equation has essentially no influence on either the proofs or the final results. Let us consider for our equation the Cauchy problem

$$u|_{t=0} = \varphi(x). \tag{6.61}$$

The solution of the problem (6.60), (6.61) is given by the formula

$$u(x, t) = \varphi(x - at), \tag{6.62}$$

The lines $x - at = \text{const}$ are the characteristics of (6.60), along which occur the propagation of perturbations (which is quite clear from (6.62)).

It is obvious that, for our problem, the optimal difference scheme is the following:

$$u_t + au_{\bar{x}} = 0, \tag{6.63}$$

$$u_h|_{t=0} = \varphi_h, \tag{6.64}$$

with $h = a\tau$ (as everywhere in this chapter, h is the step-size in x and τ the step-size in t, i.e., the points of the grid have coordinates $x = x_s = sh$, $t = t_s = s\tau$, $s = 0, \pm 1, \ldots$). When we use the finite-difference method of Fourier, we assume that $h = l/m$, where m is odd. The grid function φ_h is taken to be $\varphi(x)$ at points of the grid if $\varphi(x)$ is continuous, and, for the sake of simplicity, we shall assume that $\varphi(x)$ is smooth. The scheme (6.63), (6.64) may be denoted symbolically as ${}_{**}^{\,*}$. Its solution u_h^r, as may be seen easily, is given by

$$u_h^r(x_s) = \varphi_h(x_s - at_r) = \varphi_h(x_{s-r}), \tag{6.65}$$

so that $u_h^r(x_s)$ is equal to the value of the solution $u(x, t)$ of the problem (6.60), (6.61) at the point (x_s, t_r); consequently, in the limit as $h \to 0$, $\tau \to 0$, u_h^r yields $u(x, t)$ if the step-sizes h and τ are related by the equation $h = a\tau$. From "general considerations" (i.e., from the theory of characteristics, which in this case allowed us to write the solution explicitly by (6.62)) it is clear that if h and τ are connected by an equation $\tau/h = c$ and if we take $c > 1/a$, then the solution of the scheme (6.63)–(6.64) will not converge to $u(x, t)$ as $h \to 0$ (for u_h at (x_s, t_r) does not depend on the value of $\varphi(x)$ at the point $x = x_s - at_r$ on which the exact solution $u(x, t)$ depends at the point (x_s, t_r)). From the same arguments it follows that the solution u_h of the difference scheme

$$u_t + au_x = 0, \qquad u_h|_{t=0} = \varphi_h \tag{6.66}$$

cannot converge to $u(x, t)$ for an arbitrary relation between h and τ.

Conversely, the scheme (6.63), (6.64) and the scheme

$$u_t + \frac{a}{2}(u_{\bar{x}} + u_x) = 0, \qquad u_h|_{t=0} = \varphi_h, \tag{6.67}$$

should, it would seem, yield solutions u_h which converge to $u(x, t)$ as $h \to 0$ if τ/h is subject to the condition $\tau/h = \text{const} \le 1/a$. However, a rigorous analysis, which we now carry out, leads to other conclusions which cannot be foreseen if we take into account only the general properties of the solutions of the Cauchy problem we are studying. First, we remark that if the coefficient a in (6.60) is negative, then arguments similar to the earlier ones reject the scheme (6.63), (6.64) (for arbitrary τ/h) and the scheme (6.66) for $\tau/h > 1/|a|$. The schemes (6.66) and (6.7) seem to be good if $\tau/h = \text{const} \le 1/|a|$.

If the coefficient a in (6.60) depends on (x, t) and changes sign in the domain of (x, t), then, simply on the basis of our preliminary analysis, the schemes (6.63), (6.64), and (6.66) are unsuitable. A scheme with a claim to convergence should either take into account the size and the sign of $a(x, t)$ (and consequently be different in different domains of (x, t)) or have this property: The domain of dependence of $u_h^r(x_s)$ on φ_h should contain the domain of dependence of the exact solution $u(x_s, t_r)$ on φ.

The scheme (6.67) has this latter property for arbitrary $a = a(x, t)$ only if τ/h is subject to the condition $\tau/h \le 1/\max|a(x, t)|$.

Let us now see what a rigorous analysis yields for the schemes that we have taken. It will be no loss of generality to assume that $\varphi(x)$ is periodic with period l. Let us consider the case $a = \text{const} > 0$ (the case $a < 0$ may be considered analogously) and also, without loss of generality, let us assume that $a = 1$.

The solution $u(x, t)$ will have the same period in x, namely l, as $\varphi(x)$. In all the schemes we take φ_h at the nodes to be equal to the grid function

$$\check{\varphi}_h(x) = \sum_{k=-(m-1)/2}^{(m-1)/2} \varphi^{(k)}\mu^{(k)}(x), \tag{6.68}$$

where $\varphi^{(k)} = (1/l)\int_0^l \varphi(x)e^{-2\pi ikx/l}\,dx$. We write the solutions u_h of the difference schemes in the form

$$u_h^r = \sum_{k=-(m-1)/2}^{(m-1)/2} a_{(k)}^r \mu_h^{(k)}, \qquad r = 1, 2, \ldots, \left[\frac{T}{\tau}\right]. \tag{6.69}$$

This is possible, for we are seeking u_h which have period l. If u_h is a solution for the scheme (6.63), (6.64), then, substituting (6.69) into (6.63), (6.64) and using (5.8), we obtain

$$a_{(k)t}^r - \bar{\beta}_k(h)a_{(k)}^r = 0. \tag{6.70}$$

It follows from (6.64) that

$$a_{(k)}^0 = \varphi^{(k)}. \tag{6.71}$$

The equations (6.70) allow us to express $a_{(k)}^r$ in terms of $a_{(k)}^0$, namely,

$$a_{(k)}^{r+1} = (1 + \tau\bar{\beta}_k(h))a_{(k)}^r = \cdots = (1 + \tau\bar{\beta}_k(h))^{r+1}a_{(k)}^0, \tag{6.72}$$

where $\beta_k(h) = (e^{2\pi ikh/l} - 1)/h$. Using this expression for $\beta_k(h)$, we calculate

$$\begin{aligned} |a_{(k)}^r|^2 &= |1 + \tau\bar{\beta}_k(h)|^{2r}|a_{(k)}^0|^2 \\ &= \left|1 + 2\frac{\tau}{h}\left(\frac{\tau}{h} - 1\right)\left(1 - \cos\left(\frac{2\pi}{l}kh\right)\right)\right|^r |a_{(k)}^0|^2. \end{aligned} \tag{6.73}$$

If $\tau/h \le 1$, then $0 \le 1 - (4\tau/h)(1 - \tau/h) \le 1$ and, moreover, $|1 + (2\tau/h)(\tau/h - 1)(1 - \cos(2\pi kh/l))| \le 1$. Consequently, for all r and k,

$$|a_{(k)}^r|^2 \le |a_{(k)}^0|^2, \tag{6.74}$$

that is, the scheme (6.63), (6.64) is actually stable for arbitrary τ and h whenever $\tau/h \le 1$. The same analysis shows that, for (6.66), a part of the harmonics (i.e., the terms in the sum (6.69)) will infinitely increase for any magnitude of the ratio τ/h. We shall analyze the scheme (6.67). Let us substitute (6.69) into (6.67) and use (5.3), (5.4) (recall that we have taken $a = 1$). This leads to the equations

$$a_{(k)t}^r + i\alpha_k(h)a_{(k)}^r = 0, \tag{6.75}$$

where $\alpha_k(h) = (1/h)\sin(2\pi kh/l)$. Using (6.75) and (6.71) let us express $a_{(k)}^r$ in terms of $a_{(k)}^0$:

$$a_{(k)}^r = (1 - i\tau\alpha_k(h))a_{(k)}^{r-1} = (1 - i\tau\alpha_k(h))^r a_{(k)}^0, \tag{6.76}$$

whence

$$|a_{(k)}^r|^2 = (1 + \tau^2\alpha_k^2(h))^r|a_{(k)}^0|^2. \tag{6.77}$$

For k close to $m/4$, $\alpha_k(k) = (1/h)\sin(2\pi k/m) \approx 1/h$. For such k the expression $(1 + \tau^2\alpha_k^2(h))^{[t/\tau]}$ increases without bound as $\tau \to 0$ no matter how we fix τ/h and $t > 0$. This means that the scheme (6.67) is unstable: The solutions

u_h calculated from it will increase without bound as $h \to 0$ for a wide class of initial data if τ/h here is equal to any constant whatever.

It is not hard to see from (6.77) that, for stability (and hence for convergence) of the scheme (6.67), it is necessary and sufficient that as $h \to 0$ the ratio τ/h^2 remain bounded—a conclusion that is unexpected for equations of hyperbolic type. The sufficiency of the condition $\tau \le ch^2$ follows from the fact that

$$(1 + \tau^2\alpha_k^2(h))^{t/r} \le \left(1 + \frac{\tau^2}{h^2}\right)^{t/r} \le (1 + c\tau)^{t/r} \le e^{ct}$$

for all k,, τ, and h. The necessity of $\tau \le ch^2$ can be established in the same way it was done in subsection 2 (cf. (6.46)): from the assumption $\tau/h^2 \to \infty$ we can derive the unbounded growth of $|a_{(k)}^r|$ with $k \approx m/4$ when h and τ tend to 0 and $r = [t/\tau]$ (here t is an arbitrary fixed instant of time, $t > 0$).

We shall show schemes for the problem (6.60), (6.61) which converge for any ratio of the step-sizes h and τ. In my paper [1] (from which we take all the material presented in §§5 and 6) it was proved that such schemes converge not only for (6.60) but also for general hyperbolic systems of first order. In connection with the special case considered here, they have the form

$$u_t^r + \frac{a}{4}(u_x^{r+1} + u_{\bar{x}}^{r+1}) + \frac{a}{4}(u_x^r + u_{\bar{x}}^r) = 0, \tag{6.78$_1$}$$

$$u_h|_{t=0} = \check{\varphi}_h. \tag{6.78$_2$}$$

Equation (6.78$_1$) is written out for the layer $t = t_r = r\tau$, where $r = 0, 1, \ldots$, and should be satisfied, as everywhere in this section, at the points $x = x_s = sh$, $s = 0, \pm 1, \ldots$ with regard for the l-periodicity of $\check{\varphi}_h$ and the calculated values of u_h^r. The scheme (6.78$_i$) is implicit: for the determination of u_h^{r+1} on the layer $r + 1$ when the u_h^r are known it is necessary to solve the system of algebraic equations connecting all the values of $u_h^{r+1}(x_s)$ at the points x_s with $s = 1, \ldots, m$. To prove the stability of (6.78$_i$), we represent u_h^k in the form (6.69). Substituting (6.69) into (6.78$_1$) and using (5.3), we obtain the following relations for calculating the coefficients $a_{(k)}^r$:

$$a_{(k)t}^r + \frac{a}{2} i\alpha_k(a_{(k)}^{r+1} + a_{(k)}^r) = 0 \tag{6.79}$$

from which we determine

$$a_{(k)}^{r+1} = \frac{1 - i\dfrac{a\alpha_k}{2}\tau}{1 + i\dfrac{a\alpha_k}{2}\tau}\, a_{(k)}^r.$$

From this it follows that $|a_{(k)}^{r+1}| = |a_{(k)}^0|$ for all k and r. This guarantees that the norms $\|u_h^r\|_{2, l_h} = (\sum_{k=-k'}^{k'} |a_{(k)}^r|^2)^{1/2}$, where $k' = (m - 1)/2$, are independent of the layer r, and also that (6.78$_i$) is stable in the L_2 norm for arbitrary h and τ. We have preserved in (6.78_i) an important property of the problem (6.60), (6.61): The law of conservation of energy, which, for l-periodic solutions, looks like this:

$$\int_0^l u^2(x, t)\, dx = \int_0^l \varphi^2(x)\, dx,$$

and follows from

$$0 = \int_0^l \left(\frac{\partial u}{\partial t} + a\frac{\partial u}{\partial x}\right) u\, dx = \frac{1}{2}\frac{d}{dt}\int_0^l u^2\, dx.$$

In addition to this classical conservation law, we have the equations

$$\int_0^l (D_x^m u(x, t))^2\, dx = \int_0^l (D_x^m \varphi(x))^2\, dx, \tag{6.80}$$

where $D_x^m u$ is the m-th order derivative with respect to x (assuming that $u(x, t)$ possesses these derivatives and is l-periodic). These can be obtained from the equation

$$0 = \int_0^l D_x^m\left(\frac{\partial u}{\partial t} + a\frac{\partial u}{\partial x}\right) D_x^m u\, dx = \frac{1}{2}\frac{d}{dt}\int_0^l (D_x^m u(x, t))^2\, dx.$$

The scheme (6.78_i) preserves all these equalities, or, more precisely, their difference analogues.

Let us consider one more example of an initial-boundary value problem for equations of hyperbolic type, namely,

$$\frac{\partial^2 u}{\partial t^2} - \frac{\partial^2 u}{\partial x^2} = 0, \tag{6.81_1}$$

$$u|_{x=0} = u|_{x=\pi} = 0, \tag{6.81_2}$$

$$u|_{t=0} = \varphi(x), \qquad \left.\frac{\partial u}{\partial t}\right|_{t=0} = \psi(x), \tag{6.81_3}$$

in the domain $\{(x, t): x \in (0, \pi),\ t > 0\}$. Let $\varphi(\cdot) \in \mathring{W}_2^1(0, \pi)$ and $\psi(\cdot) \in L_2(0, \pi)$. We shall analyze the elementary difference scheme first introduced in the paper by R. Courant, K. O. Friedrichs and H. Lewy [1] and studied there in connection with the Cauchy problem. It has the form

$$u_{t\bar{t}} - u_{x\bar{x}} = 0, \tag{6.82_1}$$

$$u_h|_{x=0} = u_h|_{x=\pi} = 0, \tag{6.82_2}$$

$$u_h|_{t=0} = u_h^0, \qquad u_h|_{t=\tau} = u_h^1. \tag{6.82_3}$$

The x-axis is subdivided into intervals of length $h = \pi/m$ by the points $x_s = sh$, $s = 0, \pm 1, \ldots$, and the t-axis into intervals of length τ by the points $t_s = s\tau$, $s = 0, \pm 1, \ldots$ The equation (6.82_1) should be satisfied at the points (x_s, t_r), $s = 1, \ldots, m-1$, $r = 1, 2, \ldots$, (6.82_2) at $x = 0$ and $x = mh = \pi$

for all t_r, $r = 0, 1, \ldots$, and (6.82_3) at the points x_s, $s = 0, 1, \ldots, m$ for $t = 0$ and $t = \tau$. We shall write out the grid functions u_h^0 and u_h^1 below; they are determined by $\varphi(x)$ and $\psi(x)$. It is easy to see that (6.82_i) determines uniquely the solution u_h at the points (x_s, t_r), $s = 0, 1, \ldots, m$, $r = 0, 1, 2, \ldots$, for arbitrary h and τ. The scheme (6.82_i) is explicit; the u_h can be determined layer by layer $t = t_r$, beginning with $r = 2$. It follows from the theory of characteristics for (6.81_1) that convergence for the scheme (6.82_i) is possible only if $\tau/h \le 1$.† We shall show that the scheme (6.82_i) is actually stable for $\tau/h \le 1$ if u_h^0 and u_h^1 are chosen in a suitable way. We shall do this, having in mind stability in the energy metric.

We represent the desired solution in the form (5.23):

$$u_h^r = \sum_{k=1}^{m-1} a_{(k)}^r \chi_h^{(k)}, \qquad (6.83_1)$$

where

$$a_{(k)}^r = \frac{2}{\pi}(u_h^r, \chi_h^{(k)})_{\pi h}, \qquad r = 0, 1, \ldots, \qquad (6.83_2)$$

The superscript r denotes the number of the layer in t $(t = t_r)$, and $\chi_h^{(k)}(x) = \sin(kx)$. Let us substitute (6.83_1) into (6.82_1) and take into account the fact that $\chi_h^{(k)}$, $k = 1, \ldots, m-1$, form a complete set of linearly independent solutions of the spectral problem (5.17), (5.18). This gives us the difference equations

$$a_{(k)t\bar{t}}^r + \lambda_k a_{(k)}^r = 0, \qquad r = 1, 2, \ldots \qquad (6.84)$$

for determining the coefficients $a_{(k)}^r$, where $\lambda_k = ((2/h)\sin(kh/2))^2$, $k = 1, 2, \ldots, m-1$, $h = \pi/m$. All the $a_{(k)}^r$ are uniquely determined by (6.84) if we know $a_{(k)}^r$ for $r = 0$ and $r = 1$. As we indicated at the end of the second subsection (cf. (6.56)–(6.59)), the general solution of

$$a_{t\bar{t}} + \lambda a = 0 \qquad (6.84')$$

has the form

$$a^r = c_1(q_1)^r + c_2(q_2)^r, \qquad r = 0, 1, \ldots, \qquad (6.85)$$

where the c_j are arbitrary constants, where the q_j are the roots of the quadratic equation

$$q^2 - 2(1-\alpha)q + 1 = 0, \qquad \alpha = \tfrac{1}{2}\lambda\tau^2, \qquad (6.86)$$

and where $(q_j)^r$ is the r-th power of q_j. The general solution is given by (6.85) if $q_1 \neq q_2$. By our hypothesis that $\tau/h \le 1$, the numbers $\alpha_k = \frac{1}{2}\lambda_k\tau^2 = 2(\tau/h \sin kh/2)^2$, $k = 1, 2, \ldots, m-1$, $h = \pi/m$, lie in the interval $(0, 2)$,

† Speaking of the need for restrictions of a similar kind, we do not consider limiting processes where the ratios τ/h (or τ/h^2 in the second section) can change as $h \to 0$.

and hence the roots $q_{1,2} = 1 - \alpha \pm i\sqrt{\alpha(2 - a)}$ are distinct with $|q_{1,2}| = 1$. Let us write them in the form $q_1 = e^{i\theta}$, $q_2 = e^{-i\theta}$, where $\sin\theta = \sqrt{\alpha(2-\alpha)}$ and $\cos\theta = 1 - \alpha$. Then the general solution of (6.85) may be written either as the sum

$$a^r = c_1 e^{ir\theta} + c_2 e^{-ir\theta},$$

or as the sum

$$a^r = \hat{c}_1 \cos(r\theta) + \hat{c}_2 \sin(r\theta) \tag{6.87}$$

with arbitrary constants $\hat{c}_1, \hat{c}_2$. Since all the quantities given in the problem are real, and since the system $\chi_h^{(k)}$ into which we expanded u_h is real, the coefficients $a_{(k)}$ are also real, so that $\hat{c}_1$ and $\hat{c}_2$ may be taken real. We return now to the equations (6.84), whose general solutions may be written as

$$a_{(k)}^r = \hat{c}_{1,k} \cos(r\theta_k) + \hat{c}_{2,k} \sin(r\theta_k), \qquad r = 0, 1, \ldots, \tag{6.88}$$

where the θ_k are related to $\alpha_k = \frac{1}{2}\lambda_k \tau^2$ by

$$\sin\theta_k = \sqrt{\alpha_k(2 - \alpha_k)}, \qquad \cos\theta_k = 1 - \alpha_k, \quad \alpha_k \in (0, 2). \tag{6.89}$$

The representation (6.83_1) may be rewritten in the form

$$u_h^r = \sum_{k=1}^{m-1} [\hat{c}_{1,k} \cos(r\theta_k) + \hat{c}_{2,k} \sin(r\theta_k)]\chi_h^{(k)}, \tag{6.90}$$

where the coefficients may be uniquely determined in terms of the functions u_h^0 and u_h^1 in (6.82_3). However, we have not yet indicated the rules by which u_h^0 and u_h^1 are to be calculated in terms of $\varphi(x)$ and $\psi(x)$. The previous considerations suggest this rule for u_h^0:

$$u_h^0 = \sum_{k=1}^{m-1} \varphi_{(k)} \chi_h^{(k)} \equiv \hat{\varphi}_h, \tag{6.91}$$

where

$$\varphi_{(k)} = \frac{2}{\pi} \int_0^\pi \varphi(x) \sin kx \, dx. \tag{6.92}$$

As u_h^1 one would usually take the sum $\hat{\varphi}_h + \tau\hat{\psi}_h$, where $\hat{\psi}_h$ is calculated from $\psi(x)$ in accordance with (6.91), (6.92). However, as a calculation of the "difference energy norm" for u_h shows, this sum will not be uniformly bounded (in h) if $\tau/h = 1$. In order to obtain this, it is necessary to change somewhat the expression for u_h^1, namely, to choose u_h^1 so that formula (6.90) for the solution of (6.82_i) has the form

$$u_h^r = \sum_{k=1}^{m-1} \left[\varphi_{(k)} \cos(r\theta_k) + \frac{\tau\psi_{(k)}}{2\sin\dfrac{\theta_k}{2}} \sin(r\theta_k) \right] \chi_h^{(k)}. \tag{6.93}$$

It follows from (6.93) that u_h^0 corresponds to (6.91) and that

$$u_h^1 = \sum_{k=1}^{m=1} \left[\varphi_{(k)} \cos \theta_k + \tau \psi_{(k)} \cos \frac{\theta_k}{2} \right] \chi_h^{(k)}. \tag{6.94}$$

The quotient

$$\frac{u_h^1(x) - u_h^0(x)}{\tau} = \sum_{k=1}^{m-1} \left[-\varphi_{(k)} \frac{\alpha_k}{\tau} + \psi_{(k)} \cos \frac{\theta_k}{2} \right] \chi_h^{(k)}(x) \tag{6.95}$$

approximates $\psi(x)$ in the weak norm of $L_2(0, \pi)$. In fact, any term of (6.95) with fixed index k tends to $\psi_{(k)} \sin kx$ as $h \to 0$ (with $\tau/h \leq 1$), for $\lambda_k \to k^2$, $\alpha_k/\tau = \lambda_k \tau/2 \to 0$, and $\cos(\theta_k)/2 \to 1$, and the norms $\tau^{-1}\|u_h^1(\cdot) - u_h^0(\cdot)\|_{2,\pi_h}$ for (6.95) are uniformly bounded in h. To verify this last part, we use the orthogonality of the $\chi_h^{(k)}$ in the sense of (5.23) and (5.24) and the fact that $\lambda_k \leq k^2$:

$$\begin{aligned}
\tau^{-2}\|u_h^1 - u_h^0\|_{2,\pi_h}^2 &= \frac{\pi}{2} \sum_{k=1}^{m-1} \left[-\varphi_{(k)} \frac{\alpha_k}{\tau} + \psi_{(k)} \cos \frac{\theta_k}{2} \right]^2 \\
&\leq \pi \sum_{k=1}^{m-1} \left[\frac{1}{4} \varphi_{(k)}^2 \lambda_k^2 \tau^2 + \psi_{(k)}^2 \right] \leq \pi \sum_{k=1}^{m-1} [\varphi_{(k)}^2 \lambda_k + \psi_{(k)}^2] \\
&= 2(\hat{\varphi}_x, \hat{\varphi}_x)_{\pi_h} + 2(\hat{\psi}_h, \hat{\psi}_h)_{\pi_h} \\
&\leq \pi \sum_{k=1}^{m-1} [\varphi_{(k)}^2 k^2 + \psi_{(k)}^2] \leq 2 \left\| \frac{\partial \varphi}{\partial x} \right\|_{2,\hat{\pi}}^2 + 2\|\psi\|_{2,\hat{\pi}}^2.
\end{aligned} \tag{6.96}$$

From these two facts follows the weak convergence in $L_2(0, \pi)$ of the functions $[u_h^1(x) - u_h^0(x)]/\tau$ to $\psi(x)$. Because of the construction (cf. (6.91), (6.92)) and because of the fact that $\varphi(\cdot)$ is in $\mathring{W}_2^1(0, \pi)$ the functions $u_h^0 = \hat{\varphi}_h(x)$ converge to $\varphi(x)$ in the norm of $W_2^1(0, \pi)$, where their grid norms $[(\hat{\varphi}_h, \hat{\varphi}_h)_{\pi_h} + (\hat{\varphi}_x, \hat{\varphi}_x)_{\pi_h}]^{1/2}$ are uniformly bounded (cf. (6.96)). Because of this and (6.96), the "difference energy norms"

$$[(u_x^r, u_x^r)_{\pi_h} + (u_{\bar{t}}^r, u_{\bar{t}}^r)_{\pi_h}]^{1/2} \equiv j_h(r)$$

are uniformly bounded for the solutions u_h^r defined by (6.93).

Actually, because of (5.24) and the equations connecting α_k, λ_k, θ_k, and τ, we have

$$\begin{aligned}
(u_x^r, u_x^r)_{\pi_h} &= \frac{\pi}{2} \sum_{k=1}^{m-1} \left[\varphi_{(k)} \cos(r\theta_k) + \frac{\tau \psi_{(k)}}{2 \sin \frac{\theta_k}{2}} \sin(r\theta_k) \right]^2 \lambda_k \\
&\leq \pi \sum_{k=1}^{m-1} \left[\varphi_{(k)}^2 \lambda_k + \tfrac{1}{4} \lambda_k \tau^2 \frac{1}{\sin^2 \frac{\theta_k}{2}} \psi_{(k)}^2 \right] = \pi \sum_{k=1}^{m-1} [\varphi_{(k)}^2 \lambda_k + \psi_{(k)}^2]
\end{aligned} \tag{6.97}$$

and

$$(u_{\bar{t}}^r, u_{\bar{t}}^r)_{\pi_h} = \frac{\pi}{2}\sum_{k=1}^{m-1}\left[\varphi_{(k)}\frac{\cos r\theta_k - \cos(r-1)\theta_k}{\tau} + \frac{\tau\psi_{(k)}}{2\sin\frac{\theta_k}{2}}\frac{\sin r\theta_k - \sin(r-1)\theta_k}{\tau}\right]^2$$

$$= \frac{\pi}{2}\sum_{k=1}^{m-1}\left[-\varphi_{(k)}\frac{2}{\tau}\sin\frac{\theta_k}{2}\sin\left(r-\frac{1}{2}\right)\theta_k + \psi_{(k)}\cos\left(r-\frac{1}{2}\right)\theta_k\right]^2$$

$$\leq \pi\sum_{k=1}^{m-1}[\varphi_{(k)}^2\lambda_k + \psi_{(k)}^2]. \tag{6.98}$$

It is clear from (6.97), (6.98), and (6.96) that $j_h^2(r)$ does not exceed $4\|\partial\varphi/\partial x\|_{2,\hat{\pi}}^2 + 4\|\psi\|_{2,\hat{\pi}}^2$ for all τ and h if $\tau/h \leq 1$, i.e., we have proved the stability of the difference scheme (6.82$_i$) in the "energy norm" if $\tau/h \leq 1$ and if u_h^0 and u_h^1 are computed according to (6.91) and (6.94). If we let h tend to zero, it is not difficult to verify that the limit of u_h^r yields a solution of the problem (6.81$_i$) having finite energy norm.

We may investigate analogously other difference schemes for the problem (6.81$_i$), for example, implicit schemes in which the difference equation (6.82$_1$) is replaced by the equation

$$u_{t\bar{t}}^r - [\gamma u^{r+1} + (1-2\gamma)u^r + \gamma u^{r-1}]_{x\bar{x}} = 0, \tag{6.99}$$

where γ is a positive constant.

§7. The Dirichlet Problem for General Elliptic Equations of Second Order

We use the finite-difference method for solving the Dirichlet problem

$$\mathscr{L}u \equiv \frac{\partial}{\partial x_i}\left(a_{ij}(x)\frac{\partial u}{\partial x_j} + a_i(x)u\right) + b_i(x)\frac{\partial u}{\partial x_i} + a(x)u = f(x) + \frac{\partial f_i(x)}{\partial x_i}, \quad u|_S = 0 \tag{7.1}$$

in an arbitrary bounded domain $\Omega \subset R^n$. We assume that the coefficients of the equation are bounded, measurable functions and that they satisfy the condition

$$a_{ij}\xi_i\xi_j + (a_i - b_i)\xi_i\xi_0 - a\xi_0^2 \geq \nu_1\sum_{i=1}^n \xi_i^2, \tag{7.2}$$

where ν_1 is a positive constant and $\xi_0, \xi_1, \ldots, \xi_n$ are arbitrary real numbers. This condition can be replaced by more weakened conditions (2.3), (2.11), (2.12) of Chapter II, or even by the condition of ellipticity and the assumption that (7.1) is uniquely solvable. We shall not do this, however, for the first of these relaxations can be done by the reader for himself, while the second, on the other hand, requires some additional considerations which would divert us from the basic material.

Concerning f and f_i, we shall assume that they are elements of $L_2(\Omega)$.

Let us subdivide all of R^n by elementary cells $\omega_{(kh)}$ by the planes $x_k = mh_k$, $k = 1, \ldots, n$, $h_k > 0$, and use the notation in §§2 and 3 of this chapter. We shall construct stable difference schemes for the problem (7.1), following the principle described in the introductory paragraph. As a basis we take the space $W_2^1(\Omega)$ and the energy bound

$$\|u\|_{2,\Omega}^{(1)} \leq c_1(\|f\|_{2,\Omega} + \|\mathbf{f}\|_{2,\Omega}) \tag{7.3}$$

for the generalized solution $u(x)$ of problem (7.1) in $W_2^1(\Omega)$. This bound guarantees the stability of the problem (7.1) with respect to the variations of f and f_i, and, in particular, the validity of the uniqueness theorem for problem (7.1) in the class of generalized solutions in $W_2^1(\Omega)$. We recall that a generalized solution of (7.1) in $W_2^1(\Omega)$ is defined as an element of $\mathring{W}_2^1(\Omega)$ satisfying the integral identity

$$\begin{aligned} \mathscr{L}(u, \eta) &\equiv \int_\Omega \left[\left(a_{ij}\frac{\partial u}{\partial x_j} + a_i u\right)\frac{\partial \eta}{\partial x_i} - \left(b_i\frac{\partial u}{\partial x_i} + au\right)\eta\right] dx \\ &= \int_\Omega \left(f_i\frac{\partial \eta}{\partial x_i} - f\eta\right) dx \end{aligned} \tag{7.4}$$

for all $\eta \in \mathring{W}_2^1(\Omega)$. The bound (7.3) is easily derived from (7.4) if we set $\eta = u$ in (7.4) and then use (7.2). What we want to do is construct a difference scheme for which the difference analogue of (7.3) holds.

For this we start from (7.4) and construct an approximation to it. The integrals over Ω are replaced by sums of integrals over the cells $\omega_{(kh)}$ contained in Ω and, within each cell, u and η are replaced by piecewise-constant functions $\tilde{u}_h$ and $\tilde{\eta}_h$; the derivatives $\partial u/\partial x_i$ and $\partial \eta/\partial x_i$ are replaced by piecewise-constant interpolations of some kind of difference quotients of the same type which approximate them; for example, u_{x_i} and η_{x_i}. The requirement that u and η be in $\mathring{W}_2^1(\Omega)$ is replaced naturally by the requirement that the grid functions u_h and η_h vanish on the boundary of the domain where they are defined, i.e., on S_h. Thus, (7.4) is replaced by

$$\begin{aligned} \mathscr{L}u(u_h, \eta_h) &\equiv \int_{\Omega_h} [(a_{ij}\tilde{u}_{x_j} + a_i\tilde{u}_h)\tilde{\eta}_{x_i} - (b_i\tilde{u}_{x_i} + a\tilde{u}_h)\tilde{\eta}_h]\, dx \\ &= \int_{\Omega_h} (f_i\tilde{\eta}_{x_i} - f\tilde{\eta}_h)\, dx, \end{aligned} \tag{7.5}$$

or, what is the same thing, by

$$\mathscr{L}_h(u_h, \eta_h) = \Delta_h \sum_{\Omega_h^+} [(a_{ijh} u_{x_j} + a_{ih} u_h)\eta_{x_i} - (b_{ih} u_{x_i} + a_h u_h)\eta_h]$$

$$= \Delta_h \sum_{\Omega_h^+} (f_{ih}\eta_{x_i} - f_h \eta_h), \tag{7.6}$$

where Ω_h^+ is the set of vertices $x = (kh)$ of those cells $\omega_{(kh)}$ which belong to Ω (we recall that $\omega_{(kh)} = \{x: k_i h_i < x_i < (k_i + 1)h_i\}$ and each cell $\omega_{(kh)}$ is associated with only one of its vertices $x = (kh)$, so that the set Ω_h^+ is part of the set $\overline{\Omega}_h$). The sums $\Delta_h \sum \cdots$ can be extended to all $\overline{\Omega}_h$ if we stipulate that u_h and η_h are defined to be zero outside $\overline{\Omega}_h$ (they also vanish on S_h) and that all the remaining functions $a_{ijh}, \ldots, f_h$ are set equal to zero at those points of S_h where they are not defined by the method described below.

The grid functions a_{ijh} in (7.6) are equal at the node $(kh) = (k_1 h_1, \ldots, k_n h_n)$ to the mean $\Delta_h^{-1} \int_{\omega_{(kh)}} a_{ij}(x)\, dx$ taken over the cell $\omega_{(kh)}$. This will be the case for all coefficients and free terms in (7.1).

The "drift" of the known functions that we take (i.e., of the coefficients and the free terms) on the grid is far from being uniquely realized. For example, if these functions are continuous, then, for computational purposes, it is more convenient simply to take their values at the nodes of the grid. But, if they are discontinuous, we have to take one or another of their averages. To be definite, we shall deal with those described above.

The identity (7.6) must be satisfied by all grid functions η_h defined on $\overline{\Omega}_h$ and vanishing on S_h and outside $\overline{\Omega}_h$, so that (7.6) contains as many independent, freely varying quantities as there are "interior" points of the lattice $\overline{\Omega}_h$, i.e., as there are vertices of $\Omega_h = \overline{\Omega}_h \backslash S_h$. We transform (7.6) by our formula (2.11) for "summation by parts" into the form

$$\Delta_h \sum_{\Omega_h} [(a_{ijh} u_{x_j} + a_{ih} u_h)_{\bar{x}_i} + (b_{ih} u_{x_i} + a_h u_h)]\eta_h = \Delta_h \sum_{\Omega_h} (f_{ih\bar{x}_i} + f_h)\eta_h \tag{7.7}$$

and then use the fact that η_h is arbitrary at the points of Ω_h. This leads us to the following system of difference equations:

$$\mathscr{L}_h u_h \equiv (a_{ijh} u_{x_j} + a_{ih} u_h)_{\bar{x}_i} + b_{ih} u_{x_i} + a_h u_h = f_{ih\bar{x}_i} + f_h, \tag{7.8}$$

which should be satisfied at points of the grid Ω_h. We must couple this with the boundary condition

$$u_h|_{S_h} = 0. \tag{7.9}$$

Equations (7.8), (7.9), or, what is the same thing, (7.6), (7.9), give us the desired difference scheme. These equations form a linear algebraic system containing as many unknowns (the values of u_h) as there are equations. The unique solvability of this system, as well as the stability of the scheme, follows from *a priori* bounds for the u_h, which we now proceed to obtain. For this we set

η_h equal to u_h in (7.6), which we can do, and take into account that, by (7.2) and the structure of the grid functions $a_{ijh}, \ldots$ at the points of Ω_h^+,

$$a_{ijh}u_{x_j}u_{x_i} + (a_{ih} - b_{ih})u_{x_i}u_h - a_h u_h^2 \geq \nu_1 \sum_{i=1}^{n} u_{x_i}^2 \equiv \nu_1 u_x^2.$$

Hence

$$\nu_1 \Delta_h \sum_{\Omega_h^+} u_x^2 \leq \mathscr{L}_h(u_h, u_h) = \Delta_h \sum_{\Omega_h^+} (f_{ih}u_{x_i} - f_h u_h)$$

$$\leq \|\mathbf{f}_h\|_{2,\Omega_h^+} \|u_x\|_{2,\Omega_h^+} + \|f_h\|_{2,\Omega_h^+} \|u_h\|_{2,\Omega_h^+}, \tag{7.10}$$

where

$$\|\mathbf{f}_h\|_{2,\Omega_h^+} = \left[\Delta_h \sum_{\Omega_h^+} \sum_{i=1}^{n} f_{ih}^2\right]^{1/2}, \qquad \|v_h\|_{2,\Omega_h^+} = \left[\Delta_h \sum_{\Omega_h^+} v_h^2\right]^{1/2}.$$

On the other hand, for any grid function u_h vanishing on S_h, we have

$$\Delta_h \sum_{\Omega_h^+} u_h^2 \leq c^2 \Delta_h \sum_{\Omega_h^+} u_x^2, \tag{7.11}$$

which is the difference analogue of the Poincaré–Friedrichs inequality (cf. (3.19)). If the system (7.8), (7.9) is homogeneous, i.e., if the right-hand side of (7.8) is zero at all points of Ω_h, then, as is obvious from (7.10) and (7.11), its only solution is $u_h \equiv 0$. In other words, there can be no more than one solution of the system (7.8), (7.9). But then, by a theorem in linear algebra, this system has a solution u_h for any set of free terms, i.e., for arbitrary f_{ih} and f_h, and the bounds (7.10), (7.11) hold for u_h. From these bounds we have

$$\nu_1 \Delta_h \sum_{\Omega_h^+} u_x^2 \leq (\|\mathbf{f}_h\|_{2,\Omega_h^+} + c\|f_h\|_{2,\Omega_h^+})\|u_x\|_{2,\Omega_h^+},$$

which implies that

$$\|u_x\|_{2,\Omega_h^+} \leq \frac{1}{\nu_1}(\|\mathbf{f}_h\|_{2,\Omega_h^+} + c\|f_h\|_{2,\Omega_h^+}). \tag{7.12}$$

It follows from the definition of f_h and Cauchy's inequality that

$$\|f_h\|_{2,\Omega_h^+}^2 = \Delta_h \sum_{\Omega_h^+} f_h^2 = \Delta_h \sum_{\omega(kh) \in \Omega_h} \left(\frac{1}{\Delta_h} \int_{\omega(kh)} f\,dx\right)^2$$

$$\leq \sum_{\omega(kh) \in \Omega_h} \int_{\omega(kh)} f^2\,dx = \int_{\Omega_h} f^2\,dx \leq \|f\|_{2,\Omega}^2, \tag{7.13}$$

and, analogously,

$$\|\mathbf{f}_h\|_{2,\Omega_h^+}^2 \leq \|\mathbf{f}\|_{2,\Omega}^2 \equiv \int_\Omega \sum_{i=1}^{n} f_i^2\,dx. \tag{7.14}$$

By these inequalities, (7.12) and (7.11) guarantee the stability of the scheme we constructed in the grid norm, which corresponds to the energy norm.

They form the difference analogue of the inequality (7.3). Starting from the same principle of approximating the identity (7.4), we can construct other difference schemes possessing stability in the form (7.11), (7.12). They differ from (7.5) by the way we replace the derivatives $\partial/\partial x_i$ by the difference quotients and by the way we form the grid functions $a_{ijh}, \ldots, f_h$ from the functions $a_{ij}, \ldots, f$. For example, we can replace $\tilde{u}_{x_j}$ and $\tilde{\eta}_{x_i}$ in (7.5) by $\tilde{u}_{\bar{x}_j}$ and $\tilde{\eta}_{\bar{x}_i}$.

We go now to the proof of convergence of the scheme (7.8), (7.9) when all the h_i, $i = 1, \ldots, n$, tend to zero, that is, to the proof of the fact that the interpolations of the solutions $\{u_h\}$ of the systems (7.8), (7.9) described in §3 of this chapter converge to a function $u(x)$ which is the generalized solution of (7.1).

In this connection we shall not use the fact that the generalized solution $u(x)$ of (7.1) exists, but rather we shall prove the result by using the approximate solutions $\{u_h\}$. By §3 it follows from (7.11), (7.12), and (7.9) that the interpolations $\{u'_h(x)\}$ form a bounded set in $\overset{\circ}{W}{}_2^1(\Omega)$ (here we assume that the u_h vanish outside $\bar{\Omega}_h$). We select an arbitrary subsequence $\{u'_{h^\alpha}(x)\}$ which is weakly convergent in $L_2(\Omega)$, along with the derivatives $\{\partial u'_{h^\alpha}/\partial x_i\}$, $i = 1, \ldots, n$, to some function $u(x) \in \overset{\circ}{W}{}_2^1(\Omega)$, and its derivatives $\partial u/\partial x_i$ correspondingly, and we shall show that this limit function is a generalized solution in $W_2^1(\Omega)$ of the problem (7.1). Since (7.1) under the condition (7.2) can have at most one generalized solution in $W_2^1(\Omega)$, we shall have proved the existence of a unique limit point in $\overset{\circ}{W}{}_2^1(\Omega)$ for every set $\{u'_h(x)\}$, or, what is the same thing, the weak convergence in $L_2(\Omega)$ of $\{u'_h(x)\}$ and $\{\partial u'_h/\partial x_i\}$ to $u \in \overset{\circ}{W}{}_2^1(\Omega)$ and $\partial u/\partial x_i$, respectively. By Theorem 6.1, Chapter I, the functions $u'_h(x)$ themselves will converge strongly in $L_2(\Omega)$ to $u(x)$. Hence, it is necessary to show that the limit function $u(x)$ for $\{u'_{h^\alpha}(x)\}$ is the generalized solution of (7.1), and for this we use the assertion proved in §3 of this chapter. Lemma 3.1 and Remark 3.1 to Theorem 3.1 guarantees, for the sequence $\{h^\alpha\}$, the strong convergence in $L_2(\Omega)$ to $u(x)$ of the interpolations $\{\tilde{u}_{h^\alpha}(x)\}$ and the weak convergence in $L_2(\Omega)$ of the functions $\{\widetilde{u_{x_i}}(x)\}$ to $\partial u/\partial x_i$. Hence we can take the limit in (7.5) with $\{h^\alpha\}$ provided that we take as η_h the values on the grid $\bar{\Omega}_h$ of an arbitrary but fixed function $\eta(x)$ in $\dot{C}^\infty(\Omega)$ (because for all sufficiently small h_i the functions η_h vanish on S_h and outside $\bar{\Omega}_h$, and, as $h_i \to 0$, the interpolations formed for η, namely $\tilde{\eta}_h(x)$ and $\widetilde{\eta_{x_i}}(x)$, will converge uniformly on $\bar{\Omega}$ to $\eta(x)$ and $\partial\eta/\partial x_i$). The domain of integration Ω_h in (7.5) can be replaced by the domain Ω not depending on h, for $\tilde{\eta}_h$ and $\widetilde{\eta_{x_i}}$ vanish outside $\bar{\Omega}_h$. As a result of the limit process in (7.5) with $\{h^\alpha\}$, we have proved that the limit function $u \in \overset{\circ}{W}{}_2^1(\Omega)$ satisfies (7.4) for the function $\eta(x) \in \dot{C}^\infty(\Omega)$ that we took. Hence it follows that $u(x)$ is the generalized solution of (7.1) in $W_2^1(\Omega)$, for η is an arbitrary element of $\dot{C}^\infty(\Omega)$, and $\dot{C}^\infty(\Omega)$ is dense in $\overset{\circ}{W}{}_2^1(\Omega)$. Thus, we have proved

Theorem 7.1. *The difference scheme (7.8), (7.9) is stable in the energy norm, and the interpolations $\{u'_h(x)\}$ formed by means of the solutions $\{u_h\}$ of the*

system (7.8), (7.9) converge strongly in $L_2(\Omega)$ to the generalized solution $u \in \overset{\circ}{W}{}_2^1(\Omega)$ of the problem (7.1), and their derivatives $\{\partial u_h'/\partial x_i\}$ converge weakly in $L_2(\Omega)$ to $\partial u/\partial x_i$, $i = 1, \ldots, n$. Here it is assumed that the coefficients in (7.1) are bounded, measurable functions satisfying (7.2), that f and f_i are in $L_2(\Omega)$, and that Ω is a bounded domain.

It can be shown that the functions $\{u_h'(x)\}$ actually converge strongly to $u(x)$ in $\overset{\circ}{W}{}_2^1(\Omega)$, and we shall do this in the first subsection of §11.

§8. The Neumann Problem and Third Boundary Value Problem for Elliptic Equations

Let us consider the boundary value problem

$$\mathscr{L}u \equiv \frac{\partial}{\partial x_i}\left(a_{ij}(x)\frac{\partial u}{\partial x_j}\right) + b_i(x)\frac{\partial i}{\partial x_i} + a(x)u = f(x),$$

$$\left.\frac{\partial u}{\partial N} + \sigma(s)u\right|_S = 0, \tag{8.1}$$

in a domain Ω with boundary $S \equiv \partial\Omega$ satisfying the conditions of §5, Chapter II, where $\partial u/\partial N = a_{ij}\,\partial u/\partial x_j \cos(\mathbf{n}, \mathbf{x}_i)$. We shall assume that $\sigma(s)$ and the coefficients of $\mathscr{L}$ are bounded functions, and that $f \in L_2(\Omega)$. In addition, let

$$a_{ij}(x)\xi_i\xi_j - b_i(x)\xi_i\xi_0 - a(x)\xi_0^2 \geq \nu_1 \sum_{i=1}^{n} \xi_i^2 + \lambda\xi_0^2. \tag{8.2}$$

for arbitrary real $\xi_0, \xi_1, \ldots, \xi_n$ with positive constants ν_1 and λ. If $\sigma(s) \geq 0$, then ν_1 and λ can be arbitrary constants greater than zero. If $\sigma(s)$ is non-positive but bounded on S, then λ in (8.2) should be sufficiently large. These restrictions on $\sigma(s)$ and the coefficients of $\mathscr{L}$ should guarantee, for all $u \in W_2^1(\Omega)$, that

$$\mathscr{L}(u, u) + \int_S \sigma u^2\, ds \geq \nu_2(\|u\|_{2,\Omega}^{(1)})^2, \qquad \nu_2 > 0, \tag{8.3}$$

where

$$\mathscr{L}(u, \eta) \equiv \int_\Omega \left(a_{ij}\frac{\partial u}{\partial x_j}\frac{\partial \eta}{\partial x_i} - b_i\frac{\partial u}{\partial x_i}\eta - au\eta\right) dx, \tag{8.4}$$

which ensures the uniqueness of the solution of (8.1). These conditions, which constitute a stronger requirement than the unique solvability of (8.1), allow us to construct for (8.1) stable difference schemes for which the solutions u_h can be calculated uniquely for arbitrary step-sizes h_i.

The generalized solution $u(x) \in W_2^1(\Omega)$ of (8.1) as defined in §5, Chapter II, is an element of $W_2^1(\Omega)$ satisfying the identity

$$\mathscr{L}(u, \eta) + \int_S \sigma u \eta \, ds = -(f, \eta) \tag{8.5}$$

for all $\eta \in W_2^1(\Omega)$. If we take $\eta = u$ in (8.5), then

$$\mathscr{L}(u, u) + \int_S \sigma u^2 \, ds = -(f, u), \tag{8.6}$$

from which it follows, by virtue of (8.2), that

$$\int_\Omega (\nu_1 u_x^2 + \lambda u^2) \, dx + \int_S \sigma u^2 \, ds \leq -(f, u) \leq \|f\|_{2,\Omega} \cdot \|u\|_{2,\Omega}. \tag{8.7}$$

If $\sigma \geq 0$, then we obtain from (8.7) *a priori* bounds for the solution $u(x)$:

$$\|u\|_{2,\Omega} \leq \frac{1}{\lambda} \|f\|_{2,\Omega}, \qquad \|u_x\|_{2,\Omega}^2 \leq \frac{1}{\lambda \nu_1} \|f\|_{2,\Omega}^2. \tag{8.8}$$

Now if $\sigma \geq -c$, where $c > 0$, then we get for $u(x)$ the bound $\|u\|_{2,\Omega}^{(1)} \leq c_1 \|f\|_{2,\Omega}$ if we assume that λ is large enough (cf. §5, Chapter II). We limit ourselves to considering the case $\sigma \geq 0$, the second case being handled analogously. Thus, let $\sigma \geq 0$ and let ν_1 and λ in (8.2) be greater than zero. We subdivide R^n into cells $\omega_{(kh)}$ by the planes $x_i = k_i h_i$, $h_i > 0$, $k_i = 0, \pm 1, \pm 2, \ldots$, and we denote by $\overline{\Omega_h^*}$ the totality of cells $\bar{\omega}_{(kh)}$ covering $\bar{\Omega}$ and meeting Ω. The boundary of the closed region $\overline{\Omega_h^*}$ will be denoted by S_h^*, and $\overline{\Omega_h^*} \setminus S_h^* = \Omega_h^*$. We shall use the symbols $\overline{\Omega_h^*}$, S_h^*, and Ω_h^* to designate the set of vertices of our subdivision belonging to $\overline{\Omega_h^*}$, S_h^*, and Ω_h^*.

We shall assume that the coefficients of (8.1) and the function $f(x)$ have been extended from Ω to a somewhat larger domain and preserve the properties mentioned at the beginning of the section. We shall construct a stable finite-difference scheme starting from an approximation of (8.5). We replace the integral over Ω by sums over the cells $\omega_{(kh)} \subset \Omega_h^*$, the functions $u(x)$ and $\eta(x)$ by piecewise-constant interpolations $\tilde{u}_h(x)$ and $\tilde{\eta}_h(x)$, and the derivatives $\partial u/\partial x_i$ and $\partial \eta/\partial x_i$ by piecewise-constant functions $\tilde{u}_{x_i}$ and $\tilde{\eta}_{x_i}$ (instead of these we could take, for example, $\tilde{u}_{\bar{x}_i}$ and $\tilde{\eta}_{\bar{x}_i}$); the known functions we leave unchanged.

To the left-hand side of the summation identity, which we want to replace by the left-hand side of (8.5), we add the sum

$$J_h(u_h, \eta_h) = \Delta_h \sum_{S_h^*} \left[\theta u_h \eta_h + \sum_{i=1}^n \theta_i u_{x_i} \eta_{x_i} \right], \tag{8.9}$$

in which θ and θ_i are functions on the set of vertices in S_h^* assuming the values 0 or 1 and are defined by a rule that we shall explain a bit later. Thus, we replace (8.5) by the following identity:

$$\int_{\Omega_h^*} [a_{ij}\tilde{u}_{x_j}\tilde{\eta}_{x_i} - b_i\tilde{u}_{x_i}\tilde{\eta}_h - a\tilde{u}_h\tilde{\eta}_h]\,dx + \int_S \sigma\tilde{u}_h\tilde{\eta}_h\,ds + J_h(u_h, \eta_h) = -\int_{\Omega_h^*} f\tilde{\eta}_h\,dx. \tag{8.10}$$

which can be rewritten in the form

$$\Delta_h \sum_{\Omega_h^{*+}} [a_{ijh}u_{x_j}\eta_{x_i} - b_{ih}u_{x_i}\eta_h - a_h u_h \eta_h] + \sum_{\hat{S}_h} \sigma_h u_h \eta_h + J_h(u_h, \eta_h) = -\Delta_h \sum_{\Omega_h^{*+}} f_h \eta_h, \tag{8.11}$$

where Ω_h^{*+} is the set of points of $\overline{\Omega_h^*}$ which are vertices, with coordinates (kh), of the cells $\omega_{(kh)} \subset \overline{\Omega_h^*}$ (a similar symbol Ω_h^+, defined for $\overline{\Omega}_h$, was used in §7; again we emphasize that we associate one vertex to each cell $\omega_{(kh)} \subset \Omega_h^*: x = (kh)$). The sum $\sum_{\hat{S}_h} \sigma_h u_h \eta_h$ is another way of writing the integral $\int_S \sigma\tilde{u}_h\tilde{\eta}_h\,ds$, stressing the fact that this integral depends on the values of u_h and η_h only on some set of points of the grid; this set is denoted by $\hat{S}_h$. The grid functions $a_{ijh}, \ldots, f_h$ have been computed from $a_{ij}, \ldots, f$ as the mean over the corresponding cell $\omega_{(kh)}$, namely, $a_{ijh}|_{x=(kh)} = \Delta_h^{-1} \int_{\omega(kh)} a_{ij}(x)\,dx$ and similarly for f and the remaining coefficients of $\mathscr{L}$. The sum $\Delta_h \sum_{\Omega_h^{*+}} u_h \eta_h$ does not contain the values of u_h (nor, indeed, of η_h) at all points of $\overline{\Omega_h^*}$. These missing values of u_h we add in the form of the sum $\Delta_h \sum_{S_h^*} \theta u_h \eta_h$, which obviously contains a part of the vertices belonging to S_h^*. Moreover, the sum $\Delta_h \sum_{\Omega_h^{*+}} a_{ijh}u_{x_j}\eta_{x_i}$ in (8.11) does not contain all the difference quotients u_{x_i} (nor η_{x_i}) which can be formed from the values of u_h on $\overline{\Omega_h^*}$. These missing quotients are added in the form of the sum $\Delta_h \sum_{S_h^*} \sum_{i=1}^n \theta_i u_{x_i} \eta_{x_i}$. They are also formed from the values of u_h on part of the vertices of S_h^*. Now all terms appearing in the identity (8.11) have been described; they have been formed by means of the values of u_h and η_h at all the vertices of $\overline{\Omega_h^*}$. The values of η_h at these vertices of $\overline{\Omega_h^*}$ vary arbitrarily. If in (8.11) we write out explicitly all the u_{x_i} and η_{x_i} in the form of difference quotients of u_h and η_h and then represent the left-hand and right-hand sides of (8.11) in the form

$$\sum_{\overline{\Omega_h^*}} \mathscr{L}_h(u_h) \cdot \eta_h = \sum_{\overline{\Omega_h^*}} \mathscr{F}_h(f_h)\eta_h, \tag{8.12}$$

then, because of the independence of the values of η_h at the points of $\overline{\Omega_h^*}$, (8.12) is equivalent to the following system of equations:

$$\mathscr{L}_h(u_h) = \mathscr{F}_h(f_h), \qquad (kh) \in \overline{\Omega_h^*}. \tag{8.13}$$

The system (8.13) contains as many equations as there are vertices in the grid $\overline{\Omega_h^*}$, and in the system there appear as many unknowns as there are values of u_h at the vertices of $\overline{\Omega_h^*}$. We shall show that u_h can be determined uniquely on $\overline{\Omega_h^*}$ by means of this system. For this we set $\eta_h = u_h$ in (8.11), which is equivalent to the system (8.13), and use the fact that (8.2) implies that

$$a_{ijh}u_{x_j}u_{x_i} - b_{ih}u_{x_i}u_h - a_h u_h^2 \geq \nu_1 u_x^2 + \lambda u_h^2. \tag{8.14}$$

This yields the relation

$$\begin{aligned}\Delta_h \sum_{\Omega_h^{*+}} (\nu_1 u_x^2 + \lambda u_h^2) &+ \int_S \sigma(\tilde{u}_h)^2\, ds + J_h(u_h, u_h) \\ &\leq -\Delta_h \sum_{\Omega_h^{*+}} f_h u_h \leq \left(\Delta_h \sum_{\Omega_h^{*+}} f_h^2\right)^{1/2} \left(\Delta_h \sum_{\Omega_h^{*+}} u_h^2\right)^{1/2} \\ &\equiv \|f_h\|_{2,\Omega_h^{*+}} \|u_h\|_{2,\Omega_h^{*+}}.\end{aligned} \tag{8.15}$$

Since $\sigma \geq 0$, the integral $\int_S \sigma(\tilde{u}_h)^2\, ds$ may be discarded from the left-hand side of (8.15). If we recall the definition of the sum $J_h(u_h, u_h)$, then (8.15) implies the bound

$$\nu_2(\|u_h\|_{2,\overline{\Omega_h^*}}^{(1)})^2 \leq \|f_h\|_{2,\Omega_h^{*+}} \|u_h\|_{2,\Omega_h^{*+}}, \tag{8.16}$$

where $\nu_2 = \min(\nu_1, \lambda, 1)$, and the quantity

$$(\|u_h\|_{2,\overline{\Omega_h^*}}^{(1)})^2 \equiv \|u_h^*\|_{2,\overline{\Omega}_h^*}^2 + \sum_{i=1}^n \Delta_h \sum_{\overline{\Omega_h^*(i)}} u_{x_i}^2 \tag{8.17}$$

is equivalent to the square of the norm $\|u_h\|_{2,\overline{\Omega_h^*}}^{(1)}$ defined by (3.11) with $m = 2$. In (8.17) the symbol $\sum_{\overline{\Omega_h^*(i)}}$ denotes the summation over all points of $\overline{\Omega_h^*}$ at which the difference quotient u_{x_i} is defined for the function u_h, which is known only on $\overline{\Omega_h^*}$. The addition of the term $J_h(u_h, u_h)$ in (8.10) was due to our wish to have in (8.16) the grid norm, which is equivalent to the norm appearing in the hypotheses of Theorem 3.3 in this chapter. It follows from (8.16) that, if the system (8.13) is homogeneous, i.e., if $f_h \equiv 0$, then it is satisfied only by $u_h \equiv 0$. But then, by a theorem of linear algebra, the non-homogeneous system (8.13) has a unique solution for arbitrary free terms, i.e., for arbitrary f_h. Then (8.16) holds for the solutions of the system, which implies that our difference scheme is stable in the "energy norm," namely (cf. (7.13)),

$$\|u_h\|_{2,\overline{\Omega_h^*}} \leq \frac{1}{\nu_2}\|f_h\|_{2,\Omega_h^{*+}} \leq \frac{1}{\nu_2}\|f\|_{2,\Omega_h^*} \tag{8.18}$$

and

$$\|u_h\|_{2,\overline{\Omega_h^*}}^{(1)} \leq \frac{1}{\nu_2}\|f_h\|_{2,\Omega_h^{*+}} \leq \frac{1}{\nu_2}\|f\|_{2,\Omega_h^*}. \tag{8.19}$$

In order to carry out the limit process as $h_i \to 0$, $i = 1, \ldots, n$, we use the embedding theorems for grid functions proved in §3 of this chapter. As above, from a sequence $\{h\}$ tending to zero we must first select subsequences on which we have the convergence indicated below. But then it turns out that the convergence yields a limit function u which will be the generalized solution of (8.1); since such a solution is unique, we can conclude that every sequence $\{u_h\}$ must converge in the corresponding way to u. This part of the consideration is completely analogous to that carried out in §7; therefore we shall take a certain liberty in speaking at once of all sequences $\{u_h\}$ and their convergences instead of formally carrying out every argument on subsequences.

Thus, on the basis of the theorems of §3 it follows from the bounds (8.18), (8.19) that the interpolations $\{u_h'\}$ converge in $L_2(\Omega)$ to some function $u \in W_2^1(\Omega)$. Moreover, they converge to u on the surface S in the norm $L_2(S)$. Their derivatives $\{\partial u_h'/\partial x_i\}$, $i = 1, \ldots, n$, converge weakly in $L_2(\Omega)$ to $\partial u/\partial x_i$. Furthermore, the functions $\{\tilde{u}_h\}$ converge to u in $L_2(\Omega)$ and in $L_2(S)$, and $\{\tilde{u}_{x_i}\}$, $i = 1, \ldots, n$, converge weakly in $L_2(\Omega)$ to $\partial u/\partial x_i$. Finally, we take into account that the quantities $\|\tilde{u}_x\|_{2,\Omega_h^*}$, $\|\tilde{u}_h\|_{2,\Omega_h^*}$, and $J_h(u_h, u_h)$ are bounded by some constant c_1 which is independent of h. We shall now take the limit in (8.10) with a fixed, smooth function $\eta(x)$ defined on some domain $\tilde{\Omega} \supset \bar{\Omega}$. As η_h in (8.10) we take the values of $\eta(\cdot)$ at the vertices of $\overline{\Omega_h^*}$, assuming that the h_i are already small enough so that $\overline{\Omega_h^*} \subset \tilde{\Omega}$. The functions $\tilde{\eta}_h$ and $\tilde{\eta}_{x_i}$ will converge uniformly to η and $\partial\eta/\partial x_i$ in a domain $\tilde{\tilde{\Omega}}$ contained in $\tilde{\Omega}$ and containing all the $\overline{\Omega_h^*}$, beginning with some $h_i^0 > 0$, $i = 1, \ldots, n$. The integrals

$$j_h' \equiv \int_{\Omega_h^* \setminus \Omega} [a_{ij}\tilde{u}_{x_j}\tilde{\eta}_{x_i} - b_i\tilde{u}_{x_i}\tilde{\eta}_h - a\tilde{u}_h\tilde{\eta}_h]\,dx$$

tend to zero as $h_i \to 0$ because of the bound

$$\begin{aligned} |j_h'| &\le c[\|\tilde{u}_x\|_{2,\Omega_h^*\setminus\Omega} + \|\tilde{u}_h\|_{2,\Omega_h^*\setminus\Omega}] \\ &\quad \times [\|\tilde{\eta}_x\|_{2,\Omega_h^*\setminus\Omega} + \|\tilde{\eta}_h\|_{2,\Omega_h^*\setminus\Omega}] \\ &\le 2cc_1[\|\tilde{\eta}_x\|_{2,\Omega_h^*\setminus\Omega} + \|\tilde{\eta}_h\|_{2,\Omega_h^*\setminus\Omega}], \end{aligned}$$

and because $\text{mes}(\Omega') \to 0$ where $\Omega' = \Omega_h^* \setminus \Omega$. The same bounds show that $\int_{\Omega'} f\tilde{\eta}_h\,dx$ and $J_h(u_h, \eta_h)$ also tend to zero as $h_i \to 0$, $i = 1, \ldots, n$. But then it is clear that the limit of (8.0) will be the identity (8.5) which lies at the heart of the definition of a generalized solution $u(x)$ of the problem (8.1). We have proved the validity of (8.5) for the limit function and for all smooth η. But such functions η are dense in $W_2^1(\Omega)$, so that (8.5) is satisfied for all $\eta \in W_2^1(\Omega)$. In this regard, we call attention to the fact that, as well as the use of the embedding in $W_2^1(\Omega)$ into $L_2(S)$, we should know that the boundary surface S has a certain regularity (which was stipulated at the beginning of the section). From the analysis of the difference scheme just carried out, it is clear that it can be varied within some limits (for example, we could have replaced (8.9) by $\alpha J_h(u_h, \eta_h)$ with some $\alpha > 0$). This does not change the final result (that in the limit we obtain $u(x)$), but it does change the system (8.13). We must keep this possibility in mind when we put together a difference scheme for a specific equation (8.1) in a specific domain.

§9. Equations of Parabolic Type

We consider the first initial-boundary value problem

$$\mathcal{M}u \equiv \frac{\partial u}{\partial t} - \mathcal{L}u = f(x, t), \tag{9.1}$$

$$u|_{t=0} = \varphi(x), \qquad u|_{S_T} = 0, \tag{9.2}$$

where

$$\mathcal{L}u \equiv \frac{\partial}{\partial x_i}\left(a_{ij}(x, t)\frac{\partial u}{\partial x_j}\right) - b_i(x, t)\frac{\partial u}{\partial x_i} - a(x, t)u$$

in a bounded domain $Q_T = \Omega \times (0, T)$. We shall assume that conditions (3.3)–(3.5), Chapter III, are satisfied. The functions a_i and f_i were taken to be zero only for the sake of economy of space. For the elliptic part of (9.1) and the boundary condition we take the difference scheme which was considered in detail in §7 (we remind the reader that it is not the only one possible). We consider two substitutions for the derivative $\partial u/\partial t$: $u_{\bar{t}}$ and u_t, the first of which yields an implicit scheme, the second an explicit scheme. As we shall show below, the difference schemes to be obtained in this connection will be stable, for essentially the same ratios of the step-sizes in t and x as in the elementary representative of the equations (9.1), considered in subsection 2 of §6. Besides these two schemes, we shall analyze also a variant of an implicit scheme for higher-dimensional equations (9.1) (i.e., when $n \geq 2$) in which the algebraic systems defining the grid functions u_h are essentially simpler than the systems arising from the first implicit scheme. Such schemes converge, as the first one did, for any ratio of step-sizes in x and t, but the convergence takes place in a weaker norm than in the case of the first. At present there are many variants of difference schemes of similar type differing from one another by the way in which the higher-dimensional elliptic part is decomposed into simpler components. They are all investigated principally in the same way as the case we considered in subsection 2.

We partition the whole Euclidean space R^{n+1} of the variables (x, t) by the planes $x_i = k_i h_i$, $h_i > 0$, $k_i = 0, \pm 1, \ldots$, and $t = k_0 \tau$, $\tau > 0$, $k_0 = 0, 1, 2, \ldots$, into parallelopipeds $Q_{(k, k_0)} = \omega_{(kh)} \times (k_0 \tau, (k_0 + 1)\tau) = \{(x, t): k_i h_i < x_i < (k_i + 1)h_i,\ k_0 \tau < t < (k_0 + 1)\tau\}$, and use the notation Ω_h, $\overline{\Omega}_h$, Ω_h^+, and S_h introduced earlier in §7. To denote the functions prescribed in the grid we shall use a subscript Δ: u_Δ. In writing the difference quotients for such functions, as we did above, we shall omit the subscript Δ, so that by u_{x_i} we shall understand $(u_\Delta)_{x_i}$. Moreover, we shall agree that the index k of $u_\Delta(k)$ means that the grid function u_Δ is taken at the level $t = t_k = k\tau$. Finally, for the sake of shortening the notation, we shall often write simply $\|u_\Delta(k)\|$ in place of the grid norm $\|u_\Delta(k)\|_{2, \overline{\Omega}_h}$.

1. The First Implicit Grid Scheme. The first implicit grid scheme has the form

$$\mathcal{M}_\Delta u_\Delta(k) \equiv u_{\bar{t}}(k) - (a_{ij\Delta}(k)u_{x_j}(k))_{\bar{x}_i} + b_{i\Delta}(k)u_{x_i}(k) + a_\Delta(k)u(k) = f_\Delta(k), \tag{9.3}$$

$$u_\Delta(k)|_{S_h} = 0, \tag{9.4}$$

$$u_\Delta|_{t=0} = \varphi_h. \tag{9.5}$$

The difference equations (9.3) must be satisfied on the levels $k = 1, 2, \ldots, N = [T/\tau]$ at the "interior" points of the grid $\overline{\Omega}_h$, i.e., at the points of Ω_h; the equations (9.4) for $k = 0, 1, \ldots, N$; and the equation (9.5) at points of Ω_h, where, for φ_h, we take $\varphi_h|_{(kh)} = \Delta_h^{-1}\int_{\omega(kh)} \varphi(x)\,dx$, $\Delta_h = h_1, \ldots, h_n$. The functions $a_{ij\Delta}, \ldots, f_\Delta$ are formed from the known functions $a_{ij}, \ldots, f$ in this way:

$$a_{ij\Delta}|_{x=(kh),\,t=k_0\tau} = \Delta_h^{-1}\tau^{-1}\int_{(k_0-1)\tau}^{k_0\tau}\int_{\omega(kh)} a_{ij}(x, t)\,dx\,dt,$$

and similarly for the others. The solution of the system (9.3)–(9.5) is determined sequentially by levels $t = t_k$, beginning with $k = 1$. For each level we have to solve a linear algebraic system containing as many equations and unknowns $u_\Delta(k)$ as there are points of the lattice Ω_h. These systems have the same form as those in §7. We shall show that they are uniquely solvable for all τ less than some $\tau_0 > 0$. This will follow from essentially the same bound which guarantees the stability of the scheme that is constructed. It will hold for arbitrary ratio of step-sizes τ and h_i. The equations (9.3), (9.4) are equivalent to the summation identity

$$\Delta_h \sum_{\Omega_h^+} [u_{\bar{t}}\eta_\Delta + a_{ij\Delta}u_{x_j}\eta_{x_i} + b_{i\Delta}u_{x_i}\eta_\Delta + a_\Delta u_\Delta \eta_\Delta] = \Delta_h \sum_{\Omega_h^+} f_\Delta \eta_\Delta, \tag{9.6}$$

in which η_Δ is an arbitrary function on Ω_h which vanishes on S_h. The identities (9.6) are satisfied on all levels $t = t_k$, $k = 1, \ldots, N$. In order to prove the solvability of the system (9.3), (9.4) on the level t_k, we take the homogeneous system corresponding to it, or, what is the same thing, the identity (9.6) on the level t_k in which f_Δ and $u_\Delta(k-1)$ have been replaced by zero ($u_\Delta(k-1)$ goes into $u_{\bar{t}}(k)$). Setting $\eta_\Delta = u_\Delta$ in this identity, we obtain

$$\Delta_h \sum_{\Omega_h^+} [\tau^{-1}u_\Delta^2 + a_{ij\Delta}u_{x_j}u_{x_i} + b_{i\Delta}u_{x_i}u_\Delta + a_\Delta u_\Delta^2| = 0.$$

It follows from (3.3)–(3.5), Chapter III, that

$$\Delta_h \sum_{\Omega_h^+} [\tau^{-1}u_\Delta^2 + \nu u_x^2] \le \mu\Delta_h \sum_{\Omega_h^+} [|u_x|\cdot|u_\Delta| + u_\Delta^2]$$

$$\le \Delta_h \sum_{\Omega_h^+} \left[\nu u_x^2 + \left(\frac{\mu^2}{4\nu} + \mu\right)u_\Delta^2\right]. \tag{9.7}$$

If $\tau < (\mu^2/4\nu + \mu)^{-1} \equiv \tau_0$, then (9.7) implies that $u_\Delta(k)$ vanishes identically; that is, we have shown that, for $\tau < \tau_0$, the system (9.3), (9.4) is uniquely solvable on the level t_k (where k is arbitrary, starting with $k = 1$) for arbitrary $f_\Delta(k)$ and $u_\Delta(k-1)$. We shall now prove the stability of the system, namely, we shall derive a bound of the form (5.14), Chapter III. This is done in exactly the same way as we derived the inequalities (5.10)–(5.14) in Chapter III. We can even retain all the constants in these inequalities if we just regard $\|u\|^2_{2,\Omega_k}$ as $\|u_\Delta(k)\|^2 \equiv \Delta_h \sum_{\bar{\Omega}_h} u^2_\Delta(k)$ and $\|u_x\|^2_{2,\Omega_k}$ as the sum $\|u_x(k)\|^2 \equiv \Delta_h \sum_{\Omega_h^+} u^2_x(k)$ (in §5, Chapter III, we denoted the derivatives $\partial u/\partial x_i$ by u_{x_i}, and here u_{x_i} are the "right-hand" difference quotients; we remark that, by (9.4),

$$\sum_{\Omega_h^+} u^2_\Delta(k) = \sum_{\bar{\Omega}_h} u^2_\Delta(k) = \sum_{\Omega_h} u^2_\Delta(k)$$

and $\sum_{\Omega_h^+} u^2_x(k) = \sum_{\bar{\Omega}_h} u^2_x(k)$ if we assume that $u_\Delta(k)$ is zero outside $\bar{\Omega}_h$). Thus we have for u_Δ the bounds

$$\begin{aligned}(1 - c\tau)\|u_\Delta(k)\|^2 - \|u_\Delta(k-1)\|^2 + \|\delta u_\Delta(k-1)\|^2 \\ + \nu\tau\|u_x(k)\|^2 \le 2\tau\|f_\Delta(k)\|\,\|u_\Delta(k)\|\end{aligned} \tag{9.8}$$

and

$$\begin{aligned}\|u_\Delta(m)\|^2 + \nu\tau \sum_{k=1}^m \|u_x(k)\|^2 + \sum_{k=1}^m \|\delta u_\Delta(k-1)\|^2 \\ \le c_1[\|\varphi_\Delta\|^2 + \|f_\Delta\|^2_{2,1,Q_m^\Delta}], \qquad m = 1, \ldots, N,\end{aligned} \tag{9.9}$$

where $c = \mu^2/\nu + 2\mu$, $\tau \le (2c)^{-1}$, where the constant c_1 depends only on ν, μ, and T, where $\delta u_\Delta(k-1) = u_\Delta(k) - u_\Delta(k-1)$, and where $\|f_\Delta\|_{2,1,Q_m^\Delta} = \tau \sum_{k=1}^m (\Delta_h \sum_{\Omega_h^+} f^2_\Delta(k))^{1/2}$. These inequalities guarantee the stability of our difference scheme (without any restrictions on the step-sizes h_i and τ, except that $\tau \le (2c)^{-1}$), for the right-hand side of (9.9) cannot exceed the constant $c_2 \equiv c_1(\|\varphi\|^2_{2,\Omega} + \|f\|^2_{2,1,Q_T})$ (in this connection see the choice of φ_Δ and f_Δ and (7.13)). On the basis of these bounds and of the theorems of §3 in this chapter, we can prove the weak convergence in $W_2^{1,0}(Q_T)$ of the interpolations $u'_\Delta(x, t)$ to a generalized solution in $W_2^{1,0}(Q_T)$ of (9.1), (9.2). This is done in the same way as for elliptic equations in §7, and we leave it for the reader to do independently. We give only a few indications. The identities (9.6), valid for the levels $t = t_k$, $k = 1, \ldots, N$, imply the identity

$$\begin{aligned}\int_{Q_T} [-\tilde{u}_\Delta \tilde{\eta}_t + a_{ij}\tilde{u}_{x_j}\tilde{\eta}_{x_i} + b_i\tilde{u}_{x_i}\tilde{\eta}_\Delta + a\tilde{u}_\Delta\tilde{\eta}_\Delta]\,dx\,dt \\ - \int_\Omega \varphi\tilde{\eta}_\Delta(0)\,dx = \int_{Q_T} f\tilde{\eta}_\Delta\,dx\,dt\end{aligned} \tag{9.10}$$

for an arbitrary function $\eta_\Delta(k)$ vanishing on S_k and outside $\overline{\Omega}_h$ for all k, and vanishing on the level $t = [T/\tau]$ and above. We also assume that the function $u_\Delta(k)$ has been defined to be zero outside $\overline{\Omega}_h$ for all k (on S_h it is zero because of (9.4)). The completions of $\tilde{u}_\Delta$, $\tilde{u}_{x_j}$, $\tilde{\eta}_\Delta$, and $\tilde{\eta}_{x_i}$ are zero outside $\overline{\Omega}_h$ for all k. This makes it possible to assume that all integrations in x have been carried out on Ω. The piecewise-constant interpolations ($\tilde{\ }$) have the form: $\tilde{u}_\Delta(x, t) = u_{(kh, (k_0+1)\tau)}$ on the cell $Q_{(k, k_0)}$. We must carry out the limit process in (9.10) as $h_i \to 0$ taking as η_Δ the values at points of the grid of some smooth function $\eta(x, t)$ which vanishes in the vicinity of the lateral surface of the cylinder Q_T and its upper base. Thus, we may assume as proved the following theorem:

Theorem 9.1. *Let* (3.3)–(3.5), *Chapter III, be satisfied for the problem* (9.1), (9.2), *let* $f \in L_{2,1}(Q_T)$, *and let* $\varphi \in L_2(\Omega)$. *Then the difference scheme* (9.3)–(9.5) *defines uniquely a grid function* u_Δ *for all* $\tau \leq (2\mu^2/\nu + 4\mu)^{-1}$, *and its interpolations* $u'_\Delta(\cdot, \cdot)$ *convergence weakly in* $L_2(Q_T)$ *as* $h_i \to 0$ *and* $\tau \to 0$ *to a generalized solution* $u(\cdot, \cdot) \in W_2^{1,0}(Q_T)$ *of the problem* (9.1), (9.2). *The derivatives* $\partial u'_\Delta/\partial x_i$ *converge weakly in* $L_2(Q_T)$ *to* $\partial u/\partial x_i$.

2. The Second Implicit Scheme (the Alternating-Direction Scheme). The systems (9.3)–(9.5) defining u_Δ contain as many unknowns as there are nodes in the lattice in Ω_h. The number of equations in these systems is very large for $n \geq 2$ when h_i are small, and the systems do not split into separate blocks. The solution of such systems requires machines of immense memory, as well as a lot of time, for their solution. Because of this, other implicit difference schemes have been proposed for special classes of equations (9.1); these require the solution of significantly smaller systems.† The first such schemes were proposed by J. Douglas [1] and D. W. Peaceman and H. H. Rachford [1] for simplest elliptic and parabolic equations. These and some other schemes were investigated in the books of G. I. Marcuk [1], N. N. Yanenko [1], A. A. Samarakii [1], and in some papers. We shall give such schemes for the general parabolic equations (9.1). Like the scheme in subsection 1, they converge as τ and h_i tend to zero in an arbitrary way, but the convergence itself is "of a lower quality" than that in subsection 1 (in the sense that u_Δ converges to u in a weaker norm). This can be proved in essentially the same way as we proved convergence for the scheme (9.3)–(9.5) in subsection 1. We shall show this in connection with (9.1), (9.2) for the scheme (9.11) below. We take the partition of R^{n+1} into cells, as described at the beginning of the section, and add to it the cross-sections $t_{k+p/n} = (k + p/n)\tau, p = 1, 2, \ldots, n-1$,

† Schemes of a similar type are known in the literature under the names of alternating-direction schemes, fractional-step schemes, and splitting schemes.

$k = 0, 1, \ldots, N-1$. We shall compute the solution u_Δ on all the levels $t_{k+p/n}$ following such a scheme:

$$\frac{1}{\tau}\left[u_\Delta\left(k+\frac{p}{n}\right) - u_\Delta\left(k+\frac{p-1}{n}\right)\right] - 2\sum_{q=1}^{p-1}\left[a_{pq\Delta}(k+1)u_{x_q}\left(k+\frac{\sqrt{q}}{n}\right)\right]_{\bar{x}_p}$$
$$-\left[a_{pp\Delta}(k+1)u_{x_p}\left(k+\frac{p}{n}\right)\right]_{\bar{x}_p} + b_{p\Delta}(k+1)u_{x_p}\left(k+\frac{p}{n}\right)$$
$$+\frac{1}{n}a_\Delta(k+1)u_\Delta\left(k+\frac{p}{n}\right) = \frac{1}{n}f_\Delta\left(k+\frac{p}{n}\right), \tag{9.11}$$

for $k = 0, 1, \ldots, N-1$, $p = 1, 2, \ldots, n$, and add to it initial and boundary conditions

$$u_\Delta\left(k+\frac{p}{n}\right)\bigg|_{S_h} = 0, \qquad u_\Delta|_{t=0} = \varphi_h. \tag{9.12}$$

The equations (9.11) should be satisfied on the levels $t_{k+p/n}$ when x is an "interior node" of the lattice $\bar{\Omega}_h$ (i.e., for $x \in \Omega_h$). It is not difficult to see that on each level we must solve systems in each of which the unknowns are "linked" only to one segment parallel to some coordinate axis. The fundamental matrices of such systems are tridiagonal and are comparatively easy to invert, for example, by means of the double-sweep method (this can be found, for example, in the book by S. K. Godunov and V. S. Rjaben'kii [1]).

To prove the solvability of the systems (9.11), (9.12), as well as the stability and convergence (in some norm) of our difference process, we derive the analogue of the bounds (9.7)–(9.9).

Let us consider the homogeneous system corresponding to (9.11) on the level $t = (k + p/n)\tau$, which has the form

$$\frac{1}{\tau}u_\Delta\left(k+\frac{p}{n}\right) - \left[a_{pp\Delta}(k+1)u_{x_p}\left(k+\frac{p}{n}\right)\right]_{\bar{x}_p}$$
$$+ b_{p\Delta}(k+1)u_{x_p}\left(k+\frac{p}{n}\right) + \frac{1}{n}a_\Delta(k+1)u_\Delta\left(k+\frac{p}{n}\right) = 0, \tag{9.13}$$

from which it is clear that (9.13) splits into separate subsystems; in each of these subsystems only those nodes of the lattice of $\bar{\Omega}_h$ are interconnected which lie on some straight line l_p parallel to the x_p-axis. If we multiply (9.13) by $h_p u_\Delta(k + p/n)$ and sum over all nodes in $\bar{\Omega}_h$ lying on l_p, we obtain, after elementary manipulations,

$$\tau^{-1}h_p\sum_{l_p}u_\Delta^2\left(k+\frac{p}{n}\right) + h_p\sum_{l_p}\left[a_{pp\Delta}u_{x_p}^2\left(k+\frac{p}{n}\right)\right.$$
$$\left.+ b_{p\Delta}u_{x_p}\left(k+\frac{p}{n}\right)u_\Delta\left(k+\frac{p}{n}\right) + \frac{1}{n}a_\Delta u_\Delta^2\left(k+\frac{p}{n}\right)\right] = 0. \tag{9.14}$$

Here we used (2.4) of §2 and the vanishing of $u_\Delta(k + p/n)$ at all points of l_p except those which belong to Ω_h. By (3.3)–(3.5), Chapter III, it follows from (9.14) that $u_\Delta \equiv 0$ if $\tau < (\mu^2/4\nu + \mu/n)^{-1} \equiv \tau_0$ (cf. (9.7) above). Thus we have proved the unique solvability of all the systems (9.11) for $\tau < \tau_0$. For bounds on the solutions we multiply (9.11) by $2\tau\Delta_h u_\Delta(k + p/n)$ and sum over all nodes of $\overline{\Omega}_h$; we then transform the expression we get, by means of (5.9), Chapter III, and (2.11) of this chapter, into the following form:

$$\begin{aligned}
&\left\|u_\Delta\left(k+\frac{p}{n}\right)\right\|^2 - \left\|u_\Delta\left(k+\frac{p-1}{n}\right)\right\|^2 \\
&\quad + \left\|u_\Delta\left(k+\frac{p}{n}\right) - u_\Delta\left(k+\frac{p-1}{n}\right)\right\|^2 \\
&\quad + 2\tau\Delta_h \sum_{\Omega_h^+} \left[a_{pp\Delta}(k+1) u_{x_p}^2\left(k+\frac{p}{n}\right) \right. \\
&\quad + 2\sum_{q=1}^{p-1} a_{pq\Delta}(k+1) u_{x_p}\left(k+\frac{p}{n}\right) u_{x_q}\left(k+\frac{q}{n}\right) \\
&\quad + b_{p\Delta}(k+1) u_{x_p}\left(k+\frac{p}{n}\right) u_\Delta\left(k+\frac{p}{n}\right) \\
&\quad \left. + \frac{1}{n} a_\Delta(k+1) u_\Delta^2\left(k+\frac{p}{n}\right)\right] \\
&= \frac{2\tau}{n}\Delta_h \sum_{\Omega_h^+} f_\Delta\left(k+\frac{p}{n}\right) u_\Delta\left(k+\frac{p}{n}\right).
\end{aligned} \tag{9.15}$$

If we sum (9.15) from $p = 1$ to $p = n$ and use (3.3)–(3.5), Chapter III, we are led to the inequalities

$$\begin{aligned}
&\|u_\Delta(k+1)\|^2 - \|u_\Delta(k)\|^2 + \sum_{p=1}^n \left\|u_\Delta\left(k+\frac{p}{n}\right) - u_\Delta\left(k+\frac{p-1}{n}\right)\right\|^2 \\
&\quad + 2\nu\tau \sum_{p=1}^n \left\|u_{x_p}\left(k+\frac{p}{n}\right)\right\|^2 \\
&\le 2\mu\tau \sum_{p=1}^n \left[\left\|u_{x_p}\left(k+\frac{p}{n}\right)\right\| \cdot \left\|u_\Delta\left(k+\frac{p}{n}\right)\right\| + \frac{1}{n}\left\|u_\Delta\left(k+\frac{p}{n}\right)\right\|^2\right] \\
&\quad + \frac{2}{n}\tau \sum_{p=1}^n \left\|f_\Delta\left(k+\frac{p}{n}\right)\right\| \left\|u_\Delta\left(k+\frac{p}{n}\right)\right\| \\
&\le \tau \sum_{p=1}^n \left[\nu\left\|u_{x_p}\left(k+\frac{p}{n}\right)\right\|^2 + \left(\frac{\mu^2}{\nu} + \frac{2\mu}{n}\right)\left\|u_\Delta\left(k+\frac{p}{n}\right)\right\|^2\right] \\
&\quad + \frac{2\tau}{n} \sum_{p=1}^n \left\|f_\Delta\left(k+\frac{p}{n}\right)\right\| \left\|u_\Delta\left(k+\frac{p}{n}\right)\right\|,
\end{aligned} \tag{9.16}$$

for $k = 0, 1, \ldots, N-1$. We can derive the bounds we need from these inequalities, essentially in the same way as in §5, Chapter III, we derived the inequalities (5.12) and (5.14) (it is only necessary, to derive an inequality similar to (5.12), to use the presence of the third term on the left-hand side of (9.16)). For the sake of minor simplifications we assume that $f \in L_2(Q_T)$. Then we can bound the last term on the right-hand side of (9.16) by Cauchy's inequality by replacing

$$2\left\|f_\Delta\left(k+\frac{p}{n}\right)\right\|\left\|u_\Delta\left(k+\frac{p}{n}\right)\right\| \quad \text{by} \quad \left\|f_\Delta\left(k+\frac{p}{n}\right)\right\|^2 + \left\|u_\Delta\left(k+\frac{p}{n}\right)\right\|^2.$$

It follows from (9.16) that

$$\|u_\Delta(k+1)\|^2 - \|u_\Delta(k)\|^2 + \sum_{p=1}^{n}\left\|u_\Delta\left(k+\frac{p}{n}\right) - u_\Delta\left(k+\frac{p-1}{n}\right)\right\|^2$$
$$\leq c\tau \sum_{p=1}^{n}\left\|u_\Delta\left(k+\frac{p}{n}\right)\right\|^2 + \frac{\tau}{n}\sum_{p=1}^{n}\left\|f_\Delta\left(k+\frac{p}{n}\right)\right\|^2, \tag{9.17}$$

where $c = \mu^2/\nu + 2\mu/n + 1/n$.

Each of the terms $\|u_\Delta(k+p/n)\|^2$, $p = 1, \ldots, n-1$, can be bounded by terms on the left-hand side of (9.17) if we start from the obvious representation

$$u_\Delta\left(k+\frac{p}{n}\right) = \sum_{s=1}^{p}\left[u_\Delta\left(k+\frac{s}{n}\right) - u_\Delta\left(k+\frac{s-1}{n}\right)\right] + u_\Delta(k). \tag{9.18}$$

Actually, it follows from (9.18) that

$$\left\|u_\Delta\left(k+\frac{p}{n}\right)\right\|^2 \leq \left[\sum_{s=1}^{p}\left\|u_\Delta\left(k+\frac{s}{n}\right) - u_\Delta\left(k+\frac{s-1}{n}\right)\right\| + \|u_\Delta(k)\|\right]^2$$
$$\leq (p+1)\left[\sum_{s=1}^{p}\left\|u_\Delta\left(k+\frac{s}{n}\right) - u_\Delta\left(k+\frac{s-1}{n}\right)\right\|^2\right.$$
$$\left. + \|u_\Delta(k)\|^2\right]. \tag{9.19}$$

If we substitute these bounds into the right-hand side of (9.17) and make a rough estimate of them, we obtain

$$\|u_\Delta(k+1)\|^2 - \|u_\Delta(k)\|^2 + \sum_{p=1}^{n}\left\|u_\Delta\left(k+\frac{p}{n}\right) - u_\Delta\left(k+\frac{p-1}{n}\right)\right\|^2$$
$$\leq c\tau\frac{n(n+3)}{2}\left[\sum_{p=1}^{n}\left\|u_\Delta\left(k+\frac{p}{n}\right) - u_\Delta\left(k+\frac{p-1}{n}\right)\right\|^2\right.$$
$$\left. + \|u_\Delta(k)\|^2\right] + \frac{\tau}{n}\sum_{p=1}^{n}\left\|f_\Delta\left(k+\frac{p}{n}\right)\right\|^2. \tag{9.20}$$

Hence, for $c\,\tau[n(n+3)/2] \le \frac{1}{2}$ (we recall that $n \ge 2$ at this stage), we obtain

$$\|u_\Delta(k+1)\|^2 + \frac{1}{2}\sum_{p=1}^{n}\left\|u_\Delta\left(k+\frac{p}{n}\right) - u_\Delta\left(k+\frac{p-1}{n}\right)\right\|^2$$
$$\le \left(1 + c\tau\frac{n(n+3)}{2}\right)\|u_\Delta(k)\|^2 + \frac{\tau}{n}\sum_{p=1}^{n}\left\|f_\Delta\left(k+\frac{p}{n}\right)\right\|^2. \quad (9.21)$$

If we discard now the second term on the left-hand side of (9.21), we obtain a recurrence relation for $\|u_\Delta(k)\|^2$ by means of which we can bound $\|u_\Delta(m)\|^2$ in terms of known quantities:

$$\|u_\Delta(m)\|^2 \le c_1\left[\|\varphi_h\|^2 + \tau\sum_{k=0}^{m-1}\sum_{p=1}^{n}\left\|f_\Delta\left(k+\frac{p}{n}\right)\right\|^2\right], \quad (9.22)$$

where c_1 is defined by ν, μ, n, and T (a similar bound was given in §5, Chapter III; cf. (5.11)–(5.12)). We derive a similar bound for the non-integer levels from (9.21), (9.22), and (9.19):

$$\left\|u_\Delta\left(m+\frac{p}{n}\right)\right\|^2 \le c_2\left[\|\varphi_h\|^2 + \tau\sum_{k=0}^{m}\sum_{p=1}^{n}\left\|f_\Delta\left(k+\frac{p}{n}\right)\right\|^2\right] \equiv c_2\mathscr{F}(m+1), \quad (9.23)$$

$m = 0, 1, \ldots, N+1$, $p = 1, \ldots, n$. If we sum the inequalities (9.16) from $k = 0$ to $k = m$ and then use (9.23), we obtain the desired bound

$$\|u_\Delta(m+1)\|^2 + \sum_{k=0}^{m}\sum_{p=1}^{n}\left\|u_\Delta\left(k+\frac{p}{n}\right) - u_\Delta\left(k+\frac{p-1}{n}\right)\right\|^2$$
$$+ \nu\tau\sum_{k=0}^{m}\sum_{p=1}^{n}\left\|u_{x_p}\left(k+\frac{p}{n}\right)\right\|^2 \le c_3\mathscr{F}(m+1) \quad (9.24)$$

for $m = 0, 1, \ldots, N-1$, where $\mathscr{F}(m+1)$ is defined in (9.23). The right-hand side of (9.24) does not exceed a certain constant which is defined by $\varphi(x)$ and $f(x, t)$. This bound yields the specific stability of the scheme (9.11), (9.12) with respect to the variationof φ and f. This stability makes it easy to prove the weak convergence in $L_2(Q_{T-\delta})$, for all $\delta > 0$, of the piecewise-constant interpolations $\tilde{u}_\Delta$ of the grid functions u_Δ computed according to (9.11), (9.12) to the solution $u(x, t)$ of the problem (9.1), (9.2). The best convergence of u_Δ to $u(x, t)$ can be proved as follows: For $u(x, t)$ we construct a grid function $\hat{u}_\Delta$ (if $u(x, t)$ is the classical solution and the coefficients of $\mathscr{L}$ are smooth functions, then we can simply take as $\hat{u}_\Delta$ the values of $u(x, t)$ at the nodes of the grid), write for $\hat{u}_\Delta$ the relations (9.11) with remainders $r_\Delta(k + p/n)$ on the right-hand sides, and then from these subtract the relations (9.11) for u_Δ. Hence for $v_\Delta = \hat{u}_\Delta - u_\Delta$ we obtain the relations (9.11) whose right-hand sides will have the remainders $r_\Delta(k + p/n)$ instead of $(1/n)f_\Delta(k + p/n)$. These remainders need not be small, but the sum $\sum_{p=1}^{n} r_\Delta(k + p/n)$ tends to zero as $\tau \to 0$ and $h_i \to 0$. This permits us to obtain a bound of the form (9.24) for v_Δ whose right-hand side tends to zero as

$\tau \to 0$ and $h_i \to 0$. Thus we have proved the convergence of u_Δ to u in the norm corresponding to the left-hand side of (9.24). A similar argument was carried out rigorously for a more difficult problem in my book [15, Chapter 6, §9, Subsection 2] (the second Russian edition).

3. The Explicit Scheme. The explicit difference scheme for (9.1), (9.2) differs from (9.3) only in the replacement of $u_{\bar{t}}$ by u_t, namely,

$$u_t(k) - (a_{ij\Delta}(k)u_{x_j}(k))_{\bar{x}_i} + b_{i\Delta}(k)u_{x_i}(k) + a_\Delta(k)u_\Delta(k) = f_\Delta(k), \tag{9.25}$$

$$u_\Delta(k)|_{S_h} = 0, \qquad k = 0, 1, \ldots, N, \quad u_\Delta|_{t=0} = \varphi_h. \tag{9.26}$$

The space R^{n+1} is partitioned into elementary cells in the same way as in subsection 1. The equation (9.25) should be satisfied for $t = t_k = k\tau$, $k = 0, 1, \ldots, N - 1$, with $x \in \Omega_h$. It is more convenient to form the grid functions $a_{ij\Delta}, \ldots, f_\Delta$ from $a_{ij}, \ldots, f$ as

$$a_{ij\Delta}|_{(kh, k_0\tau)} = \Delta_h^{-1}\tau^{-1}\int_{k_0\tau}^{(k_0+1)\tau}\int_{\omega(kh)} a_{ij}(x, t)\,dx\,dt$$

and similarly for the others. If we take (9.25) at one such point (x, t_k), we define u_Δ at $(x, t_k + \tau)$ from (9.25) if the values of u_Δ are already known at the nodes of the level $t = t_k$. Thus the approximate solution u_Δ can be determined uniquely quite simply from (9.25), (9.26).

In order that these solutions u_Δ yield in the limit a solution $u(x, t)$ of (9.1), (9.2), it is necessary to impose on the step-size in t a restriction of the form $\tau \le c_i h_i^2$ for certain c_i. The need for such a restriction was shown in §6, subsection 2, where we considered a simple one-dimensional parabolic equation of the form (9.1). We assume, for the sake of incidental simplification, that all the h_i are equal to h, and show that, for $\tau = ch^2$ with $c < (4n\mu)^{-1}$, where μ is taken from the condition (3.5), Chapter III, the solutions u_Δ yield in the limit as $h \to 0$ the solution $u(x, t)$ of (9.1), (9.2).

Thus, let (3.3)–(3.5), Chapter III, be satisfied, and let $\varphi \in L_2(\Omega)$, $f(\cdot, \cdot) \in L_{2,1}(Q_T)$. We derive for the solutions u_Δ of (9.25), (9.26) the "energy" bounds below. For this we multiply (9.25) by $2\tau h^n u_\Delta(k + 1)$ and sum the result over all the nodes of Ω_h (or, what is the same thing, over the nodes of $\bar{\Omega}_h$). Then, by means of (5.9) of Chapter III and (2.11) of this chapter, we transform this equality into the form

$$\begin{aligned}
&\|u_\Delta(k + 1)\|^2 - \|u_\Delta(k)\|^2 + \|\delta u_\Delta(k)\|^2 \\
&\quad + 2\tau h^n \sum_{\Omega_h^+}[a_{ij\Delta}(k)u_{x_j}(k)u_{x_i}(k + 1) + b_{i\Delta}(k)u_{x_i}(k)u_\mu(k + 1) \\
&\quad + a_\Delta(k)u_\Delta(k)u_\Delta(k + 1)] - 2\tau h^n \sum_{\Omega_h^+} f_\Delta(k)u_\Delta(k + 1),
\end{aligned} \tag{9.27}$$

where $\delta u_\Delta(k) = u_\Delta(k + 1) - u_\Delta(k)$.

We write the sum $j_1(k) \equiv 2a_{ij\Delta}(k)u_{x_j}(k)u_{x_i}(k+1)$ as

$$j_1(k) = a_{ij\Delta}(k)u_{x_j}(k)[u_{x_i}(k) + \delta u_{x_i}(k)]$$
$$+ a_{ij\Delta}(k)u_{x_i}(k+1)[u_{x_j}(k+1) - \delta u_{x_j}(k)] = a_{ij\Delta}(k)u_{x_j}(k)u_{x_i}(k)$$
$$+ a_{ij\Delta}(k)u_{x_j}(k+1)u_{x_i}(k+1) - a_{ij\Delta}(k)\delta u_{x_j}(k)\delta u_{x_i}(k),$$

where $\delta u_{x_i}(k) = u_{x_i}(k+1) - u_{x_i}(k)$. We substitute this expression for $j_1(k)$ into (9.27), transpose the third term of $j_1(k)$ and terms with $b_{i\Delta}$ and a_Δ to the right-hand side, after which we bound the left-hand and right-hand sides of our equality from below and from above, using (3.3)–(3.5) of Chapter III,

$$\|u_\Delta(k+1)\|^2 - \|u_\Delta(k)\|^2 + \|\delta u_\Delta(k)\|^2 + v\tau(\|u_x(k)\|^2 + \|u_x(k+1)\|^2) \le \mu\tau[\|\delta u_x(k)\|^2 + 2\|u_x(k)\|\,\|u_\Delta(k+1)\| + 2\|u_\Delta(k)\|\,\|u_\Delta(k+1)\|] + 2\tau\|f_\Delta(k)\|\,\|u_\Delta(k+1)\|. \tag{9.28}$$

We write out δu_{x_i} explicitly at an arbitrary node (x, t_k) of our lattice and bound its modulus:

$$|\delta u_{x_i}(x, k)| = \frac{1}{h}|\delta u_\Delta(x + he_i, k) - \delta u_\Delta(x, k)|$$
$$\le \frac{1}{h}(|\delta u_\Delta(x + he_i, k)| + |\delta u_\Delta(x, k)|),$$

whence it follows that

$$\|\delta u_x(k)\|^2 \le h^n \sum_{\Omega_h^+} \sum_{i=1}^{n} \frac{2}{h^2}(|\delta u_\Delta(x + he_i, k)|^2 + |\delta u_\Delta(x, k)|^2)$$
$$\le \frac{4n}{h^2}\|\delta u_\Delta(k)\|^2. \tag{9.29}$$

We substitute this bound into the right-hand side of (9.28), and replace $2\mu\|u_x(k)\|\,\|u_\Delta(k+1)\|$ in (9.28) by the larger quantity $(v/2)\|u_x(k)\|^2 + (2\mu^2/v)\|u_\Delta(k+1)\|^2$. As a result of this and of the reduction of similar terms in (9.28), we have

$$\|u_\Delta(k+1)\|^2 - \|u_\Delta(k)\|^2 + \varepsilon\|\delta u_\Delta(k)\|^2 + \frac{v}{2}\tau\|u_x(k)\|^2 + v\tau\|u_x(k+1)\|^2 \le \mu\tau\left[\left(1 + \frac{2\mu}{v}\right)\|u_\Delta(k+1)\|^2 + \|u_\Delta(k)\|^2\right] + 2\tau\|f_\Delta(k)\|\,\|u_\Delta(k+1)\|, \tag{9.30}$$

in which $\varepsilon = 1 - 4n\mu h^{-2}\tau$. Let us impose on the ratio $h^{-2}\tau$ the condition

$$1 - 4n\mu h^{-2}\tau \equiv \varepsilon, \tag{9.31}$$

where ε is some number in (0, 1) which is fixed in all subsequent considerations. From (9.30) we deduce the desired bound in a well-known way (cf. §5, Chapter III):

$$\|u_\Delta(m)\|^2 + \nu\tau \sum_{k=0}^{m} \|u_x(k)\|^2 + \varepsilon \sum_{k=0}^{m} \|\delta u_\Delta(k)\|^2 \le c\left[\|u_\Delta(0)\|^2 + \left(\tau \sum_{k=1}^{m} \|f_\Delta(k)\|\right)^2\right], \tag{9.32}$$

in which c is defined by ν, μ, and T, and does not depend on $u_\Delta(k)$, nor on the choice of the grid. This inequality, which indicates the stability of the difference scheme and allows us to prove the convergence of the solutions u_Δ of the difference equations (9.25), (9.26) to the solution of (9.1), (9.2), was derived under the condition (9.31). It forces us to take the time step-sizes τ quite small, which leads to the need for long computation to determine u_Δ at some instant $t_1 > 0$ if h is small.

We shall not carry out here the proof of the fact that the interpolations u'_Δ of the approximate solutions u_Δ converge to the solution $u(x, t)$ of the problem (9.1), (9.2), for this is done according to the same plan as in §7 (see the explanation at the end of the first subsection of this section).

On this note we conclude our analysis of difference schemes for parabolic equations. We remark only that the method set forth here for studying difference schemes can be applied to many other implicit schemes. By using it, we can establish the convergence of difference schemes for the second and third boundary value conditions if their "elliptic part" has the form described in §8 and if the derivative with respect to t is replaced according to one of the methods described here.

§10. Equations of Hyperbolic Type

In this section we shall consider the Cauchy problem and the first initial-boundary value problem for hyperbolic equations of the type

$$\mathscr{L}u = \frac{\partial^2 u}{\partial t^2} - \frac{\partial}{\partial x_i}\left(a_{ij}(x, t)\frac{\partial u}{\partial x_j}\right) + a_i(x, t)\frac{\partial u}{\partial x_i} + a(x, t)u = f(x, t), \tag{10.1}$$

assuming, for the sake of simplification, that the coefficients are continuous functions of (x, t). In addition, the conditions must be satisfied:

$$\nu\xi^2 \le a_{ij}\xi_i\xi_j \le \mu\xi^2, \qquad \nu, \mu = \text{const} > 0, \tag{10.2}$$

$$\left(\sum_{i=1}^{n} a_i^2\right)^{1/2}, \quad |a| \le \mu_1, \quad |a_{ijt}\xi_i\eta_j| \le \mu_2|\xi||\eta|, \tag{10.3}$$

where μ_1, μ_2 are constants, $\xi_1, \ldots, \xi_n, \eta_1, \ldots, \eta_n$ are arbitrary real numbers, and

$$f \in L_{2,1}(Q_T), \qquad \varphi(x) = u|_{t=0} \in W_2^1(\Omega), \qquad \psi(x) = \frac{\partial u}{\partial t}\bigg|_{t=0} \in L_2(\Omega) \quad (10.4)$$

in those domains Q_T of the space (x, t) where the problem is treated. (If the coefficients satisfy weaker requirements than those enumerated here, say, the conditions of Theorem 3.2 of §3, Chapter IV, then, in difference schemes for them, we need to use some averaged values of coefficients as we did in §§7 and 9 of this chapter.)

As we showed in §§1 and 2, Chapter IV, the solution of the Cauchy problem for (10.1) can be reduced to the solution of the first initial-boundary value problem for (10.1). However, this reduction, while relieving us of the need to prove separately the solvability of the Cauchy problem, has only a methodical significance. To actually find the solution of the Cauchy problem in the general case, it is disadvantageous, for it requires solving a more difficult problem.

In this section we shall find separately the solutions of that and of another problem by means of finite-difference approximations. However, the proof of convergence for the Cauchy problem cannot be carried out independently and must be reduced to the proof of convergence for an initial-boundary value problem.

1. The Cauchy Problem. We partition the half-space $(R^{n+1})^+ = \{(x, t): t \geq 0\}$ by the planes $x_i = kh_i$, $h_i > 0$, $k = 0, \pm 1, \pm 2, \ldots, t = k\tau$, $k = 0, 1, 2, \ldots$, into elementary cells, and at their vertices we determine the solutions u_Δ of some difference equations. For the sake of simplicity we shall assume in this section that $h_i = h$, $i = 1, \ldots, n$. We replace (10.1) by the difference equation

$$\mathscr{L}_\Delta u_\Delta = u_{t\bar{t}} - (a_{ij} u_{x_j})_{\bar{x}_i} + a_i u_{x_i} + a u_\Delta = f_\Delta, \qquad (10.5)$$

and require that it be satisfied at all vertices lying in the planes $t = k\tau$, $k = 1, \ldots$. We couple this with the following conditions at the vertices of the grid which lie on the planes $t = k\tau$ $(k = 0, 1)$:

$$u_\Delta|_{t=0} = \varphi_h, \qquad u_\Delta|_{t=\tau} = \varphi_h + \tau\psi_h. \qquad (10.6)$$

The functions f_Δ, φ_h, and ψ_h are formed from f, φ, and ψ, for example, as their (Steklov) means (cf. §9). It is easy to see that the system (10.5), (10.6) enables us to determine uniquely the solution u_Δ at all vertices of the lattice; in addition, the computational process is very simple and does not require the solutions of algebraic systems: each equation of (10.5) contains only one unknown whose coefficient is equal to τ^{-2}. Furthermore, it is clear from (10.5) that the solution u_Δ is connected, at some point of the lattice $(x^0, t^0 = k\tau)$, with its values only at points belonging to a pyramid—the "pyramid of influence" $Q^\Delta(x^0, t^0)$ with vertex at (x^0, t^0) and base on the plane $t = 0$. The base of this pyramid is a cube $\Omega^\Delta(x^0)$ with center x^0 and edge equal to

$2kh$, and its faces lie in the coordinate planes. If we recall the basic property of hyperbolic equations on the finite propagation speed of perturbations and the finiteness of the influence domain (cf., §§1 and 2, Chapter IV), we see without difficulty that, for the convergence of the solutions u_Δ of (10.5), (10.6) to the solution of the Cauchy problem for (10.1), it is necessary that the pyramids of influence for (10.5), with vertex at an arbitrary point (x^0, t^0), contain (or if only their limits as $h, \tau \to 0$ contain) the cone of influence for (10.1) with the same vertex (x^0, t^0). This condition imposes on the step-sizes of τ and h a restriction of the form $\tau/h \leq c$. A common majorant for the propagation speed of perturbations is given by $\mu^{-1/2}$ for all equations (10.1). We shall prove that if we take c smaller than $(2\mu n)^{-1/2}$ or $n^{-1/2}\nu^{1/2}\mu^{-1}$, then (10.5), (10.6) will yield solutions u_Δ which converge to the solution of the Cauchy problem as $h, t \to 0$.

The proof of these assertions is based on a bound for u_Δ which is the difference analogue of the energy bound (2.15) of Chapter IV. This bound can be obtained for u_Δ in an arbitrary "influence pyramid" $Q^\Delta(x^0, t^0)$ in terms of f_Δ in it and φ_h and ψ_n on its base $\Omega^\Delta(x^0)$. However, we proceed instead as follows: We set f_Δ equal to zero outside $Q^\Delta(x^0, t^0)$, we set φ_h equal to the Steklov mean of $\varphi(x)\zeta(x)$—where $\zeta(x)$ is a smooth function equal to 1 on $\Omega^\Delta(x^0)$ and in an h-neighborhood of $\Omega^\Delta(x^0)$ and equal to zero at a certain distance from $\Omega^\Delta(x^0)$—and we set ψ_h equal to zero outside $\Omega^\Delta(x^0)$. To these altered functions f_Δ, φ_h, ψ_h corresponds a solution $\hat{u}_\Delta$ of equations (10.5), (10.6) which coincides with the earlier u_Δ on $Q^\Delta(x^0, t^0)$ and at large distances from $Q^\Delta(x^0, t^0)$ the grid function $\hat{u}_\Delta$ vanishes. Hence $\hat{u}_\Delta$ can be considered as an approximate solution of the initial-boundary value problem in some parallelopiped $\Pi^\Delta(x^0, t^0)$, of height $t^0 = k\tau$, containing $Q^\Delta(x^0, t^0)$ and such that, on its lateral surface, $\hat{u}_\Delta$ is zero. Thus, these $\hat{u}_\Delta$ originate as solutions of the first initial-boundary value problem with zero boundary conditions by means of the difference scheme (10.5), (10.6). In the next subsection we shall obtain for them the difference analogue of the energy bound. From this estimate follows the uniform boundedness of u_Δ in Q^Δ. By means of this the convergence of u_Δ to the solution of the Cauchy problem for (10.1) can be proved in the same way as it was above in §§7–9 for equations of other types.

2. The First Initial-Boundary Value Problem. In order to find a solution $u(x, t)$ of the problem

$$\mathscr{L}u = f, \qquad u|_{t=0} = \varphi(x), \qquad \left.\frac{\partial u}{\partial t}\right|_{t=0} = \psi(x), \qquad u|_{S_T} = 0 \quad (10.7)$$

in the cylinder $Q_T = \Omega \times (0, T)$, we use a difference scheme analogous to that studied in sub-section 3 of §6 of this chapter, for the equation of a vibrating string. It coincides, in its principal part, with the scheme of the preceding subsection for the solution of the Cauchy problem. Let the conditions (10.2)–(10.4) be satisfied for (10.7), and let $\varphi(x) \in \mathring{W}_2^1(\Omega)$. In this subsection we use the same notation Ω_h, S_h, $\overline{\Omega}_h$, Ω_h^+, and $\|u_\Delta\|$, $\|u_x\|$ as we did in §§7–9.

We recall that, as long as the functions $u_\Delta(k)$ appearing in our considerations vanish on S_h for all k, the sum $\|u_\Delta(k)\|^2 \equiv \Delta_h \sum_{\Omega_h^+} u_\Delta^2(k)$ will be equal to the sums $\Delta_h \sum_{\bar{\Omega}_h} u_\Delta^2(k)$ and $\Delta_h \sum_{\Omega_h} u_\Delta^2(k)$, and

$$\|u_x(k)\|^2 \equiv \Delta_h \sum_{\Omega_h^+} \sum_{i=1}^{n} u_{x_i}^2(k) = \Delta_h \sum_{\bar{\Omega}_h} u_x^2(k),$$

if we stipulate that $u_\Delta(k)$ vanish outside $\bar{\Omega}_h$. Just as in the preceding subsection, we shall assume for simplicity that all $h_i = h, i = 1, \ldots, n$.

We form for $\varphi(x)$ grid functions φ_h in such a way that φ_h vanishes on S_h and outside Ω_h and so that the interpolations $\tilde{\varphi}_h(\cdot)$ and $\tilde{\varphi}_{x_i}(\cdot)$ converge weakly in $L_2(\Omega)$ to $\varphi(\cdot)$ and $\partial\varphi/\partial x_i$, respectively (in particular, in such a way that $\|\varphi_h\|$ and $\|\varphi_x\|$ are uniformly bounded); for the construction of such φ_h, see the end of §4 of this chapter.

As ψ_h and f_Δ we may take, for example, the same grid functions as we did in subsection 3, §9, for φ and f.†

We define the approximate solutions u_Δ as grid functions satisfying the difference equations (10.5) at the vertices $x_k \in \Omega_h$ at the levels t_k, $k = 1, 2, \ldots, N$. To this we add the equations (10.6) at the points $x_k \in \Omega_h$, along with the equations

$$u_\Delta(k)|_{S_h} = 0, \qquad k = 0, 1, \ldots, N. \tag{10.8}$$

It is easily seen that these equations determine u_Δ uniquely, where the calculation of the u_Δ does not require the solution of any algebraic systems, that is, the scheme is explicit. For u_Δ we have the bound

$$h^n \sum_{\Omega_h^+} (u_{\bar{t}}^2(m) + u_x^2(m) + u^2(m)) \leq ch^n \sum_{\Omega_h^+} (\psi_h^2 + \varphi_h^2 + \varphi_x^2) + c\left[\tau \sum_{k=1}^{m-1} \left(h^n \sum_{\Omega_h^+} (f_\Delta(k))^2\right)^{1/2}\right]^2, \tag{10.9}$$

in which c is defined by the constants ν, μ, μ_1, μ_2 from (10.2), (10.3) and by the height T, and where m is an arbitrary integer less than $N + 2 \equiv [T/\tau] + 2$ if τ/h is only less than $(2\mu n)^{-1/2}$ or $n^{-1/2}\nu^{1/2}\mu^{-1}$. This bound is the difference analogue of the energy inequality which we mentioned above. It implies the stability of our difference scheme, and provides a basis for the proof of the convergence of the u_Δ to the solution $u(x, t)$ of the problem (10.7). The derivation of (10.9) is sufficiently complicated that we first clarify it by the example of the wave operator. Thus, let $u(x, t)$ be a solution of the problem

$$\mathscr{L}_0 u = \frac{\partial^2 u}{\partial t^2} - \sum_{i=1}^{n} \frac{\partial^2 u}{\partial x_i^2} = f, \qquad u|_{t=0} = \varphi, \frac{\partial u}{\partial t}\bigg|_{t=0} = \psi, \qquad u|_{S_T} = 0, \tag{10.10}$$

† As will be clear from what follows, it is enough for us to know $f_\Delta(k)$ only for $x \in \Omega_h$, and we use $f_\Delta(k)$ on part of S_h only for writing the norm $\|f_\Delta\|$ in the same way. At points of S_h, for example, we may assume that $f_\Delta(k)$ vanishes.

and u_Δ the solution of the corresponding difference equations

$$\mathscr{L}_{0\Delta}u_\Delta = u_{t\bar{t}} - u_{x_i\bar{x}_i} = f_\Delta, \qquad u_\Delta|_{t=0} = \varphi_h,$$
$$u_\Delta|_{t=\tau} = \varphi_h + \tau\psi_h, \qquad u_\Delta|_{S_h} = 0. \tag{10.11}$$

Let us multiply both sides of the first of the equations of (10.11), taken at the level t_k, by $\tau h^n(u_t(k) + u_{\bar{t}}(k))$, sum the result over all nodes of Ω_h^+ and over all k from 1 to m, and then transform the terms with $u_{x_i\bar{x}_i}$ by means of (2.11), our formula for "summation by parts," recalling that the quantity $u_t(k) + u_{\bar{t}}(k)$ vanishes on S_h. Thus we obtain

$$\begin{aligned}\tau\sum_{k=1}^{m} h^n \sum_{\Omega_h^+} u_{t\bar{t}}(k)(u_t(k) + u_{\bar{t}}(k))\\ + \tau\sum_{k=1}^{m} h^n \sum_{\Omega_h^+} u_{x_i}(k)(u_{tx_i}(k) + u_{\bar{t}x_i}(k))\\ = \tau\sum_{k=1}^{m} h^n \sum_{\Omega_h^+} f_\Delta(k)(u_t(k) + u_{\bar{t}}(k)).\end{aligned} \tag{10.12}$$

We can verify by direct calculation that the following equalities hold:

$$\tau\sum_{k=1}^{m} u_{t\bar{t}}(k)(u_t(k) + u_{\bar{t}}(k)) = \tau\sum_{k=1}^{m} (u_{\bar{t}}^2(k))_t = u_{\bar{t}}^2(m+1) - u_{\bar{t}}^2(1), \tag{10.13}$$

$$2\tau u_{x_i}(k)u_{\bar{t}x_i}(k) = u_x^2(k) - u_x^2(k-1) + (\delta u_x(k-1))^2, \tag{10.14}$$

where $\delta v(k) = v(k+1) - v(k)$, and

$$2\tau u_{x_i}(k)u_{tx_i}(k) = u_x^2(k+1) - u_x^2(k) - (\delta u_x(k))^2 \tag{10.15}$$

(in connection with (10.14) and (10.15), see (2.7) and (2.8) in this chapter). Because of (10.13)–(10.15), we can represent (10.12) as

$$\begin{aligned}\|u_{\bar{t}}(m+1)\|^2 + \tfrac{1}{2}\|u_x(m+1)\|^2 + \tfrac{1}{2}\|u_x(m)\|^2 - \tfrac{1}{2}\|\delta u_x(m)\|^2\\ - \|u_{\bar{t}}(1)\|^2 - \tfrac{1}{2}\|u_x(1)\|^2 - \tfrac{1}{2}\|u_x(0)\|^2 + \tfrac{1}{2}\|\delta u_x(0)\|^2\\ = \tau\sum_{k=1}^{m} h^n \sum_{\Omega_h^+} f_\Delta(k)(u_t(k) + u_{\bar{t}}(k)).\end{aligned} \tag{10.16}$$

We write the sum

$$j_2(m) = \tfrac{1}{2}\|u_x(m+1)\|^2 + \tfrac{1}{2}\|u_x(m)\|^2 - \tfrac{1}{2}\|\delta u_x(m)\|^2$$

in the form

$$\begin{aligned}j_2(m) &= h^n \sum_{\Omega_h^+}\sum_i u_{x_i}(m+1)u_{x_i}(m)\\ &= h^n \sum_{\Omega_h^+} [u_x^2(m+1) - u_{x_i}(m+1)\delta u_{x_i}(m)],\end{aligned}$$

and then estimate $\|\delta u_x(m)\|$, recalling the definitions of the symbols $(\ )_x$ and δ, in such a way that:

$$\|\delta u_x(m)\| = \left(h^n \sum_{\Omega_h^+} \sum_{i=1}^n [u_{\bar{t}}(x + he_i, m + 1) - u_{\bar{t}}(x, m + 1)]^2\right)^{1/2} \frac{\tau}{h}$$

$$\leq 2\sqrt{n}\frac{\tau}{h}\|u_{\bar{t}}(m + 1)\|. \tag{10.17}$$

Hence

$$j_2(m) \geq \|u_x(m + 1)\|^2 - 2\sqrt{n}\frac{\tau}{h}\|u_x(m + 1)\| \cdot \|u_{\bar{t}}(m + 1)\|. \tag{10.18}$$

Because of (10.18), it follows from (10.16) that

$$\|u_{\bar{t}}(m + 1)\|^2 + \|u_x(m + 1)\|^2 - 2\sqrt{n}\frac{\tau}{h}\|u_x(m + 1)\|\,\|u_{\bar{t}}(m + 1)\|$$

$$\leq \tfrac{1}{2}\|u_x(1)\|^2 + \tfrac{1}{2}\|u_x(0)\|^2 - \tfrac{1}{2}\|\delta u_x(0)\|^2 + \|u_{\bar{t}}(1)\|^2$$

$$+ \tau \sum_{k=1}^m h^n \sum_{\Omega_h^+} f_\Delta(k)(u_t(k) + u_{\bar{t}}(k))$$

$$\leq h^n \sum_{\Omega_h^+} u_x(1)u_x(0) + \|\psi_h\|^2 + \tau \sum_{k=1}^m \|f_\Delta(k)\|(\|u_t(k)\| + \|u_{\bar{t}}(k)\|)$$

$$\leq \|\varphi_x\|^2 + 2\sqrt{n}\frac{\tau}{h}\|\varphi_x\|\,\|\psi_h\| + \|\psi_h\|^2$$

$$+ 2\tau \sum_{k=1}^m \|f_\Delta(k)\| \cdot \max_{1 \leq k \leq m+1} \|u_{\bar{t}}(k)\|. \tag{10.19}$$

For these bounds we wrote $j_2(0) \equiv h^n \sum_{\Omega_h^+} u_x(1)u_x(0)$ in the form $\|u_x(0)\|^2 + h^n \sum_{\Omega_h^+} \delta u_x(0)u_x(0)$ and bounded the last term by means of (10.17), taking $m = 0$ in (10.17). If τ and h are subject to the condition

$$\sqrt{n}\frac{\tau}{h} \leq 1 - \varepsilon \quad \text{for some} \quad \varepsilon \in (0, 1), \tag{10.20}$$

then (10.19) implies that

$$\|u_{\bar{t}}(m + 1)\|^2 + \|u_x(m + 1)\|^2 \leq 2\varepsilon^{-1}(j_0 + \mathscr{F}_\Delta(m) \max_{1 \leq k \leq m+1} \|u_{\bar{t}}(k)\|), \tag{10.21}$$

where $m \geq 1$,

$$j_0 = \|\varphi_x\|^2 + \|\psi_h\|^2, \tag{10.22$_1$}$$

and

$$\mathscr{F}_\Delta(m) \equiv \|f_\Delta\|_{2,1,Q_m^\Delta} \equiv \tau \sum_{k=1}^m \|f_\Delta(k)\|. \tag{10.22$_2$}$$

We introduce the notations

$$\max_{1\le k\le m} \|u_{\bar t}(k)\| = z(m), \qquad \max_{1\le k\le m} \|u_x(k)\| = y(m). \tag{10.23}$$

For $z(m)$ it follows from (10.21) that

$$\begin{aligned} z^2(m+1) &\le 2\varepsilon^{-1}[j_0 + \mathscr{F}_\Delta(m)z(m+1)] \\ &\le 2\varepsilon^{-1}\left[j_0 + \frac{\varepsilon}{4}z^2(m+1) + \varepsilon^{-1}\mathscr{F}_\Delta^2(m)\right], \end{aligned}$$

whence

$$z^2(m+1) \le 4\varepsilon^{-1}[j_0 + \varepsilon^{-1}\mathscr{F}_\Delta^2(m)]. \tag{10.24}$$

From (10.24) and (10.21) we obtain the desired "energy" bound,

$$\begin{aligned} &\max_{1\le k\le m} [\|u_{\bar t}(k)\|^2 + \|u_x(k)\|^2] \\ &\quad\le 2\varepsilon^{-1}[j_0 + \mathscr{F}_\Delta(m-1)2\varepsilon^{-1/2}\cdot\sqrt{j_0 + \varepsilon^{-1}\mathscr{F}_\Delta^2(m-1)}] \\ &\quad\le 2\varepsilon^{-1}(2j_0 + 3\varepsilon^{-1}\mathscr{F}_\Delta^2(m-1)). \end{aligned} \tag{10.25}$$

Remark 10.1. For $n = 1$ in the problem (10.10), the stability of the difference scheme (10.11) holds when $\tau/h \le 1$, which is somewhat less restrictive than (10.20). This can be shown if we regroup the sum of the first four terms on the left-hand side of (10.16) in the corresponding way. But we shall not do this here since such a regrouping achieves this goal only for special classes of the equations considered in this subsection.

We shall now prove (10.9) for the general case, i.e., for the solutions u_Δ of the system

$$\mathscr{L}_\Delta u_\Delta(k) = f_\Delta(k), \tag{10.26}$$

$$u_\Delta|_{t=0} = \varphi_h, \qquad u_\Delta|_{t=\tau} = \varphi_h + \tau\psi_h, \qquad u_\Delta(k)|_{S_h} = 0, \tag{10.27}$$

where $\mathscr{L}_\Delta$ was defined in (10.5). Just as in the case of the system (10.11), we multiply (10.26) by $\tau h^n(u_t(k) + u_{\bar t}(k))$, sum the result over all nodes of Ω_h^+ and over k from 1 to m, and transform the two principle terms into forms like (10.13) and (10.12):

$$\begin{aligned} &\|u_{\bar t}(m+1)\|^2 - \|u_{\bar t}(1)\|^2 + \tau\sum_{k=1}^{n} h^n \sum_{\Omega_h^+} [a_{ij}(k)u_{x_j}(k)(u_{tx_i}(k) + u_{\bar t x_i}(k)) \\ &\qquad + (a_i(k)u_{x_i}(k) + a(k)u_\Delta(k))(u_t(k) + u_{\bar t}(k))] \\ &\quad = \tau\sum_{k=1}^{m} h^n \sum_{\Omega_h^+} f_\Delta(k)(u_t(k) + u_{\bar t}(k)). \end{aligned} \tag{10.28}$$

We transform the sum

$$j_1(m) = \tau \sum_{k=1}^{m} a_{ij}(k) u_{x_j}(k)(u_{tx_i}(k) + u_{\bar{t}x_i}(k))$$

in the following way:

$$j_1(m) = \sum_{k=1}^{m} a_{ij}(k) u_{x_j}(k)[u_{x_i}(k+1) - u_{x_i}(k-1)]$$

$$= \sum_{k=1}^{m} a_{ij}(k) u_{x_i}(k+1) u_{x_j}(k) - \sum_{k=1}^{m} a_{ij}(k) u_{x_i}(k) u_{x_j}(k-1)$$

$$= \sum_{k=1}^{m} a_{ij}(k+1) u_{x_i}(k+1) u_{x_j}(k) - \sum_{k=1}^{m} a_{ij}(k) u_{x_i}(k) u_{x_j}(k-1)$$

$$- \sum_{k=1}^{m} \delta a_{ij}(k) u_{x_i}(k+1) u_{x_j}(k) = a_{ij}(m+1) u_{x_i}(m+1) u_{x_j}(m)$$

$$- a_{ij}(1) u_{x_i}(1) u_{x_j}(0) - \sum_{k=1}^{m} \delta a_{ij}(k) u_{x_i}(k+1) u_{x_j}(k).$$

Hence (10.28) assumes the form

$$\|u_{\bar{t}}(m+1)\|^2 + h^n \sum_{\Omega_h^+} a_{ij}(m+1) u_{x_i}(m+1) u_{x_j}(m)$$

$$= \|u_{\bar{t}}(1)\|^2 + h^n \sum_{\Omega_h^+} a_{ij}(1) u_{x_i}(1) u_{x_j}(0)$$

$$+ \tau \sum_{k=1}^{m} h^n \sum_{\Omega_h^+} \left\{ \frac{\delta a_{ij}(k)}{\tau} u_{x_i}(k+1) u_{x_j}(k) \right.$$

$$\left. - [a_i(k) u_{x_i}(k) + a(k) u_\Delta(k)][u_t(k) + u_{\bar{t}}(k)] \right\}$$

$$+ \tau \sum_{k=1}^{m} h^n \sum_{\Omega_h^+} f_\Delta(k)[u_t(k) + u_{\bar{t}}(k)]. \tag{10.29}$$

Let us write the expression

$$j_2(m) = a_{ij}(m+1) u_{x_i}(m+1) u_{x_j}(m)$$

in the form of a sum

$$j_2(m) = a_{ij}(m+1) u_{x_j}(m+1) u_{x_i}(m+1)$$
$$- a_{ij}(m+1) u_{x_j}(m+1) \delta u_{x_i}(m).$$

Below, in Remark 10.2 to the proof of the inequality (10.50), we shall require another representation for $j_2(m)$, namely,

$$j_2(m) = \tfrac{1}{2}[a_{ij}(m+1) u_{x_i}(m+1) u_{x_j}(m+1)$$
$$+ a_{ij}(m) u_{x_i}(m) u_{x_j}(m) - a_{ij}(m+1) \delta u_{x_i}(m) \delta u_{x_j}(m)$$
$$+ \delta a_{ij}(m) u_{x_i}(m) u_{x_j}(m)]. \tag{10.30}$$

But now we return to (10.29), and bound the terms appearing in it, namely,

$$j_3(m) = \tau \sum_{k=1}^{m} h^n \sum_{\Omega_h^+} \frac{\delta a_{ij}(k)}{\tau} u_{x_i}(k+1) u_{x_j}(k),$$

$$j_4(m) = -\tau \sum_{k=1}^{m} h^n \sum_{\Omega_h^+} [a_i(k) u_{x_i}(k) + a(k) u_\Delta(k)][u_t(k) + u_{\bar{t}}(k)]$$

by applying (10.3) in the following way:

$$\begin{aligned} |j_3(m)| &\le \mu_2 \tau \sum_{k=1}^{m} h^n \sum_{\Omega_h^+} |u_x(k+1)| |u_x(k)| \\ &\le \mu_2 \tau \sum_{k=1}^{m} \|u_x(k+1)\| \, \|u_x(k)\|, \end{aligned} \tag{10.31}$$

$$\begin{aligned} |j_4(m)| &\le \mu_1 \tau \sum_{k=1}^{m} h^n \sum_{\Omega_h^+} (|u_x(k)| + |u_\Delta(k)|)(|u_t(k)| + |u_{\bar{t}}(k)|) \\ &\le \mu_1 \tau \sum_{k=1}^{m} (\|u_x(k)\| + \|u_\Delta(k)\|)(\|u_t(k)\| + \|u_{\bar{t}}(k)\|). \end{aligned}$$

The last term in (10.29),

$$j_5(m) = \tau \sum_{k=1}^{m} h^n \sum_{\Omega_h^+} f_\Delta(k)[u_t(k) + u_{\bar{t}}(k)],$$

is bounded as follows:

$$|j_5(m)| \le \tau \sum_{k=1}^{m} \|f_\Delta(k)\| (\|u_t(k)\| + \|u_{\bar{t}}(k)\|). \tag{10.32}$$

Finally, we give a lower bound for $h^n \sum_{\Omega_h^+} j_2(m)$ by applying (10.2), (10.3), and (10.17):

$$\begin{aligned} h^n \sum_{\Omega_h^+} j_2(m) &\ge \nu \|u_x(m+1)\|^2 - \mu \|u_x(m+1)\| \, \|\delta u_x(m)\| \\ &\ge \nu \|u_x(m+1)\|^2 - 2\mu \sqrt{n} \frac{\tau}{h} \|u_x(m+1)\| \, \|u_{\bar{t}}(m+1)\|. \end{aligned} \tag{10.33}$$

It follows from (10.29)–(10.33) that

$$\begin{aligned} &\|u_{\bar{t}}(m+1)\|^2 + \nu \|u_x(m+1)\|^2 - 2\mu \sqrt{n} \frac{\tau}{h} \|u_x(m+1)\| \, \|u_{\bar{t}}(m+1)\| \\ &\le h^n \sum_{\Omega_h^+} j_2(0) + \tau \sum_{k=1}^{m} [\mu_2 \|u_x(k+1)\| \, \|u_x(k)\| + (\mu_1 \|u_x(k)\| \\ &\quad + \mu_1 \|u_\Delta(k)\| + \|f_\Delta(k)\|)(\|u_t(k)\| + \|u_{\bar{t}}(k)\|)] + \|u_{\bar{t}}(1)\|^2 \\ &\le \mu \|u_x(1)\| \, \|u_x(0)\| + \sum_{k=1}^{m+1} \|u_x(k)\|^2 (\mu_2 + \mu_1) \\ &\quad + \tau \mu_1 \sum_{k=1}^{m} \|u_\Delta(k)\|^2 + \tau \sum_{k=1}^{m+1} 2\mu_1 \|u_{\bar{t}}(k)\|^2 \\ &\quad + \tau \sum_{k=1}^{m} \|f_\Delta(k)\| (\|u_t(k)\| + \|u_{\bar{t}}(k)\|) + \|u_{\bar{t}}(1)\|^2. \end{aligned} \tag{10.34}$$

If τ and h are subject to the condition

$$\frac{\mu}{\sqrt{\nu}}\sqrt{n}\frac{\tau}{h} = 1 - \varepsilon, \tag{10.35}$$

where ε is some number in (0, 1), then the left-hand side of (10.34) will not be less than $\varepsilon\|u_{\bar{t}}(m+1)\|^2 + \varepsilon\nu\|u_x(m+1)\|^2$, whence

$$\begin{aligned}\|u_{\bar{t}}(m+1)\|^2 + \nu\|u_x(m+1)\|^2 \le \varepsilon^{-1}\Big[&\mu(\|\varphi_x\| + \tau\|\psi_x\|)\|\varphi_x\| \\ &+ \|\psi_h\|^2 + (\mu_1 + \mu_2)\tau\sum_{k=1}^{m+1}\|u_x(k)\|^2 + \mu_1\tau\sum_{k=1}^{m}\|u_\Delta(k)\|^2 \\ &+ 2\mu_1\tau\sum_{k=1}^{m+1}\|u_{\bar{t}}(k)\|^2 + 2\|f_\Delta\|_{2,1,Q_m^\Delta}\max_{1\le k\le m+1}\|u_{\bar{t}}(k)\|\Big],\end{aligned} \tag{10.36}$$

where

$$\|f_\Delta\|_{2,1,Q_m^\Delta} = \tau\sum_{k=1}^{m}\|f_\Delta(k)\|. \tag{10.37}$$

From the obvious equality

$$u_\Delta(k) = u_\Delta(0) + \tau\sum_{l=1}^{k}u_{\bar{t}}(l)$$

we obtain the following bounds for $\|u_\Delta(k)\|$:

$$\begin{aligned}\|u_\Delta(k)\| &\le \|u_\Delta(0)\| + \tau\sum_{l=1}^{k}\|u_{\bar{t}}(l)\| \\ &\le \|u_\Delta(0)\| + \left(\tau\sum_{l=1}^{k}\|u_{\bar{t}}(l)\|^2\right)^{1/2}\sqrt{k\tau}\end{aligned} \tag{10.38}$$

and

$$\|u_\Delta(k)\|^2 \le 2\|u_\Delta(0)\|^2 + 2k\tau^2\sum_{l=1}^{k}\|u_{\bar{t}}(l)\|^2. \tag{10.39}$$

Let us substitute (10.39) into the right-hand side of (10.36) and bound it as follows:

$$\begin{aligned}&\|u_{\bar{t}}(m+1)\|^2 + \nu\|u_x(m+1)\|^2 \\ &\le \varepsilon^{-1}\Big[2\mu\|\varphi_x\|^2 + 2\|\psi_h\|^2 + 2\mu_1 m\tau\|\varphi_h\|^2 \\ &\quad + (\mu_1 + \mu_2)\tau\sum_{k=1}^{m+1}\|u_x(k)\|^2 + (2(m\tau)^2\mu_1 + 2\mu_1)\tau\sum_{k=1}^{m+1}\|u_{\bar{t}}(k)\|^2 \\ &\quad + 2\|f_\Delta\|_{2,1,Q_m^\Delta}\max_{1\le k\le m+1}\|u_{\bar{t}}(k)\|\Big].\end{aligned} \tag{10.40}$$

We introduce the notation

$$\left.\begin{aligned}&\max_{1\le k\le m}\|u_{\bar t}(k)\| = z(m), \qquad \nu^{1/2}\max_{1\le k\le m}\|u_x(k)\| = y(m),\\&2\mu\|\varphi_x\|^2 + 2\|\psi_h\|^2 + 2\mu_1 m\tau\|\varphi_h\|^2 = c(m),\\&(\mu_1+\mu_2)\nu^{-1} = c_1,\ 2(m\tau)^2\mu_1 + 2\mu_1 = c_2(m),\\&\|f_\Delta\|_{2,1,Q_m^\Delta} = \mathscr{F}_\Delta(m).\end{aligned}\right\} \tag{10.41}$$

Since (10.40) holds for arbitrary m, beginning with $m = 0$, it follows from (10.40) that

$$\begin{aligned}z^2(m+1) + y^2(m+1) &\le 2\varepsilon^{-1}[c(m) + c_1\tau(m+1)y^2(m+1)\\&\quad + c_2(m)(m+1)\tau z^2(m+1) + 2\mathscr{F}_\Delta(m)z(m+1)]\\&\le 2\varepsilon^{-1}\Big\{c(m) + c_1\tau(m+1)y^2(m+1)\\&\quad + \left[\frac{\varepsilon}{8} + c_2(m)\tau(m+1)\right]z^2(m+1) + 8\varepsilon^{-1}\mathscr{F}_\Delta^2(m)\Big\}.\end{aligned} \tag{10.42}$$

If m satisfies the condition

$$\max\{2\varepsilon^{-1}c_1\tau(m+1);\ \tfrac{1}{4} + 2\varepsilon^{-1}c_2(m)\tau(m+1)\} \le \tfrac{1}{2} \tag{10.43}$$

we obtain from (10.42) the bound

$$z^2(m+1) + y^2(m+1) \le 4\varepsilon^{-1}[c(m) + 8\varepsilon^{-1}\mathscr{F}_\Delta^2(m)], \tag{10.44}$$

and from this and (10.39) we have

$$\begin{aligned}&w^2(m+1) + y^2(m+1) + z^2(m+1)\\&\qquad \le 2\|\varphi_h\|^2 + [1 + 2(\tau(m+1))^2]4\varepsilon^{-1}[c(m) + 8\varepsilon^{-1}\mathscr{F}_\Delta^2(m)],\end{aligned} \tag{10.45}$$

where

$$w(m) = \max_{1\le k\le m}\|u_\Delta(k)\|. \tag{10.46}$$

The bound (10.45) is derived for m satisfying (10.43), in which c_1 and $c_2(m)$ depend only on ν, μ_1, μ_2, and $m\tau$. From the initial data φ and ψ we use in (10.45) the same norms as we get for $u_\Delta(k)$ on the left-hand side, i.e., from $\varphi = u|_{t=0}$ we use the quantity $\|\varphi_h\|^2 + \|\varphi_x\|^2$, and from $u_t|_{t=0}$ the quantity $\|\psi_h\|$. Hence if m_1 is the largest value of m for which (10.43) is satisfied, and hence for which (10.45) holds, and if $\tau(m_1+1) < T$, then the level $t_1 = m_1\tau$ can be taken as the starting level, and, starting from it, we can deduce (10.45) for the level $t_2 \le \min\{(2m_1+1)\tau; T\}$. If t_2 turns out to be less than T, then it is possible to rise to the height $(m_1+1)\tau$ from $t_2 = \tau + 2m_1\tau$, and so on, until we exhaust all of T. It is clear that, by a finite number of steps satisfying (10.43), we exhaust all of T and arrive at the bound

$$\begin{aligned}&\max_{1\le k\le m}\|u_{\bar t}(k)\|^2 + \max_{1\le k\le m}\|u_x(k)\|^2 + \max_{1\le k\le m}\|u_\Delta(k)\|^2\\&\qquad \le c(m\tau;\varepsilon)[\|\psi_h\|^2 + \|\varphi_x\|^2 + \|\varphi_h\|^2 + \|f_\Delta\|^2_{2,1,Q_m^\Delta}],\end{aligned} \tag{10.47}$$

which holds for all $m \leq T/\tau + 1$. Here $c(m\tau: \varepsilon)$ depends only on $\nu, \mu, \mu_1, m\tau$ and on ε from (10.35). The inequality (10.47) is the difference analogue of the energy inequality; it guarantees the stability of our difference scheme with a fixed ε from (0, 1) (in those norms appearing in it). It can be used to establish the convergence of the solutions u_Δ of the difference equations (10.5), (10.6), (10.8) to a solution of the problem (10.1), (10.4). This can be done in the same way as it was done in §7 for elliptic equations, where in this case, instead of (10.47), it is enough to have the inequality

$$\tau \sum_{k=1}^{[T/\tau]+1} h^n \sum_{\Omega_h^+} [u_{\bar{t}}^2(k) + u_x^2(k) + u_\Delta^2(k)] \leq Tc(T, \varepsilon)[\|\psi_h\|^2 + \|\varphi_x\|^2 + \|\varphi_h\|^2 + \|f_\Delta\|^2_{2,1,Q_T^\Delta}], \tag{10.48}$$

which is the corollary of (10.47). We shall not carry out this argument in detail, for it does not differ appreciably from the corresponding arguments in earlier sections. We shall describe only the basic steps and the results obtained. We can deduce from (10.48) the uniform boundedness of the norms in $W_2^1(Q_T)$—and hence the weak compactness in $W_2^1(Q_T)$—of the continuous multilinear interpolations u'_Δ of the solutions u_Δ of the systems (10.5), (10.6), (10.8) for arbitrary h and τ such that τ/h satisfies (10.35) with a fixed $\varepsilon \in (0, 1)$. Furthermore, it can be proved that if some subsequence of $\{u'_\Delta\}$ converges weakly in $W_2^1(Q_T)$ to a function $u(x, t)$, then $u(x, t)$ is the generalized solution of (10.7) in $W_2^1(Q_T)$. From these two facts and the uniqueness theorem (Theorem 3.1), if, naturally, its hypotheses are satisfied, it follows that all sequences of $\{u'_\Delta\}$ converge weakly in $W_2^1(Q_T)$, and strongly in $L_2(Q_T)$, to the solution $u(x, t)$. By this argument we prove simultaneously with the calculation of the approximate solutions of (10.7) the theorem of existence of a generalized solution in $W_2^1(Q_T)$.

Remark 10.2. A bound of the form (10.47) can be obtained under a condition on τ/h differing from (10.35), namely, under the condition

$$\sqrt{2\mu n}\,\frac{\tau}{h} = 1 - \varepsilon, \tag{10.49}$$

where ε is an arbitrary number in (0, 1). For this we must bound $h^n \sum_{\Omega_h^+} j_2(m)$ from below, starting from the representation (10.30) for $j_2(m)$ and the estimate (10.17) in such a way that

$$\begin{aligned} h^n \sum_{\Omega_h^+} j_2(m) &\geq \frac{\nu}{2}\|u_x(m+1)\|^2 + \frac{\nu}{2}\|u_x(m)\|^2 \\ &\quad - \frac{\mu}{2}\|\delta u_x(m)\|^2 - \frac{\mu_2}{2}\tau\|u_x(m)\|^2 \geq \frac{\nu}{2}\|u_x(m+1)\|^2 \\ &\quad + \frac{\nu}{2}\|u_x(m)\|^2 - 2\mu n\left(\frac{\tau}{h}\right)^2\|u_{\bar{t}}(m+1)\|^2 - \frac{\mu_2\tau}{2}\|u_x(m)\|^2. \end{aligned} \tag{10.50}$$

The method of finite differences allows us to observe the increase of smoothness of generalized solutions of (10.7) in $W_2^1(Q_T)$ with the increase in smoothness of all the data of the problem and in their order of compatibility. This is tedious enough in practice, but transparent in principle. Such an investigation was undertaken by us and is presented in our book [4]; it enabled us to clarify those actual connections which take place in the problem (10.7) between the smoothness of the data and the smoothness of the solution. We remark, incidentally, that this way was the first by which one could investigate successfully the solvability of (10.7) in various function spaces under reasonable hypotheses on the data of the problem. The method of finite differences gives a comparatively simple method of actually computing the solution on a finite time-interval.

The method is suitable for determining the solutions of initial-boundary value problems under other classical boundary conditions. Moreover, without any essential changes, it can be used to investigate the so-called strongly hyperbolic systems. Such systems are of the type

$$\mathbf{u}_{tt} - \mathscr{L}\mathbf{u} = \mathbf{f},$$

where $\mathscr{L}$ is a strongly elliptic operator with symmetric principal part. To these systems belong a number of important systems of mathematical physics, for example the system of the dynamical theory of elasticity. We shall consider some of these systems in §11.

In [LA 4] there is for (10.7) an implicit difference scheme, which is convergent for arbitrary step-size ratios in x and t. A special case of this is the scheme (6.99) for $\gamma = \frac{1}{2}$, which we considered in the third subsection of §6. This scheme is suitable not only for the equations (10.7), but also for the most general hyperbolic equations containing terms of the form $\partial^2 u/\partial t\, \partial x_i$. We describe it for the problem

$$\frac{\partial^2 u}{\partial t^2} + 2\sum_{i=1}^{n} a_{0i}(x,t)\frac{\partial^2 u}{\partial t\, \partial x_i} - \sum_{i,j=1}^{n} a_{ij}(x,t)\frac{\partial^2 u}{\partial x_i\, \partial x_j} + \sum_{i=1}^{n} a_i(x,t)\frac{\partial u}{\partial x_i} + a_0(x,t)\frac{\partial u}{\partial t} + a(x,t)u = f(x,t), \tag{10.51}$$

$$u|_{S_T} = 0, \qquad u|_{t=0} = \varphi(x), \qquad \left.\frac{\partial u}{\partial t}\right|_{t=0} = \psi(x). \tag{10.52}$$

We introduce the notations

$$\frac{\tilde{\Delta}^2 u}{\Delta t\, \Delta x_i} = \tfrac{1}{4}(u_t + u_{\bar{t}})_{x_i} + \tfrac{1}{4}(u_t + u_{\bar{t}})_{\bar{x}_i}$$

and

$$\frac{\tilde{\Delta}^2 u(x,t)}{\Delta x_i\, \Delta x_j} = \tfrac{1}{2}u_{x_i\bar{x}_j}(x, t+\tau) + \tfrac{1}{2}u_{x_i\bar{x}_j}(x, t-\tau).$$

Then (10.51) can be replaced by

$$u_{t\bar{t}} + 2\sum_{i=1}^{n} a_{0i}\frac{\bar{\Delta}^2 u}{\Delta t\,\Delta x_i} - \sum_{i,j=1}^{n} a_{ij}\frac{\bar{\Delta}^2 u}{\Delta x_i\,\Delta x_j} + \sum_{i=1}^{n} a_i u_{\bar{x}_i} + a_0 u_{\bar{t}} + au = f_\Delta, \tag{10.53}$$

which should be satisfied for $x \in \Omega_h$ and $t = t_k$, $k = 1, \ldots, N$. Conditions (10.27) are to be adjoined to it. It has been proved that this scheme is stable (and hence, in the limit, gives the solution of the problem (10.51), (10.52)) for arbitrary size of the ratio $\Delta t/\Delta x_i$ (where the Δx_i can, as everywhere, be chosen to be different). For the solutions u_Δ of the scheme we have an inequality of the form (10.47) which can be derived much more easily than the inequality (10.47) for the scheme (10.26), (10.27). We remark that the scheme (10.26), (10.27) yields for the problem (10.51), (10.52) a non-convergent process for arbitrary ratio-size $\Delta t/\Delta x$ = const (if, naturally, $a_{0i} \not\equiv 0$).

§11. Strong Convergence, Systems, Diffraction Problems

1. On Strong Convergence. In the passage to the limit from solutions u_Δ of a difference equation to solutions of the corresponding boundary value problem, we confined ourselves, in this chapter, to the proof of weak convergence in the energy norms. Thus, for example, it was proved for the Dirichlet problem (7.1) in §7 that the interpolations $u'_\Delta(\cdot)$ converge strongly in $L_2(\Omega)$ to a solution $u(\cdot)$ and that their derivatives $\partial u'_\Delta(\cdot)/\partial x_i$ converge weakly in $L_2(\Omega)$ to $\partial u/\partial x_i$. We shall show it is possible, under the same assumptions on the data of the problem, to prove the strong convergence in $L_2(\Omega)$ of the $\tilde{u}_{x_i}$ to $\partial u/\partial x_i$. This is done principally in the same way as in Galerkin's method (see the end of §8, Chapter II). To be precise, we take a solution $u(x)$ of (7.1). Since it is in $\overset{\circ}{W}{}_2^1(\Omega)$, and since $\dot{C}^\infty(\Omega)$ is dense in $\overset{\circ}{W}{}_2^1(\Omega)$, there exists a sequence $u^m(\cdot)$ in $\dot{C}^\infty(\Omega)$, $m = 1, 2, \ldots$, for which $\|u^m - u\|_{2,\Omega} + \|\partial u^m/\partial x - \partial u/\partial x\|_{2,\Omega} \le 1/m$. Let the approximate solutions u_h be calculated for the grids Ω_h with $h = (h_1, \ldots, h_n) \to 0$. We shall assume, for the sake of notational brevity, that the elementary cells $\omega_{(kh)}$ are of the same edge-length, $h_1 = \cdots = h_n \equiv h$, and that the index k of h is used to denote the number of the grid Ω_{h_k} in the sequence of contracting grids $\{\Omega_{h_k}\}$, so that, as $k \to \infty$, the step-size or mesh of the grid decreases to zero. For each function $u^m(x)$, which has been extended to be zero outside Ω, the grid function $u^m_{h_k}$ which is equal to it at points of the grid with mesh h_k not exceeding some $h(m)$ vanishes on S_{h_k} and outside $\bar{\Omega}_{h_k}$, and

$$\|u^m - \tilde{u}^m_{h_k}\|_{2,\Omega} + \left\|\frac{\partial u^m}{\partial x} - \tilde{u}^m_{h_k x}\right\|_{2,\Omega} \le \frac{1}{m}.$$

By the triangle inequality,

$$\|u - \tilde{u}_{h_k}^m\|_{2,\Omega} + \left\|\frac{\partial u}{\partial x} - \tilde{u}_{h_k x}^m\right\|_{2,\Omega} \leq \frac{2}{m} \quad \text{where} \quad h_k \leq h(m). \tag{11.1}$$

Let us take the integral identities satisfied by $u(x)$ and u_h,

$$\mathcal{M}(u, \eta) \equiv \mathcal{L}(u, \eta) - \mathcal{F}(f_i, f, \eta) = 0, \tag{11.2}$$

where

$$\mathcal{L}(u, \eta) = \int_\Omega \left(a_{ij}\frac{\partial u}{\partial x_j}\frac{\partial \eta}{\partial x_i} + a_i \frac{\partial \eta}{\partial x_i} u - b_i \frac{\partial u}{\partial x_i}\eta - au\eta\right)dx,$$

and

$$\mathcal{F}(f_i, f, \eta) = \int_\Omega \left(f_i \frac{\partial \eta}{\partial x_i} - f\eta\right) dx,$$

and

$$\mathcal{M}_h(u_h, \eta_h) \equiv \mathcal{L}_h(u_h, \eta_h) - \mathcal{F}_h(f_i, f, \eta_h) = 0, \tag{11.3}$$

where

$$\mathcal{L}_h(u_h, \eta_h) = \int_\Omega (a_{ij}\tilde{u}_{hx_i}\tilde{\eta}_{hx_j} + a_i\tilde{u}_h\tilde{\eta}_{hx_i} - b_i\tilde{u}_{hx_i}\tilde{\eta}_h - a\tilde{u}_h\tilde{\eta}_h)\, dx,$$

and

$$\mathcal{F}_h(f_i, f, \eta_h) = \int_\Omega (f_i\tilde{\eta}_{hx_i} - f\tilde{\eta}_h)\, dx.$$

(Here and below we shall write the subscript h on u_{hx_i} in order to emphasize the step-size for which we calculate u_{x_i}. Above, we omitted it.) It follows from (11.1)–(11.3) and from our assumptions about the data of the problem, mentioned in §7, that

$$|\mathcal{M}_{h_k}(u_{h_k}^m, u_{h_k}^m)| \leq \frac{c}{m} \quad \text{for} \quad h_h \leq h(m). \tag{11.4}$$

Using (11.3) and the fact that $h_k \leq h(m)$, we calculate the expression

$$\begin{aligned}
&\mathcal{L}_{h_k}(u_{h_k} - u_{h_k}^m, u_{h_k} - u_{h_k}^m) \\
&= \mathcal{L}_{h_k}(u_{h_k}, u_{h_k}) - \mathcal{L}_{h_k}(u_{h_k}, u_{h_k}^m) - \mathcal{L}_{h_k}(u_{h_k}^m, u_{h_k}) \\
&\quad + \mathcal{L}_{h_k}(u_{h_k}^m, u_{h_k}^m) = \mathcal{F}_{h_k}(f_i, f, u_{h_k}) - \mathcal{F}_{h_k}(f_i, f, u_{h_k}^m) \\
&\quad + \left[-\mathcal{L}_{h_k}(u_{h_k}, u_{h_k}^m) + \int_\Omega (a_i + b_i)(\tilde{u}_{h_k x_i}^m \tilde{u}_{h_k} - \tilde{u}_{h_k x_i}\tilde{u}_{h_k}^m)\, dx\right] \\
&\quad + [\mathcal{M}_{h_k}(u_{h_k}^m, u_{h_k}^m) + \mathcal{F}_{h_k}(f_i, f, u_{h_k}^m)] \\
&= \mathcal{F}_{h_k}(f_i, f, u_{h_k} - u_{h_k}^m) + \int_\Omega (a_i + b_i)(\tilde{u}_{h_k x_i}^m \tilde{u}_{h_k} - \tilde{u}_{h_k x_i}^m \tilde{u}_{h_k}^m \\
&\quad + \tilde{u}_{h_k x_i}^m \tilde{u}_{h_k}^m - \tilde{u}_{h_k x_i}\tilde{u}_{h_k}^m)\, dx + \mathcal{M}_{h_k}(u_{h_k}^m, u_{h_k}^m).
\end{aligned} \tag{11.5}$$

By (7.2), as well as by (11.4) we have from (11.5) that

$$\nu_1 \|\tilde{u}_{h_k x} - \tilde{u}^m_{h_k x}\|^2_{2,\Omega} \le \mathscr{F}_{h_k}(f_i, f, u_{h_k} - u^m_{h_k})$$
$$+ \int_\Omega (a_i + b_i)[\tilde{u}^m_{hx_{k}i}(\tilde{u}_{h_k} - \tilde{u}^m_{h_k}) + (\tilde{u}^m_{h_k x_i} - \tilde{u}_{h_k x_i})\tilde{u}^m_{h_k}]\, dx + \frac{c}{m}. \quad (11.6)$$

The right-hand side tends to zero when $m \to \infty$ and $h_k \le h(m) \to 0$, as we know the convergence in $L_2(\Omega)$ of $\tilde{u}_{h_k}$ and $\tilde{u}^m_{h_k}$ to u, the weak convergence of $\tilde{u}_{h_k x_i}$ and $\tilde{u}^m_{h_k x_i}$ to $\partial u/\partial x_i$, and the uniform boundedness of $\|\tilde{u}^m_{h_k}\|_{2,\Omega}$ and $\|\tilde{u}^m_{h_k x}\|_{2,\Omega}$. From this and from (11.1) it follows that

$$\|u - \tilde{u}_{h_k}\|_{2,\Omega} + \left\| \frac{\partial u}{\partial x} - \tilde{u}_{h_k x} \right\|_{2,\Omega} \to 0$$

when $h_k \to 0$.

2. Strongly Elliptic Systems. These systems were considered in §1, Chapter V. Here we shall restrict ourselves to second-order systems

$$(\mathscr{L}\mathbf{u})_k \equiv \frac{\partial}{\partial x_i}\left(a^{kl}_{ij}\frac{\partial u_l}{\partial x_j}\right) + b^{kl}_i \frac{\partial u_l}{\partial x_i} + c^{kl} u_l = f_k,$$
$$k, l = 1, \ldots, N, \qquad i, j = 1, \ldots, n, \quad (11.7)$$

where it is assumed that we have the inequalities

$$a^{kl}_{ij}\gamma_{jl}\gamma_{ik} \ge \nu \sum_{i=1}^n \sum_{k=1}^N \gamma^2_{ik}, \qquad \nu = \text{const} > 0, \quad a^{kl}_{ij} = a^{kl}_{ji} \quad (11.8)$$

for arbitrary real γ_{ik}. This condition (11.8) is more restrictive than the condition (1.2), Chapter V, of strong ellipticity. It facilitates in an essential way the proof of stability of the difference scheme that we shall construct to find the solution $\mathbf{u} = (u_1, \ldots, u_N)$ of the system (11.7) satisfying the boundary condition

$$\mathbf{u}|_S = 0. \quad (11.9)$$

This scheme leads to the solution of the problem (11.7), (11.9), and in the general case, when (11.8) is replaced by the condition (1.2), Chapter V. However, the proof of this is rather long, and we shall not present it here. On the other hand, an extension of the arguments below to the case of systems of arbitrary order $2m$ (to which, in particular, the general elliptic equation of order $2m$ is related) does not present difficulty.

We partition the space R^n where x is defined by hyperplanes $x_i = k_i h_i$, $h_i > 0$, $k_i = 0, \pm 1, \pm 2, \ldots$, and use the notations Ω_h, $\overline{\Omega}_h$, S_h, Ω_h^+ introduced earlier for these sets. We replace (11.7) by a system of difference equations

$$(a^{kl}_{ijh} u_{lx_j})_{\bar{x}_i} + b^{kh}_{ih} u_{lx_i} + c^{kl}_h u_{lh} = f_{kh}, \quad (11.10)$$

which must be satisfied at all vertices of Ω_h, and the boundary condition by the inequalities

$$u_{lh}|_{S_h} = 0. \tag{11.11}$$

The grid functions $a_{ijh}^{kl}, \ldots, f_{kh}$ are formed from the $a_{ij}^{kl}, \ldots, f_k$ by the same rule that we used in §7. To the condition (11.8) we adjoin the assumption that

$$a_{ij}^{kl}\gamma_{jl}\gamma_{ik} - b_i^{kl}\gamma_{il}\gamma_{0k} - c^{kl}\gamma_{0l}\gamma_{0k} \geq \nu_1 \sum_{i=1}^{n} \sum_{k=1}^{N} \gamma_{ik}^2, \qquad \nu_1 > 0. \tag{11.12}$$

This inequality (11.12) must be satisfied for all real γ_{ik}, $i = 0, 1, \ldots, n$, $k = 1, \ldots, N$. It guarantees the uniqueness theorem for the problem (11.7), (11.9). Moreover, we assume as everywhere that the coefficients of (11.7) are bounded functions and that $\mathbf{f} \in L_2(\Omega)$. We shall show the unique solvability of the system (11.10), (11.11). For this we multiply (11.10) by u_{kh} and sum the result over k from 1 to N over all points of Ω_h (or, what is the same thing, over all points of $\overline{\Omega}_h$, assuming that the extension of u_{kh} is zero outside $\overline{\Omega}_h$). After this, we use (2.11) for a representation of the principal terms. This gives the equality

$$\Delta_h \sum_{\Omega_h^+} [a_{ijh}^{kl} u_{lx_j} u_{kx_i} - b_{ih}^{kl} u_{lx_i} u_{kh} - c_h^{kl} u_{lh} u_{kh}] = -\Delta_h \sum_{\Omega_h^+} f_{kh} u_{kh}. \tag{11.13}$$

From this and from (11.12) it follows that

$$\nu_1 \Delta_h \sum_{\Omega_h^+} \sum_{i=1}^{n} \sum_{k=1}^{N} (u_{kx_i})^2 \leq -\Delta_h \sum_{\Omega_h^+} f_{kh} u_{kh}. \tag{11.14}$$

If (11.10) is homogeneous, i.e., if $f_{kh} = 0$, then it is clear from (11.14) that $u_{kx_i} = 0$ for all k and i, whence (11.11) implies that the unique solution u_{kh} of this system is $u_{kh} = 0$. Indeed, we have shown that the system (11.10), (11.11) is uniquely solvable for arbitrary f_{kh}. Moreover, it follows from (11.14) and from (3.19), which holds for functions u_{kh}, that

$$\Delta_h \sum_{\Omega_h^+} \sum_{k=1}^{N} \left[\sum_{i=1}^{n} (u_{kx_i})^2 + (u_{kh})^2 \right] \leq c\Delta_h \sum_{\Omega_h^+} \sum_{k=1}^{N} (f_{kh})^2, \tag{11.15}$$

with the constant c independent of h; this guarantees the stability of the difference scheme in the "energy" norm. On the basis of this, we can prove the convergence of $\mathbf{u}_h$ to a solution $\mathbf{u}$ of the problem (11.7), (11.9) in the same way as we did in §7 for the case of one equation.

3. The Equations of the Theory of Elasticity. Let us consider the problem of equilibrium of an isotropic elastic body under the conditions of rigid fastening, i.e., the problem of determining the elastic displacement vector $\mathbf{u} = (u_1, u_2, u_3)$ as a solution of the system

$$(\mathscr{L}\mathbf{u})_i \equiv \sum_{k=1}^{3} \frac{\partial \tau_{ik}(\mathbf{u})}{\partial x_k} = -f_i(x), \qquad i = 1, 2, 3, \tag{11.16}$$

satisfying the boundary conditions

$$\mathbf{u}|_S = 0. \tag{11.17}$$

We recall (§1, Chapter V) that

$$\tau_{ik}(\mathbf{u}) = 2\mu\varepsilon_{ik}(\mathbf{u}) + \delta_i^k \lambda \sum_{l=1}^{3} \varepsilon_{ll}(\mathbf{u}), \qquad \varepsilon_{ik}(\mathbf{u}) = \frac{1}{2}\left(\frac{\partial u_i}{\partial x_k} + \frac{\partial u_k}{\partial x_i}\right), \tag{11.18}$$

and that λ and μ are positive constants. Let $\mathbf{f} \in L_2(\Omega)$, where Ω is a bounded domain in R^3. We associate with the problem (11.16), (11.7) the difference scheme

$$\sum_{k=1}^{3} (\tau_{ik}^h(\mathbf{u}_h))_{\bar{x}_k} = -f_{ih}, \tag{11.19}$$

in which $\tau_{ik}^h(\mathbf{u}_h)$ is connected with $\varepsilon_{ik}^h(\mathbf{u}_h)$ by (11.18), f_{ih} is formed from f_i in the same way as in §7, and $\varepsilon_{ik}^h(\mathbf{u}_h) = \frac{1}{2}(u_{ix_k} + u_{kx_i})$. The equations (11.19) must be satisfied at all points of Ω_h, and at points of S_h we must have

$$\mathbf{u}_h|_{S_h} = 0. \tag{11.20}$$

To prove the unique solvability of (11.19), (11.20) and the stability of the scheme that we choose, we multiply (11.19) by u_{ih} and then sum the equations we get from $i = 1$ to $i = 3$ and over all points of $\overline{\Omega}_h$. (Here, as in all problems we have treated with the first boundary condition, it is convenient to assume that the functions u_{ih} have been defined to be zero outside $\overline{\Omega}_h$ and to write all sums $\Delta_h \sum$ as sums over the points of $\overline{\Omega}_h$.) After this, we use (2.11) to represent the resulting equality in the form

$$\Delta_h \sum_{\overline{\Omega}_h} \tau_{ik}^h(\mathbf{u}_h) u_{ix_k} = \Delta_h \sum_{\overline{\Omega}_h} f_{ih} u_{ih}. \tag{11.21}$$

Since τ_{ik}^h and ε_{ik}^h are invariant under a permutation of the indices i and k, we can rewrite (11.21) as

$$\Delta_h \sum_{\overline{\Omega}_h} \tau_{ik}^h(\mathbf{u}_h)\varepsilon_{ik}^h(\mathbf{u}_h) = \Delta_h \sum_{\overline{\Omega}_h} f_{ih} u_{ih}. \tag{11.22}$$

Since the constants λ and μ are positive, the left-hand side of (11.22) is non-negative:

$$\tau_{ik}^h(\mathbf{u}_h)\varepsilon_{ik}^h(\mathbf{u}_h) = 2\mu \sum_{i,k=1}^{3} (\varepsilon_{ik}^h(\mathbf{u}_h))^2 + \lambda\left(\sum_{i=1}^{3} \varepsilon_{ii}^h(\mathbf{u}_h)\right)^2. \tag{11.23}$$

We shall show that it is equivalent to the quantity $(\|\mathbf{u}_h\|^{(1)}_{2,\bar{\Omega}_h})^2$. For this we use again the formula (2.11) for summation by parts, assuming that u_{kh} is zero outside $\bar{\Omega}_h$ by extension. Thus

$$\begin{aligned}\Delta_h \sum_{\bar{\Omega}_h} \varepsilon^h_{ik}(\mathbf{u}_h)\varepsilon^h_{ik}(\mathbf{u}_h) &= \tfrac{1}{4}\Delta_h \sum_{\bar{\Omega}_h}\sum_{i,k} [(u_{ix_k})^2 + 2u_{ix_k}u_{kx_i} + (u_{kx_i})^2] \\ &= \tfrac{1}{2}\Delta_h \sum_{\bar{\Omega}_h}\sum_{i,k} [(u_{ix_k})^2 - u_{ix_k\bar{x}_i}u_k] \\ &= \tfrac{1}{2}\Delta_h \sum_{\bar{\Omega}_h}\sum_{i,k} [(u_{ix_k})^2 + u_{i\bar{x}_i}u_{k\bar{x}_k}] \\ &\geq \tfrac{1}{2}\Delta_h \sum_{\bar{\Omega}_h}\sum_{i,k} (u_{ix_k})^2 \equiv \tfrac{1}{2}\|\mathbf{u}_x\|^2_{2,\bar{\Omega}_h}.\end{aligned} \tag{11.24}$$

It follows from (11.22)–(11.24) and (11.20) that the homogeneous system (11.19), (11.20) has only the zero solution, so that the non-homogeneous system (11.19), (11.20) is uniquely solvable for arbitrary f_{ih}. Moreover, these relations, together with (3.19), Chapter VI, imply the bound

$$\|\mathbf{u}_h\|^2_{2,\Omega_h} + \|\mathbf{u}_x\|^2_{2,\bar{\Omega}_h} \leq c\|\mathbf{f}\|^2_{2,\Omega_h} \leq c_1, \tag{11.25}$$

which guarantees the stability of the scheme (11.19), (11.20) and allows us to prove that, in the limit as $h_i \to 0$, $\mathbf{u}_h$ gives us the solution $\mathbf{u}$ of the problem (11.16), (11.17).

Other boundary value problems for (11.16) can be investigated in the same spirit, but it becomes more complicated to derive the inequality (11.14). The arguments of this subsection can be extended easily to the case of anisotropic media, where we can allow the coefficients to have discontinuities of the first kind, In particular, the coefficients λ and μ in (11.18) can be arbitrary positive piecewise-constant functions of $x \in \bar{\Omega}$.

4. Non-stationary Boundary Problems for Systems of Parabolic and Hyperbolic Type. The classical initial-boundary value problems for the so-called strongly parabolic systems can be investigated by the method of finite differences, basically in the same way as we did for one parabolic equation in §9. These systems are of the form

$$\frac{\partial \mathbf{u}}{\partial t} - \mathscr{L}\mathbf{u} = \mathbf{f}, \tag{11.26}$$

where $\mathscr{L}$ is a strongly elliptic operator (cf. §2, Chapter V). The investigations of §10, related to one hyperbolic equation of second order, can be extended without essential changes to the so-called strongly hyperbolic systems, which are of the form

$$\frac{\partial^2 \mathbf{u}}{\partial t^2} - \mathscr{L}u = \mathbf{f}, \tag{11.27}$$

where $\mathscr{L}$ is a strongly elliptic operator with symmetric principal part (cf. §2, Chapter V). The system of the dynamical theory of elasticity, for example, belongs to this class of systems. The first initial-boundary value problem for (11.26) and (11.27) was considered in the thesis by E. M. Chistjakova [1].

5. Equations of Schrödinger Type. Such equations (or systems) are of the form

$$\frac{\partial u}{\partial t} = i\mathscr{L}(t)u + f(t), \tag{11.28}$$

where $\mathscr{L}$ is a family of symmetric operators depending on a parameter t, and $i = \sqrt{-1}$. We shall set down only a general plan for studying them by the method of finite differences, not specifying the type of operator $\mathscr{L}$. We denote by H the complex Hilbert space in which $\mathscr{L}(t)$ acts. We shall assume that, for arbitrary t (in that interval of the t-axis which interests us), $\mathscr{L}(t)$ is self-adjoint and that its domain of definition $\mathscr{D}(\mathscr{L})$ is dense in H and independent of t. We consider the Cauchy problem for (11.28), where, for economy of space, we shall assume that $f \equiv 0$. Thus, it is required to find a function $u(t)$ for $t \geq 0$, with values in H, which satisfies the equations

$$\frac{du}{dt} = i\mathscr{L}(t)u, \qquad u|_{t=0} = \varphi. \tag{11.29}$$

We take for (11.29) the difference scheme

$$\frac{u_h(t+h) - u_h(t)}{h} = \frac{i}{2}\mathscr{L}(t)[u_h(t+h) + u_h(t)], \qquad u_h(0) = \varphi_h. \tag{11.30}$$

The equations (11.30) should be satisfied at the points $t = t_k = kh$, $k = 0, 1, \ldots$. They determine u_h uniquely at these grid points, for we have from (11.30) that

$$\left(E - \frac{ih}{2}\mathscr{L}(t_k)\right)u_h(t_{k+1}) = \left(E + \frac{ih}{2}\mathscr{L}(t_k)\right)u_h(t_k), \qquad k = 0, 1, \ldots,$$

and, by the self-adjointness of $\mathscr{L}(t)$, the operator $E - (ih/2)\mathscr{L}(t_k)$ has a bounded inverse, so that

$$u_h(t_{k+1}) = \left(E - \frac{ih}{2}\mathscr{L}(t_k)\right)^{-1}\left(E + \frac{ih}{2}\mathscr{L}(t_k)\right)u_h(t_k). \tag{11.31}$$

Here E is the identity operator in H. The scheme (11.30) preserves the main property of (11.29), that the norm of its solutions, $\|u(t)\|$, does not depend on t. That this property holds is clear from (11.31): the operator $(E - (ih/2)\mathscr{L}(t_k))^{-1}(E + (ih/2)\mathscr{L}(t_k)) \equiv A(t_k)$ is the Cayley transform of the self-adjoint operator $\mathscr{L}(t_k)$ and is hence isometric on $\mathscr{D}(\mathscr{L})$, i.e., $(A(t_k)v, A(t_k)w) = (v, w)$ for all v and w in $\mathscr{D}(\mathscr{L})$. If $\varphi \in \mathscr{D}(\mathscr{L})$, then (11.31) determines $u_h(t_{k+1})$ for all $k = 0, 1, \ldots$ and $u_h(t_{k+1}) \in \mathscr{D}(\mathscr{L})$. If φ is an arbitrary element of H, then (11.31) must be replaced by

$$u_h(t_{k+1}) = \hat{A}(t_k)u_h(t_k), \tag{11.32}$$

where $\hat{A}(t_k)$ is the isometric extension of $A(t_k)$ to all of H. In any of these cases, we have

$$\|u_h(t_{k+1})\| = \|u_h(t_k)\|, \tag{11.33}$$

which shows the stability of the difference scheme (11.30) in the norm of H. Using these relationships, we can take the limit as $h \to 0$ and prove for the problem (11.29) the existence of at least one generalized solution $u(t)$ in the space $H_1 = L_2((0, T); H)$. By H_1 we mean the Hilbert space whose elements are functions $v(\cdot)$ with values in H, which are defined and measurable on $t \in [0, T]$, and which have finite norm

$$\|v\|_{H_1} = \left[\int_0^T \|v(t)\|_H^2 \, dt\right]^{1/2}.$$

Under certain hypotheses on the smoothness of $\mathscr{L}(t)$, as functions of t, we can establish the uniqueness of such solutions of the problem (11.29) (see the papers [LA 8]). This is done in roughly the same way as it was in §2, Chapter III, for parabolic equations and in §6, Chapter IV, for hyperbolic equations. The method of finite differences allows us to observe how properties of generalized solutions of (11.29) improve in connection with improvement of φ and of the smoothness of $\mathscr{L}(t)$ in t (this is done by the same plan which was described at the end of §4, Chapter IV). For abstract equations (11.29), when the space H is not specified, "improvement" has to be understood as the property that φ and $u(t)$ belong to $\mathscr{D}(A)$ or $\mathscr{D}(A^\alpha)$ for some α.

In the scheme (11.30) that we considered, only the derivative with respect to t was approximated. We can approximate $\mathscr{L}(t)$ in (11.30) along with the derivative $\partial/\partial t$ so that the approximations of $\mathscr{L}(t)$ are self-adjoint operators in those spaces where they are defined. If H is a function space with elements $u(x)$ and $\mathscr{L}(t)$ is a symmetric differential operator containing some derivatives of $u(x)$ with respect to x_i, with coefficients depending on (x, t), then we can take as $\mathscr{L}_h(t)$ some kind of symmetric difference operator approximating $\mathscr{L}(t)$ and acting on grid functions satisfying the boundary conditions corresponding to those of the problem.

6. On Diffraction Problems. We showed in §4, Chapter V, that diffraction problems can be included in the class of problems considered in Chapters II–IV. In particular, the difference schemes treated in this chapter can be used to solve them. It is necessary only to keep in mind that whenever we have two or more different media, the coefficients in the integral identities corresponding to these problems will have discontinuities on the interfaces of these media, and hence the piecewise-constant interpolations of the grid functions corresponding to them (these should be uniformly bounded functions) will converge to them, not uniformly, but only almost everywhere. But such convergence is sufficient for passage to the limit as $h_i \to 0$ (cf. [LA 5, 9]).

For some particular classes of problems of diffraction type, a more detailed analysis has been conducted (see, for example, [SA 1]).

§12. Approximation Methods

The desire to connect Galerkin's method with the finite-difference methods has led to the construction of some other finite-dimensional approximations for solutions of boundary value problems. But again this only turned out to be possible because of a non-classical interpretation of boundary value problems which reduced the solution of these problems to the finding of functions satisfying a corresponding integral identity and belonging to a properly chosen function space. We shall show how these new approximations originated by the example of the problem

$$\Delta u = f(x), \qquad u|_S = 0. \tag{12.1}$$

whose generalized solution $u(\cdot)$ in $W_2^1(\Omega)$ should be in $\mathring{W}_2^1(\Omega)$ and satisfy

$$[u, \eta] \equiv \int_\Omega \frac{\partial u}{\partial x_i} \cdot \frac{\partial \eta}{\partial x_i}\, dx = -\int_\Omega f \cdot \eta \, dx \equiv -(f, \eta) \tag{12.2}$$

for all $\eta(\cdot)$ in $\mathring{W}_2^1(\Omega)$.

In Galerkin's method (in its original form, stemming from the works of Galerkin) we choose a sequence of finite-dimensional spaces $\mathfrak{M}_N$, where N is the dimension of $\mathfrak{M}_N$, approximating the space $\mathring{W}_2^1(\Omega)$ in which (12.2) is considered. These spaces $\mathfrak{M}_N$ are subspaces of $\mathring{W}_2^1(\Omega)$, and $\mathfrak{M}_N \subset \mathfrak{M}_{N+1}$. In each of the $\mathfrak{M}_N$ we seek a function u^N satisfying (12.2) in which $\eta(x)$ is an arbitrary element of $\mathfrak{M}_N$.

To define such a function we introduce a basis $\{\varphi_k\}$, $k = 1, \ldots, N$, in $\mathfrak{M}_N$ and seek u^N in the form $u^N(x) = \sum_{k=1}^N c_k^N \varphi_k(x)$. To determine the coefficients c_k^N, $k = 1, \ldots, N$, we obtain from (12.2) a system of N linear algebraic equations from which c_k^N can be found. From the arguments of §8, Chapter II, related to Galerkin's method, it is easy to see that we can discard the requirement that $\mathfrak{M}_N$ be contained in $\mathfrak{M}_{N+1}$. In fact, in the first part of the reasoning where we prove the unique solvability of the systems defining c_k^N and give a bound for the norms $\|u^N\|_{2,\Omega}^{(1)}$, this requirement is not used. In the second part, where we take the limit as $N \to \infty$, it is necessary only to bring in the following modification. Let the sequence $\{u^{N_k}\}$ converge weakly in $\mathring{W}_2^1(\Omega)$ to $u(x)$ (i.e., u^{N_k} and $\partial u^{N_k}/\partial x_i$ converge weakly in $L_2(\Omega)$ to u and $\partial u/\partial x_i$, respectively). We take an arbitrary $\eta(\cdot)$ in $\mathring{W}_2^1(\Omega)$ and a sequence $\{\eta^{N_k}\}$ converging strongly to η in $\mathring{W}_2^1(\Omega)$ such that $\eta^{N_k} \in \mathfrak{M}_{N_k}$ for all k. This can be done because, by assumption, $\{\mathfrak{M}_{N_k}\}$ approximates $\mathring{W}_2^1(\Omega)$ in the norm of $\mathring{W}_2^1(\Omega)$. Now the pair u^{N_k} and η^{N_k} satisfy (12.2), and by taking limits as $k \to \infty$, we obtain (12.2), for u and η. Thus we have shown that the condition $\mathfrak{M}_N \subset \mathfrak{M}_{N+1}$ may be discarded. Within each $\mathfrak{M}_N$ the basis $\{\varphi_k\}$, $k = 1, \ldots, N$, can be chosen arbitrarily; the function $u^N(\cdot)$ does not depend on the basis, for as is easy to see, there is only one function in $\mathfrak{M}_N$ which satisfies (12.2) for arbitrary η in $\mathfrak{M}_N$. This remark allows us to construct finite-dimensional approximations to the problem (12.1), which can naturally be classified as approximations of finite-difference type. Here the approximate solutions u_h are determined by their values at the vertices of Ω_h, where the u_h may be computed as solutions of linear algebraic systems, which connect the values of u_h only at a number of neighboring points. Such approximations for (12.2) can be constructed in the following way: We choose a partition of R^n into cells and the corresponding grids Ω_h (recall that the closed region $\overline{\Omega}_h \subset \overline{\Omega}$). For each grid function η_h defined on the vertices of $\overline{\Omega}_h$ and vanishing on S_h we construct an interpolation belonging to $\mathring{W}_2^1(\Omega)$. Among the interpolations we defined in §3 of this chapter, the interpolation $\eta_h'(x)$ possesses this property, and we consider this one, in order to be definite, and we take as $\mathfrak{M}_h$ the totality of all such interpolations $\eta_h'(x)$. The dimension of $\mathfrak{M}_h$ is equal to the number of vertices belonging to Ω_h, and the sets $\mathfrak{M}_{h(m)}$, $h(m) \to 0$, can be considered as the finite-dimensional subspaces of $\mathring{W}_2^1(\Omega)$; they approximate $\mathring{W}_2^1(\Omega)$ in the norm of this space. Actually, the functions of $\dot{C}^\infty(\Omega)$ are dense in $\mathring{W}_2^1(\Omega)$, and every function $\eta(\cdot)$ in $C^\infty(\Omega)$ can be approximated in the norm of $\mathring{W}_2^1(\Omega)$ by functions $\eta_h'(\cdot)$ in $\mathfrak{M}_h$ if we choose as η_h (for sufficiently small h) the values of $\eta(x)$ at the vertices of $\overline{\Omega}_h$. The approximate solutions $u_h'(\cdot) \in \mathfrak{M}_h$ are determined from the identity

$$[u_h', \eta_h'] = -\int_\Omega f\eta_h' \, dx, \tag{12.3}$$

where η_h' is an arbitrary element of $\mathfrak{M}_h$. As a basis in $\mathfrak{M}_h$ we can take the totality of functions, each of which is an interpolation of the form $(\ )_h'$ of a grid function which is 1 at some vertex of Ω_h and 0 at all the remaining vertices of $\overline{\Omega}_h$. We denote these functions as $\{\varphi_{k,h}'(x)\}$, where (kh) is the coordinate of that vertex of Ω_h where the grid function $\varphi_{k,h}$ has value 1 (at the other vertices of $\overline{\Omega}_h$, $\varphi_{k,h}$ vanishes). The grid function u_h corresponding to the desired solution $u_h'(x)$ can be written as a sum $\sum_{(kh)\in\Omega_h} u_h(kh)\varphi_{k,h}$, where by $u_h(kh)$, we mean the value of u_h at the vertex $x = kh$. If we substitute the sum $\sum_{(kh)\in\Omega_h} u_h(kh)\varphi_{k,h}$ for $u_h'(x)$ in (12.3) and take as $\eta_h'(x)$ one of the functions $\varphi_{k_0,h}'(x)$, then (12.3), after we perform all integrations in it, turns out to be a linear algebraic system relating the value of $u_k(k_0 h)$ to the values of $u_h(kh)$ at those points (kh) which are vertices of the cells $\omega_{(kh)}$ with $(k_0 h)$ as one vertex. Each such system (for fixed h) has as many equations as the number of vertices of Ω_h, and along with the equalities $u_h|_{S_h} = 0$, determine the difference scheme. Its unique solvability, as well as its stability and convergence (more precisely, the convergence of $u_h'(x)$ to $u(x)$), follows from what was said concerning the extended interpretation of Galerkin's method described above.

If we replace the interpolation $(\ )_h'$ by some other possible interpolation of the grid functions (i.e., one which yields an element of $\mathring{W}_2^1(\Omega)$ for the grid function η_h), then we get another convergent difference scheme for (12.1). For example, for two-dimensional problems (i.e., for $n = 2$), it is convenient to use another interpolation η_h instead of $\eta_h'(x)$, namely, we partition each of the cells $\omega_{(kh)}$ by the line $x_2 = (h_2/h_1)(x_1 - k_1h_1) + k_2h_2$ into equal triangles and in each of them we interpolate η_h by a linear function. As a result, with the grid function η_h, which is zero on S_h, is associated a continuous, piecewise linear function $\hat{\eta}_h(x)$, which is zero outside Ω_h. These interpolations $\hat{\eta}_h(\cdot)$ belong to $\mathring{W}_2^1(\Omega)$. If we use these instead of $\eta_h'(x)$, (12.3) reduces to the elementary difference scheme $u_{x_i\bar{x}_i} = f_h$, $u_h|_{S_h} = 0$. This property was first noted by R. Courant [1].† N. A. Strelkov [1] extended this result to the case of arbitrary $n > 2$; that is, he constructed piecewise-linear simplicial approximations $\hat{\eta}_h(x)$ which give for (12.1) the difference scheme $\sum_{i=1}^n u_{x_i\bar{x}_i} = f_h$, $u_h|_{S_h} = 0$. We shall not cite other realizations of the described method for constructing difference schemes, except to remark that the method is applicable to all the boundary value problems and initial-boundary value problems considered in this book. In a series of papers it was called the variational-difference method, but it seems to us that it might better have been called the projection-difference method, for the method described comprises not only those problems which can be solved by direct methods of variational calculus, but also all of those problems to which Galerkin's method is applicable in various forms.

For equations with variable coefficients the difference schemes that can be obtained by this method usually turn out to be more complicated than those that were considered in §§7–10. We shall look at how the approximations that are constructed by the difference schemes of §§7–10 (we call these schemes classical) differ from approximations arising from various realizations of the projection-difference method described just now. We take as an example the problem (7.1) with $a_i = f_i \equiv 0$ and its corresponding difference scheme (7.8), (7.9), or, what is the same thing, the scheme defined by the identity

$$\int_\Omega (a_{ij}\tilde{u}_{x_j}\tilde{\eta}_{x_i} - b_i\tilde{u}_{x_i}\tilde{\eta}_h - a\tilde{u}_h\tilde{\eta}_h)\,dx = -\int_\Omega f\tilde{\eta}_h\,dx, \tag{12.4}$$

† In addition, Courant expressed ideas close to those described above. However, for a long time these were not widely known and were not realized in rigorous mathematical investigations. In practice these and other realizations (which were not infrequently called "the finite-element method") appeared in various works of engineers without any theoretical justification. Apparently the first paper after [CR 1], in which difference equations arose as Euler equations for the solution of a variational problem by the Ritz method, was the paper of L. A. Oganesjan [1], in which the author carried out a rigorous investigation of the Neumann problem for the biharmonic equation. For this point on, there began to appear works by mathematicians who analyzed and used various approximation methods in connection with equations of elliptic type (see the papers of Oganesjan + K^0, Dem'janovich [1], and the books of D'jakonov [1], Cea [1], Aubin [1], and others.)

in which u_h is the unknown and η_h is an arbitrary grid function on $\overline{\Omega}_h$, satisfying the condition

$$u_h|_{S_h} = 0, \qquad \eta_h|_{Sh} = 0 \tag{12.5}$$

(cf. (7.5), (7.5)). For the problem (7.1) one of the realizations of the projection-difference method will be the following: The approximate solution v_h', which is an interpolation of the form $(\ )_h'$ of the grid function v_h, can be defined by the identity

$$\int_\Omega \left(a_{ij} \frac{\partial v_h'}{\partial x_j} \frac{\partial \eta_h'}{\partial x_i} - b_i \frac{\partial v_h'}{\partial x_i} \eta_h' - a v_h' \eta_h' \right) dx = - \int_\Omega f \eta_h' \, dx, \tag{12.6}$$

in which η_h' is an interpolation of the form $(\ )_h'$ from the arbitrary grid function η_h satisfying (12.5). The unknown function v_h' also vanishes on S_h, so that

$$v_h'|_{S_h} = 0, \qquad \eta_h'|_{S_h} = 0. \tag{12.7}$$

The essential difference between the approximations $\tilde{u}_h$ obtained by the scheme (12.4), (12.5) and the approximations v_h' obtained by the scheme (12.6), (12.7) consists of the fact that the v_h' are in the space $\mathring{W}_2^1(\Omega)$ which contains the solution u of (7.1), but the $\tilde{u}_h$ do not belong to this space, and, moreover, the $\tilde{u}_{x_i}(x)$ are not the derivatives of $\tilde{u}_h(x)$. Thus, in the scheme (12.4), (12.5) we go outside the space $\mathring{W}_2^1(\Omega)$ and approximate its elements u and their gradients $\partial u/\partial x = (\partial u/\partial x_1, \ldots, \partial u/\partial x_n)$ by piecewise-constant functions $\tilde{u}_h$ and $\tilde{u}_x = (\tilde{u}_{x_1}, \ldots, \tilde{u}_{x_n})$ in such a way that $\tilde{u}_h$ and $\tilde{u}_x$ converge in $L_2(\Omega)$ (strongly or weakly) to u and $\partial u/\partial x$, respectively. Thus, the method of finite differences in its classical form (e.g., in the form (12.4), (12.5)) is not contained as a special case in Galerkin's method, even in its extended form as described in this subsection. But the investigation of the method of finite differences, given in §7, is not difficult to axiomatize, translating it to the language of operator theory, and giving an approximation method of solving operator equations $Au = f$ in a Hilbert space (or Banach space) which is different from what is called Galerkin's method for such equations (in connection with this, see Cea [1]).

Supplements and Problems

1. We have described in §10 an implicit scheme (10.53) for hyperbolic equations of general form (10.51) which is convergent for Δx and Δt tending independently to zero. Explicit schemes of the type (10.5) for these equations are divergent if $\Delta t/\Delta x = O(1)$. Verify that this is true, e.g., for the equation

$$\frac{\partial^2 u}{\partial t^2} + \frac{\partial^2 u}{\partial t \, \partial x} - \frac{\partial^2 u}{\partial x^2} = 0$$

with periodic boundary conditions. Schemes of the type (10.5) for this equation have the form

$$u_{t\bar{t}} + \frac{\Delta}{\Delta x} u_{\bar{t}} - u_{x\bar{x}} = 0,$$

where $\Delta/\Delta x(\cdot) = \alpha(\cdot)_x + (1-\alpha)(\cdot)_{\bar{x}},\ \alpha \in [0, 1].$

2. The solution of the Cauchy problem for symmetric hyperbolic systems (25) considered in Section 10 of the supplements to Chapter V can be obtained with the help of an implicit scheme of the type (6.78_1), (6.78_2), namely

$$\mathbf{u}_t^r + \frac{1}{4}\sum_{k=1}^{n} A_k(\mathbf{u}_{x_k}^{r+1} + \mathbf{u}_{\bar{x}_k}^{r+1} + \mathbf{u}_{x_k}^r + \mathbf{u}_{\bar{x}_k}^r) \tag{1}$$

$$+ B\mathbf{u}^r = \mathbf{f}^r, \qquad r = 0, 1, \ldots, \quad \mathbf{u}^0 = \boldsymbol{\varphi}, \tag{2}$$

(as in the entire Chapter VI, all summands are calculated at the same point of the grid; the upper index "r" shows the number of the layer $t = r\tau$). Join to equations (1), (2) the boundary conditions 2R-periodic with respect to x_k, $k = 1, \ldots, n$, and prove that the solutions $\mathbf{u}_h$ of this scheme converge to a solution $\mathbf{u}$ of the 2R-periodic initial-boundary value problem for (25), when $h = \Delta x_k$ and $\tau = \Delta t$ tend to zero in an arbitrary way.

3. To find solutions of the problem described in **2** we can also employ some explicit schemes. For example, we can use the scheme

$$u_t + \frac{1}{2}\sum_{k=1}^{n} A_k(\mathbf{u}_{x_k} + \mathbf{u}_{\bar{x}_k}) + B\mathbf{u} - \varepsilon\sum_{k=1}^{n} \mathbf{u}_{x_k\bar{x}_k} = \mathbf{f}, \qquad \varepsilon > 0, \tag{3}$$

with the artificial "viscosity" $\varepsilon\sum_{k=1}^{n} \mathbf{u}_{x_k\bar{x}_k}$. Clarify under which conditions on $\Delta t/\Delta x$ and ε the 2R-periodic solutions u_h of (3) converge to a 2R-periodic solution u of (25). To begin with, it is useful to consider the simplest representative of the system (25), the equation

$$\frac{\partial u}{\partial t} + a\frac{\partial u}{\partial x} = 0. \tag{4}$$

For (4), the scheme (3) has the form

$$u_t + \frac{a}{2}(u_x + u_{\bar{x}}) - \varepsilon u_{x\bar{x}} = 0, \qquad \varepsilon > 0. \tag{5}$$

Equality (5) evaluated at the point (x, t) connects the values of u_h at four points, namely

$$u_h(x, t+\tau) = u_h(x, t)\left(1 - 2\varepsilon\frac{\tau}{h^2}\right) + u_h(x+h, t)\left(\varepsilon\frac{\tau}{h^2} - a\frac{\tau}{2h}\right)$$

$$+ u_h(x-h, t)\left(\varepsilon\frac{\tau}{h^2} + a\frac{\tau}{2h}\right). \tag{6}$$

The sum of the coefficients of the three terms on the right-hand side equals one. If they are non-negative then

$$\min\{u_h(x, t); u_h(x + h, t); u_h(x - h, t)\}$$
$$\leq u_h(x, t + \tau) \leq \max\{u_h(x, t); u_h(x + h, t); u_h(x - h, t)\}, \qquad (7)$$

and therefore the scheme is stable in the norm of the space C. Non-negativeness of the coefficients in (6) imposes some restrictions on τ, h and ε; these restrictions show that $\tau/h = O(1)$. Examine what restrictions arise in the framework of Fourier difference analysis for the case when a is constant. First it is useful to prove the statement of Section 10 of the Supplements to Chapter V. The scheme (3) and (5) has to be completed by some boundary conditions for u_h and u_h. The most suitable conditions are those of periodicity with respect to all x_k.

4. Of the possible boundary conditions for hyperbolic systems of the first order we considered only periodic conditions. The description of the boundary conditions generating correct initial-boundary value problems for multidimensional systems is an extensive job (see [FR 7], [GV 1], and others). The problem simplifies for systems with one space variable. To solve these it is convenient to apply the difference method. Consider for the beginning the following problems in the domain $Q = \{(x, t): 0 < x < 1, 0 < t < \tau\}$:

$$\frac{\partial u}{\partial t} + \frac{\partial u}{\partial x} = f, \qquad u|_{t=0} = 0, \qquad u|_{x=0} = 0; \qquad (8)$$

$$\left.\begin{aligned} &\frac{\partial u_1}{\partial t} + \frac{\partial u_1}{\partial x} + \sum_{k=1}^{2} b_{1k}(x, t)u_k = f_1, \\ &\frac{\partial u_2}{\partial t} - \frac{\partial u_2}{\partial x} + \sum_{k=1}^{2} b_{2k}(x, t)u_k = f_2, \\ &u_k|_{t=0} = \varphi_k, k = 1, 2; \qquad u_1|_{x=0} = 0, u_2|_{x=1} = 0; \end{aligned}\right\} \qquad (9)$$

and

$$\left.\begin{aligned} &\frac{\partial u_1}{\partial t} + a_1(x, t)\frac{\partial u_1}{\partial x} + \sum_{k=1}^{2} b_{1k}(x, t)u_k = f_1, \\ &\frac{\partial u_2}{\partial t} - a_2(x, t)\frac{\partial u_2}{\partial x} + \sum_{k=1}^{2} b_{2k}(x, t)u_k = f_2, \\ &u_k|_{t=0} = \varphi_k, k = 1, 2; \qquad u_1|_{x=0} = 0, u_2|_{x=1} = 0, \end{aligned}\right\} \qquad (10)$$

with $a_k(x, t) \geq a_0 > 0, k = 1, 2$. Compose for them the difference equations of the type (1) and (3) accompanied by the equations approximating the initial and boundary relations.

In Chapter VI we have formulated many other problems which can be solved by the difference method or by some of its variations.

Bibliography

[ADN] AGMON, S., DOUGLIS, A., AND NIRENBERG, L.

1. Estimates near the boundary for solutions of elliptic partial differential equations satisfying general boundary conditions, I, II. *Comm. Pure Appl. Math.*, **12**, No. 4 (1959), 623–727; **17**, No. 1 (1964), 35–92.

[AK] AKILOV, G. P. AND KANTOROVICH, L. V.

1. *Functional Analysis in Normed Spaces.* Gos. Izdat. Fiz. Mat. Lit., Moscow, 1959, 684 pp.

[AR] ARONSZAJN, N.

1. Boundary value of functions with the finite Dirichlet integral. *Conference on Partial Differential Equations.* Univ. of Kansas, 1955, pages 77–93.

[AU] AUBIN, I. P.

1. Behaviour of the error of the approximate solutions of boundary value problems for linear elliptic equations by the Galerkin and finite difference methods. *Ann. Scuola Norm. Super. Pisa,* **21**, No. 4 (1967), 599–638.

[BB] BABICH, V. M.

1. On an extension of functions. *Uspehi Math. Nauk,* **8**, No. 2 (1953), 111–113.

[BS] BABICH, V. M. AND SLOBODEČKII, L. N.

1. On the boundedness of the Dirichlet integral. *Dokl. Akad. Nauk SSSR,* **106**, No. 4 (1956), 604–606.

[BA] BARRAR, R. B.

1. Some estimates for solutions of parabolic equations. *J. Math. Anal. and Appl.*, **3**, No. 2 (1961), 373–397.

[BE] BERNSTEIN, S. N.

1. *Investigation and Integration of Elliptic Differential Equations with Partial Derivatives of Second Order*. Har'kov, 1908, 164 pp.
2. *Collection of Works*, Vol. III, Izdat. Akad. Nauk SSSR, Moscow, 1960, 439 pp.

[BHK] O'Brien, G. G., Hyman, M. A., and Kaplan, S.
1. A study of the numerical solutions of partial differential equations. *J. Math. Phys.*, **29**, No. 4 (1951), 223–251.

[BR] Browder, F. F.
1. The Dirichlet problem for linear elliptic equations of arbitrary even order with variable coefficients, *Proc. Nat. Acad. Sci. USA*, **38** (1952), 230–235.
2. The Dirichlet and vibration problems for linear elliptic differential equations of arbitrary order. *Proc. Nat. Acad. Sci. USA*, **38** (1952), 741–745.
3. Strongly elliptic systems of differential equations. Ann. Math. Studies, No. 33, Princeton, 1954, pages 15–41.
4. Non-linear elliptic boundary value problems. *Bull. Amer. Math. Soc.*, **69** (1963), 862–874.
5. Non-linear monotone operators and convex sets in Banach spaces. *Bull. Amer. Math. Soc.*, **71** (1965), 780–785.

[BY] Byhovskii, E. B.
1. Solution of the mixed problem for Maxwell's system in the case of the ideal conductive boundary. Vestnik Leningrad. Univ., No. 13, 1957, pages 50–66.

[CC] Caccioppoli, R.
1. Limitazioni integrali per soluzioni di un'equazioni lineare ellitica a derivate parziali. *Giorn. Mat. Battaglini*, **80** (1950–51), 186–212.

[CE] Cea, I.
1. Approximation variationnelle des problemes aux limites. *Ann. Inst. Fourier Grenoble*, **14**, No. 2 (1964), 345–444.

[CH] Chistjakova, E. M.
1. Application of the difference method to the solving of initial-boundary value problems for some parabolic and hyperbolic systems. Thesis for a candidate's dissertation. Pedagog. Inst., Leningrad, 1956, 9 pp.

[CR] Courant, R.
1. Variational methods for the solutions of problems of equilibrium and vibration. *Bull. Amer. Math. Soc.*, **49**, No. 1 (1943), 1–23.

[CFL] Courant, R., Friedrichs, K. O., and Lewy, H.
1. Über der partiellen Differentialgleichungen der mathematischen Physik. *Math. Ann.*, **100** (1928), 32–74.

[CHI] Courant, R. and Hilbert, D.
1. *Methoden der Mathematischen Physik*. Bd I and II. Springer-Verlag, Berlin, 1931 and 1937, 469 pp., 549 pp. (*Methods of Mathematical Physics*. II. Interscience, New York–London, 1962.)

[DE] Dem'janovich, J. K.
1. On estimates of the rate of convergence of some projective methods for elliptic equations. *J. Numer. Mat. and Mat. Physics*, **8**, No. 1 (1968), 79–96.

[DY] Derguzov, V. I. and Yacubovich, V. A.
1. Existence of solutions of Hamiltonian equations with unbounded operator coefficients. *Dokl. Akad. Nauk SSSR*, **151**, No. 6 (1963), 1264–1267.

[DK] D'jakonov, E. G.
1. Difference methods of solving of boundary value problems. Izdat. Moscow Univ., Part. I, 1971, 242 pp; Part II, 1972, 225 pp.

[DG] DOUGLAS, J.
1. On the numerical integration of $u_{xx} + u_{yy} = u_t$ by implicit methods. *J. Soc. Indust. Appl. Math.*, **3**, No. 1 (1955), 42–65.

[DN] DOUGLIS, A. AND NIRENBERG, L.
1. Interior estimates of elliptic systems of partial differential equations. *Comm. Pure Appl. Math.*, **8** (1955), 503–538.

[EI] EĬDEL'MAN, S. D.
1. *Parabolic Systems.* Izdat. Nauka, Moscow, 1964, 443 pp.

[ED] EĬDUS, D. M.
1. On the mixed problem of the theory of elasticity. *Dokl. Akad. Nauk SSSR*, **76** No. 2 (1951), 181–184.

[EV] EVANS, G. C.
1. *Fundamental Points of Potential Theory*, Rice Inst., Houston, Texas, No. 7, 1920, pages 252–359.

[FA] FAEDO, S.
1. Un nuovo metodo per l'analisi esistenziale e quantativa dei problemi di propagatione. *Ann. Scuola Norm. Pisa*, **1** (1949), 1–40.

[FI] FICHERA, G.
1. Sulle equazioni differenziali lineari ellittico-paraboliche. *Atti Acc. Naz. Lincei. Mem. Ser.*, **8**, 5 (1956), 1–30.
2. On a unified theory of boundary value problems for elliptic-parabolic equations of second order. *Boundary Value Problems in Differential Equations.* Wisconsin Univ. Press, 1960, pages 97–120.
3. Existence theorems in elasticity. Boundary value problems of elasticity with unilateral constraints. *Handbuch der Physik.* Springer-Verlag, Berlin, **VIa**/2, 1972, pages 347–424.

[FO] FOMIN, B. N.
1. Parametrical resonance in elastic systems with infinitely many degrees of freedom. Vestnik Leningrad. Univ., No. 13, 1965, pages 73–87.

[FM] FRIEDMAN, A.
1. *Partial Differential Equations of Parabolic Type.* Prentice-Hall, Englewood Cliffs, NJ, 1964, 347 pp.

[FR] FRIEDRICHS, K. O.
1. Die Randwert—und Eigenwertprobleme aus der Theorie der elastischen Platten (Anwendung der direkten Methoden der Variationsrechnung). *Math. Ann.*, **98** (1928), 205–247.
2. Spektraltheorie halbbeschrankten Operatoren und ihre Anwendung auf Spektralzerlegung von Differentialoperatoren. *Math. Ann.*, **109** (1934), Part I, 465–487; Part II, 685–713.
3. On differential operators in Hilbert spaces. *Amer. J. Math.*, **61**, No. 2 (1939), 523–544.
4. The identity of weak and strong extensions of differential operators. *Trans. Amer. Math. Soc.*, **55**, No. 1 (1944), 132–151.
5. On the boundary value problems of the theory of elasticity and Korn's inequality. *Ann. of Math.*, **48** (1947), 441–471.
6. On the differentiability of the solutions of linear elliptic differential equations. *Comm. Pure Appl. Math.*, **6**, No. 3 (1953), 299–326.
7. Symmetric hyperbolic linear differential equations. *Comm. Pure Appl. Math.*, **7**, No. 2 (1954), 345–392.

[GA] GAGLIARDO, E.

1. Caratterizzioni delle tracce sulla frontiera relative ad alcune classi di funzioni in n variabili. *Rend. Sem. Mat. Univ. Padova*, **27** (1957), 284–305.
2. Proprieta di alcune classi di funzioni in pia variabili. *Richerche Mat.*, **7**, No. 1 (1958), 102–137; **8**, No. 1 (1959), 24–51.

[GD] GÅRDING, L.

1. Dirichlet's problem for linear elliptic differential equations. *Math. Scand.*, **1** (1953), 55–72.
2. *Cauchy's Problem for Hyperbolic Equations.* Univ. of Chicago Lecture Notes, 1957. (Russian translation with an additional chapter on hyperbolic systems. Gos. Isdat. Foreign. Lit., Moscow, 1961, 122 pp.)

[GŠ] GELFAND, I. M. AND ŠILOV, G. E.

1. *Some Questions of Theory of Differential Equations*, Vol. 2. Gos. Isdat. Fiz. Mat. Lit., Moscow, 1958, 274 pp.

[GV] GINDIKIN, S. G. AND VOLEVICH, L. R.

1. Method of energetical estimates in mixed problem. *Uspehi. Mat. Nauk*, **35**, No. 5 (1980), 55–120.

[GR] GODUNOV, S. K. AND RJABEN'KII, V. S.

1. *Introduction to the Theory of Difference Schemes.* Gos. Isdat. Fiz Mat. Lit., Moscow, 1962, 340 pp.

[GO] GOLOVKIN, K. K.

1. Two theorems of imbedding. *Dokl. Akad. Nauk SSSR*, **134**, No. 1 (1960), 19–20.
2. Two classes of inequalities for sufficiently smooth functions of *n* variables. *Dokl. Akad. Nauk SSSR*, **138** (1961), 22–25.

[GN] GREEN, J. W.

1. An expansion method for parabolic differential operators. *J. Res. Nat. Bur. Standards*, **51** (1953), 127–132.

[GT] GÜNTER, N. M.

1. Sur les integrales des Stieltjes et leur applications aux problemes fondamentaux de la physique mathematique. *Trudy Fiz.-Mat. Inst. Akad. Nauk SSSR*, **1** (1932), 1–494.
2. On the smoothing of functions and associated problems. *Uchenie Zapiski Leningrad. Univ. ser. Mat.*, **17** (1937), 51–78.
3. On the statement of some problems of mathematical physics. *Uchenie Zapiski Leningrad. Univ. ser. mat.*, **10** (1940), 12–26.
4. On the problem of small oscillations of string. *Uchenie Zapiski Leningrad. Univ. ser. mat.*, **15** (1948), 23–74.

[GU] GUSEVA, O. V.

1. On boundary value problems for strong elliptic systems. *Dokl. Akad. Nauk SSSR*, **102** (1955), 1069–1072.

[HA] HADAMARD, J.

1. *Le Problème de Cauchy et les Équations aux Dérivées Partielles Linéaires Hyperboliques.* Hermann et Cie, Paris, 1932.

[HI] HILBERT, D.

1. Über das Dirichlet Prinzip. *Jber. Deutsch. Math. Verein.*, **8** (1900), 184–188.

[HPH] HILLE, F. AND PHILLIPS, R. S.

1. *Functional Analysis and Semigroups*, rev. edn. Amer. Math. Soc. Publ., No. 31, 1957, 808 pp.

[HO] Hopf, E.

1. Elementare Betrachtungen über die Lösungen partieller Differentialgleichungen zweiter Ordnung vom elliptischen Typus. *Sitzungber. Preuss. Akad. Wiss.*, **19** (1927), 147–152.
2. Über den functionalen, insbesondere den analytischen Charakter, der Lösungen elliptischen Differentialgleichungen zweiter Ordnung. *Math. Z.*, **34** (1931), 194–233.
3. Über die Anfangswertaufgabe für die hydrodynamischen Grundgleichungen. *Math. Nach.*, **4** (1950–1951), 213–231.

[HR] Hörmander, L.

1. *Linear Partial Differential Operators.* Springer-Verlag Berlin, 1963, 287 pp.

[IL] Il'in, V. P.

1. Some inequalities in functional spaces and their applications to investigation of convergence in variational methods. *Trudy Mat. Inst. Steklov*, **53** (1959), 64–127.
2. Some integral inequalities and their applications in theory of differential equations with many variables. *Mat. Sb.*, **54**, No. 3 (1961), 331–380.

[JO] John, F.

1. On integration of parabolic equations by difference methods. *Comm. Pure Appl. Math.*, **5** (1952), 155–211.
2. Derivatives of continuous weak solutions of elliptic equations. *Comm. Pure Appl. Math.* **6**, No. 3 (1953), 327–333.
3. Derivatives of solutions of linear elliptic partial differential equations. *Ann. Math. Studies*, **33** (1954), 53–61.

[KM] Kamynin, L. J. and Maslenikova, V. N.

1. Boundary estimates for solutions of the third boundary problem for parabolic equations. *Dokl. Akad. Nauk SSSR*, **153**, No. 3 (1963), 526–529.

[KA] Kato, T.

1. Integration of the equation of evolution in a Banach space, *J. Math. Soc. Japan*, **5** (1953), 208–304.
2. On linear differential equations in Banach spaces. *Comm. Pure Appl. Math.*, **9** (1956), 479–486.
3. Abstract evolution equations of parabolic type in Banach and Hilbert spaces. *Nagoya Math. J.*, **19** (1961), 93–125.

[KE] Keldyš, M. V.

1. On some cases of degeneracy of an elliptic equation on the boundary of the domain. *Dokl. Akad. Nauk SSSR*, **77** (1951), 181–183.

[KL] Knowles, I. K.

1. On Saint-Venant's principle in the two-dimensional linear theory of elasticity. *Arch. Rat. Mech. Anal.*, **21** (1966), 1–22.
2. A Saint-Venant's principle for a class of second-order elliptic boundary value problems. *ZAMP*, **18** (1967), 478–490.

[KN] Kohn, J. J. and Nirenberg, L.

1. Degenerate elliptic-parabolic equations of second order. *Comm. Pure Appl. Math.*, **20** (1967), 797–872.

[KKS] Krasnoselskiĭ, M. A., Krein, S. G., and Sobolevskiĭ, P. E.

1. On differential equations with unbounded operational coefficients in Hilbert space. *Dokl. Akad. Nauk SSSR*, **112** (1957), 990–993.

[KR] Krein, S. G.

1. *Linear Differential Equations in Banach Space.* Isdat. Nauka, Moscow, 1967, 464 pp.

[LA] LADYZHENSKAYA, O. A.
1. On integration of the Cauchy problem for hyperbolic systems by the difference method. Thesis for a candidate's dissertation. Leningrad Univ., March 1949; *Uchenie Zapiski Leningrad. Univ. ser. math.*, **23** (1952), 192–246; *Dokl. Akad. SSSR*, **88**, No. 4 (1953), 607–610.
2. On Fourier's method for wave equations. *Dokl. Akad. Nauk SSSR*, **75**, No. 6 (1950), 765–768.
3. On the closure of an elliptic operator. *Dokl. Akad. Nauk SSSR*, **79**, No. 5 (1951), 723–725.
4. *The Mixed Problem for the Hyperbolic Equation*. Gos. Izdat. Tehn.-Teor. Lit., Moscow, 1953, 279 pp.
5. On integration of the general diffractional problem. *Dokl. Akad. Nauk SSSR*, **96**, No. 3 (1954), 433–436.
6. On solvability of classical boundary value problems for equations of parabolic and hyperbolic types. *Dokl. Akad. Nauk SSSR*, **97**, No. 3 (1954), 395–398.
7. A simple proof of solvability of classical boundary value problems and eigenvalue problem for elliptic equations. Vestnik Leningrad Univ., No. 11, 1955, pages 23–29.
8. (a) On solvability of operational equations of different types. *Dokl. Akad. Nauk SSSR*, **102** (1955), 207–210; (b) On integration of non-stationary operational equations. *Mat. Sb.*, **39**, No. 4 (1956), 491–524; (c) On non-stationary operational equations and their applications to linear problems of mathematical physics. *Mat. Sb.*, **45**, No. 2 (1958), 123–158.
9. The difference method in the theory of partial differential equations. *Uspehi Mat. Nauk*, **12**, No. 5 (1957), 123–149.
10. On equations with small parameter by highest derivatives in linear partial differential equations. Vestnik Leningrad Univ., No. 7, 1957, pages 111–120.
11. On integral estimates, convergence, approximate methods, and solutions in functionals for elliptic operators. Vestnik Leningrad Univ., No. 7, 1958, pages 60–69.
12. The solvability "in the large" of the first value problem for quasilinear parabolic equations. *Dokl. Akad. Nauk SSSR*, **107** (1956), 636–639; *Trudy Moscow Mat. Soc.*, **7** (1958), 149–177.
13. Study of the Navier–Stokes equations in the case of stationary motion of incompressible fluids. *Uspehi Mat. Nauk*, **13** (1958), 219–220; *ibid.*, **14** (1959), 75–97.
14. The solvability "in the large" of the boundary value problem for the Navier–Stokes equations in the case of two space variables. *Dokl. Akad. Nauk SSSR*, **123** (1958), 427–429; *Comm. Pure Appl. Math.*, **12** (1959), 427–433.
15. *Mathematical Problems in the Dynamics of Viscous Incompressible Flow*. Gos. Izdat. Fiz.-Mat., Moscow, 1961, 203 pp. (first Russian edn.); English transl. Gordon & Breach, New York–London, 1963, 184 pp.; German transl. Akademie-Verlag, Berlin, 1965, 180 pp.; Izdat. Nauka, Moscow, 1970, 288 pp. (second Russian edn.).

[LS] LADYZHENSKAYA, O. A. AND SOLONNIKOV, V. A.
1. The solvability of some non-stationary problems of magneto-hydrodynamics for viscous incompressible flow. *Trudy Mat. Inst. Steklov*, **59** (1960), 115–173.
2. Some problems of vector analysis and generalized statements of boundary value problems for the Navier–Stokes equations. *Zap. Nauch. Sem. Leningrad. Otdel. Mat. Inst. Steklov*, **59** (1976), 84–116.
3. On the solvability of boundary and initial-boundary value problems for the Navier–Stokes equations in a domain with non-compact boundaries. Vestnik Leningrad Univ., No. 13, 1977, pages 39–47.
4. On finding the solutions which have the infinite Dirichlet integral for the stationary

Navier–Stokes equations. *Zap. Nauch. Sem. Leningrad. Otdel. Mat. Inst. Steklov*, **96** (1980), 117–160.

[LSS] Ladyzhenskaya, O. A. and Stupjalis, L.
1. On equations of mixed type. Vestnik Leningrad. Univ. ser. Mat., Mech. Astronomy, No. 19, 1965, pages 38–46.
2. Boundary value problems for equations of mixed type. *Trudy Mat. Inst. Steklov*, **116** (1971), 101–136.

[LSU] Ladyzhenskaya, O. A., Solonnikov, V. A., and Ural'ceva, N. N.
1. *Linear and Quasi-linear Equations of Parabolic Type*. Izdat. Nauka, Moscow, 1967, 736 pp.

[LU] Ladyzhenskaya, O. A. and Ural'čeva, N. N.
1. *Linear and Quasi-linear Equations of Elliptic Type*. Izdat. Nauka, Moscow, 1964, 538 pp. (first Russian edn.); 1973, 576 pp. (second Russian edn.).

[LV] Ladyzhenskaya, O. A. and Višik, M. I.
1. Boundary value problems for partial differential equations and certain classes of operator equations. *Uspehi. Mat. Nauk* **11**, No. 6 (1956), 89–152 (Russian); English transl. *Amer. Math. Soc.*, (2) **10** (1958), 223–281.

[LD] Landis, E. M.
1. *Equations of the Second Order of Elliptic and Parabolic Types*. Izdat. Nauka, Moscow, 1971, 287 pp.

[LX] Lax, P. D.
1. On the Cauchy problem for hyperbolic equations and the differentiability of solutions of elliptic equations. *Comm. Pure Appl. Math.*, **8** (1955), 615–633.

[LM] Lax, P. D. and Milgram, A. N.
1. Parabolic equations, Ann. Math. Studies, Princeton, No. 33, 1954, pages 167–189.

[LB] Lebesgue, H.
1. Sur le problème de Dirichlet. *Rend. Circ. Math. Palermo*, **24** (1907), 371–402.

[LR] Leray, J.
1. Essai sur les mouvements plans d'un liquide visqueux que limitent des parois. *J. Math. Pures Appl.*, serie 9, **13** (1934), 331–418.
2. *Hyperbolic Differential Equations*, Part 1–3. Inst. Advanced Study, Princeton, 1952, 238 pp.

[LN] Levinson, N.
1. The first boundary value problem for $\varepsilon\Delta u + Au_x + Bu_y + Cu = \mathbf{D}$ with small ε. *Ann. of Math.*, **51**, No. 2 (1950), 428–445.

[LI] Lions, J. L.
1. *Equations Differentielles-Operationneles et Problemes aux Limites*. Springer-Verlag, Berlin, 1961, 292 pp.
2. *Quelques Methodes de Resolution des Problemes aux Limites Non-lineares*. Dunod Gauthier-Villars, Paris, 1969, 554 pp.

[LE] Lewy, H.
1. Über den analytischen Charakter der Lösungen elliptischer Differentialgleichungen Göttingen Nach. (1927) 178–186.
2. Neuer Beweis des analytischen Charakters der Lösungen elliptischer Differentialgleichungen. *Math. Ann.*, **101** (1929), 609–619.

[LT] LÜSTERNIK, L. A.
1. The Dirichlet problem. *Uspehi. Mat. Nauk SSSR*, **8** (1941), 115–124.

[LTS] LÜSTERNIK, L. A. AND SOBOLEV, V. I.
1. *Elements of Functional Analysis*, Izdat. Nauka, Moscow, 1965, 520 pp.

[LTV] LÜSTERNIK, L. A. AND VIŠIK, M. I.
1. Regular degeneration and boundary layer for linear differential equations with small parameter. *Uspehi. Mat. Nauk*, **12**, No. 5 (1957), 3–122.

[MA] MARCHUK, G. I.
1. *Numerical Methods of Calculations of Nuclear Reactors.* Atomizdat, Moscow, 1961, 667 pp.

[MI] MIHLIN, S. G.
1. On a convergence of Galerkin's method. *Dokl. Akad. Nauk SSSR*, **61**, No. 2 (1948), 197–199.
2. Some sufficient conditions of a convergence of Galerkin's method. *Uchenie Zapiski Leningrad. Univ. ser. mat.*, **135**, No. 18 (1950).
3. *Direct Methods in Mathematical Physics.* Gos. Izdat. Tehn.-Teor. Lit., Moscow–Leningrad, 1950, 428 pp.
4. *The Problem of the Minimum of the Quadratic Functional.* Gos. Izdat. Techn.-Teor. Lit., Moscow–Leningrad, 1952, 216 pp.
5. Degenerate elliptic equations. Vestnik Leningrad Univ., No. 8, 1954, pages 19–48.

[MR] MILGRAM, A. AND ROSENBLOOM, P. C.
1. Harmonic forms and heat conduction: Part I, Closed Riemannian manifolds; Part II, Heat distribution on complexes and approximation theory. *Proc. Nat. Acad. Sci. USA*, **37** (1951), 180–184; 435–438.

[MN] MINTY, G. J.
1. Monotone (non-linear) operators in Hilbert space. *Duke Math. J.*, **29** (1962), 341–346.
2. On a monotonicity method for the solution of non-linear equations in Banach spaces. *Proc. Nat. Acad. Sci. USA*, **50** (1963), 1038–1041.

[MR] MIRANDA, C.
1. *Partial Differential Equations of Elliptic Type* (second rev. edn.). Springer-Verlag, Berlin–Heidelberg–New York, 1970, 370 pp.

[MY] MORREY, C. B.
1. A class of representation of manifolds, I. *Amer. J. Math.*, **55** (1933), 683–707; II. *ibid.*, **56** (1934), 275–293.
2. *Multiple Integrals in the Calculus of Variations.* Springer-Verlag, Berlin–Heidelberg–New York, 1966, 506 pp.

[NI] NIRENBERG, L.
1. Remarks on strongly elliptic partial differential equations. *Comm. Pure Appl. Math.*, **8** (1955), 648–674.
2. Estimates and existence of solutions of elliptic equations. *Comm. Pure Appl. Math.*, **9** (1956), 509–520.
3. On elliptic partial differential equations. *Ann. Sc. Norm. Sup. Pisa, ser. III*, **13**, F.II (1959), 115–162.

[OD] ODQVIST, F. K. G.
1. Über die Randwertaufgaben der Hydrodynamik zäher Flüssigkeiten. *Math. Z.*, **32** (1930), 329–375.

[OG] OGANESJAN, L. A.
1. Numerical calculations of plates Sbornik "Solution of engineering problems with the help of EVM". Leningrad ČBTI (1963), 84–97.

[ORS] OGANESJAN, L. A., RIVKIND, V. J., AND SAMOKISH, B. A.
1. Variational-difference schemes in problems of electronical optik. *Trudy the Third All-Union Seminar on Electronical Optik.* Izdat. Nauka Novosibirsk, 1970, pages 163–168.

[OR] OGANESJAN, L. A. AND RUHOVETZ, L. A.
1. On variational-difference schemes for linear elliptic equations of second order in a two-dimensional region with a piece-smooth boundary. *J. Numer. Mat. and Mat. Phys.*, **8**, No. 1 (1968), 97–114.

[OL] OLEJNIK, O. A.
1. On equations of elliptic type with a small parameter in the highest derivatives. *Mat. Sb.*, **31** (1952), 104–117.
2. On equations of elliptic type which degenerate on the boundary of the region. *Dokl. Akad. Nauk SSSR*, **87** (1952), 885–888.
3. On linear second-order equations with non-negative characteristic form. *Mat. Sb.*, **69** (1966), 111–140.

[PR] PEACEMEN, D. W. AND RACHFORD, H. H.
1. The numerical solution of parabolic and elliptic differential equations. *J. Soc. Industr. Appl. Math.*, **3**, No. 1 (1955), 28–42.

[PT] PETROWSKIĬ, I. G.
1. On the Cauchy problem for the system of equations with partial derivatives. *Mat. Sb.*, **2** (1937), 815–868.
2. On analyticity of solutions of equations with partial derivatives. *Mat. Sb.*, **5**, No. 1 (1939), 6–68.
3. New proof of existence of solution of Dirichlet's problem by the difference method. *Uspehi. Mat. Nauk*, **8** (1941), 161–170.
4. *Lectures on Equations with Partial Derivatives.* Gos. Isdat. Fis. Mat. Lit., Moscow, 1961, 400 pp.

[PE] PETROX, G. I.
1. The application of Galerkin's method to the problem of stability of a viscous flow, *Prikl. Math. and Mech.*, **4**, No. 3 (1940), 3–12.

[RI] RICHTMAYER, R. D.
1. *Difference Methods for the Initial Value Problem.* New York–London, 1957, 238 pp.

[RN] RIESZ, F. AND SZ-NAGY, B.
1. *Lecons d'Analyse Fonctionnelle.* Akadémiai Kiadó, Budapest, 1952, 448 pp.

[RK] RIVKIND, V. YA.
1. On estimates of the rate of convergence of solutions of difference equations to solutions of elliptic equations with discontinuous coefficients, and on one numerical method of solving Dirichlet's problem, *Dokl. Akad. Nauk SSSR*, **149**, No. 6 (1963), 1264–1267.
2. An approximate method of solving Dirichlet's problem and estimates of the rate of convergence of solutions of difference equations to solutions of the elliptic equations with discontinuous coefficients. Vestnik Leningrad Univ., No. 13, 1964, pages 37–52.

3. On estimates of the rate of convergence in homogeneous difference schemes for elliptic and parabolic equations with discontinuous coefficients "Problems of Mat. Analysis", Leningrad Univ., No. 1, 1966, pages 110–119.

[RF] RJABEN'KIĬ, V. S. AND FILIPPOV, A. F.
1. *On the Stability of Difference Equations.* Gos. Tech. Izdat., Moscow, 1956, 171 pp.

[RO] ROTHE, E.
1. Wärmeleitungs gleichungen mit nichtconstanten koeffizienten. *Math. Ann.*, **104** (1931), 340–362.

[SA] SAMARSKIĬ, A. A.
1. *Introduction to the Theory of Difference Schemes.* Izdat. Nauka, Moscow, 1971, 552 pp.

[SR] SCHAUDER, J.
1. Über lineare elliptische Differentialgleichung zweiter Ordnung. *Math. Z.*, **38** (1934), 251–282.
2. Hyperbolische Differentialgleichungen. *Fund Math.*, **24** (1935), 213–246.

[SZ] SCHWARTZ, L.
1. *Theorie des Distributions.* t.1. Hermann, Paris, 1950, 148 pp.; t.2, 1951, 169 pp.

[SI] ŠILOV, G. E.
1. *Mathematical Analysis. Special Course.* Gos. Izdat. Fiz. Mat. Lit., Moscow, 1961, 436 pp.
2. *Mathematical Analysis. Second Special Course.* Izdat. Nauka, Moscow, 1965, 327 pp.

[SN] SMIRNOV, M. M.
1. *Degenerate Elliptic and Hyperbolic Equations.* Izdat. Nauka, Moscow, 1966, 292 pp.

[SM] SMIRNOV, V. I.
1. *A Course on Higher Mathematics*, Vol. IV. Gos. Izdat. Fis. Mat. Lit., Moscow, 1958, 812 pp.
2. *A Course on Higher Mathematics*, Vol. V. Gos. Izdat. Fis. Mat. Lit., Moscow, 1959, 655 pp.

[SO] SOBOLEV, S. L.
1. Methode nouvelle a resoudre le problème de Cauchy pour les equations lineaires hyperboliques normales, *Mat. Sb.*, **1** (1936), 39–72.
2. On a boundary value problem for polyharmonic equations. *Mat. Sb.*, **2** (1937), 467–500.
3. On a theorem of functional analysis. *Mat. Sb.*, **4** (1938), 471–497.
4. On estimates of some sums for functions determined on a mesh. *Izv. Akad. Nauk SSSR*, **4** (1940), 5–16.
5. *Applications of Functional Analysis in Mathematical Physics.* Izdat. Leningrad Gos. Univ., 1950, 255 pp.

[SB] SOBOLEVSKII, P. E.
1. Generalized solutions of differential equations of first order in Hilbert space. *Dokl. Akad. Nauk SSSR*, **122** (1958), 994–996.
2. On parabolic equations in Banach space. *Trudy Moscow Mat. Soc.*, **10** (1961), 237–250.

[SK] SOLOMJAK, M. Z.
1. Analyticity of a semigroup generated by an elliptic operator in L_p spaces. *Dokl. Akad. Nauk SSSR*, **127** (1959), 37–39.

2. On differential equations in Banach spaces. *Izv. Hight. Uchebn. Zaved., Mat.*, **1** (1960), 198–209.

[SL] SOLONNIKOV, V. A.

1. Linear differential equations with small parameters in the terms of highest order. *Dokl. Akad. Nauk SSSR*, **199** (1958), 454–457.
2. On estimates of Green's tensor for certain boundary problems. *Dokl. Akad. Nauk SSSR*, **130** (1960), 988–991.
3. Some stationary boundary value problem for magneto-hydrodynamics. *Trudy Mat. Inst. Steklov*, **59** (1960), 174–187.
4. *A priori* estimates for certain boundary value problems. *Dokl. Akad. Nauk SSSR*, **138**, No. 4 (1961), 781–784.
5. Estimates for solutions of general boundary value problems for elliptic systems. *Dokl. Akad. Nauk SSSR*, **151** (1963), 783–785.
6. On general boundary value problems for systems of differential equations of elliptic and parabolic types. *Proceedings of the Soviet–American Symposium on Partial Differential Equations*. Izdat. Nauka, Novosibirsk, 1963, pages 246–252.
7. On general boundary value problems for elliptic Douglis–Nirenberg's systems: I. *Izv. Akad. Nauk SSSR*, **28** (1964), 665–706; II. *Trudy Mat. Inst. Steklov*, **92** (1966), 233–297.
8. On boundary value problems for general linear parabolic systems. *Dokl. Akad. Nauk SSSR*, **151**, No. 1 (1964), 56–59.
9. On boundary value problems for linear parabolic systems of differential equations of general type. *Trudy Mat. Inst. Steklov*, **83** (1965), 3–162.

[SP] SOLONNIKOV, V. A. AND PILEČKAS, K. I.

1. On some spaces of solenoidal vectors and solvability the boundary value problem for the Navier–Stokes system in regions with non-compact boundaries. *Zap. Nauch. Sem. LOMI*, **73** (1977), 136–151.

[ST] STRELKOV, N. A.

1. Simplicial expansions of mesh functions and their applications to the solving of problems of mathematical physics. *J. Numer. Mat. and Mat. Physics*, **11**, No. 4 (1971), 969–981.

[SS] STUPJALIS, L.

1. Spectral problem for equations of mixed type. Vestnik Leningrad Univ., No. 13, 1967, pages 86–102.
2. Boundary value problems of elliptic-hyperbolic equations. *Trudy Mat. Inst. Steklov*, **125** (1973), 211–229.
3. Initial boundary value problems for equations of mixed type. *Trudy Mat. Inst. Steklov*, **127** (1975), 115–145.

[TO] TONELLI, L.

1. *Opera Scelte*, Roma; vol. 1, 1960, 604 pp.; vol. 2, 1961, 534 pp.; vol. 3, 1962, 508 pp.; vol. 4, 1963, 333 pp.

[TP] TOUPIN, R. A.

1. Saint-Venant's principle. *Arch. Rat. Mech. and Anal.*, **18** (1965), 83–96.

[TR] TRUDINGER, N. S.

1. On imbeddings into Orlicz spaces and some applications. *J. Math. Mech.*, **17** (1967), 473–483.

[VI] VIŠIK, M. J.

1. The method of orthogonal projection for self-adjoint differential equations. *Dokl. Akad. Nauk SSSR*, **56** (1947), 115–118.

2. The method of orthogonal projection and direct decomposition in the theory of elliptic differential equations. *Mat. Sb.*, **25** (1949), 189–234.
3. On strongly elliptic systems of differential equations. *Mat. Sb.*, **29** (1951), 615–676.
4. Boundary value problems for elliptic equations degenerating on the boundary. *Mat. Sb.*, **35**, No. 3 (1954), 513–568.
5. Cauchy problem for equations with operational coefficients, mixed boundary value problem for systems of differential equations, and an approximate method of their solution. *Mat. Sb.*, **39**, No. 1 (1956), 51–148.
6. Mixed boundary value problems for equations containing the first time-derivative and an approximate method for their solution. *Dokl. Akad. Nauk SSSR*, **99** (1954), 189–192.

[WE] WEYL, H.
1. The method of orthogonal projection in potential theory. *Duke Math. J.*, **7** (1940), 411–444.

[WI] WIENER, N.
1. The operational calculus. *Math. Ann.*, **95** (1926), 557–584.

[ZA] ZAREMBA, S.
1. Sur un problème toujours possible comprenant à titre de cas particuliers, le problème de Dirichlet et celui de Neumann. *J. Math. Pures appl.*, **6** (1927), 127–163.

[YA] YANENKO, N. N.
1. *Method of Fractional Steps for Many-Dimensional Problems of Mathematical Physics*. Izdat. Nauka, Novosibirsk, 1967, 195 pp.

Index

Printed in the United States
by Baker & Taylor Publisher Services